Cohesive Properties of Semiconductors under Laser Irradiation

NATO ASI Series

Advanced Science Institutes Series

A Series presenting the results of activities sponsored by the NATO Science Committee, which aims at the dissemination of advanced scientific and technological knowledge, with a view to strengthening links between scientific communities

The Series is published by an international board of publishers in conjunction with the NATO Scientific Affairs Division

A	Life Sciences	Plenum Publishing Corporation
B	Physics	London and New York
C	Mathematical and Physical Sciences	D. Reidel Publishing Company Dordrecht and Boston
D	Behavioural and Social Sciences	Martinus Nijhoff Publishers The Hague/Boston/Lancaster
E	Applied Sciences	
F	Computer and Systems Sciences	Springer-Verlag Berlin/Heidelberg/New York
G	Ecological Sciences	

Series E: Applied Sciences – No. 69

Cohesive Properties of Semiconductors under Laser Irradiation

edited by

Lucien D. Laude

Professor of Solid State Physics
Université de l'Etat
Mons, Belgium

1983 **Martinus Nijhoff Publishers**
The Hague / Boston / Lancaster
Published in cooperation with NATO Scientific Affairs Division

Proceedings of the NATO Advanced Study Institute on Cohesive Properties of Semiconductors under Laser Irradiation, Corsica, France, July 19-30, 1982

Library of Congress Cataloging in Publication Data
NATO Advanced Study Institute on Cohesive Properties
   of Semiconductors under Laser Irradiation (1982 :
   Cargèse, Corsica)
   Cohesive properties of semiconductors under laser
irradiation.

   (NATO ASI series. Series E, Applied sciences ;
no. 69)
   Proceedings of the NATO Advanced Study Institute on
Cohesive Properties of Semiconductors under Laser
Irradiation, Cargèse, Corsica, France, July 19-30, 1982.
   "Published in cooperation with NATO Scientific Affairs
Division."
   1. Semiconductors--Effect of radiation on--Congresses.
2. Laser beams--Congresses. I. Laude, Lucien D.
II. Title. III. Series: NATO advanced science institutes
series. Series E, Applied sciences ; no. 69.
QC611.6.R3N37   1982   537.6'22   83-11377
ISBN-13: 978-94-009-6892-9   e-ISBN-13: 978-94-009-6890-5
DOI: 10.1007/978-94-009-6890-5

ISBN-13: 978-94-009-6892-9 (this volume)
ISBN 90-247-2689-1   (series)

Distributors for the United States and Canada: Kluwer Boston, Inc., 190 Old Derby Street, Hingham, MA 02043, USA

Distributors for all other countries: Kluwer Academic Publishers Group, Distribution Center, P.O. Box 322, 3300 AH Dordrecht, The Netherlands

PREFACE

The impact of Materials Science in our environment has probably never been as massive and decisive as it is today. In every aspect of our lives, progress has never been so dependent on the techniques involved in producing ever more sophisticated materials in ever larger quantities, nor so demanding for technologists to imagine novel processes and circumvent difficulties, or take up new challenges. Every technique is based on a physical process which is put into practice and optimized. The better we know that process, the better the optimization, and more powerful the technique. Laser processing of materials is inscribed in that context.

As soon as powerful coherent light sources were made available, it was realized that such intense sources of energy could be used to "heat, melt and crystallize" materials, i.e., to promote phase transitions in atomic systems. As early as 1964, attempts in that direction were made but received very little (if any) attention. Reasons for this lack of interest were several. For one thing, laser technology was not fully developed, so that the process offered poor reliability and no versatility. Also, improving the existing techniques was believed to be sufficient to meet the needs of the time, and there was no real motivation to explore new ways. Finally, and more important, the fundamentals of the physics behind the scenes were, and continue to be, way out of the running stream. Therefore, on the fundamental side, the gap between actual problems and available theoretical machinery was just too large for initiatives to show up towards understanding the phenomenology of laser processing.

The situation in the early seventies did not improve until it was realized that laser annealing of semiconductors might open new horizons in material processing. Not surprisingly, the motivation to resume investment in that field came from the industry. Ion-implantation had been discovered to produce extremely precise doping profiles in Si crystals, which were required in the development of microelectronics. However, implanted materials were so severely damaged that a way of re-crystallizing the altered region which would not affect the implanted profiles was mandatory. Whilst oven-annealing was unable to meet this requirement, ruby-laser irradiation demonstrated the ability to do so. Immediately, an important research activity developed aiming at technically controlling every step of the ion-implantation/laser annealing technique. At first, results of that activity attracted considerable interest from the Si industry. However, as industrial requirements became more astringent, some of them (e.g., on the dislocation concentration) could not be met by laser annealing, and industrial interests would fade out rapidly. Would this lead to the vanishing of the whole activity in laser-processing? Fortunately, no. The implanted Si laser annealing era may well be discontinued now, as it has served to establish the field of laser-solid interactions. Examples have been multiplied over the past few years which demonstrate the versatility of laser processing. As a result, an increasing number of solid state physicists and chemists are sensitized to the novel possibilities offered by the laser source in promising domains such as non-equilibrium phase-transitions, dynamics of nucleation, or synthesis of complex materials. Development of the field is then open to the wildest variety of problems dealing with the cohesion of matter. At this point, and considering the risk of seeing this field desaggregating *before* reaching its plain maturity, it appears essential to preserve its *unity* by properly delineating it and setting its rules.

The task may appear to be impossible. Both from experimental and theoretical points-of-views, laser-solid interactions do not resemble the classical figures one may find in textbooks. What we are concerned with here is the physics of systems which are brought out of equilibrium under circumstances (laser irradiations of various sorts) forbidding the usual perturbation schemes. Instabilities are produced which perturb cohesion so deeply that conventional approaches must make room for new concepts. The purpose of this book is to convince the reader of the importance of finding new concepts; a challenge worth the effort and experience.

The first step along the path would be an assessment of the present situation, limited to the most reliable material but historically complete. As mentioned above, this field has already endured its ups-and-downs. With history working on such a long-time basis, one may assume the field to now be in quasi-equilibrium, whatever the ups-and-downs to come. After introducing the field as it stands today, the series of lectures is subdivided into three groups: i) the first relates to the thermodynamic aspects of phase transitions, progressively bringing the concepts to the limit of their validity and further into the hydrodynamics field, and, therein, instabilities; problems dealing with cohesion are treated here in detail, in nucleation, crystallization, dislocations and fusion, in relation with laser processing; ii) the second part is concerned with the optical properties of semiconductors under intense irradiation. Models are put forward in an attempt to understand the involved physical mechanisms; iii) the last section exemplifies some of the latest developments of laser processing.

Professor Bertolotti presents a comprehensive review of the present state-of-the-art, not missing the current arguments on laser melting which attracted, attracts, and will attract (?) interest. Another important part of the assessment is given by Professor May on the lasers to be used (the tooling) and their problems, one of which is the speckle (or punctuation) effects in a laser beam, which are considered as being either detrimental or beneficial, depending on the users. Common thinking would tend to favor the former (a clean, flat beam), but this presentation may provide some ideas to many on shaping-up a laser beam. Professor Rimini's lecture serves two purposes: i) to finish up the assessment; and, (ii) to open the discussion on the fundamentals by a detailed and well-substantiated presentation of what is called the "thermal" model of laser anneal- ing. On doing so, and assuming light to be instantaneously absorbed, E. Rimini shows to what extent diffusion equations may be used to describe laser-irradiated implanted materials behaviours.

The problems of the applicability of such an approach is ad- dressed immediately after by Dr. Dewel through the transport theory and the concepts of hydrodynamics. Unmistakably, instabilities are around the corner, but before progressing in that direction, a fur- ther look at equilibrium states is given by Prof. Rosenberger in an impressive lecture on phase diagrams. We are all used to such repre- sentations, but only for stable systems. What happens to a phase diagram in a destabilized system is almost unpredictable, but cer- tainly produces lively discussions! Prof. Binder addresses the crucial problem of the dynamics of phase transition in a lecture that is probably at the core of this meeting. Experimental evidence on laser-induced phase transitions are just too few at the moment to confront or confirm these views, but there are reasons to believe that spinodal decomposition and/or Bénard instabilities might play an important role in laser-induced phase transformations. The pro- blem of instabilities is again raised by Prof. Müller-Krumbhaar after a general presentation on crystal growth. One striking fact is that very little is known on the melt, i.e., not enough to build up a unifying theory of melting. Most probably, lasers may provide the means to explore this field, particularly in tracking solid- liquid interfaces, a problem which is surprisingly (and convincingly) studied in ice melting by Dr. Bilgram. The state-of-the-art in nucleation theory is presented by Dr. Germain and co-workers, a field in which lasers may again provide for a discrete and most reliable tool. Finally, Prof. Bok presents a model of melting based on the instability of the electron-hole plasma, together with some views on the kinetics of crystallization of amorphous Si, a subject which, again, is shared by many of the participants to the Institute. This lecture introduces, in fact, the third part of the Proceedings.

Up to this point, and following Prof. Rimini's lecture, one is solely concerned with the thermodynamics of laser processing. Prof. Ulbrich introduces the other aspect of the problem, i.e., the optical absorption which precedes an eventual phase transformation, concentrating on hot electron behaviour and non-linear effects. Quite logically, Prof. Voos follows with the various recombination mechanisms governing energy transfer to the lattice and the life-time of the excited carriers. Prof. Smirl develops extensively this latter subject into the dense plasmas produced in the picosecond regime, together with an impressive instrumental background. Impressive also is the presentation of Prof. Compaan on his time-resolved Raman experiments. Questions were raised at the meeting on interpreting Raman line shift vs. phonon populations, i.e., lattice temperature. The picosecond regime is again exemplified by Dr. Kurz, who cannot find evidence for the long-lasting electron plasma predicted by Dr. Van Vechten. The latter gives an extensive survey of experimental facts which apparently do not find satisfactory interpretation without the above plasma, and one should not be blamed for systematically raising unanswered questions. Very hot discussions took place in Cargèse on these items. Heat or light? The average answer was "undecided", as usual, but light was clearly first and essential. Finally, in the context of semi-conductors, proper consideration is to be paid to defects or inhomogeneities. Prof. Schröter gives a remarkable lecture on dislocations, which is followed by Dr. Wautelet's presentation on such defects being optically excited. Both lectures find connections with the two trends developed in these proceedings and provide for valuable approaches to the problem of laser-assisted nucleation. Last, but not least, Dr. von Allmen's lecture concentrates on the behaviour of interfaces under laser irradiation, a domain to experience fast developments in the future. These Proceedings terminate with a series of short presentations aimed at introducing some of the latest developments in laser processing.

These, and many other subjects which were discussed in a series of ebullient round-table seminars along the course of this Institute, open up this field into domains which were unpredicted just a few years ago. Towards the end of the two-week session, it was the general impression among all participants that we probably stand at the even of a renaissance in Materials Science, one in which given structures could be tailored at will, using a more and more sophisticated laser tooling. It will probably be some time before reaching this prospect, which is in sight, however. For one thing, the Institute helped to discern clearly the limits of existing theoretical approaches and the directions along which works are urgently needed within the next few years.

In the early stage of the preparation of this Institute, the advice and support of Professor J. Friedel have been extremely useful in setting up the programme. We would like to thank him sincerely for his help. Prof. Binder and Dr. Gaspard also provided valuable comments, as well as Profs. Bertolotti, Cardona, Dagonnier, Dr. Dewel, Prof. A.A. Lucas, and Dr. Van Vechten. The collaboration of Drs. Rod Andrews and Michel Wautelet during the preparation stage of the Institute has been essential to the success of the meeting; and appreciation is also extended to them for their invaluable assistance in editing these Proceedings. The efficient assistance of Miss M.F. Hanseler in the local organization at the Institut d'Etudes Scientifiques de Cargèse is gratefully acknowledged, and particular thanks are due to Mrs. Irène Doison-Bednarz for her dedication in skillfully typing the manuscript.

L.D. LAUDE

January, 1983

x

Acknowledgements

The organizers benefitted from the co-sponsorhship of the following:

> *The European Physical Society*
> *IBM-Belgium and IBM-Europe*
> *The US Army European Research Office, London*
> *The USAF-EOARD, London*
> *Cie Thomson-CSF, France*
> *The National Science Foundation, Washington*
> *Fonds National de la Recherche Scientifique, Brussels*
> *The Ministry for Science Policy, Brussels*
> *Université de l'Etat, Mons*

The organizers of this NATO Advanced Research Institute are grateful to all these co-sponsors for their support.

Particular thanks go to Drs. Mario di Lullo and Craig Sinclair of NATO Scientific Affairs Division for their encouragement and assistance.

CONTENTS

K. BINDER :

STATICS AND DYNAMICS OF PHASE TRANSITIONS : A BRIEF
INTRODUCTION

H. MÜLLER-KRUMBHAAR :

THEORY OF CRYSTAL GROWTH

J.H. BILGRAM :

DYNAMICAL PROCESSES DURING SOLIDIFICATION

XIV

P. GERMAIN and K. ZELLAMA :

NUCLEATION IN CONDENSED MATTER

K. ZELLAMA, P. GERMAIN, P.A. THOMAS and A. GHEORGHIU :

TRANSIENT BULK INDUCED NUCLEATION IN AMORPHOUS GROUP IV
SEMICONDUCTORS

M. COMBESCOT and J. BOK :

CRYSTALLINE, AMORPHOUS AND LIQUID SILICON

XVIII

D. PRIBAT :

EFFECTS OF PULSED LASER IRRADIATION ON THE ELECTRICAL PROPERTIES
OF GaAS

G. LANGOUCHE :

LASER ANNEALING OF SEMICONDUCTORS STUDIED BY MÖSSBAUER SPECTROSCOPY

PRESENTATION ON REORDERING PROCESSES IN LASER IRRADIATED
SEMICONDUCTORS

Mario Bertolotti

Istituto di Fisica - Facoltà di Ingegneria
Università di Roma, Roma, Italy and
Gruppo Nazionale Elettronica Quantistica e
Plasmi of CNR, Italy.

1. HISTORICAL INTRODUCTION

The Study of the effects of laser irradiation on semiconductors started very early in 1964 [1] . It was immediately appreciated the possibility of producing a large number of excess carriers by irradiating with energies larger than the band gap. The first researches were in fact on the enhanced reflectivity produced by the extra carriers introduced by laser irradiation in semiconductor mirrors to be used as passive Q-switches in ruby and Nd lasers [2] [3] .

Several researchers in the USA, USSR, and Italy started to examine different aspects of the problem. Attention was given to the change in reflectivity, to mechanical damage, etc. [4] . Although people understood quite readily that melting of a thin surface layer was possible, it was not clearly appreciated that there was a range of laser density power in which surface is melted and extended mechanical damage is not produced. The presence of a large mechanical damage in the form of crack lines or, by increasing the laser power density, of craters, discouraged in fact applications. To make clear this point Figs. 1 and 2 show two sequences, obtained by increasing the laser power density over the surface of a Ge and Si sample, respectively, which show the appearance of crack lines clearly connected to the cleavage habit of the surface and, by increasing the power density, the formation of a crater at the spot center, as seen with an optical microscope [5] .

Electron microscope observation at the boundary of the damaged regions is shown in Fig. 3 where several different structures are visible as produced by a ruby laser on a Ge sample [6] .

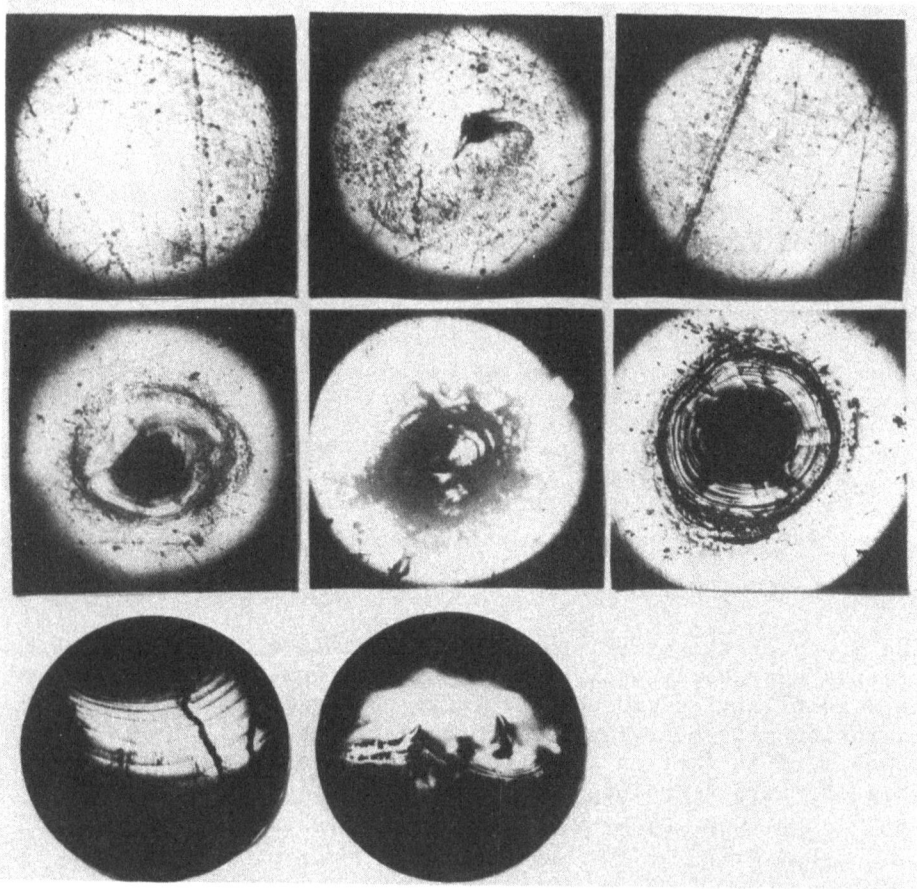

Fig. 1 Optical Microscope micro-photographs of ruby laser irradiated germanium samples at increasing powers. The last microphotograph is a side view of protuberance produced after melting (work [5]).

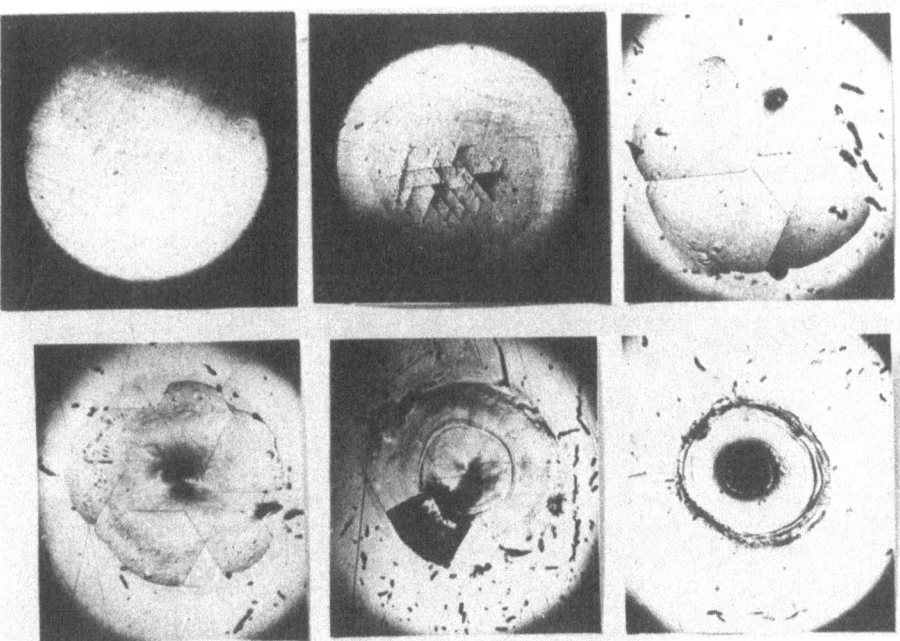

Fig. 2 Optical microscope microphotographs of ruby laser irradiated silicon samples at increasing powers (work [5]).

The laser irradiated surface appears as a flat background from
which characteristic patterns stream out (Fig. 3). Pseudo-hexagonal
structures, regularly arranged, cover the surface of this region
(Fig. 3). The main characteristics of these hexagonal cells is
the presence of a relief formation at their center. At times these
relief formations have a geometrical shape which depends upon the
crystallographic orientation of the surface. This shape tends to
have a triangular symmetry on (111) surface (Fig. 3) and a rectangular symmetry on the (100) surface.

Sometimes the pseudo-hexagonal structures merge to form
parallel and quasi-regular structures (channels) (Fig. 3). The distance of one channel from the other is typically of the order of
0.7-0.9 μm, in the figure.

The structures were associated with expitaxial regrowth. Channels
we know today are produced by interference between the incoming
laser field and a surface wave propagating along the sample surface
[7] [8] . Initially a weak surface plasma wave is excited by

4

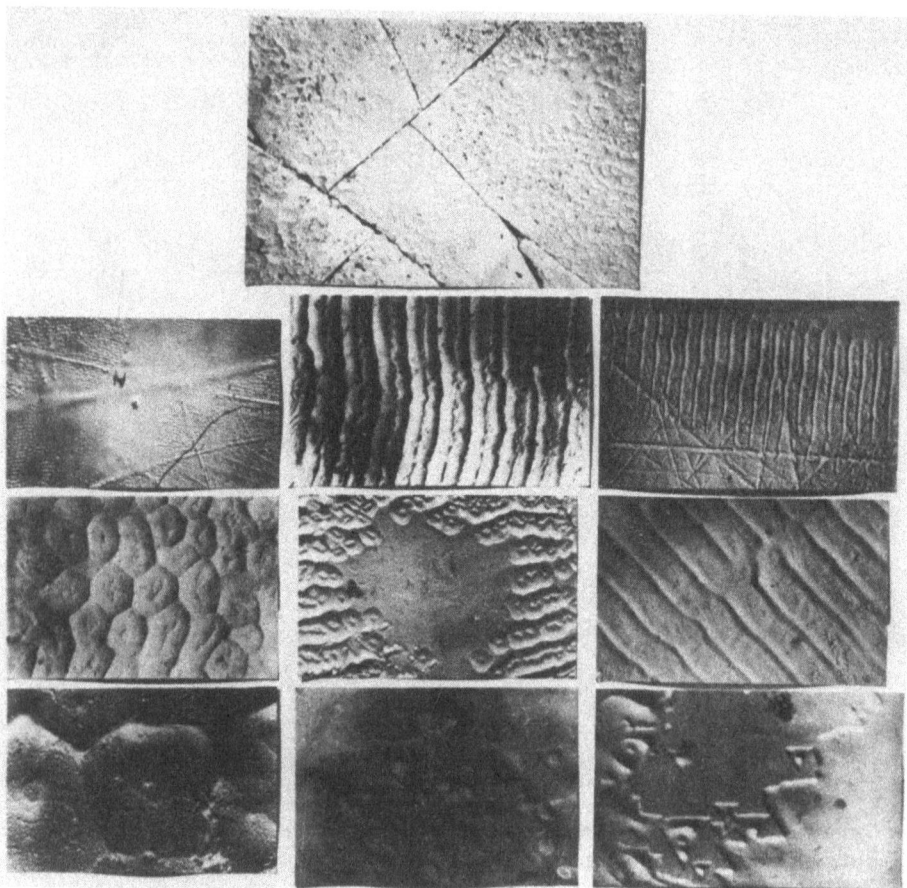

Fig. 3 Electron microscope microphotographs of a ruby laser irradiated germanium surface showing different produced structures
(work [6]).

surface roughness-induced scattering from the incident field. The spatial modulation of the optical intensity resulting from the interference between the incident wave and the surface plasma wave promotes the growth of a periodic structure (Channels) which increases the scattering into the surface wave. Ehrlich [7] has found that there is a positive gain coefficient increasing both the amplitude of the surface structure and the surface plasma wave intensity for spatial frequencies ranging from approximately the incident light wavevector to the surface wavevector. The process, which is analogous to stimulated Raman scattering, where the grating structure plays the role of the material excitation and the surface wave corresponds to the Stokes beam, is the first demonstration of an exponentially growing instability involving surface electromagnetic waves.

Starting from 1968 some courageous researchers notwhistanding the existence of the mechanical damage tryed to use laser melting for alloying and doping semiconductors [9] and it is remarkable that Marquardt and Giurliani already in 1974 had understood the possibility of changing dopant concentration by laser irradiation [10].

For some reason or the other the research however was not pushed further while, shortly after, several soviet groups considered anew laser effects on implanted semiconductors claiming good electrical recovery of irradiated samples and perhaps recrystallization [11].

In 1977 clear evidence of structural reordering under laser irradiation was obtained in Rome [12-15] and a simple melting model was proposed to explain the obtained results [16].

As soon as these results were presented at some Conferences in the United States [17] [18] many laboratories started research. The term *laser annealing* came in use rather improperly, and is nowadays of large use.

It was immediately shown that reordering can be produced in at least two different ways.

In the first case, melting of a thin surface layer was hypotized with subsequent fast regrowth [16]. In the second case, it was shown that treating with a c.w. laser of sufficient power density for a time of the order of a millisecond was sufficient to regrow a thin layer of material in solid state by just the same thermally activated process which occurs in ordinary furnace annealing [19] [20] [21].

It was also readily shown that reordering is possible in desordered materials produced for ex. by ion implantation as well as in truly *amorphous* materials obtained by glow discharge, CVD, or other

techniques [22] . Also polycrystalline material was made to turn
to perfect crystal, which was a further evidence of melting [23] .

Incidentally one may wonder whether the *coherence* of lasers is
important in laser annealing or not. Present evidence is that only
first-order coherence may be useful allowing good focusing and well
controlled coupling of radiation with electrons, which however is
not necessary in many cases. Annealing has been obtained with
flashlamps [24] or sun light [25] and with electron beams [26] as
well, all methods which are equally effective in producing a number
of excited carrier pairs in the bands.

2. GENERAL MECHANISM OF LASER ANNEALING

Very schematically we may describe the laser annealing process
in the following way [16] . Laser radiation is absorbed in the
semiconductor. If the photon energy is larger than band-gap the
main absorption channel is through the production of excess electron-
hole pairs. If the photon energy is lower than band-gap, absorption
is through free carriers in the band (of course these absorb also
in the previous case and may give contribution [27]), or impurities
in the band gap. Non-linear processes, like two-photon-absorp-
tion have not been proved .until now of being of importance in the
studied cases [28] . Coupling of laser radiation to the material
is a very important point and is considered further in this volume
[29] . In the melting model energy is assumed to be rapidly converted into heat in the lattice. If the deposited energy is large
enough melting occurs, otherwise simple heating is produced.

In the first case a thin layer of surface material is melted
which then starts to regrow. The way this regrowing occurs determines the characteristics of the regrown material. The most important parameter is the cooling velocity.

Several calculations have been performed on the basis of this
model, which is particularly used in pulsed laser annealing [30] .
In this case melting starts during the laser pulse and progresses
towards the interior. Then, after the end of the laser pulse it
stops and cooling starts.
Cooling is extremely fast and it occurs a few hundred nanoseconds
after the pulse end, starting from the back, if 20-30 nsec pulses
are used.
Cooling rates of $10^8 - 10^{10}$°K/sec are produced. Such high cooling
rates correspond to very high regrowth velocities; the melt front
moving typically at speeds of the order of 10m/sec during melting
and returning to the surface at 1-2 m/sec during solidification.
A non-equilibrium growing situation is therefore created which will
be discussed in this volume [31] .

Regrowth following melting by picosecond pulses in cases where a very shallow layer is molten, results in even higher cooling rates and refreezing interface velocities larger than 10 m/sec. If cooling rate is of the order of 10^{13}°K/sec or more in Si the material has no time to grow crystalline and an amorphous phase can form instead [32] . Apparently the high interface velocities exceed the rate at which atoms can re-arrange as a crystal at the freezing interface, in this case [33] .

If cooling rate is lower than this value the material can re-order when cooling and, if the substrate is able to give it the just hint, a perfect single crystal is produced.

In this model impurities can largely diffuse in the liquid phase over sensible distances giving rise to redistribution effects and other important phenomena which will be treated in this volume [34] . This melting regrowth mode is usually called *Liquid Phase Epitaxy* (LPE).

The other way of producing annealing is by using a scanning c.w. laser (for ex. an Ar laser) over the sample surface. In this case the various parameters can be adjusted so that not enough energy for melting is given but temperature is made to increase somehow below the melting temperature of the material. The annealing proceeds by an accelerated solid-phase epitaxial process similar to conventional oven annealing. The amorphous – crystalline interface moves towards the surface at a rate which increases rapidly at elevated temperatures. Empirical studies [35] of this regrowth velocity show that it can be written in the form of:

$$V = V_0 \exp{(-E_a / kT)} .$$

The values of the constants V_0 and E_a depend of course on the material and the crystalline regrowth orientation. For <100> Si they are $V_0 = 2.9 \times 10^7$ cm/sec and $E_a = 2.7$ eV [36] .

For example, at 900°C, the regrowth velocity is expected to be approximately 7.5 cm/sec, which is adequate to regrow an amorphous layer of 1000 Å in about 1.3 ms.

More recent measurements [37] in As-implanted (100) Si give somehow different values $E_a = 2.6$ eV and $V_0 = 5.75 \times 10^7$ cm/sec.

Re-ordering in solid phase is what is nowadays called *Solid Phase Epitaxy* (SPE). In this case of course, no redistribution of dopants can occur due to the very short time the high temperature is maintained and to the very low diffusion constant ($D \simeq 10^{-11}$ cm^2/sec) in solid state.

This very rough overview of the general processes of laser annealing immediately shows the novelty of the laser treatment.

The very short time during which regrowing occurs makes it possible to avoid heating and consequent contamination of the material which is not directly interested in the treatment (substrate, nearby structures, etc.). The possibility of focalizing the laser beam over very small areas makes also possible very punctual treatments as can be required in the construction of integrated devices.

In the melting case, impurities can diffuse rapidly into the melt and a redistribution occurs after the laser treatment, which can be avoided in solid phase laser annealing.

In both cases, but especially in the former one, cooling rates are extremely fast, and allow to obtain metastable situations in which dopants at concentration values largerly exceeding the thermo-dynamical limit are obtained. Other phenomena can occur in the liquid phase as segregation, solute trapping, cell formation [38] etc.

The properties of the new technique, and the fact that in most cases the treatment can be made just in the open air, seem to open new possibilities of speeding up construction, reduce costs and obtain better devices in some cases.

Lasers have a great flexibility and can be chosen according to which results are requested.

To make an example, shortening the laser wavelength makes the radiation to be absorbed in a thinner and thinner surface layer whose temperature increases the thinner is the layer. Cooling rates become accordingly higher and higher and it is possible to reach that value of $10^{13}°C/sec$ which allows to obtain the single-crystal amorphous transition [39] .

Conversely increasing wavelength increases the laser interested region and makes possible to treat deeper layers of material.

If wavelength is increased beyond the absorption edge of the material, this becomes in principle transparent. However, desordered materials like the amorphous ones have very different absorption coefficients than the corresponding crystalline one, so it may happen than Ge, which is transparent at the CO_2 laser radiation (λ = 10.6 µm) in the single crystal form, is not if amorphous and can be annealed via energy transfer through free electron absorption and in band states.

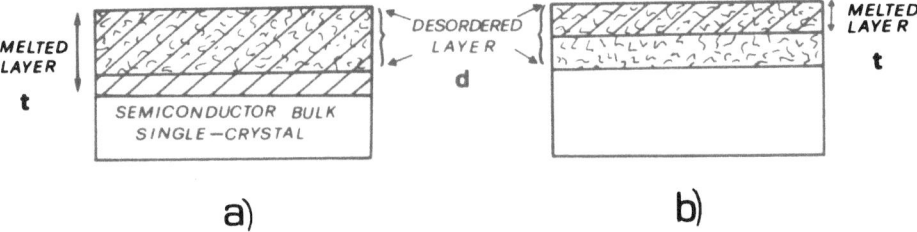

<u>Fig. 4</u> Explains the amorphous to single crystal (a) or the amorphous to polycrystal (b) transition in the liquid phase epitaxy.

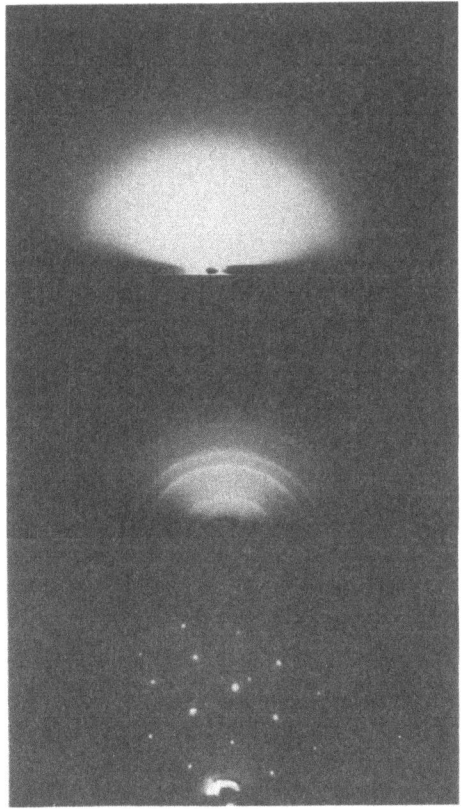

<u>Fig. 5</u> RHEED patterns of a Silicon sample as implanted (top) irradiated with a ruby laser power of 0.6 J/cm^2 (middle) and after 4 J/cm^2 (bottom).

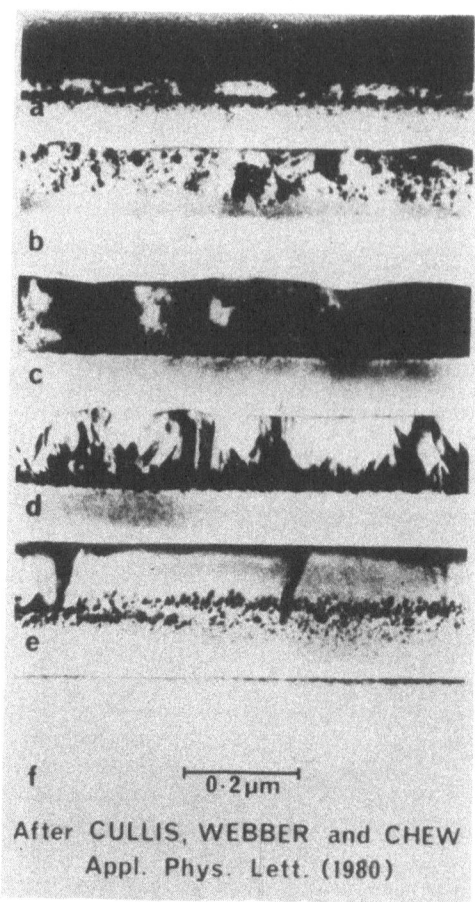

Fig. 6 Transmission electron micrographs of cross sections of (001) Si specimens implanted with 4.10^{15}, 150 KeV As^+/cm^2 (a) as implanted, (b) 0.20 J/cm^2, (c) 0.35 J/cm^2, (d) 0.85 J/cm^2, (e) 1.00 J/cm^2 and (f) 1.20 J/cm^2 (layer surfaces face, upwards) (from [40]).

3. EXAMPLES

A few examples of laser effects are now presented. To start with let us consider a single crystal with an amorphous surface layer of depth d. According to the value of the pulsed laser energy a thickness t of material is molten. If t < d (Fig. 4) upon re-growing the material regrows as polycrystal; if instead t > d regrowth starts from a well ordered substrate and takes place epitaxially. A threshold therefore exists which must be overcome to obtain single crystal regrowth, which depends on the initial amorphous layer thickness. Fig. 5 [13] shows an example of this as seen via RHEED pattern, and Fig. 6 [40] shows the development of grains as the laser fluence is increased.

The laser power density necessary to have this is obtained by calculating the expected thickness of melted material [18][15] .

Calculation is not simple because one must take into account changes with temperature of most thermal and optical parameters of the material during irradiation. There is a lot of numerical calculations, and they are discussed also here [34] . Perhaps one of the most interesting quantities is the thickness of the molten material as a function of the laser energy density. This quantity is reported for Si in several cases [30][41][42][43] .

In the case of SPE the laser is scanned over the semiconductor surface which increases its temperature. If the laser has a Gaussian intensity distribution :

$$I = \frac{P}{\Pi W_o^2} \exp\ (-r^2/W_o^2),\qquad\qquad(2)$$

where r is distance from the center of the beam, P is the total absorbed power, and W_o is the laser beam waist, the temperature rise at the sample surface can be found as :

$$\theta(r) = (\frac{P}{2\sqrt{\Pi}W_o K_o})\ \exp\ (-r^2/2W_o^2)\ I\ (r^2/2W_o^2)\qquad\qquad(3)$$

where $I_o(x)$ is the modified Bessel function of order zero and K_o is the thermal conductivity assumed constant. If k is assumed to be temperature dependent with a law :

$$k(T) = A/(T-T_k)\ ,\qquad\qquad(4)$$

with f.e. A = 299 W/cm and T_k = 99°K, one obtains the true temperature :

$$T(r) = T_k + (T_o - T_k)\ \exp\ [\theta(r)/(T_o-T_k)]\ ,\qquad\qquad(5)$$

being T_o the back-surface temperature of the sample.

This gives a maximum temperature at the spot center:

$$T_{max} = T_k + (T_o - T_k) \exp (P/2\sqrt{\Pi} \ W_o A) \qquad (6)$$

and a reacted-layer thickness (in case of a single scan at velocity v):

$$z \simeq R_o \ t_{eff} \ \exp (-E_a/ \ kT_{max}), \qquad (7)$$

where the effective annealing time $t_{eff} \simeq (2 \ W_o/v)f$; f is a "dwell-time reduction factor" of the order of 0.2-0.4, for Si, $E_a = 2.35$ eV is the activation energy, and $R_o = 3.2 \times 10^{14}$ Å/sec [44].

In c.w. laser annealing an interesting regime can be obtained, known as *explosive crystallization*. The term refers to the rapid and self-sustaining crystallization of an entire amorphous film after crystallization is initiated at any point of the film.

Initiation can be by a pinprick or by a localized heat pulse from a laser, for example.

The velocities of propagation of the crystallization front have been measured to be lm/sec. This phenomenon has been studied histo-rically for a number of amorphous materials, but most extensively in the 1970's by Mineo et al. [45] and more recently has been found in laser annealing [46][47].

The mechanism of explosive crystallization involves the release of the energy of crystallization from a small region triggered by the initiating disturbance. This energy heats the adjacent material which in turn crystallizes and drives the reaction outward. The high speed of propagation of the crystallization front was first associated exclusively with melting. It was demonstrated [48] that for a-Ge the explosive crystallization proceeds via an intermediate melting or liquid phase process.

The phenomenon can however occur either in the liquid or in the solid phase; which phase dominates depends on the laser scan speed, the proximity of the amorphous material melting isotherm and the magnitude of the "thermal kick" provided by fluctuations in laser power or at localized regions of higher absorption [49]. By adjusting the laser scan velocity and/or laser output power it is possible to have explosive crystallization via the liquid phase or in solid state [49]. The two regimes are clearly distinguished by their apparence. In the liquid phase assisted phenomenon the crystallization is characterized by the appearance of crystal bounda-ries in the form of crescents where the surface morphology is very

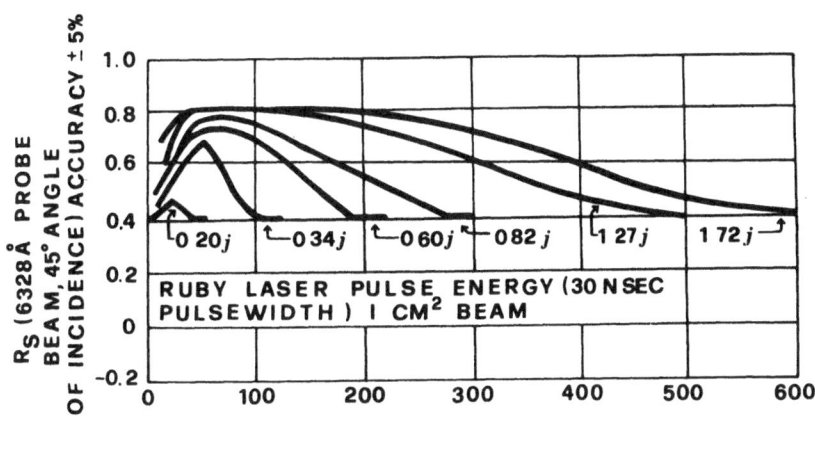

T, NANOSECONDS

Fig. 7 Time changes in reflectivity of Ge irradiated with a Q-switched ruby laser (from [3]).

rough. In solid phase explosive crystallization the surface is essentially as smooth as in ordinary SPE.

In order to eliminate the periodic crescent boundaries it is necessary to increase the scan speed above 100 cm/s.SP explosive crystallization occurs first and if the laser scan speed is adjusted to sustain the growth front.

4. TO BE MELTED OR NOT TO BE ?

Until now we have presented annealing results basing their explanation on a simple thermal model. From a theoretical point of view, the simple melting theory has been questioned [50] and there has been much debate as to whether the laser pulse energy is transferred quickly to phonons in the semiconductor, thus producing melting, or whether the absorbed energy creates instead a hot, dense plasma while the lattice temperature remains low [51] . A discussion of this second model is presented in this volume by Van Vechten. We discuss here some relevant experiments which have been made to understand what is really happening.

One of the simplest experiment is to look at the change in reflectivity during laser irradiation.

14

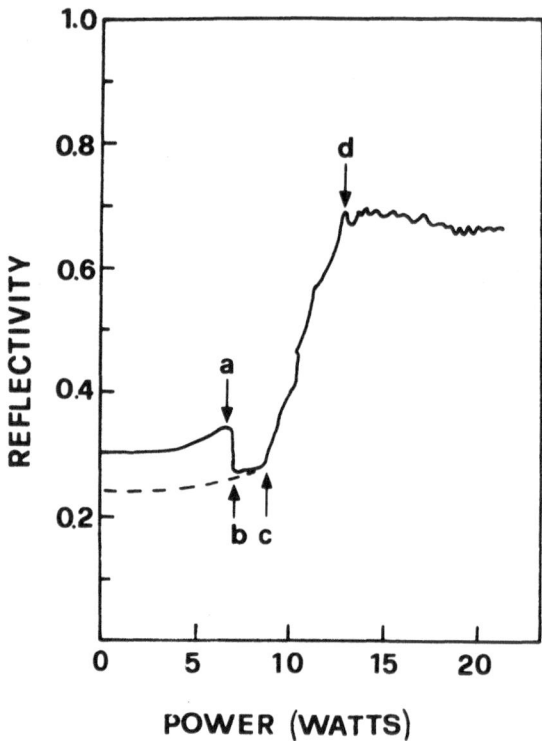

<u>Fig. 8</u> He-Ne laser surface reflectivity from ion implanted silicon
during cw Ar laser annealing at different power levels (from [52]).

Such experiments were already done since 1964, as already noted.
Fig. 7 [3] shows the results obtained in germanium irradiated with
a Q-switched ruby laser of t_D = 30 nsec. The same experiment was
repeated 15 years later by Auston and collaborators obtaining substantially identical results. In 15 years only time resolution
was increased and now we can fully appreciate the time evolution of
the reflectivity signal.

In Auston's experiment the optical reflectivity of the sample
was measured during annealing in both c.w. or pulsed laser irradiation. We will first discuss the c.w. case. When c.w. laser annealing is performed in SPE (on Si irradiated with an Ar laser) reflectivity changes as a function of laser power as shown in Fig. 8 [52] .
Below a given laser power the measured reflectivity is that of the
amorphous layer. Increasing the power, the thickness of the amorphous layer decreases as annealing from the rear interface progresses.

At 6.5. W (beam waist 50 μm, scan speed 0.5 cm/sec) the reflectivity goes through a peak (point a in the figure) which is produced by constructive interference between the regrowth interface and the front surface of the crystal. Increasing the laser power further results in a drop in reflectivity, indicating that the thermal anneal has proceeded to the surface at point b. A similar reflectivity scan is shown by the dashed line for a previously annealed sample; the interference effect is obviously absent. As the power is increased between points c and d, a large change in reflectivity is observed produced by silicon melting. The absolute values of reflectivity below point a, between points b and c, and above point d, agree well with those of a-Si, crystalline Si · and metallic liquid Si at the Ar wavelength, respectively.

The onset of melting at point c corresponds to a peak surface temperature of 1685° K. Using this value and calculation of the surface temperature as a function of laser power, the peak surface temperature at the annealing threshold (point b) was estimated to be 1100°K which agrees quite well with predictions.

Taking instead the annealing power fixed and monitoring reflectivity changes as a function of time, the method can be used to study directly the kinetics of SPE, as shown in Fig. 9 [37].

As the amorphous film crystallizes epitaxially from the substrate, the reflected intensity of the probe laser (He-Ne) varies in time due to alternating constructive and destructive interference between reflected light from the wafer surface and light reflected from the advancing epitaxial interface. Since the temporal variation in reflectivity is caused by a change in thickness of the amorphous film, these data can be used to deduce the absolute position of the amorphous/crystalline interface as well as its temporal variation by comparing the measured reflectivity as a function of time with the calculated reflectivity as a function of film thickness. This method, therefore allows the growth rate to be determined as a function of the position of the interface.

Reflectivity measurements were made also in pulsed laser annealing [53][54][55]. Fig. 10 [54] shows a typical time-resolved reflectivity measurement for the case of 5300 Å annealing of an implanted Si sample. The incident pulse had a duration of 30 ns and energy of 2.75 J/cm^2. As the laser turns on, the initial reflectivity of the amorphous layer R_a rises with the increasing temperature of the sample surface to a value R'_a when the surface melts and an abrupt increase in reflectivity occurs due to the metallic nature of liquid silicon. The reflectivity remains high at the value R_e as the liquid-solid interface penetrates more deeply into the sample and reverses its direction some time after the annealing laser pulse has ended. As the liquid-solid interface proceeds towards the surface, the reflectivity decreases to R'_c of the hot solid. The surface is

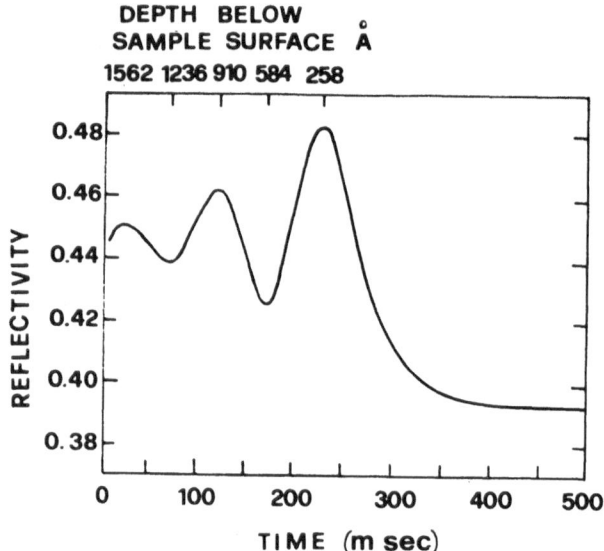

**DEPTH BELOW
SAMPLE SURFACE Å**
1562 1236 910 584 258

Fig. 9 Time variation of the reflected He-Ne probe laser intensity
during laser-induced SPE. Incident Ar laser power 8.5 W; laser
spot radius 83 µm, substrate temperature, 673°K. Upper scale shows
the absolute position of the crystal/amorphous interface below the
sample surface (from [37]).

melted for a total time τ. The fall time τ_f is the time required
for the regrowth front to move through one optical skin depth and
is a measure of the regrowth velocity. As the crystal cools further,
the room temperature value R_c of the crystalline phase is obtained.
A better information could be reached if a direct temperature measurement could be done during irradiation. Raman scattering seems
able to assess lattice temperature. The principle of the measure
is very simple; the ratio of Stokes to anti-Stokes Raman scattering
counting rates is given by :

$$\frac{R_s}{R_{As}} = \frac{(\alpha_L + \alpha_{AS})\,\omega_S^3\,\sigma(\omega_L,\omega_S)\,e^{\hbar\omega_o/kT}}{(\alpha_L + \alpha_S)\,\omega_{AS}^3\,\sigma'(\omega_L,\omega_{AS})} \qquad (8)$$

where ω_L, ω_S, ω_{AS} and ω_o are the laser, Stokes anti-Stokes and phonon frequencies respectively; $\sigma(\omega_L,\omega_S)$ and $\sigma'(\omega_L,\omega_{AS})$ are the Stokes and anti-Stokes Raman scattering cross-section; the α's are the

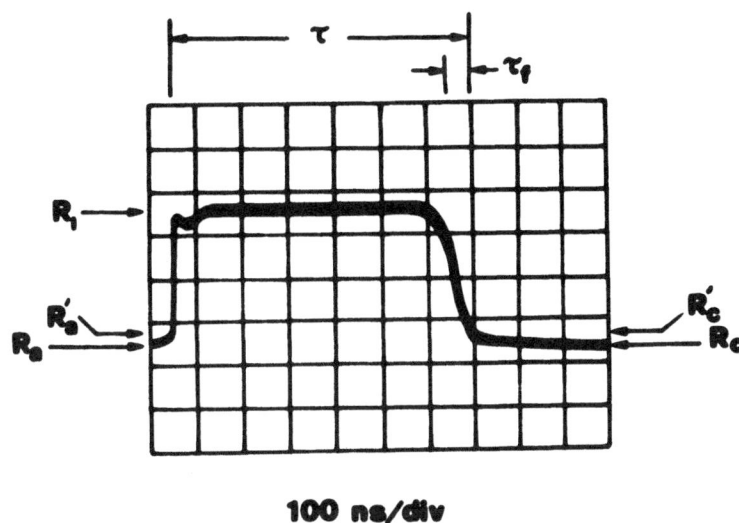

100 ns/div

Fig. 10 He-Ne laser surface reflectivity from silicon following a 30 ns pulse from the second harmonic of a Nd-YAG (from [54]).

absorption coefficients at laser, anti-Stokes and Stokes frequencies and the term $\exp(\hbar\omega_o/ kT)$ follows directly from the ratio $[n(\omega_o)+1]$ $/ n(\omega_o)$ where $n(\omega_o)$ is the phonon occupation probability at the lattice temperature T. This exponential term assumes thermal equilibrium in phonon population.

A measurement of the ratio R_S/R_{AS}, corrected for further errors as the ones due to differences in spectrometer and detector efficiency at the Stokes and anti-Stokes frequencies, is able therefore to give directly the temperature of the lattice at each time.

Compaan and Lo have performed a series of very beautiful measurements with this method both in the case of SPE [56] and LPE [57] [58][59] .

Fig. 11 [56] shows the peak temperatures as a function of Ar laser power in the case of SPE together with three theoretical curves. Data were obtained from further correcting Eq. 8 because the spectrometer response was a spatial integration of signal. It was necessary to deconvolute also the effect of the radial temperature

18

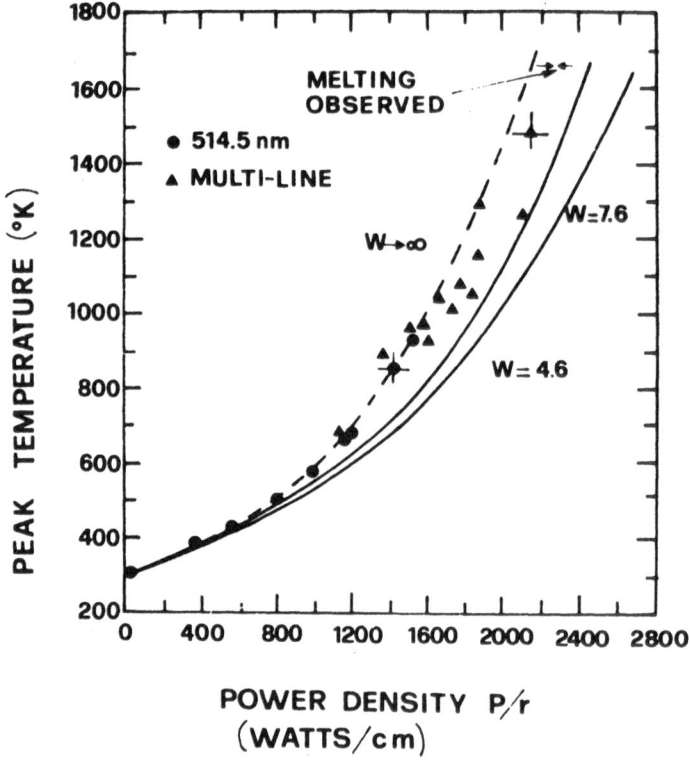

POWER DENSITY P/r
(WATTS/cm)

Fig. 11 Peak temperature versus absorbed laser power density.
Solid curves are calculated peak temperatures using room-temperature
absorption constants and parameters appropriate for 514.5 μm (W=4.6)
and multi-line excitation from on Ar laser (W=7.6). Dashed curves
show peak temperatures in the limit W → ∞ (from [56]).

profile from the observed R_{AS}/R_S using calculated temperature profiles.

In Fig. 11 the theoretical predicted peak temperature depends
on the ratio W = rα of the spot radius r to the light absorption
length (1/α). Due to the large increase in α with temperature the
better theoretical curve is the one with W → ∞.

When silicon is uniformly heated, anharmonic effects produce
a decrease in the Raman frequency at a rate of approximatively
$2cm^{-1}/100°C$ (60). Under laser heating, the semiconductor is free
to expand only in the direction normal to the surface. Parallel to

surface the heated region is "clamped" by the surrounding unheated material.

The position, width and shape of the Raman peaks are consistent with effects to be expected in nonuniform heated silicon.

The deformation of the heated surface during c.w. laser heating can be measured also with a simple interferometric method [61] and the vertical displacement of the surface and its spatial profile as a function of time can be measured.

The Raman method of temperature measurement has also been applied by Compaan and Lo to pulsed laser annealing [57][58].

Their results which were an averaging over a 10 nsec period have given a temperature well below the melting point of silicon [57].

Repeated measurements by von der Linde and Wartmann with 1 ns time resolution and well-defined spatial resolution give instead different results [62]. The reasons of the discrepancy may be found in the use of UV radiation as probing which avoids spatial averaging of the temperature under the probed surface. This point has also been raised by Wood and collab. [63].

Other evidence of the temperature reached by surface upon annealing comes from X-ray diffraction and electron and ion thermal emission.
Time resolved X-ray diffraction measurements of lattice strain in silicon during pulse-laser annealing have been made with nanosecond resolution by using synchrotron radiation by Larson et al. [64]. Analyses of the strain in pure and B-implanted Si in terms of temperature indicate high temperatures and evidence for near-surface melting in qualitative agreement with the melting model.

These measurements provide direct information on the time duration and depth distribution of near-surface lattice strains associated with laser annealing of Si. The extended Bragg scattering intensity that is found in the vicinity of Bragg reflections as a result of Bragg-like scattering from near surface strains was used.

Transforming the strain distribution so derived into lattice temperatures give the temperature distributions as a function of depth as shown in Fig. 12.

Information about the electron and lattice temperatures during and following laser irradiation with 20 nsec pulses at 5320 Å is provided also by the emission of charged particles from the silicon surface [32].

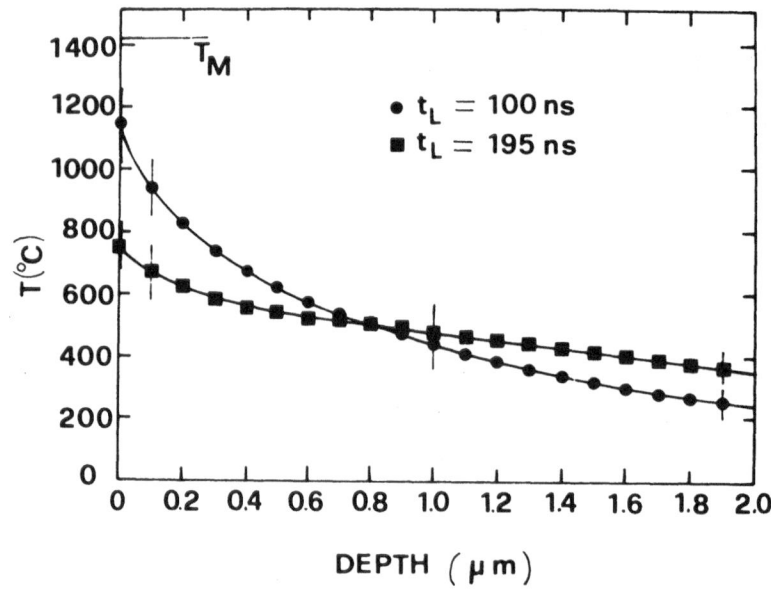

<u>Fig. 12</u> Temperature as a function of depth as derived from strain profiles determined from X-ray scattering at 100 and 195 ns after laser pulses (15 ns, ruby, 1.35 J/cm^2) (from [64]).

At the threshold for phase transformation (0.2J/cm^2 in this case) no emission from Si is observable. If all the absorbed laser energy remained in the electron-hole plasma, 10^{22} carriers/cm^3 would be created with an excess kinetic energy of about 1 eV. Fast Auger recombination would reduce the carrier density and increase the electron temperature.

Thus the electron temperature should excess 10^4°K, if the lattice remained cool, since diffusion is negligible during the picosecond pulse. In that case thermionic emission of electrons from the hot plasma during the laser pulse should already have exceeded 10^{11} charges, five orders of magnitude above the detection limit of instrumentation which was 10^6 charges. Applications of the Richardson-Dushman equation shows that the electron temperature remained below 5000°K during the 20 ps pulse. This indicates that a considerable fraction of the absorbed laser energy was transferred to the lattice during the pulse. The data are consistent with thermal equilibrium between carriers and lattice. At 0.2 J/cm^2, the lattice just reaches the melting point. For energy fluences higher than 0.26 J/cm^2, positive and negative particles are emitted in equal amount. This is an indication that evaporation of Si atoms from the surface is taking place, a fraction of which is ionized.

Other experiments are listed in reference [65] . The presence of a plasma of carriers with a recombination time shorter than some tens ps has also been proved together with convincing evidence that above a definite threshold (0.2 J/cm^2, for 25 ps pulses in Si) liquid silicon is produced [66] .

A number of other experiments have been made which on the contrary claims for a non-melting interpretation.

Among these, apart Compaan and Lo, Raman measurements we referred to previously, experiments in which contemporary measurement of surface reflectivity and sample transparency were made during laser irradiation [59][67] .

In some case the absence of the expected transparency total fading during the liquid phase was found.

This result has however been questioned [63] and will be discussed in this volume by A. Compaan. It is also worthwhile to point out that evidence for carriers plasma formation followed by melting has been recently obtained and is discussed in this volume.

From all the data it seems reasonable to assume that a dense plasma of carriers is produced by the powerful laser pulse, which subsequently recombines in times of the order 1-10 psec transferring energy to the lattice in these short times. Subsequent melting occurs which is the final mechanism responsible for the observed reordering processes [91] .

5. IONIZATION ENHANCED ANNNEALING AND LOW POWER EFFECTS

Notwithstanding the growing experimental evidence in support of a melting model of pulsed laser annealing at the high laser fluences, non-thermal effects can be found instead when low laser fluences are used.

These effects can be evidenced when point defects or agregates of defects are considered.

The interaction between defects and laser light is at present not yet completely understood but we may find roughly speaking three different types of effects :
a) in case of laser melting upon re-solidification beneath the solidified layer new defects have been found which have been undoubtely introduced by the laser treatment [68][69][70];
b) when laser is used to heat the material for some long time, usual thermally activated annealing of defects can occur [68] . For

example, a sample of P-doped Si was bombarded with 1 MeV electrons and exposed to laser beams under a variety of conditions. A c.w. beam from a Nd : YAG laser induced annealing of A (Oxygen-vacancy) and E (Phosphor-vacancy) centers, consistent with an instantaneous rise in sample temperature up to 350°C. A Q-switched Nd:YAG laser produced no defect annealing at any incident power below that producing surface damage [71] .
These results were consistent with the conclusion of damage annealing under pure thermal effects;
c) finally there is growing evidence of laser annealing of defects which is not connected to thermally activated effects but seems rather to be associated with the laser produced ionization.

Electronic stimulation of defect processes is a phenomenon known from some years and for what concerns here can be roughly divided into [71] *configurational processes* which relate to a change in local electronic configuration and a modification of the defect structure and reactivity, and *vibrational processes* which refer to the excitation of the defect into an energetic vibrational state by an electronic transition at a localized state. These last processes are also known as *recombination enhancement*.

The phenomenon of recombination-enhanced defect migration is now a well recognized important factor for defects in many III-V semiconductors where many of the defects produced by radiation damage have been found to exhibit enhanced recovery under minority-carrier injection conditions [72] .

The mechanism for this enhanced migration is that the electronic energy released upon carrier capture and recombination supplies part of the energy necessary for the defect to make a diffusional jump [73] .

A few years ago migration of room-temperature-stable defect in Si was reported [74] under injection conditions. This observation shows that also in semiconductors with relatively small band-gap, injection-enhanced motion is possible not only for defects stable at cryogenic temperatures (vacancies, interstitials, etc.) [75] but also for defects stable at room temperature.

The process observed by Watkins et al. [74] uses approximatively 0.9 eV to assist the motion with the migration rate enhanced by a factor of 10^8 under injection conditions at room temperature. The defect was identified as interstitial Al. Energy is provided either by electrical carrier injection or by laser illumination. Charge state effects belong instead to the class of configurational processes and are another class of electronic enhancement of defect reaction mechanisms [76] .

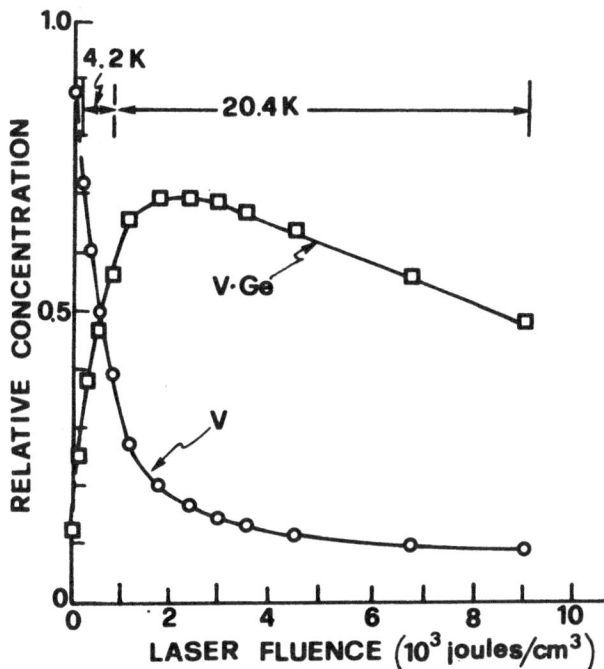

Fig. 13 Nd-YAG laser annealing of vacancies and vacancy-Ge pairs in p-type silicon at 20.4°K (from [78]).

In the case of diffusion the normal ionization-enhanced diffusion mechanism occurs when the defect exhibits a diffusion coefficient which has two distinct different activation energies corresponding to its two possible charge states. These normal ionization enhanced diffusion mechanisms appear to be quite common in silicon [77] .

An example of laser annealing of point defects is shown in Fig. 13 [78] . Samples of p-Si doped with Ge were first irradiated at 20.4°K with electrons to produce isolated vacancies (V) and vacancy-Ge pairs (V-Ge). Following the intensities of the EPR spectra the samples were then irradiated with 1.064 μm light from a Nd-YAG laser while keeping the sample at 20.4° K. The chosen laser photon energy (1.17 eV) is ideally matched to the band gap of Si; the absorption coefficient at this temperature (0.5-1 cm^{-1}) assuring uniform electron-hole pair excitation throughout the bulk of the sample. In Fig. 13 it is shown that as a function of illumination time, the vacancies convert quickly to V-Ge pairs through an ionization-enhanced migration of the vacancies to produce the pairs, and also, at the same time, the reverse reaction.

Charge state effects were also found in GaAs [79] .

Experimentally, even without any sure assignement of the phenomenon to a determinate mechanism there is growing evidence of the effects of ionization on annealing properties of defects. Ionization produced by low energy electron irradiation has been proved to be effective in changing annealing properties of point defects [80] .

Evidence has been also found of the effect of ionization (produced again by an electron beam) on crystal growth [81] .

Results on more complicated defects have been obtained with Si implanted with non-amorphizing doses of B-ions [82] .

The existence of extended boron-vacancy (B-V) complexes in regions with a high density of intrinsic point defects which are predominantly of the vacancy type was previously hypotized by Frank and Berry [83] , to explain the existence of a negative or reverse annealing in the 500-700° C range. Laser treatment has been shown to affect these complexes.

The results by J. Suski et al. [82] show that the electrical activity of implanted Si layers increases gradually with the increase of pulse energy density thus showing the lack of annealing energy threshold. Final hole concentration is limited by the number of electrically inactive extended B-V pairs that cannot be annealed by the applied energy densities of laser pulses (between 0.4 and 1 J/cm^2 i.e. below melting threshold) (see Fig. 14). In the figure the laser energy is increasing from 0.4 to 0.7 in curves from 5 to 1 showing the gradual disappearance of the reverse annealing.

These results show that post-implantation defects anneal with some activation energy under laser illumination. Previous measurements by Rzewuski et al. [84] of activation energy for thermal annealing of compensating defects and B-V pairs gave 3.2 and 5.6 eV, respectively. Because of the very high values of these energies, purely thermal processes cannot account for the laser pulse annealing. The recovery of electrical parameters of B-implanted layers was therefore attributed to annihilation of vacancy clusters associated with ionization effects, presumably recombination-enhanced annealing.

Reversible photostructural and photoconductivity changes have been discussed by M. Wautelet and M. Failly-Lovato [85] using simple models involving electron-phonon couplings.

Finally there is growing evidence of reordering produced under low power irradiation.

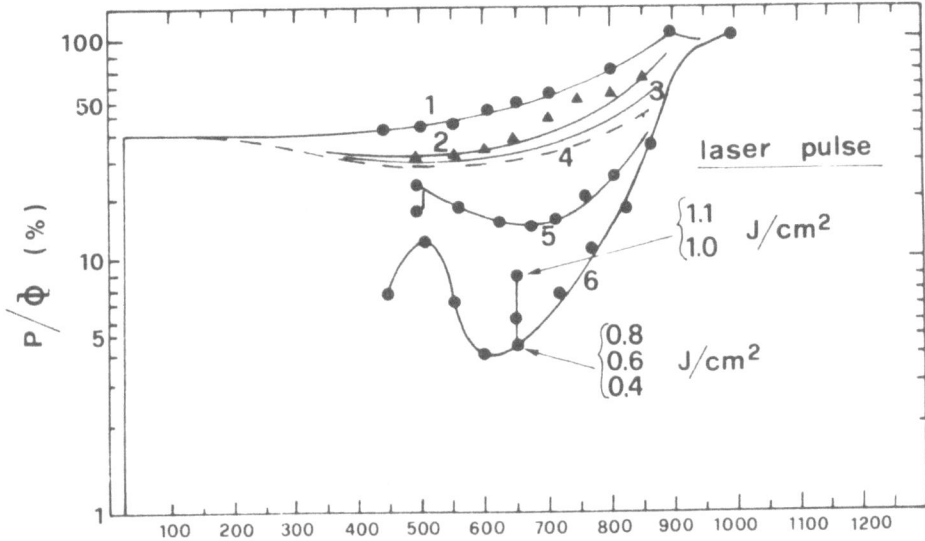

Fig. 14 Fraction of free carrier concentration versus annealing temperature for B implanted laser annealed Si samples. Curve 1 : as implanted; Curves 2 and 5:preannealed samples to 500°C; Curves 3 and 4:preannealed samples to 550°C Curve 6:sample annealed in dark. Laser pulses applied to samples : 1 through 3, 0.7 J/cm²; 4,0.56 J/cm²; 5,0.4 J/cm².

Laude et al. [86] found crystallization of a-Ge films irradiated with a dye-laser, with pulses of 10^{-6}s, repetition rate 16.7 Hz, $h\nu$ = 2.08 eV and power density about 30 Kw/cm², irradiated for several hours. The temperature increase was calculated as negligible.

More recently he and collab. have found evidence of laser ordering in Ge and Si under short pulses at low power in the range 0.01 to 0.15 J/cm² with star formation [87] .

The star formation may be viewed as the product of a nucleation event in the film, followed by a very rapid growth. Most of the stars show a fringe of polycrystals at their uppermost border. The rise of these polycrystals and their distribution would indicate that they might be the results of a thermally-induced crystallization initiated by the latent heat being liberated by the a-x transition within the larger crystallites belonging to the stars.

The radial distribution of these larger crystallites is a clear evidence that their growth is initiated around a given site in the film where ordering would develop preferentially due to local non-equilibrium conditions. The origin of the trigger of nucleation is not clear and may not be connected to a simple thermal kick. In any case important modification in the electronic structure of solids under laser irradiation may be produced [88] as discussed further in this volume.

Vitali et al. [89] also have demonstrated that reordering can be produced gradually under repetitive low power pulse irradiation with repetition frequencies as low as 10^{-2}Hz in Ge. RHEED analysis of the sample increasing the number of superposed pulses shows that the reordered fraction of material is increasing continuously with the increasing number of pulses until the whole layer is completely reordered. Reordering depends on surrounding material and not simply on the underlying layer. This can be seen in the different behaviour found in solid state annealing in the low power pulsed regime. In this case true amorphous material, obtained for example by glow-discharge, regrows single-crystal starting from the bottom at the single-crystal amorphous interface. On the contrary desordered ion implanted material can grow from surface using the surrounding material of the still ordered matrix. One possible explanation of this behavior can be found in the production of high stresses associated with the very large thermal gradients produced during laser annealing.

The interaction of ionizing radiation and defects in semiconductors appear still to need deeper study to be fully understood and can deserve interesting results.

REFERENCES

1. For a general review see Bertolotti, M. and Vitali, G. in Current Topics in Materials Sciences, vol. 8 ed. E. Kaldis, Amsterdam, North-Holland Pu. Co., 1982, p. 95.
2. Carmichael, C.H. and Simpson, G.N., Nature 202 (1964) 787;
    Birnbaum, M., J. Appl. 36 (1965) 657;
    Bell, M.I., Laser and Their Applications IEEE London 1964, pp. 34-1, 34-2;
    Birnbaum, M. and Stocker, T.L., Brit. J. Appl. Phys. 17 (1966) 461;
    Birnbaum, M. and Stocker, T.L., IEEE J. Quant. Electr. QE-2 (1966) 184;
    Birnbaum, M. and Stocker, T.L., J. Appl. Phys. 39 (1968) 6032.
3. Sooy, W.R., Geller, M. and Bortfeld, D.P., Appl. Phys. Lett. 5 (1964) 54.
4. See [1] for biliographic list.

5. Bertolotti, M., de Pasquale, F., Marietti, P., Sette, D. and
   Vitali, G., J. Appl. Phys. 38 (1967) 4088.
6. Bertolotti, M., Marietti, P., Sette, D., Stagni, L. and Vitali,
   G., Rad. Eff. 1 (1969) 161.
7. Ehrlich, D.J., Brueck, S.R.J. and Tsao, J.Y. XIIth Intern.
   Quantum Electr. Conference Munich, June 22-25, 1982;
   Brueck, S.R.J. and Ehrlich, D.J., Phys. Rev. Lett. 48 (1982)
   1678.
8. Young, J.F., Sipe, J.E., Gallant, M.I., Preston, J.S., van Driel,
   H.M., in Laser and Electron-Beam Interactions with Solids,
   B.R. Appleton and G.K. Celler eds., New York, North Holland,
   1982, p. 233; see also Keilmann, F. and Bai, Y.H., Appl. Phys.
   A (1982);
   Fauchet, M.P. and Siegman, A.E., Appl. Phys. Lett. 40 (1982)
   824, but see also Young, J.F., Sipe, J.E., Preston, J.S. and
   van Driel, H.M., Appl. Phys. Lett. 41 (1982) 261.
9. Fairfield, J.M. and Schwuttke, G.H., Solid State Electr. 11
   (1968) 1175;
   Harger, F.E. and Cohen, M.I., Solid State Electr. 13 (1970)
   1103;
   see also US patent 3 585 088, June 15, 1971 under the names
   Schwuttke, G.H., Ross, R.F. and Howard, J.K. .
10. Marquardt, C.L., Giuliani, J.F. and Fraser, F.W., Rad. Eff. 23,
    (1974) 135;
    Giuliani, J.F. and Marquardt, C.L., J. Appl. Phys. 45 (1974)
    4993.
11. Klimenko, A.G., Klimenko, E.A. and Donin, V.I., Kvant. Elec-
    tron. 2 (1975) 2356 and Soviet J. Quant. Electron. 5 (1976)
    1289;
    Shtyrkov, E.I., Khaibullin, I.B., Zaripov, M.M., Galyatudinov,
    M.F. and Bayazitov, R.M., Fiz. Tekh. Poluprovódu. 9 (1975)
    2000 and Soviet Phys. Semicond. 9 (1976) 1309;
    Shtyrkov, E.I., Khaibullin, I.B., Galyatudinov, M.F. and Zari-
    pov, M.M., Opt. Spektrosk. 38 (1975) 1031.
12. Vitali, G., Bertolotti, M., Foti, G., Rimini, E. in Proc. 7th
    Int. Conf. on Amorphous and Liquid Semiconduc. Edinburgh June
    27-July 1, 1977 ed. W.E. Spear (Centre for Industrial Consul-
    tancy and Liaison, Univ. Edinburgh, 1977, p. 24).
13. Vitali, G., Bertolotti, M., Foti, G., Rimini, E., Phys. Lett.
    63A (1977) 351.
14. Foti, G., Rimini, E., Vitali, G. and Bertolotti, M., Appl.
    Phys. 14 (1977) 189.
15. Foti, G., Rimini, E., Bertolotti, M. and Vitali, G., Phys.
    Lett. 65A (1978) 430.
16. The model as it was first used (see [17],[18]) was fully ex-
    plained later in the paper Bertolotti, M., Vitali, G., Rimini,
    E. and Foti, G., J. Appl. Phys. 50 (1979) 259. The Catania
    group worked also some computer calculations which were publis-
    hed nearly contemporarily; see Baeri, P., Campisano, S.U.,
    Foti, G. and Rimini, E., J. Appl. Phys. 50 (1979) 788.

17. Foti, G., Rimini, E., Bertolotti, M. and Vitali, G., Proc. of the First USA-USSR Seminar on Ion Implantation, compiled by Stromberg, R.L. and Corbet, J.W. (Dept. Phys.,State Univ. of New York at Albany, July 1977) p. 99.

18. Foti, G., Rimini, E., Bertolotti, M. and Vitali, G., in Thin Film Phenomena Interfaces and Interactions, eds. J.E.E. Baglin and J.M. Poate, vol. 78-2 (The Electrochem. Soc. Inc., Princeton, N.J. 1978), p. 88.

19. Williams, J.S., Brown, W.L., Leamy, H.J., Poate, J.M., Rodgers, J.W., Rousseau, D., Rozgonyi,G.A., Shelnutt, J.A. and Sheng, T.T., Appl. Phys. Lett. 33 (1978) 542.

20. Auston, D.H., Golovchenko, J.A., Smith, P.R., Surko, C.M. and Venkatsan, T.N.C., Appl. Phys. Lett. 33 (1978) 539.

21. Gat, A., Gibbons, J.F., Magee, T.J., Peng, J., Deline, V.H., Williams, P. and Evans, C.A.,Jr., Appl. Phys. Lett. 32 (1978) 278.

22. Lau, S.S., Tseng, W.F., Nicolet, M.A., Mayer, J.W., Eckardt, R.C., Wagner, R.J., Appl. Phys. Lett. 33 (1978) 130;
Hoonhout, D., Kerkdik, C.B., Saris, F.W., Phys. Lett. 66A (1978) 145;
Rovesz, P., Farkas, G., Mczey, G., Gyulai, J., Appl. Phys. Lett. 33 (1978) 431;
Bertolotti, M., Vitali, G. and Spear, W.E., in Laser Solid Interactions and Laser Processing, 1978, eds. S.D. Ferris, H.J. Leamy, J.M. Poate, New York, Am. Inst. Phys., 1979, p.492.

23. Vitali, G., Bertolotti, M., Foti, G. and Rimini, E., Appl. Phys. 17 (1978) 111.

24. Cohen, R.L., Williams, J.S., Feldman, L.C. and West, K.W., Appl. Phys. Lett. 33 (1978) 751;
Bomke, H.A., Berkowitz, H.L., Harmatz, M., Kronenberg, S. and Lux, R., Appl. Phys. Lett. 33 (1978) 955 .

25. Lau, S.S., von Allmen, M., Golecki, I., Nicolet, M.A., Kennedy, E.E. and Tseng, W.F., Appl. Phys. Lett. 35 (1979) 327.

26. See e.g. Laser and Electron Beam Processing of Electronics Materials, eds.C.L.Anderson, G.K. Keller and G.A. Rozgonyi (New York, The Electrochemical Soc. Princeton, 1979).

27. Lietoila, A. and Gibbons, J.F., Appl. Phys. Lett. 34 (1979) 332.

28. The mechanisms of coupling beam energy to a solid can be found f.e. in M. von Allmen in Laser and Electron Beam Processing of Materials, eds. C.W. White and P.S. Peercy, New York, Academic Press, 1980, p. 6 or von Allmen, in Physical Processes in Laser Irradiated Materials, ed. M. Bertolotti, Plenum Press in print.

29. See lectures by Voos, M., Dewel, G. and Smirl, A.L. this volume.

30. Last calculations are by Wood, R.F. and Giles, G.E., Phys. Rev. B 23 (1981) 2923 ; Lietoila, A., Gibbons, J.F., Appl. Phys. Lett. 40 (1982) 624 and J. Appl. Phys. 53 (1982) 3024;

Bryant, G., Kelly, P., Ritchie, D., Braunlich, P., Schmid, A., Phys. Rev. B25 (1982) 2587.

31. See lectures by Binder, K. and Rimini, E., this volume.
32. Liu, J.M., Yen, R., Kurz, H. and Bloembergen, Appl. Phys. Lett. 39 (1981) 755;
    Yen, R., Liu, J.M., Kurz, H. and Bloembergen, N., Appl. Phys. A 27 (1982) 153;
    see also lecture by Kurz, H.,this volume.
33. Spaepen, F. and Turnbull, D., Laser-Solid Interactions and Laser Processing, ed. S.D. Ferris, H.J. Leamy and J.M. Poate, (New York, Am. Inst. Phys., 1979) p. 73.
34. See for ex. Rimini, E., this volume.
35. See for ex. Csepregi, L., Mayer, J.W. and Sigmon, T.W., Phys. Lett. A54 (1975) 157.
36. Nishi, H., Sakurai, T. and Furu, T., J. Electrochem. Soc. 125 (1978) 461.
37. Kokorowski, S.A., Olson, G.L. and Hess, L.D., J. Appl. Phys. 53 (1982) 921.
38. Narayan, J., Naramoto, H. and White, C.W., J. Appl. Phys. 53 (1982) 912; C.W. White in Laser and Electron-Beam Interactions with Solids, eds. B.R. Appleton and G.K. Celler, (New York, North Holland, 1982), p. 109;
    Wood, R.F., Kirkpatrick, J.R. and Giles, G.E., Phys. Rev. B23 (1982) 5555;
    Wood, R.F., Phys. Rev. B25 (1982) 2786.
39. See [30] and also Liu, P.R., Yen, R., Bloembergen, N. and Hodgson, R.T., Appl. Phys. Lett. 34 (1979) 864;
    Tsu, R., Hodgson,R.T., Tan, T.Y. and Baglin, J.E., Phys. Rev. Lett. 42 (1979) 1356;
    Cullis, A.G., Webber, H.C. and Chew, N.G. in Laser and Electron beam Interactions with Solids, eds. Appleton and G.K. Celler (New York, North Holland, 1982), p. 131.
    Bloembergen, N., Kurz, H., Liu, J.M., Yen, R., ibidem p. 3,
    Yen, R., Liu, J.M., Kurz, H., Bloembergen, N., ibidem p. 37.
40. Cullis, A.G., Webber, H.C. and Chew, N.G., Appl. Phys. Lett. 36 (1980) 547.
41. Bertolotti, M. and Sibilia, C., IEEE J. Quant. Electr. QE-17 (1981) 1980.
42. Bell, R.O., Toulemonde, M. and Siffert, P., Appl. Phys. 19 (1979) 313.
43. Baeri, P. in Laser and Electron-Beam Interactions with Solids eds. B.R. Appleton and G.K. Celler (New York, North Holland, 1982) p. 151.
44. Gold, R.B. and Gibbons, J.F., J. Appl. Phys. 51 (1980) 1256.
45. Mineo, A., Matsuda, A., Kurosu, T., Kikuchi, M., Solid State Comm. 13 (1973) 329;
    Takamori, T., Messier, R., Roy. R., Appl. Phys. Lett. 20 (1972) 201 and J. Mat. Sciences 8 (1973) 1809;
    Messier,R.,Takamori, T.,Roy,R., Solid State Comm.16 (1974) 311;
    Matsuda, A., Mineo, A., Kurosu, T., Kikuchi, M., Solid State

Comm. 13 (1973) 1165;
Kikuchi, M., Matsuda, A., Kurosu, T., Mineo, A., Callanan, M.J.,
Solid State Comm. 14 (1974) 731.

46. Fan, J.C.C., Zeiger, H.J., Gale, R.P. and Chapman, R.C.,
Appl. Phys. Lett. 36 (1980) 158.

47. Gold, R.B., Gibbons, J.F., Magee, T.J., Peng, J., Ormond, R.,
Deline, V.R., Evans, C.E.,Jr., in Laser and Electron Beam
Processing of Materials, ed. C.W. White and P.S. Peercy,
(New York, Academic, 1980), p. 221.

48. Leamy, H.J., Brown, W.L., Celler, G.K., Foti, G., Gilmer, G.H.,
Fan, J.C.C., Appl. Phys. Lett. 38 (1981) 137;
Gilmer, G.H. and Leamy, H.J. in Laser and Electron Beam Proces-
sing of Materials, ed. C.W. White and P.S. Peercy (New York,
Academic, 1980) p. 227.

49. Auvert, G., Bensahel, D., Perio, A., N'Guyen, V.T., Rozgonyi,
G.A., Appl. Phys. Lett. 39 (1981) 724.

50. Van Vechten, J.A., Tsu, R., Saris, F.W., Phys. Lett. 74A (1979)
422;
Van Vechten, J.A., Tsu, R., Saris, F.W., Hoonhout, D., Phys.
Lett. 74A (1979) 417;
Van Vechten, J.A., J. de Phys. 41 (1980) C-15.

51. Yoffa, E.J. in Laser and Electron Beam Processing of Materials,
(New York, Academic Press, 1980) p.59; Phys. Rev. B21 (1980)
2415; J. de Phys. 41 (1980) C-15.

52. Auston, D.H., Golovchenko, J.A., Smith, P.R., Surko, C.M.,
Appl. Phys. Lett. 33 (1978) 539.

53. Auston, D.H., Surko, C.M., Venkatesan, T.N.C., Slusher, R.E.
and Golovchenko, J.A. , Appl. Phys. Lett. 33 (1979) 437.

54. Auston, D.H., Golovchenko, J.A., Simons, A.L., Surko, C.M.
and Venkatesan, T.N.C., Appl. Phys. Lett. 34 (1979) 777.

55. Other reflectivity measurements of this kind can be found in
Murakami, K., Kawabe, M., Gamo, K., Namba, S., Aoyagi, Y.,
Int. Conf. on Semicond. Phys.Kyoto, 1980;
Murakami, K., Kawabe, M., Gamo, K., Namba, S., Aoyagi, Y.,
Phys. Lett. 70A (1979) 332;
Gamo, K., Murakami, K., Kawabe, M., Namba, S., Aoyagi, Y.,
in Laser and Electrons Beam Solid Interactions and Materials
Processing, J.F. Gibbons, L.D.Hess, T.W. Sigmon, eds. (New
York, North Holland, 1981), p. 97.

56. Lo, H.W. and Compaan, A., J. Appl. Phys. 51 (1980) 1565.

57. Lo, H.W. and Compaan, A., Phys. Rev. Lett. 44 (1980) 1604.

58. Lo, H.W. and Compaan, A., Appl. Phys. Lett. 38 (1980) 179.

59. Compaan, A., Aydinli, A., Lee, M.C., Lo, H.W., in Laser and
Electron-Beam Interactions with Solids, B.R. Appleton and
Celler, G.K.,(New York, North Holland, 1981), p. 43.

60. See Compaan, A. and Lo, H.W. in Laser and Electron Beam Processing of Materials ed. C.W. White and P.S. Peercy, (New York,
Academic Press, 1980) p. 71.

61. Bertolotti, M., Ferrari, A., Jani, P., Sibilia, C., in press.

62. von der Linde, V. and Wartman , G., XII IGEC, Munich, 1982.
63. Wood, R.F., Rasolt, M., Jellison, G.E.,Jr. in Laser and Elec-
    tron-Beam Interactions with Solids, B.R. Appleton and G.K.
    Celler eds., (New York, North Holland, 1982), p. 61;
    Wood, R.F., Lowndes, D.H., Giles, G.E., ibidem p. 67;
    Wood, R.F., Lowndes, D.H., Jellison, G.E.,Jr. and Modine, F.A.
    Appl. Phys. Lett. 41 (1982) 287.
64. Larson, B.C., White, C.W., Noggle, T.S. and Mills, D.,
    Phys. Rev. Lett. 48 (1982) 337.
65. Stritzker, B., Pospieszcyk, A., Tagle, J.A., Phys. Rev. Lett.
    47 (1981) 356;
    Nathan, M.I., Hodgson, R.T., Yoffa, E.J., Appl. Phys. Lett.
    36 (1980) 512;
    Galvin, G.J., Thompson, M.O., Mayer, J.W., Hammond, R.B.,
    Paulter, N., Peercy, P.S., Phys. Rev. Lett. 48 (1982) 33.
66. von der Linde, D., Fabricius, N., Observation of an electronic
    plasma in picosecond laser annealing of Si - in press.
67. Lee, M.C., Lo, H.W., Aydinli, A.A., Compaan, A., Appl. Phys.
    Lett. 38 (1981) 499;
    Aydinli, A., Lo, H.W., Lee, M.C., Compaan, A., Phys. Rev. Lett.
    46 (1981) 1640;
    Laude, L.D., Andrew, R. and Baufay, L., in Physical Processes
    in Laser Materials Interactions 1980, ed. M. Bertolotti,(New York,
    Plenum Press in print),
    see also Andrew, R. and Lovato, M., J. Appl. Phys. 50 (1979)
    1142.
68. Kimerling,L.C.and Benton, J.L., in Laser and Electron Beam
    Processing of Materials, eds. C.W. White and P.S. Peercy
    (New York, Academic Press, 1980) p. 385.
69. Benton, J.L., Kimerling , L.C., Moller, G.L., Robinson, D.A.H.,
    Celler, G.K., in Laser-Solid Interactions and Laser Processing
    1978, (AIP Conf. Proc. No.50), New York, 1979, p. 543;
    Johnson, N.M., Gold, R.B., Lietoila, A., Gibbons J.F., ibidem
    p. 550;
    Wang, K.L., Liu, Y.S., Kirkpatrick, C.G., Possin, G.E., ibidem
    p. 569;
    Benton, J.C., Doherty, C.J., Ferris, S.D., Kimerling, L.C.,
    Leamy, H.J. and Celler, G.K., in Laser and Electron Beam Pro-
    cessing of Materials, eds. C.W. White and P.S. Peercy,
    (New York, Academic Press, 1980) p. 430;
    Street, R.A., Johnson, N.M., Lietoila, A., ibidem, p. 435;
    Brower, K.L. and Peercy, P.S., ibidem, p. 441;
    Fan, Z.K., Ho, V.Q. and Sugano, T., Appl. Phys. Lett. 40 (1982)
    418.
70. Mooney, P.M., Young, R.T., Karins, J., Lee, Y.H., Corbett, J.W.
    Phys. Stat. Sol. A 48 (1978) K31.
71. Kimerling, L.C., Inst. Phys. Conf. Ser. 46 (1979) 56.
72. See for ex., Kimerling, L.C. and Long, D.V., Inst. Phys. Conf.
    Ser. 23 (1975) 589;
    Lang, D.V. and Kimerling, L.C., Appl. Phys. Lett. 28 (1976) 248.

73. See for ex. :
    Kimerling, L.C., Solid State Electron. 21 (1978) 1391;
    Bourgoin, J.C. and Corbett, J.W., Rad. Eff. 36 (1978) 157;
    Stoneham, A.M., Phil. Mag. 36 (1977) 983;
    Dean, P.J. and Choyke, W.J., Adv. Phys. 26 (1977) 1;
    Weeks, J.D., Tully, J.V. and Kimerling, L.C., Phys. Rev. 12B
    (1975) 3286.

74. Troxell, J.R., Chatterjee, A.P., Watkins, G.D., Kimerling, L.C.,
    Phys. Rev. B19 (1979) 5336.

75. Gregory, B.L., J. Appl. Phys. 36 (1965) 3765;
    Watkins, G.D., Phys. Rev. B12 (1975) 5824;
    Watkins, G.D., Troxell, J.R. and Chatterjee, A.P., Inst. Phys.
    Conf. Ser. 46 (1979) 16;
    Kimerling, L.C., Blood, P., Gibson, W.M., Inst. Phys. Conf.
    Ser. 46 (1979) 273.

76. Watkins, G.D., Radiation Effects in Semiconductors (New York,
    Plenum, 1968) p. 115.

77. Kimerling, L.C. and Carnes, C.P., J. Appl. Phys. 42 (1971)
    352, see also [71].

78. Watkins, G.D., Troxell, J.R. and Chatterjee, A.P., Harris, R.D.,
    Radiation Physics of Semiconductors and Related Materials,
    1979, (Tbilisi State Univ. 1980) p.97;
    Inst. Phys. Conf. Ser. 46 (1979) 16.

79. Pons, D., Inst. Phys. Conf. Ser. 59 (1980) 269.

80. Rzewuski, H., Suski, J., Krynick, J., Rad. Eff. 39 (1978);
    Suski, J., Csepregi, L., Gyulai, J., Rzewuski, H., Werner, Z.
    Rad. Eff. 29 (1976) 137;
    Suski, J., Krymcki, J., Rzewuski, H., Gyulai, J., Rad. Eff.
    30 (1976) 125;
    Suski, J., Krymcki, J., Rzewuski, H., Gyulai, J., Loferski,
    J.J., Rad. Eff. 35 (1978) 13.
    Kraut'chiuskii, A., Rzewuski, H., Suski, J., Werner, Z.,
    Rad. Eff. 14 (1972) 277;
    Phys. Stat. Sol. (a) 13 (1972) 661.

81. Suski, J., Rzewuski, H., Rad. Eff. 40 (1979) 81;
    Blosse, A., Gallati, N.H. and Bourgoin, J.C., Inst. Phys.
    Conf. Ser. 59 (1980) 521.

82. Suski, J., Rzewuski, H. and Grotzschel, R., Inst. Phys. Conf.
    Ser. 59 (1980) 485.

83. Frabk, W.F.J. and Berry, B.S., Rad. Eff. 21 (1974) 105.

84. Rzewuski, H., Suski, J. and Krynicki, J., Rad. Eff. 39
    (1978) 123.

85. Wautelet, M. and Failly-Lovato, M. in Physical Processes in
    Laser Materials Interactions, ed. M. Bertolotti, (Plenum
    Press, New York, in press);see also Wautelet, M. in this vo-
    lume.

86. Laude, L.D., Lovato, M., Martin, M.C., Wautelet, M., Phys. Rev.
    Lett. 39 (1977) 1565.
    Lovato, M., Wautelet, M., Laude, L.D., Appl. Phys. Lett. 34
    (1979) 160.

87. Laude, L.D., Andrew, R. and Baufay, L., in Physical Processes in Laser Materials Interactions, 1980, ed. M. Bertolotti, (New York, Plenum Press, in print); see also Andrew, R. and Lovato, M., J. Appl. Phys. 50 (1979) 1142.

88. Wautelet, M. and Laude, L.D., Phys. Stat. Sol. b89 (1978) 275; Tzoar, N. and Gersten, J.I., Phys. Rev. B12 (1975) 1132; Van Vechten, J.A. and Wautelet, M., Phys. Rev. B23 (1981) 5543; Wautelet, M. and Van Vechten, J.A., Phys. Rev. B23 (1981) 5551.

89. Bertolotti, M., Vitali, G. and Spear, W.E. in Laser and Electron Beam Processing of Materials, eds. C.W. White and P.S. Peercy, (New York, Academic Press, 1980) p. 189; Vitali, G., Phys. Lett. 78A (1980) 387; Vitali, G., Bertolotti, M., Zammit, U. and Marinelli, M., Phys. Lett. 89A (1982) 199; Bertolotti, M., Vitali, G, Zammit, U. and Marinelli, M., in Laser and Electron Beam Interactions with Solids, (New York, North Holland, 1982) p. 203.

90. Yen, R., Liu, J.M., Kurz, H., Bloembergen, N., Appl. Phys. A27, (1982) 153.

91. This view is confirmed by several evidences of the existence of a fast recombining plasma. See e.g. Lui, J.M., Kurz, H. and Bloembergen, N., in Picosecond Phenomena III ed. Eisenthal, K.B., Hochstrassen, R.M., Kaiser, W., Laubereau, A., (New York, Springer-Verlag, 1982) p. 332; von der Linde, D. and Fabricius, N., ibidem p. 336; see also Lietoila, A. and Gibbons, J.F. [30].

LASERS AND SPECKLE PATTERNS

Marie MAY

Institut d'Optique et Université P. et M. Curie
Tour 13, 4 Place Jussieu, 75230 Paris Cedex 05, France.

In the infrared, visible and ultraviolet region, there are three main kinds of sources :
- thermal sources which radiate at virtually all wavelengths (e.g. solid hot bodies and hot gases under high pressure),
- thermal sources giving line spectra (gase discharge tubes under low pressure),
- lasers.

In typical thermal source, light is emitted by excited atoms which are, in general, unrelated to each other. In fact, light emitted by a given atom must be represented by trains of waves. The profile of each wave is that of a simple sine curve whose length is limited in space. If an atom is excited several times, it can emit several consecutive wavetrains generally far apart (compared with their duration) and are emitted randomly in time. The wavetrains emitted by a single atom therefore bear no constant phase relation with each other.

A beam of light is the superposition of a large number of elementary wavetrains emitted by atoms and molecules. Consequently, there is a random variation of phase and amplitude in a beam of light generated by the superposition of many independant waves of random phase but the same frequency.

The length of time for which light beams can be considered as simple harmonic vary greatly from one source to another. At one end, there is the "white light" which contains all the radiations of the visible spectrum. The simple harmonic property persists for about 10^{-14}s. In a monochromatic beam, e.g. one of the spectrum line for a mercury discharge lamp, the time for which the wave is a simple sine wave, is of the order of $10^{-11} - 10^{-10}$ s. This can be expressed in terms of the length of the wavetrain which is sinusoidal in form i.e. a few millimeters to a few tens of a millimeter.

Laser light contains even longer stretches of purely sinusoidal waves. This is because the light emitting atoms are constrained to emit photons which are in phase with each other. In stabilized lasers, the light may be truly simple harmonic for times as long as 10^{-6} s so that the wavetrains are about 1000 m long.

The first section of this article is devoted to a brief recall of the concept of temporal and spatial coherence of a light source. Lasers are the most coherent sources actually available. The different types of lasers are enumerated in section II. A laser-illuminated rough surface shows a peculiar granular appearance called "speckle". The last section of this article deals with the statistical theory of the speckle effect and describes some fundamental properties of the speckle patterns generated by a laser-illuminated rough surface in any plane of the surrounding space.

1. INTERFERENCE PHENOMENON AND COHERENCE

The time τ for which the wavetrain remains simple harmonic and the corresponding distance L are quantities which determine the possibility of getting certain interference effects.

Let S_1 and S_2 be two light point sources emitting spherical waves (Fig. 1). Let us assume that a wavetrain emitted by an atom

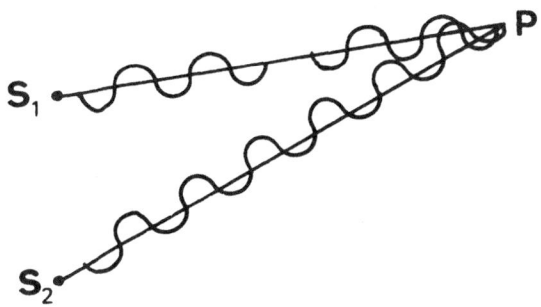

Fig. 1 The sources S_1 and S_2 generate a time-dependent amplitude at point P.

of S_1 and a wavetrain emitted by an atom of S_2 both superpose at a point P of the surrounding space. The light amplitude at P is the sum of the two incident amplitudes and the intensity is the square modulus of the amplitude. The phase relationship between the fields at point P being only constant for times comparable to τ, the intensity at point P is therefore constant during the same duration. A time later, the phase of each source is changed at random. As a consequence, the intensity distribution of the interference field varies randomly. This variation is more rapid than the response time for the fastest detector and so, the detector records a time-averaged interference field. Over the long term, constant intensity, given by the sum of the intensities of the two waves, is observed. The sources S_1 and S_2 are regarded as *incoherent* since they are unable to generate a time-independent interference phenomenon. Interference phenomena only occur with light from the same source, split and suitably recombined. Under these conditions, the phase fluctuations are the same in both interfering beams.

Suppose now than the two point sources S_1 and S_2 are identical images of the same point source S. At time t, S_1 and S_2 simultaneously emit two identical sinusoidal wavetrains whose length is L and whose duration is τ. They both propagate with the same velocity and, respectively, reach the point P one at time $t_1 = t + S_1P/C$ and the other at time $t_2 = t + S_2P/C$ where C is the light velocity in vacuo. The interference phenomenon occurs at point P only if these two identical wave trains may superpose, i.e. if :

$$t_2 - t_1 = \frac{S_2P - S_1P}{C} < \tau . \tag{1}$$

It is obvious that higher is the value of τ (or L), higher is the ability of the point source S to give rise to an interference phenomenon. The value of τ is directly related to the range $\Delta\nu$ covered by the distribution of frequency of the light, by the relationship :

$$\tau = \frac{1}{\Delta\nu} . \tag{2}$$

Let δ be the value of the optical path difference involved at point P by the geometrical arrangement represented in Fig. 1. Interference there occurs only if :

$$\delta = S_2P - S_1P < L = c\tau . \tag{3}$$

When δ is close to zero, the two wavetrains are quite superposed and the interference phenomenon will be highly constrasted. As the optical path difference approaches L, the wavefronts are gradually less and less superposed and the contrast of the fringes gradually diminishes. When the fringes have high contrast, the light is said to be *temporally coherent*. Light may be assumed to be highly coherent when $\delta < L/10$. The light is incoherent when the optical path difference exceeds L and partially coherent for the remaining values of the optical path difference. The length L of the wavetrain is called the coherence length of the source and the corresponding time τ is called the coherence time of the source. It is obvious that higher is the value of L (i.e. higher is the monochromaticity of the point source), higher is the temporal coherence of the source.

To introduce the concept of spatial coherence, let us consider the classical Young's experiment in which an extended quasimonochromatic source S, placed in the focal plane of the collimator L, illuminates two identical secondary pinholes (Fig. 2) T_1 and T_2. This source may be regarded as the juxtaposition of an infinite number of atoms. Since the waves from one atom bear no definite relation

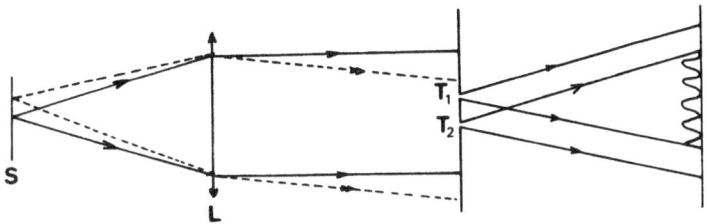

Fig. 2 The extended monochromatic source S involves an infinite number of fringe systems laterally shifted one with respect to another in observation plane.

with the wave from any other atom, the light emitted by one atom is incoherent with that emitted by any other atom. Each atom of the source generates a collimated beam and gives its own interference fringe system in plane II. The different fringe systems are laterally displaced one with respect to the other and the intensity distribution in plane II is the sum of the intensities of each of the systems. If the size of the source is not small enough, the intensity superposition of all the fringe systems gives rise to a uniform irradiance. Finally, the contrast of the fringe system obtained in the plane II depends on the size of the illuminating source, i.e. on its *spatial coherence*.

Finally, any thermal source can be made to achieve any desired degree of space or time coherence by limiting its extent with a pinhole or its wavelength range with a filter or monochromator. However, the coherent source is made extremely dim with both operations.

Laser light differs somewhat in nature from one type of laser to another. However, there are several remarkable features which are displayed to varying degrees by all the lasers :
- Most laser beams are highly coherent. The beam is spatially coherent;
- Laser light is quasimonochromatic with a highly narrow frequency bandwidth. It is temporally coherent;
- Laser can deliver a high radiant power;
- Laser light is available as a continuous wave or in the form of pulses.

2. LASERS

2.1 Spontaneous emission – Stimulated emission

As seen above, in a conventional thermal source, energy is pumped into the reacting atoms which are consequently raised in excited states. Each of them can then drop back spontaneously to the ground state emitting the absorbed energy in the form of a randomly directed photon. Atoms, in this kind of source, radiate independently. The photons in the emitted beam bear no particular phase relationship with each other.

Let us now imagine that light impinges on an atomic system in which the atoms are maintained in an excited state characterized by an energy E_2 (population inversion). As pointed out by Einstein in 1917, an excited atom of the medium can then revert to a lower state E_1 by emitting a photon exactly like the stimulating photon. This process is known as stimulated emission. The emerging photon will carry off the energy difference $h\nu$ between the initial higher state E_2 and the final lower state E_1, i.e. :

$$E_2 - E_1 = h\nu \ . \tag{4}$$

The incident electromagnetic wave which triggers an excited atom into stimulated emission, must have the frequency ν. The emitted photon is in the same radiation mode as the incident wave and tends to add to it, increasing its flux density.

The stimulated emission is a key to the operation of the laser. However, the atoms of the active medium must somehow be excited into an upper state leaving the lower state all but empty. This process is called "population inversion" and is performed by suitably supplying energy (electrical, chemical, optical ...) to the medium.

2.2 Basic schema of a laser

A laser consists of a fluorescing material (rod or tube of gas or liquid) in a suitable optical cavity generally composed of two mirrors facing each other (Fig. 3). One mirror is partially transparent and both are parallel to one another and perpendicular to the axis of the rod. The optical length of the cavity thus formed is d, the reflectance of the partially transmitting mirror is R. An additional energy is suitably supplied to the material to maintain its atoms in an excited state. The medium is thus amplifying and let G be the intensity gain of the rod.

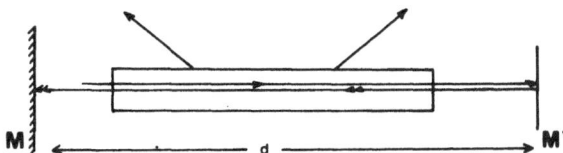

Fig. 3 Basic schema of a laser.

Initially, the only light emitted is that arising from fluorescence or spontaneous emission. The fluorescent emission is not directional but some of this light will travel along the cavity axis. A wave packet emitted along the axis by a single atom undergoes many reflections from the mirrors. After each round trip in the cavity, the net gain is G^2R and exceeds 1 if there is no dominant competitive processes such as scattering. Then, only waves that travel parallel to the axis continuously grow almost without limit. A fraction of that wave escapes through the partially reflecting mirror and the emerging useful beam is a powerful directional beam.

The wave packet emitted along the axis of the cavity has a coherence length which is great compared to d. Consequently, it takes on a standing wave configuration determined by the distance d between the mirrors. The cavity resonates when there is an integer number m of half-wavelengths spanning the region between the mirrors. Thus :

$$m \frac{\lambda}{2} = d \qquad (5)$$

and

$$\nu_m = m \frac{v}{2d} \tag{6}$$

where v is the velocity of the waves inside the active medium.
There are therefore an infinite number of possible oscillatory axial
cavity modes, each with a distinctive frequency ν_m. Consecutive
modes are separated by a constant difference :

$$\Delta\nu = \frac{v}{2d} .$$

The resonant modes of the cavity are considerably narrower in frequency than the bandwith of the spontaneous atomic transition. In
fact the radiative transitions make available a relatively broad range
of frequencies out of which the cavity will select and amplify only
certain narrow bands and, if desired, even only one such band. This
is the origin of the extreme quasi-monochromaticity of the laser.
In addition to the longitudinal modes of oscillations, transverse
electric and magnetic modes can be sustained. The lowest order of
T E M_{oo} transverse mode is the most widely used because the flux
density is ideally Gaussian over the cross section of the beam,
there is no phase shift in the electric field across the beam and
the angular divergence of the beam is the smallest.

There are many sorts of lasers available on the market. The
most important among them are the solid state, gaseous, semiconductor and organic dye lasers.

2.3 Solid state lasers

The first of them is the ruby laser whose active medium is a
rod of aluminium oxide (Al_2O_3) with a small concentration (0,03%
by weight) of chromium oxide (Cr_2O_3) impurity. The Cr^{+++} ions are
responsible for the emission of light by the crystal and the corresponding wavelength is 694 nm. These lasers are optically pumped
by means of a flash tube which generates an intense burst of light
lasting for a few milliseconds. The laser may emit 2J or more each
time the lamp is pulsed. If the pulse width is 1ms, this corresponds to an average power of a few kilowatts.The spikes have a duration of a few microseconds, so the peak power is tens or hundred of
kilowatts. The output of the laser is in general multimode both
with regards to spectral and transverse mode. The pulsed ruby laser
is successfully used for precision welding and drilling of metals,
for drilling industrial diamonds, for reparing detached retinas in
ophtalmology and for holography and photography.

The ruby laser is often Q-switched for high peak power. If
the parameters are properly adjusted, then a single giant pulse
will be emitted. The pulse may have a peak power up to 100 MW and
a full width of 10 or 20 ns.

The neodymium laser is also optically pumped. The active Nd^{+++} ion may be incorporated into several hosts in particular certain glasses and a crystal called Y A G (yttrium aluminium garnet). The Nd: YAG and Nd:glass lasers oscillate at about the same frequency. Their output is in the near infrared at 1.06 μm.

The Nd:YAG laser is most often used in a quasi-continuous fashion, i.e. repetitively pulsed at a high rate. The peak output power is of the order of kilowatts.

Nd:glass lasers on the other hand are normally operated in single pulses just as the ruby laser is. The glass laser may be pulsed or Q-switched and is highly resistant to damage from high power density. The laser may be associated with one or more amplifiers to produce very high peak powers. The Q-switched Nd:glass laser is often induced to operate in a mode-locked fashion emitting a train of pulses, each with a duration of the order of 10 ns.

Nd:YAG laser is also currently mode-locked but the pulses are 5 or 10 times longer. A single one of these pulses may be isolated and then amplified. Pulses with peak power in excess of 10^6MW have been obtained. They are used in laser fusion programs and in fluorescence spectroscopy.

2.4 Gas lasers

Gas lasers operate from the far-infrared to the ultraviolet. They oscillate generally in continuous operation; the power output is lower than in solid state lasers but their spatial and temporal coherences are better.

The helium-neon lasers are electrically continuously pumped with a d.c. power supply. The active medium is a gas mixture of about 5 parts helium to each part of neon. They emit powers ranging from 0.3 to 100 milliwatts in the T E M_{oo} transverse mode. The output of the helium-neon laser is continuous and stable and remains constant to about 1%. The spectral width of the 633 nm neon line is about 1500 Hz.

Argon ion lasers lase mainly in the green, blue green and violet in either pulsed or continuous operation. The Krypton ion laser is very similar to the argon laser. It produces a strong red line among others. These lasers provide continuous output powers of a hundred of milliwatts or more.

The most important argon laser line has a wavelength of 515.5 nm. Other lines (particularly the blue line at 488 nm) may be selected by rotating a prism inside the cavity. All these lines can be emitted simultaneously if broad band reflectors are used and the prism is removed. The output is then usually several watts but it has gone till 150 W c.w. This laser is similar in some respect to the He-Ne lasers, although it differs in its greater power, shorter wavelength , broader linewidth and higher price. Its main applications are the study of light diffusion and Raman spectroscopy. Argon ion laser can be mode-locked for short,high power pulses.

The molecular CO_2 laser oscillates at 10.6 μ in the infrared and is operated either continuously, pulsed or Q-switched. A small continuous CO_2 laser is able to emit a fraction of a Watt and can heat most materials to incandescence in a short time. The output cw power increases with the length of the discharge tube. A discharge tube nearly two hundred meters long can generate 10 kW c.w. These lasers are used for cutting and welding metals.

The helium-cadmium laser operates continuously at 442 nm in the blue and at 325 nm in the ultraviolet. This last line emitted with a c.w. power of a few milliwatts is the shortest wavelength that can be obtained in a continuous operation.

The H_2O vapor and H C N Lasers are both low power and far-infrared lasers. Their principal lines are respectively for H_2O : 28 μm and 118 μm and for H C N:373 μm.

Hydrogen and Deuterium fluoride lasers oscillate in a pulsed operation at various wavelengths between 5 and 6 μm in the infrared. The output power is relatively high.

The pulsed nitrogen laser is a high power source in the UV portion of the spectrum at 337 nm. This laser generates 10^{-3} J in pulses lasting 10 ns.

All the lasers described above emit at one or several constant frequencies since they are defined by transitions between discrete atomic or molecular levels. The following lasers are tunable ones. These devices are remarkable in that they can be continuously tuned over a range of wavelengths sometimes very spread out. This is their main advantage.

2.5 Semiconductor lasers

Transitions characterizing the semiconductor lasers occur between the conduction and valence bands. The stimulated emission results in the immediate vicinity of the p-n junction. Such lasers are available between 0.6 μm and more than 30 μm and the active medium may be GaAs, InSb, Pb $S_{1-x}Se_x$. The value of the output wavelength is precisely adjusted either by temperature variation ($d\lambda/dT \simeq 4 \times 10^{-5}$ μm/K) either by the current generated by the power supply ($d\nu/di \simeq 32$ MHz/A) , either by applying a magnetic field perpendicularly to the junction.

A remarkable feature of these lasers is that their output power is directly related to the intensity of the supplying current. The emerging beam may therefore be amplitude-modulated up to high frequencies (several hundreds of MHz). They are commonly used in optical telecommunications.

However, they present some serious drawbacks. The main of them is the heating of the p-n junction due to the very high current density (thousands of A/cm^2). The c.w. operation is therefore possible only at low temperatures with an output power between 10 and 50 mW. At room temperature, the normal operation is a pulsed one corresponding to an output power of 50 µJ in 1 µs. The spatial and temporal coherences of such beams are weak.

2.6 Organic dye lasers

A great many fluorescent dye solutions as e.g. the fluoreceins, coumarins and rhodamins have been made to lase at frequencies from the infrared to the ultraviolet. They have usually been pulsed although c.w. operations have been obtained. There exist so many organic dyes that it is theoretically possible to obtain such a laser oscillating at any frequency in the visible. These devices can be tuned continuously over a range of wavelengths of about 700 Å although a pulsed system tunable over 1700 Å exists. The primary beam is tuned internally by changing the concentration or the length of the dye cell or by adjusting a diffraction grating reflector at the end of the cavity. Several multicolor dye laser systems, which can easily be switched from one dye to another and thereby operate over a very broad frequency range, are available commercially. For example, an alcohol solution of rhodamin 6G, continuously pumped with an argon-ion laser, lases at wavelengths from 560 nm to 660 nm with an output power of 50 mW c.w.

Moreover, extremely short duration subpicosecond pulses are of considerable interest. They serve as probe on an atomic scale or to study chemical processes which occur in an interval of from 10^{-9} s to 10^{-13} s, e.g. molecular life times.

3. LASER SPECKLE PATTERNS

A rough surface, illuminated with highly coherent light whose wavelength is smaller than it depth roughness, shows a granular appearance called speckle effect. Each elementary area of the surface scatters light-waves which are coherent to each other. Their superposition in any plane of the space gives rise to an interference pattern which appears chaotic and unordered with a very fine microscopic structure. In fact, this phenomenon had been already observed at the end of the last century by Exner [1] who described the radially granular appearance of the Airy pattern displayed by .

a coherently illuminated glass plate covered with small particles. Afterwards, speckle was observed in the holographic reconstructions of three dimensional objects and was then considered as a source of noise; besides, it stood to reason than a speckle pattern contains some informations on the object itself. Since that time, many applications were found as well from the study of object deformations or optical processing of information to astronomy. Some of them will be described here after a short study of the first and second order statistics of a speckle pattern.

3.1 Statistical study of a speckle pattern

Most scattering surfaces are very rough on the scale of optical wavelengths. When such a surface O is illuminated by a coherent beam of highly monochromatic light, the amplitude distribution at any distant point P is given by the coherent superposition of the wavelets coming from the independent elementary areas of the surface (Fig. 4).

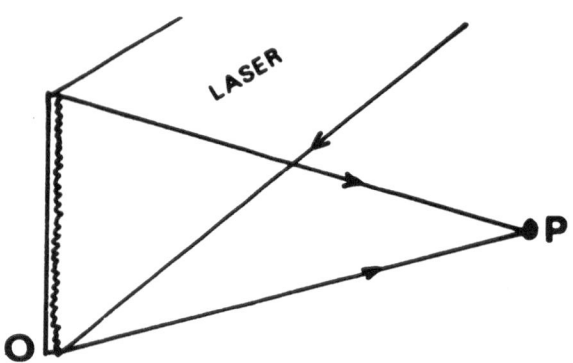

Fig. 4 Formation of a speckle in free space propagation.

The resulting intensity distribution at P is a function of the optical path travelled by each of the wavelets. Due to the roughness of the surface, these optical paths differ by several wavelengths and the dephased wavelets give rise to a granular pattern of intensity called speckle pattern. When the surface under consideration is imaged by a lens (Fig. 5) the speckle at P is generated by the coherent superposition of all the wavelets coming from a single resolution cell. If the microscopic structure of the object is resolved by the imaging system, there is no speckle structure in the image plane. The height fluctuations of the surface are randomly distributed around a mean value. Consequently, the complex

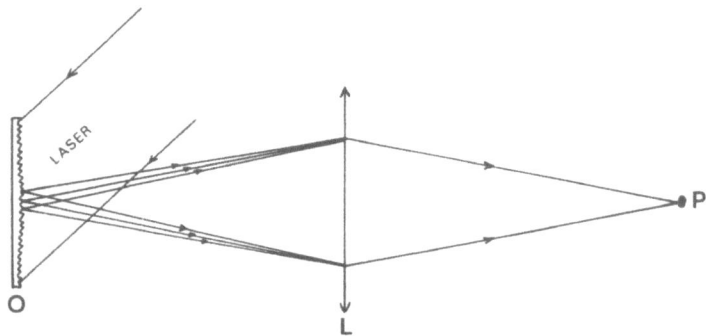

Fig. 5 Formation of a speckle in the image plane of the object.

amplitude distribution of a speckle pattern is randomly distributed and needs to be statistically studied. The presentation of this section follows that of Goodman [2].

3.1.1 First order statistics of a polarized speckle pattern

a) Complex amplitude

As seen above, a speckle, in the case of free space propagation as well as in the case of image formation, is generated by the coherent superposition of a great number N of dephased wavelets arising from different elementary scattering areas of the rough surface. The light amplitude (x,y,z) generated at a point P characterized by its rectangular coordinates x,y,z (Fig. 4) may be written as :

$$u(x,y,z) = \sum_{p=1}^{N} \frac{1}{\sqrt{N}} O_p (x,y,z) =$$

$$\sum_{p=1}^{N} \frac{1}{\sqrt{N}} |O_p(x,y,z)| \exp[j\Phi_p(x,y,z)] . \qquad (7)$$

Let us assume that:
- the elementary scattering areas are randomly distributed over the area of O with an uniform probability,
- the amplitude $|O_p|$ generated by the pth elementary area is statistically independent of the corresponding phase Φ_p and that $|O_p|$ and Φ_p are statistically independent of the other elementary amplitudes and phases,
- the phases Φ_p are uniformly distributed over the interval $-\Pi$ to Π (since the surface under consideration is rough compared to the wavelength scale),

- the polarization of the scattered beam is that of the incident beam.

Let $u^{(r)}$ and $u^{(i)}$ be the real and imaginary parts,respectively,of $u(x,y,z)$. We have :

$$u^{(r)}(x,y,z) = \frac{1}{\sqrt{N}} \sum_{p=1}^{N} \mid 0_p(x,y,z) \mid \cos \Phi_p(x,y,z) \;,$$

$$u^{(i)}(x,y,z) = \frac{1}{\sqrt{N}} \sum_{p=1}^{N} \mid 0_p(x,y,z) \mid \sin \Phi_p(x,y,z) \;.$$

$$(8)$$

With the above assumptions, the average values of $u^{(r)}$ and $u^{(i)}$ are given by :

$$<u^{(r)}> = \frac{1}{\sqrt{N}} \sum_{p=1}^{N} <\mid 0_p \mid \cos \Phi_p> = \frac{1}{\sqrt{N}} \sum_{p=1}^{N} <\mid 0_p \mid> <\cos \Phi_p> = 0 \;,$$

$$<u^{(i)}> = \frac{1}{\sqrt{N}} \sum_{p=1}^{N} <\mid 0_p \mid \sin \Phi_p> = \frac{1}{\sqrt{N}} \sum_{p=1}^{N} <\mid 0_p \mid> <\sin \Phi_p> = 0$$

$$(9)$$

and

$$<(u^{(r)})^2> = \frac{1}{N} \sum_{p=1}^{N} \sum_{q=1}^{N} <\mid 0_p \mid \mid 0_q \mid> <\cos \Phi_p \cos \Phi_q> = \frac{1}{N} \sum_{p=1}^{N} \frac{<\mid 0_p \mid^2>}{2} \;,$$

$$<(u^{(i)})^2> = \frac{1}{N} \sum_{p=1}^{N} \sum_{q=1}^{N} <\mid 0_p \mid \mid 0_q \mid> <\sin \Phi_p \sin \Phi_q> = \frac{1}{N} \sum_{p=1}^{N} \frac{<\mid 0_p \mid^2>}{2} \;,$$

$$(10)$$

$$<u^{(r)} u^{(i)}> = \frac{1}{N} \sum_{p=1}^{N} \sum_{q=1}^{N} <\mid 0_p \mid \mid 0_q \mid> <\cos \Phi_p \sin \Phi_q> = 0 \;.$$

$$(11)$$

The real and imaginary parts of the amplitude at point P have there-fore zero mean value (Eqs. 9), identical variances (Eqs. 10) and are uncorrelated (Eq. 11). If the number N of elementary areas becomes infinite, $u^{(r)}$ and $u^{(i)}$ from the central limit theorem (3) are asymptotically Gaussian. The joint probability density function is given asymptotically by a circular density function :

$$P_{ri}(u^{(r)},u^{(i)}) = \frac{1}{2\Pi\sigma^2} \exp \left[- \frac{(u^{(r)})^2 + (u^{(i)})^2}{2\sigma^2} \right] \qquad (12)$$

where

$$\sigma^2 = \lim_{N \to \infty} \frac{1}{N} \sum_{p=1}^{N} \frac{<\mid 0_p \mid^2>}{2} \;. \qquad (13)$$

The convergence of the process to a Gaussian form will generally depend on the statistics of the surface itself. However, in many practical situations, the number of scatterers is so large that the problem of convergence has no importance.

b) Intensity and phase

Let I and Θ be the respective intensity and phase of the resulting amplitude. They are defined by:

$$I = [u^{(r)}]^2 + [u^{(i)}]^2 \text{ and } \tan \Theta = \frac{u^{(i)}}{u^{(r)}} . \qquad (14)$$

The joint-probability density function may be written as:

$$P_{I,\Theta} (I,\Theta) = \frac{1}{4\Pi\sigma^2} \exp [- \frac{I}{2\sigma^2}] ; \; I \geqslant o; \; - \Pi < \Theta < \Pi \qquad (15)$$

and the probability density functions for intensity and phase are respectively given by :

$$P_I(I) = \int_{-\Pi}^{\Pi} P_{I,\Theta} (I,\Theta) \, d\Theta = \begin{cases} \frac{1}{2\sigma^2} \exp (-\frac{I}{2\sigma^2}) & I \geqslant 0 \\ 0 \text{ otherwise} \end{cases} \qquad (16)$$

and by :

$$P_\Theta(\Theta) = \int_o^{+\infty} P_{I,\Theta} (I,\Theta) \, dI = \begin{cases} \frac{1}{2\Pi} & -\pi \leqslant \Theta \leqslant \pi , \\ 0 \text{ otherwise} . \end{cases} \qquad (17)$$

It may be easily demonstrated that the second moment $\langle I^2 \rangle$ and the variance σ_I^2 of I are respectively given by :

$$\langle I^2 \rangle = 2 \langle I \rangle^2 \qquad (18)$$

and by :

$$\sigma_I^2 = \langle I^2 \rangle - \langle I \rangle^2 = \langle I \rangle^2 . \qquad (19)$$

The standard deviation σ_I being equal to the average value of the intensity, the contrast of a polarized speckle pattern defined by the ratio $C = \sigma_I / \langle I \rangle$ is always unity.

These results have been obtained considering any Fresnel plane of the object. They are identical in the Fourier plane of the object.

They are suitable for image plane provided that the elementary cell of the object generating a speckle in image plane contains a large number of scatterers. In practical situations, this condition is likely satisfied. However, the numerical aperture of the imaging system must be smaller than 0.1.

3.1.2 Second order statistics

The first order statistics of a speckle pattern does not suffice to describe another fundamental property of a speckle pattern, i.e. the coarseness of its spatial structure. This property is defined by the knowledge of the average size of the speckles forming the speckle pattern.

a) Autocorrelation function of a speckle pattern.

A rough object O, bounded by an aperture P, is illuminated by a coherent beam of laser light (Fig. 6). In the Fresnel-Kirch-hoff approximation, the amplitude distribution at a point M lying in a plane at distance z from the object is proportional to :

$$u(x,y) = \frac{j}{\lambda z} \exp \left(j\frac{2\pi}{\lambda}z\right) \iint P(\xi,\eta) \, O(\xi,\eta) \, \exp\left[j\frac{\pi}{\lambda z}[(x-\xi)^2+(y-\eta)^2]\right] d\xi d\eta$$

$$(20)$$

where $O(\xi,\eta)$ is the random amplitude distribution in the object plane, $P(\xi,\eta)$ is the amplitude transparency of the aperture P and where (ξ,η) and (x,y) are the respective rectangular coordinates of the object and the observation plane Π. The irradiance in plane Π, given by :

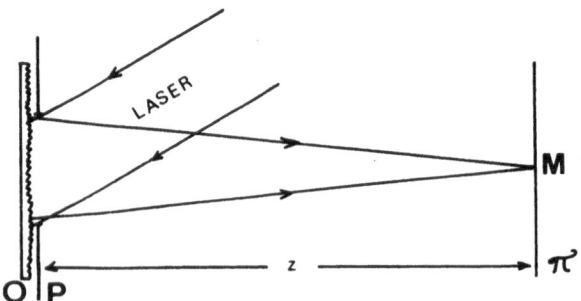

Fig. 6 Speckle pattern generated in the Fresnel plane Π by the object O limited by the aperture P.

$$I(x,y) = u(x,y) \, u^*(x,y) \tag{21}$$

is a random distribution of intensity and its autocorrelation function, defined by :

$$R(x,x+x_o;y,y+y_o) = \langle I(x,y), I^*(x+x_o,y+y_o) \rangle \,, \tag{22}$$

may be written as :

$$R(x,x+x_o;y,y+y_o) = (\lambda z)^{-4} \quad x$$

$$\int \ldots \int \langle O(\xi_1,\eta_1) O^*(\xi_2,\eta_2) O^*(\xi_3,\eta_3) O(\xi_4,\eta_4) \rangle \quad x$$

$$P(\xi_1,\eta_1) P^*(\xi_2,\eta_2) P^*(\xi_3,\eta_3) P(\xi_4,\eta_4) \quad x$$

$$\exp\left[j\frac{\Pi}{\lambda z}(\xi_1^2+\eta_1^2-\xi_2^2-\eta_2^2-\xi_3^2-\eta_3^2+\xi_4^2+\eta_4^2) \right] \quad x$$

$$\exp\left\{ - j\frac{2\pi}{\lambda z}\left[x(\xi_1-\xi_2)+y(\eta_1-\eta_2)-(x+x_o)(\xi_3-\xi_4)-(y+y_o)(\eta_3-\eta_4) \right] \right\} x$$

$$d\xi_1 d\xi_2 d\xi_3 d\xi_4 d\eta_1 d\eta_2 d\eta_3 d\eta_4 \quad .$$

$$\tag{23}$$

$O(\xi,\eta)$ is assumed to be a circular Gaussian random variable and the Wang-Uhlenbeck theorem may therefore be applied [4] i.e. :

$$\langle O(\xi_1,\eta_1) O^*(\xi_2,\eta_2) O^*(\xi_3,\eta_3) O(\xi_4,\eta_4) \rangle =$$

$$\langle O(\xi_1,\eta_1) O^*(\xi_2,\eta_2) \rangle \langle O^*(\xi_3,\eta_3) O(\xi_4,\eta_4) \rangle +$$

$$\langle O(\xi_1,\eta_1) O^*(\xi_3,\eta_3) \rangle \langle O^*(\xi_2,\eta_2) O(\xi_4,\eta_4) \rangle . \tag{24}$$

Moreover, the irregularities of the surface O are very small compared to the size of the illuminated aperture P. Consequently :

$$\langle O(\xi_i,\eta_i) \, O^*(\xi_j,\eta_j) \rangle = I_o \delta (\xi_i-\xi_j,\eta_i-\eta_j) \tag{25}$$

where δ is the Dirac distribution and where I_o is constant. The autocorrelation function of the intensity distribution of plane Π may thus be rewritten as :

$$R(x,x+x_o;y,y+y_o)=$$

$$(\lambda z)^{-4} I_o^2 \int \ldots \int |P(\xi_1,\eta_1)|^2 |P(\xi_2,\eta_2)|^2 d\xi_1 d\xi_2 d\eta_2 d\eta_2$$

$$+ (\lambda z)^{-4} I_o^2 \int \ldots \int |P(\xi_1,\eta_1)|^2 |P(\xi_2,\eta_2)|^2 \quad \times$$

$$\exp\left[j\frac{2\pi}{\lambda z}[x_o(\xi_1-\xi_2)+y_o(\eta_1-\eta_2)]\right] d\xi_1 d\eta_1 d\xi_2 d\eta_2 \qquad (26)$$

where I_o is the constant irradiance of the incident beam. Finally the function R is independent of the coordinates (x,y) and is only a function of x_o and y_o. It is given by :

$$R(x_o,y_o) = [<I(x,y)>]^2 \left\{ 1 + \left| \frac{\int\int |P(\xi,\eta)|^2 \exp[j\frac{2\pi}{\lambda z}(\xi x_o+\eta y_o)] \, d\xi d\eta}{\int\int |P(\xi,\eta)|^2 d\xi d\eta} \right|^2 \right\}$$

$$(27)$$

where :

$$<I(x,y)> = (\lambda z)^{-2} I_o \int\int |P(\xi,\eta)|^2 d\xi d\eta \qquad (28)$$

represents the mean irradiance in plane Π.
$R(x_o,y_o)$ is thus given by the sum of two terms : a constant one $<I(x,y)>^2$ and a term proportional to the square modulus of the normalized Fourier transform of the irradiance of the aperture P limiting the object O. For the special case of a circular aperture with radius R, we have :

$$R(x_o,y_o)=R(r_o) = [<I(x,y)>]^2 \left\{ 1 + \left| \frac{2 \, J_1(\frac{2\Pi}{\lambda} \frac{Rr_o}{2z})}{\frac{2\Pi}{\lambda} \frac{Rr_o}{2z}} \right|^2 \right\}$$

$$(29)$$

where $r_o = (x_o^2 + y_o^2)^{1/2}$.

Fig. 7 illustrates the variations of $R(r_o)$ versus r_o. Practically, $R(r_o)$ is given by the overlap area between the function $I(r)$ and itself laterally shifted through r_o. It seems reasonable to assume that $I(r)$ and $I(r+r_o)$ are no more correlated for the value of r_o corresponding to the first zero minimum of the function $2J_1(z)/z$. The average speckle size (diameter) in plane Π is thus defined by :

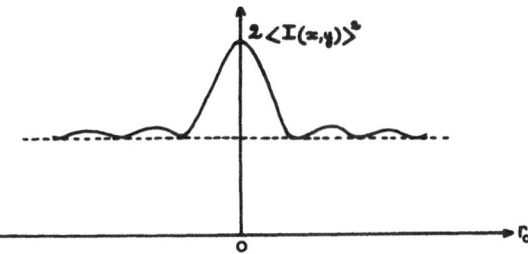

<u>Fig. 7</u> Autocorrelation function of a speckle pattern.

$$\sigma = \frac{1.22 \ \lambda z}{2R} \ . \tag{30}$$

Let us now consider that the object is imaged in plane (x,y) by a lens L. We assume that the speckle size in the plane of the pupil P of the lens is extremely small compared to the diameter of the pupil. Moreover, the fine structure of the object cannot be resolved by the lens. Under these conditions, the results are exactly the same as for free space geometry. The autocorrelation function of intensity in the image plane is given by Eq.(27) in which P represents now the amplitude transmittance of the pupil of the lens. As pointed out by Goodman, the autocorrelation function is independent of the aberrations of the imaging lens since the phase of the pupil is cancelled. The mean diameter of the speckles in image plane is thus equal to :

$$\sigma = \frac{1.22 \ \lambda z}{2N.A.} \tag{31}$$

where N.A. is the numerical aperture of the imaging lens.

b) Power spectral density.

From the Wiener-Kintchine theorem, the power spectrum is de-fined as the Fourier transform of the autocorrelation function $R(x_0,y_0)$ i.e. :

$$G_I(u,v) = \int\int R(x_o,y_o) \exp\left[j \frac{2\Pi}{\lambda} (ux_o + vy_o)\right] dx_o dy_o$$

$$(32)$$

where (u,v) are the angular coordinates of Fourier plane. In Fresnel case as in imaging case, $G_I(u,v)$ is :

$$G_I(u,v) = [<I(x,y)>]^2 \left\{ \delta(u,v) + \frac{\int\int |P(\xi,\eta)|^2 |P(\xi-zu,\eta-zv)|^2 d\xi d\eta}{[\int\int |P(\xi,\eta)|^2 d\xi d\eta]^2} \right\}.$$

$$(33)$$

The power spectral density of the speckle pattern consists of a δ function component at zero frequency ($u=v=o$) plus a spread-out component having the shape of the normalized autocorrelation function of the intensity distribution incident on the diffuse object O.

Let us suppose now that the random intensity distribution $I(x,y)$ of plane Π is recorded on a photographic plate H. After exposure, the plate is processed under the usual conditions of linearity and illuminated by a parallel beam of monochromatic light (Fig. 8). H is a partially diffusing plate which splits the incident beam into two beams : a directly transmitted beam and a scattered beam. The directly transmitted beam is focused at point S' by the lens L. The average intensity distribution in the Fourier plane of H (i.e. in back focal plane of L) is given by Eq. 33. The Fourier spectrum of H is thus formed by the image point S' and by a random intensity distribution whose average value is given by the autocorrelation function of the aperture P bounding the object O (or the imaging lens) during the recording process.

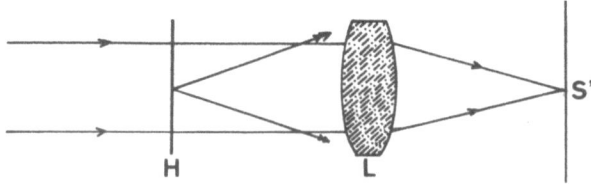

H   L

<u>Fig. 8</u> The photographic record of a speckle pattern splits the incident beam into a directly transmitted beam and a scattered beam.

It may be deduced from the previous considerations that some average properties of speckle patterns do not depend on the structure of the object by which they are generated. They are only function of the shape and of the angular size of the pupil limiting either the object in free space geometry either the imaging system in imaging geometry. These results are only applicable to speckle patterns generated by objects extremely rough on the scale of an optical wavelength. The contrast of speckle patterns generated by surface with an r.m.s. roughness that is less than a wavelength depends on the value of the roughness [5]. For very smooth surfaces, the contrast of the speckle pattern is near zero. It asymptotically approaches unity as the surface roughness increases. This is due to the fact that the light scattered from a rough surface has two components : a specular one and a diffuse one whose relative amplitudes are depending on the surface roughness [6]. Moreover, it has been demonstrated [7] that the statistics of the diffusely transmitted beam lying in image plane of a smooth object is Gaussian but non-circular.

In the following sections, we shall only consider speckle patterns with Gaussian circular statistics. For detailed study with non-circular functions, the reader is referred to the papers of Crane [8] and Miller et al. [9].

3.2 Variation of a speckle pattern with a displacement of the object

As described in the previous section, a rough object illumination by a laser beam gives rise to a random intensity distribution in any plane of the surrounding space. This intensity distribution consists of grains or "speckles" whose mean size depends only on the pupil of the optical system of observation. A translation (or a deformation) of the object involves an alteration of the corresponding speckle pattern which is studied here in some peculiar cases. The object is assumed to be trans-illuminated by a spherical laser wave and the observation plane is either a Fresnel plane, either the image plane, either the Fourier plane.

3.2.1 Random intensity distribution generated by a laser-illuminated rough object

The rough object O, characterized by its rectangular coordinates (ξ, η), is trans-illuminated by the laser point source S situated at a distance s.

a) Fresnel plane

The resulting speckle pattern is observed in a Fresnel plane π at a distance z from O (Fig. 9). The light amplitude at a point M(x,y) of plane π is proportional to:

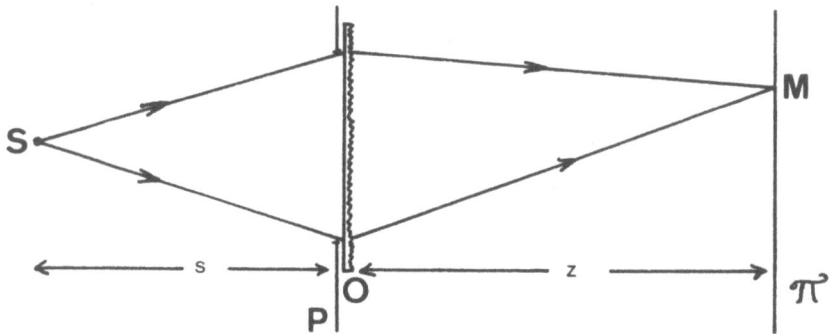

Fig. 9 Speckle pattern generated in Fresnel plane Π by an object O train-illuminated by a laser point source.

$$u_F(x,y) = \iint \exp [j \frac{\Pi}{\lambda s} (\xi^2 + \eta^2)] O(\xi,\eta) P(\xi,\eta) \quad x$$

$$\exp [j \frac{\Pi}{\lambda z} [(x-\xi)^2 + (y-\eta)^2]] d\xi d\eta \qquad (34)$$

where $P(\xi,\eta)$ is the amplitude transparency of the aperture P limiting the object. The intensity distribution in plane Π is therefore given by :

$$I_F(x,y) = |\iint O(\xi,\eta) P(\xi,\eta) \exp[j \frac{\Pi}{\lambda} \frac{s+z}{sz} (\xi^2 + \eta^2)] \quad x$$

$$\exp [- j \frac{2\Pi}{\lambda z} (\xi x + \eta y)] d\xi d\eta|^2 . \qquad (35)$$

b) Image plane

The object O is imaged in plane Π by means of the lens L (Fig. 10); p and p' which represent the respective distances of the lens from the object plane on the one hand and the image plane on the other hand are related to the focal length f of L by the classical formula :

$$\frac{1}{p} + \frac{1}{p'} = \frac{1}{f} . \qquad (36)$$

The pupil P of the imaging system is placed between the lens (assumed to be infinite) and the image plane Π at a distance d from L. It is referred to the rectangular coordinates (X,Y).

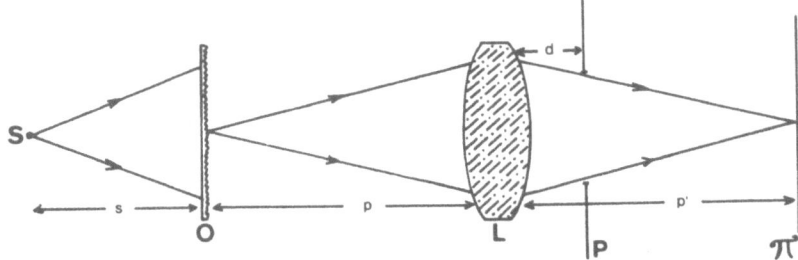

<u>Fig. 10</u> Speckle pattern lying in image plane Π of an object O imaged by the lens L through the pupil P.

The intensity distribution at a point M (x,y) of Π may be written as :

$$I_i(x,y) = \left| \iint \tilde{O}(X,Y) P(X,Y) \exp \left[-j \frac{2\pi}{\lambda(p'-d)} (xX+yY) \right] dXdY \right|^2$$

(37)

where :

$$\tilde{O}(X,Y) = \iint O(\xi,\eta) \exp \left[j \frac{\pi}{\lambda s} \left(1 + \frac{p'(f-d)s}{p(f'-d)f} \right) (\xi^2+\eta^2) \right] \times$$

$$\exp \left[-j \frac{2\pi}{\lambda} \frac{(\xi X+\eta Y)p'}{(p'-d)p} \right] d\xi d\eta$$

(38)

is a random Gaussian circular function which represents the amplitude distribution of the speckle generated by the object O at a point (X,Y) of the plane of the pupil P.

C) Fourier plane of the object

The observation plane is the Fourier plane of the object O (Fig. 11) i.e. the image plane of the source S formed by the lens L. It is situated at a distance z from L given by $(1/p+s)+(1/z)= 1/f$.
It may be easily demonstrated that the irradiance at a point M(x,y) of plane Π is proportional to :

$$I_\infty(x,y) = \left| \iint O(\xi,\eta) P(\xi,\eta) \exp \left[-j \frac{2\pi}{\lambda f} \frac{(p+s-f)}{s} (\xi x+\eta y) \right] d\xi d\eta \right|^2$$

(39)

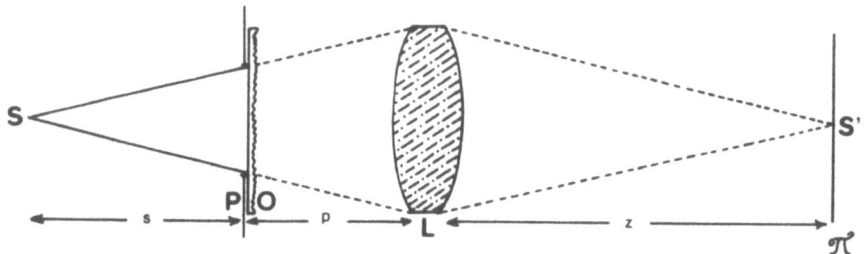

Fig. 11   Speckle pattern generated by the object O in the image plane of the illuminating point source.

provided that the aperture of the lens L is infinite.

3.2.2 Variation of I(x,y) with a lateral translation of O

The object is laterally shifted parallel to the ξ axis through an amount ξ_0. In each of the planes defined above occurs a new intensity distribution.

a) Fresnel plane

The amplitude at a point M (x,y) of plane Π, after the translation of the object, is given by Eq. 34 in which $O(\xi,\eta)$ is now replaced by $O(\xi-\xi_0,\eta)$. It may be rewritten as:

$$u'_F(x,y) = \iint \exp\left[j\frac{\Pi}{\lambda s}[(\xi+\xi_0)^2+\eta^2]\right] O(\xi,\eta) P(\xi+\xi_0,\eta) \quad \times$$

$$\exp\left[j\frac{\Pi}{\lambda z} \{[(x-\xi_0)-\xi]^2 + (y-\eta)^2\}\right] d\xi d\eta \quad (40)$$

and the corresponding irradiance in plane Π is thus given by :

$$I'_F(x,y) = \left| \iint O(\xi,\eta) P(\xi+\xi_0,\eta) \exp\left[j\frac{\Pi}{\lambda} \frac{s+z}{sz} (\xi^2+\eta^2)\right] \quad \times \right.$$

$$\left. \exp\left[- j\frac{2\Pi}{\lambda z} \{\xi(x- \xi_0\frac{s+z}{s}) + \eta y\}\right] d\xi d\eta \right|^2 .$$

$$(41)$$

The comparison between Eqs. 35 and 41 shows that the translation of the object O through ξ_0 in the ξ axis direction involves a translation in the same direction of the speckle pattern lying in plane Π, through an amount x_0 equal to :

$$x_o = \xi_o \frac{z+s}{s} \quad . \tag{42}$$

However, the corresponding speckles remain identical and consequently fully correlated if ξ_o is extremely small with respect to the size of the pupil P. Let us note that, if the object O is illuminated by a parallel beam of light (s=∞); $x_o = \xi_o$. The amount of translation suffered by the speckle pattern is thus equal to that of the object in every Fresnel plane of the surrounding space.

b) Image plane

This case is similar to the previous one. A translation of the object through ξ_o involves a translation of the speckles lying in the plane of the pupil through:

$$X_o = \frac{(p+s)(f-d)+fd}{fs} \xi_o \quad . \tag{43}$$

The intensity distribution in the image plane is then proportional to :

$$I_i' (x,y) = \left| \iint \tilde{O}(X,Y) P(X+X_o,Y) \exp\left[-j \frac{2\pi}{\lambda(p'-d)}\{X(x+\frac{\xi_o p'}{p})+Yy\}\right]dXdY \right|^2. \tag{44}$$

The speckles of the image plane are translated through :

$$x_o = - \xi_o \frac{p'}{p} \tag{45}$$

and remain fully correlated as long as X_o is negligible compared with the size of the pupil of lens L.

c) Fourier plane

The intensity distribution of the Fourier plane, given by :

$$I'_\infty(x,y) = \left| \iint O(\xi,\eta) P(\xi+\xi_o,\eta) \exp\left[-j \frac{2\pi}{\lambda f} \frac{p+s-f}{s}\{(\xi+\xi_o)x+\eta y\}\right]d\xi d\eta \right|^2, \tag{46}$$

shows that the translation of O involves a linear phase shift in Fourier plane and a decorrelation of the speckle pattern if ξ_o is not very small in comparison with the size of the aperture P. The speckles remain fixed.

3.2.3 Variation of I(x,y) with an axial translation of 0

a) Three dimensional structure of a speckle

It has been shown [10] that each speckle of the space has a
three dimensional structure which may be represented by a circular
ellipsoid whose major axis is pointed towards the center of the
object (Fig. 12). The mean size of each ellipsoid depends on its
position and on the size of the pupil limiting the object. The projection on z axis of the major axis of the ellipsoid lying at a
mean distance z from the object plane is given by the depth of
focus of a perfect lens, bounded by the same aperture as the object,
the focal length of which being equal to z. It may be assumed that
the speckle brightness and the phase distribution remain almost
constant inside each ellipsoid. The speckles lying in a plane are
the section of those ellipsoids by the plane under consideration.

Let us study the change involved into a speckle pattern by a
shift of the object 0 parallel to z axis through an amount ε.

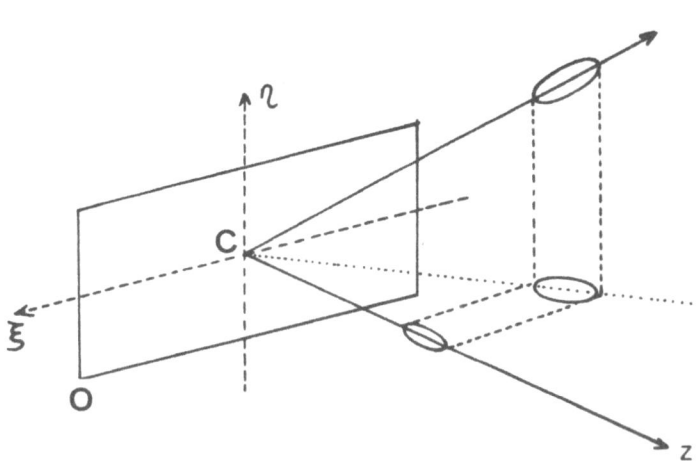

<u>Fig. 12</u> Three dimensional structure of a speckle in free space
propagation.

b) Fresnel plane.

The irradiance of plane Π is now given by Eq. 35 in which s is replaced by $(s+\varepsilon)$ and z by $(z-\varepsilon)$ (Fig. 13).

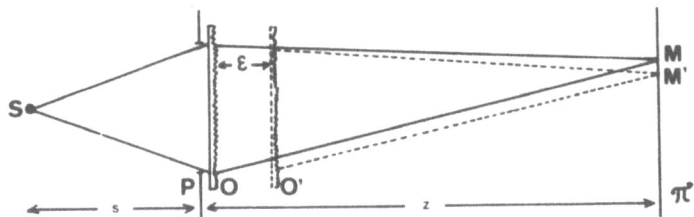

<u>Fig. 13</u> Evolution of the speckle pattern lying in plane Π with an axial shift of 0.

It can be assumed that ε is small enough for $(s+\varepsilon)^{-1}$ and $(z-\varepsilon)^{-1}$ to be respectively written as :

$$(s+\varepsilon)^{-1} \sim \frac{1}{s}(1-\frac{\varepsilon}{s}) \;\; ; \;\; (z-\varepsilon)^{-1} \sim \frac{1}{z}(1+\frac{\varepsilon}{z}). \tag{47}$$

Under these conditions, we have :

$$I'_F(x,y) = \left| \iint 0(\xi,\eta) P(\xi,\eta) \exp\left[j \frac{\Pi}{\lambda} \frac{s+z}{sz} (\xi^2+\eta^2) \right] \times \right.$$
$$\exp\left[j \frac{\Pi}{\lambda} \frac{s^2-z^2}{s^2 z^2} \varepsilon(\xi^2+\eta^2) \right] \times$$
$$\left. \exp\left[- j \frac{2\Pi}{\lambda z} \{\xi(1+\frac{\varepsilon}{z})x + \eta(1+\frac{\varepsilon}{z})y\} \right] d\xi d\eta \right|^2 .$$

$$\tag{48}$$

The axial translation of 0 involves, on the one hand, both a radial shift of the speckles and a magnification due to the change of angular size of the object in plane Π and, on the other hand, a decorrelation of the corresponding speckles due to the term :
$\exp[j\Pi\varepsilon(s^2-z^2)(\xi^2+\eta^2)/\lambda^2 s^2 z^2]$ in Eq. 48 [11] . This decorrelation term can be neglected if the maximum value of $\Pi\varepsilon(s^2-z^2)(\xi^2+\eta^2)/\lambda s^2 z^2$ is very much less than Π, i.e. if :

$$\varepsilon \ll \frac{s^2 z^2}{R^2(s^2-z^2)} \lambda \tag{49}$$

where R is the radius of the circular aperture limiting O.

In the particular case when $s=z=z_0$ and ε^2/z_0^2 is negligible in comparison with unity [12] , Eqs (35) and (48) become respectively:

$$I_F(x,y) = \left| \iint O(\xi,\eta)P(\xi,\eta) \exp\left[j \frac{2\Pi}{\lambda z_0} (\xi^2+\eta^2)\right] \right. \times$$

$$\left. \exp\left[- j \frac{2\Pi}{\lambda z_0} (\xi x+\eta y)\right] d\xi d\eta \right|^2 ,$$

$$I'_F(x,y) = \left| \iint O(\xi,\eta) P(\xi,\eta) \exp\left[j \frac{2\Pi}{\lambda z_0} (\xi^2+\eta^2)\right] \right. \times$$

$$\exp\left[- j \frac{2\Pi}{\lambda z_0}\{\xi(1+ \frac{\varepsilon}{z_0})x + \eta(1+ \frac{\varepsilon}{z_0})y\}\right] d\xi d\eta \Big|^2$$

$$= I_F\left[x(1 + \frac{\varepsilon}{z_0}), y (1+ \frac{\varepsilon}{z_0})\right] .$$

$$\tag{50}$$

With this particular geometrical configuration, $I_{F_2}(x,y)$ and $I'_F(x,y)$ remain fully correlated as long as the amount $\varepsilon 2/z^2_0$ may be neglected with comparison to unity.

·c) Image plane

After the axial shift of the object O through ε (Fig. 10), the amplitude distribution in the plane of pupil P is given by:

$$\tilde{O}'(X,Y) = \iint O(\xi,\eta) \exp\left[j \frac{\Pi}{\lambda s} (\xi^2+\eta^2)\{ 1 + \frac{p's(f-d)}{pf(p'-d)} \}\right] \times$$

$$\exp\left[j \frac{\Pi\varepsilon}{\lambda p^2} (\xi^2+\eta^2)\{ \frac{s^2-p^2}{s^2} + \frac{p'^2 d^2}{p^2(p'-d)^2} - \frac{2 p'd}{p(p'-d)\cdot} \}\right] \times$$

$$\exp\left[- j \frac{2\Pi}{\lambda} \frac{p'}{p(p'-d)} (1 + \frac{\varepsilon}{p} - \varepsilon \frac{p'd}{p^2(p'-d)}) (\xi X+\eta Y)\right] d\xi d\eta$$

$$= \tilde{O}_\varepsilon\left[X(1 + \frac{\varepsilon}{p} - \frac{\varepsilon p'd}{p^2(p'-d)}), Y(1 + \frac{\varepsilon}{p} - \frac{\varepsilon p'd}{p^2(p'-d)})\right]$$

$$\tag{51}$$

and it may be seen by comparison with Eq. 38 that the corresponding speckles suffer a decorrelation, a radial shift and a magnification. The intensity distribution in image plane may thus be written as :

$$I'_i(x,y) = \left| \iint \widetilde{O}'(X,Y)P(X,Y)\exp\left[j\frac{\pi\epsilon}{\lambda}\frac{p'^2}{p^2}\frac{(X^2+Y^2)}{(p'-d)^2} \right] \times \right.$$

$$\left. \exp\left[-j\frac{2\pi xX}{\lambda(p'-d)} \right] dXdY \right|^2 \qquad (52)$$

which may be rewritten as :

$$I'_i(x,y) = \left| \iint \widetilde{O}_\epsilon(X,Y)P\left[X\left(1 - \frac{\epsilon}{p} + \frac{\epsilon p'd}{p^2(p'-d)}\right), Y\left(1 - \frac{\epsilon}{p} + \frac{\epsilon p'd}{p^2(p'-d)}\right) \right] \right.$$

$$\times \quad \exp\left[j\frac{\pi\epsilon}{\lambda}\frac{p'^2}{p^2}\frac{(X^2+Y^2)}{(p'-d)^2} \right] \times$$

$$\left. \exp\left[-j\frac{2\pi}{\lambda(p'-d)}\left\{ X\left(1 - \frac{\epsilon}{p} + \frac{\epsilon p'd}{p^2(p'-d)}\right)x + Y\left(1 - \frac{\epsilon}{p} + \frac{\epsilon p'd}{p^2(p'-d)}\right)y \right\} \right] dXdY \right|^2.$$

$$(53)$$

The phenomena are the same as those described in free space geometry [13]. The corresponding speckles of the image plane suffer both a decorrelation, a radial shift and a magnification. The decorrelation can be reduced by reducing the effective numerical aperture of the imaging lens.

Let us consider the particular case when the pupil P of the imaging lens is lying in the focal image plane of L(d=f) [14]. Under these conditions, the axial shift suffered by the speckles at a distance $r = (x^2 + y^2)^{1/2}$ from the center of the field is given by :

$$\Delta r = \frac{\epsilon}{p}\left(\frac{p'f}{p'(p'-f)} - 1\right) r = 0 . \qquad (54)$$

The speckles generated in the image plane by an elementary cell of O before and after the axial shift of the object are superimposed. Moreover, they are decorrelated. The radial shift of the corresponding speckle pattern is removed due to the fact that the optical center of the imaging is the back-focus of lens L.

d) Fourier plane

The irradiance of the Fourier plane, after the axial shift of O through ϵ is given by Eq. 39 in which s and p are respectively replaced by $(s + \epsilon)$ and $(p - \epsilon)$, i.e. :

$$I'_{\infty}(x,y) =$$

$$\left| \int\!\!\int O(\xi,\eta)\ P(\xi,\eta)\exp\left[-j\ \frac{2\pi}{\lambda f}\ \frac{p+s-f}{s}\ \{\xi(1-\frac{\varepsilon}{s})x + \eta(1-\frac{\varepsilon}{s})y\}\right]d\xi d\eta \right|^2$$

$$= I_{\infty}[x(1-\frac{\varepsilon}{s}),y(1-\frac{\varepsilon}{s})]. \tag{55}$$

An axial translation of the object involves thus a radial shift and
a magnification of the corresponding speckles of the Fourier plane. The
two-speckle patterns remain quite correlated. If the object is
illuminated by a parallel beam of laser light ($s=\infty$), the intensity
distributions of the Fourier plane before and after the shift of O
are identical and given by :

$$I_{\infty}(x,y) = I'_{\infty}(x,y) = \left| \int\!\!\int O(\xi,\eta)P(\xi,\eta)\ \exp\left[-j\ \frac{2\pi}{\lambda f}(\xi x + \eta y)\right]d\xi d\eta \right|^2 . \tag{56}$$

In this case the speckles remain fixed without any decorrelation.
The translation of the object involves only a quadratic variation
of phase in the Fourier plane.

In fact, the variation of a speckle pattern with a shift (lateral or axial) of the diffuse object is strongly subordinated to the
localization of the observation plane, the size and the localization
of the pupil of the optical set-up and to the geometrical shape of
the illuminating beam.

3.2.4 Variation of I(x,y) with the inclination of the incident beam

We suppose here that the diffuse object O is trans-illuminated
with a parallel beam of laser light and that the observation plane is
a Fresnel plane at a distance z from O (Fig. 14).

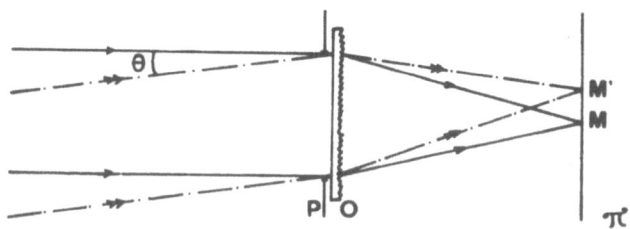

<u>Fig. 14</u> Evolution of the speckle pattern lying in plane Ⅱ with a
variation of the incidence of the illuminating beam.

In normal incidence, the irradiance of plane Π is proportional to :

$$I(x,y) = \left| \iint O(\xi,\eta)P(\xi,\eta) \exp\left[j\frac{\Pi}{\lambda}(\xi^2+\eta^2)\right] \right. \times$$
$$\left. \exp\left[-j\frac{2\Pi}{\lambda}(\xi x+\eta y)\right] d\xi d\eta \right|^2. \tag{57}$$

If now, the incident beam is inclined through an angle Θ onto the normal to surface O, the amplitude of the incident beam is thus given by :

$$u_o = \exp\left[j\frac{2\Pi}{\lambda}\xi\sin\Theta\right] . \tag{58}$$

Moreover, the amplitude transmittance of O depends on the optical path $\Delta(\xi,\eta)$ travelled by the incident beam inside the object O. It is obvious (Fig. 15) that $\Delta(\xi,\eta)$ is a function of the angle of incidence. Consequently $O(\xi,\eta)$ varies versus Θ.

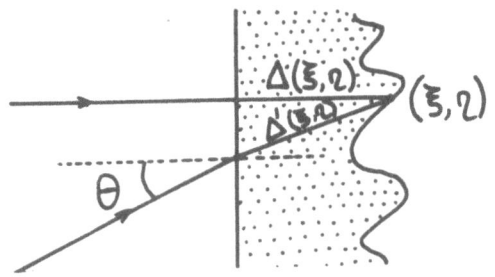

<u>Fig. 15</u> Optical path travelled by a light beam inside O vs the value of Θ.

The intensity distribution in plane Π is now given by :

$$I'(x,y) = \left| \iint \exp\left[j\frac{2\Pi}{\lambda}\xi\sin\Theta\right] O'(\xi,\eta)P(\xi,\eta) \right. \times$$
$$\left. \exp\left[j\frac{\Pi}{\lambda z}(\xi^2+\eta^2)\right] \exp\left[-j\frac{2\Pi}{\lambda z}(\xi x+\eta y)\right] d\xi d\eta \right|^2$$

$$= \left| \iint O'(\xi,\eta)P(\xi,\eta) \exp\left[j\frac{\Pi}{\lambda z}(\xi^2+\eta^2)\right] \right. \times$$
$$\left. \exp\left[-j\frac{2\Pi}{\lambda z}\{\xi(x-z\sin\Theta)+\eta y\}\right] d\xi d\eta \right|^2 . \tag{59}$$

If Θ is small enough, $O'(\xi,\eta)$ is identical to $O(\xi,\eta)$. Consequently, a very small change in the incidence of the illuminating beam involves only a translation of the speckle pattern through $z \sin\Theta$ in any plane of the surrounding space [15] but the image plane.

3.2.5 Variation of $I(x,y)$ with a change of the illuminating wavelength

The object O is illuminated in normal incidence by a parallel beam of laser light as shown in Fig. 14. Let us assume that O is a phase object whose intensity transmittance is equal to unity. Its amplitude transmittance at point (ξ,η) is therefore given by (Fig.15):

$$O(\xi,\eta) = \exp [j \frac{2\Pi}{\lambda} n e (\xi,\eta)] \tag{60}$$

where $e(\xi,\eta)$ is the random thickness of O at point (ξ,η) n is its refractive index and λ is the wavelength of the illuminating beam.

The irradiance of the Fresnel plane π is given by Eq. 57 where $O(\xi,\eta)$ depends on the value of the incident wavelength λ. We suppose now that λ is changed into λ'. The new irradiance in plane π may be written as :

$$I'(x,y) = \left| \int\int O'(\xi,\eta) P(\xi,\eta) \exp [j \frac{\Pi}{\lambda} \frac{\lambda}{\lambda'z} (\xi^2+\eta^2)] \right. \times$$

$$\left. \exp [-j \frac{\Pi}{\lambda z} \{\frac{\lambda}{\lambda'} \xi + \frac{\lambda}{\lambda'} \eta y\}] d\xi d\eta \right|^2 . \tag{61}$$

The speckles of plane Π suffer a decorrelation, a magnification and a radial shift with the change of wavelength. The decorrelation is due, on the one hand, to the phase variations involved by the random optical thickness of O and, on the other hand, to the quadratic phase factor $\exp [j \frac{\Pi}{\lambda} \frac{\lambda}{\lambda'} z(\xi^2+\eta^2)]$ [16].

By comparing Eqs.57 and 61 it may be seen that the speckles generated by the wavelength λ and lying in the Fresnel plane at the distance:

$$z' = \frac{\lambda'}{\lambda} z \tag{62}$$

from O, are identical in all respect to those generated by the wavelength λ' in plane Π, but the decorrelation due to the roughness of O.

The results are exactly the same in the image plane. A change in the illuminating wavelength involves a radial shift, a magnification and a decorrelation between the corresponding speckles [17].

3.3 Detection of in-plane deformation by speckle photography

The technique of double exposure speckle photography could afford the detection of displacements or deformations suffered by a diffuse object. All the methods which were perfected are grounded on the fundamental following experiment [18,19].

A photographic plate H is twice exposed to the speckle pattern generated in any plane but the Fourier plane by a laser-illuminated diffuse object. Between the two exposures, the object is laterally translated in its own plane. Let x_o be the lateral translation of the speckle pattern in the plane of the photographic plate. The total recorded irradiance is :

$$\mathcal{J}(x,y) = I(x,y) + I(x-x_o,y) .\qquad(63)$$

The photographic plate is processed to operate on the linear portion of the (t–E) curve. Its amplitude transmittance is therefore given by :

$$t(x,y) = a-b[\ I(x,y) + I(x-x_o,y)]\qquad(64)$$

where a and b are constants.
After processing, the plane is illuminated by a parallel beam of laser light. The amplitude distribution in its Fourier plane (Fig. 16) is given by the Fourier transform of t (x,y):

$$U(u,v) = a\delta(u,v) - b\ \tilde{I}\ (u,v)\ [\ 1 + \exp\ (i\ \frac{2\pi}{\lambda}\ u\ x_o)]\qquad(65)$$

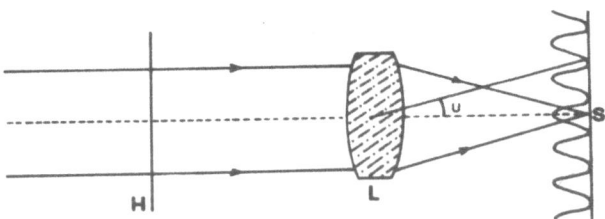

Fig. 16 Observation of the Fourier plane of the photographic record H of two laterally-shifted speckle patterns.

where $\tilde{I}(u,v)$ is the Fourier transform of $I(x,y)$ and where (u,v) are angular coordinates. The first term is the directly transmitted light and corresponds (in Fourier plane) to the direct image of the source at infinity. This image is located at the back-focus of lens L and, since it has very small size, we shall not take it into consideration while discussing the phenomena in the focal plane. The intensity of the Fourier plane may therefore be written as :

$$\mathcal{J}(u,v) = b^2 \left| \tilde{I}(u,v) \right|^2 \cos^2 \left[\frac{\Pi u x_o}{\lambda} \right] . \tag{66}$$

$\left| \tilde{I}(u,v) \right|^2$, the power spectrum of $I(x,y)$, is limited by the auto-correlation function of the pupil of the optical geometry used during the recording process. The $\cos^2$ term represents a system of Young's fringes whose contrast is maximum and whose angular spacing is λ/x_o. Let us assume that H records twice the irradiance of a Fresnel plane at a distance z from the object O bounded by a circular aperture P whose radius is R. The mean irradiance of the Fourier plane of H is represented in Fig. 17. There is no light in the Fourier plane of H at the points (u,v) such that :

$$(u^2 + v^2)^{1/2} \geqslant \frac{2R}{z} . \tag{67}$$

The fringes are visible in the Fourier plane of H if:

$$\frac{\lambda}{x_o} \leqslant \frac{2R}{z} , \quad \text{i.e. } x_o \geqslant \frac{z}{2R} . \tag{68}$$

Consequently, when the two speckle patterns recorded by H are laterally shifted through an amount x_o less than the mean speckle size, there are no observable fringes in the Fourier plane of the photographic plate.

The first method to detect in-plane deformations by speckle photography was proposed in 1971 [20]. A photographic plate H successively records the intensity distribution lying in the image plane of the laser-illuminated object O, before and after a deformation. Moreover, H is laterally shifted through an amount x_H between the exposures. The photographic plate H is therefore a record of two laterally displaced speckle patterns generated by the same object whose degree of correlation depends on the deformation undergone by the object between the exposures. The total irradiance recorded by H may be written as :

$$\mathcal{J}(x,y) = I(x,y) + I'(x-x_H,y) . \tag{69}$$

As described above, H is processed to operate under the usual conditions of linearity and its amplitude transparency is given by :

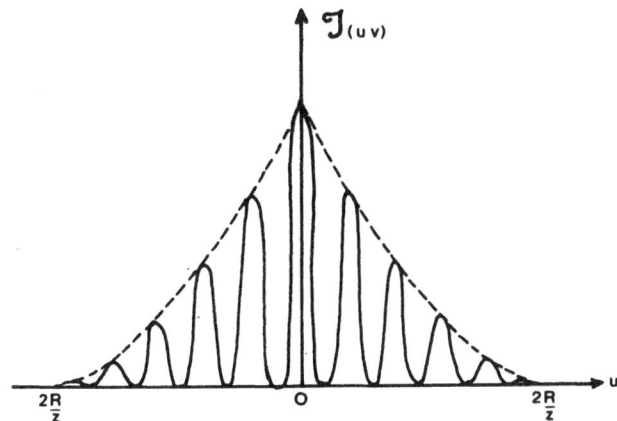

<u>Fig. 17</u> Power spectrum of the intensity distribution recorded by H.

$$t(x,y) = a - b [I(x,y)+I'(x-x_H,y)].$$ (70)

After processing, H is illuminated by a parallel beam of laser light
(Fig. 18) and displays in its Fourier plane an amplitude distribution
proportional to the Fourier transform of Eq. 70 :

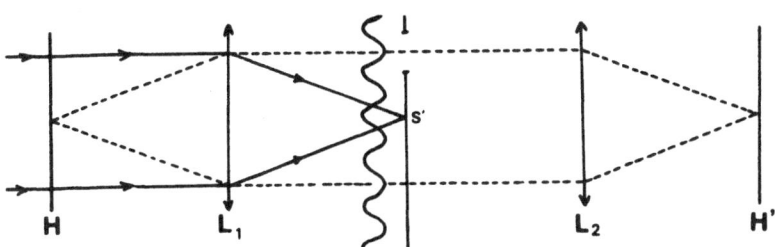

<u>Fig. 18</u> Detection of in-plane deformations by filtering the Fourier
plane of the doubly-exposed plate H.

$$\underline{t}(u,v) = b[\ \tilde{I}(u,v) + \tilde{I}'(u,v)\ \exp\ (j\ \frac{2\Pi}{\lambda}\ u\ x_H)] \tag{71}$$

which may be rewritten as :

$$\underline{t}(u,v) = b\tilde{I}(u,v)\ \{1+\exp\ [\ j\ \frac{2\Pi}{\lambda}\ ux_H]\ \} + b[\ \tilde{I}'(u,v)-\tilde{I}(u,v)]\quad x$$

$$\exp\ [\ j\ \frac{2\Pi}{\lambda}\ ux_H\ ]. \tag{72}$$

The first term of Eq. 72 represents the light scattered in the Fourier plane by the areas of signal O which suffered no deformation in between the exposures. This amplitude distribution is modulated by a system of Young's fringes with maximum contrast. However, the areas of O and O' which are not identical scatter a random background in the Fourier plane of the photographic record, given by the second term of Eq. 72. This background causes a decrease in the contrast of the Young's fringes. Finally, the plate H displays a system of Young's fringes, the contrast of which is not maximum. The maxima of the fringe system (corresponding to the values of u for which the exponential term is equal to +1) provide information concerning both O and O'. On the contrary, in the minima of the fringe system corresponding to the values of u for which the exponential term is equal to -1, there is only information about the difference of the Fourier transforms of O and O'. Consequently, if the maxima of the Fourier plane are removed by an aperture diaphragm which only lets a minimum pass, the amplitude retrieved in the image plane H' of the photographic plate is given by :

$$u(x,y) = [\ I'(-x,-y)- I(-x,-y)]\ \otimes\ \tilde{F}(-x,-y) \tag{73}$$

where $\tilde{F}(-x,-y)$ is the Fourier transform of the filtering slit. The parts of the object which have suffered the deformation are only retrieved in the image plane of the photographic plate with an irradiance proportional to the amount of deformation.

If the deformation given to the object corresponds in the image plane to a translation of the corresponding speckle through an amount less than the average speckle size, it cannot be detected without an additional modulation of the image plane by a ground glass [21].

4. CONCLUSIONS

This paper is mainly devoted to some fundamental properties of speckle patterns. These random intensity distributions are generated by the coherent superposition of the wavelets scattered by a diffuse object. We studied only the scattering surfaces whose local irregularities in depth are on the scale of optical wavelengths.

It has been then demonstrated that the contrast of the speckle patterns is always equal to unity. The mean speckle size is independent of the microstructure of the object but strongly related to the pupil of the optical set-up and to the localization of observation plane. In the same way, the evolution of a speckle pattern with a lateral or longitudinal translation of the object is only a function of the same parameters. However, the microstructure of the object plays a predominant part in the decorrelation of the speckles involved either by a change of the incidence or by a change of the wavelength of the incident beam.

REFERENCES

1. Exner, K, Sber. Akad. Wiss. Wien, 76 (1977) 522.
2. Goodman, J.W., in "Laser Speckle and Related Phenomena" ed. J.C. Dainty, vol 9 of Topics in Applied Physics (Springer-Verlag, Berlin, Heidelberg, New York, 1975).
3. Middleton, D. : Introduction to Statistical Communication Theory (McGraw Hill Book Co, New York 1960).
4. Wang, M.C. and Uhlenbeck, G.E., Revs. Modern. Phys. 17 (1945) 332.
5. Fuji, H. and Asakura, T., Opt. Commun. 11 (1974) 35.
6. Pedersen, H.M., Opt. Commun. 12 (1974) 156.
7. Goodman, J.W., Opt. Commun. 14 (1975) 324.
8. Crane, R.B., J. Opt. Soc. Am. 60 (1970) 1658.
9. Miller, M.G., Schneiderman, A.M. and Kellen, P.F., J. Opt. Soc. Am. 65 (1975) 779.
10. Eliasson, B. and Mottier, F.M., J.Opt. Soc. Am. 61 (1971) 559.
11. Mendez, J.A. and Roblin, M.L., Opt. Commun. 11 (1974) 245.
12. Dzialowski, Y., May, M., Opt. Commun. 16 (1976) 334.
13. Archbold, E. and Ennos, A.E., Opt. Acta, 19 (1972) 253.
14. Dzialowski, Y., May, M. and Shaw, R., Opt. Commun. 21 (1977) 282.
15. Debrus, S. and Grover, C.P., Opt. Commun. 13 (1971) 340.
16. Mendez, J.A. and Roblin, M.L., Opt. Commun. 13 (1975) 142.
17. Tribillon, G., Opt. Commun. 11 (1974) 172.
18. Burch, J.M. and Tokarski, J.M.J., Opt. Acta, 15 (1968) 101.
19. Debrus, S., Françon, M., Mallick, S.L., May, M. and Roblin, M.L., C.R. Acad. Sci. Paris, 267 (1968) 1332.
20. Butters, J.N. and Leendertz, J.A., J. Phys. E. Sci. Instrum. 4 (1971) 272.
21. Françon, M., Koulev, P. and May, M., Opt. Commun. 12 (1974) 63.

PULSED LASER IRRADIATION OF SEMICONDUCTORS : THERMAL DESCRIPTION

E. Rimini

Istituto di Struttura della Materia – 57 Corso Italia
95129 Catania, Italy.

1. INTRODUCTION

The use of pulsed laser or electron beams in the nanosecond duration regime and with energy densities of the order of 1 Joule/cm^2 allows to deposite a large amount of energy in short times into the near surface region [1] . Under suitable conditions the irradiation leads to the melting of the surface to a depth of several thousand angstroms. During solidification liquid phase epitaxial regrowth can occur from the underlying single crystal substrate. The irradiation implies also physical conditions far from thermodynamic equilibrium. High quenching rates $\sim 10^9$K/s, high thermal gradients $\sim 10^8$K/cm and velocity of the liquid-solid interface of several meters per second characterize for instance the thermal behaviour for laser irradiation of semiconductors [2] .

A tremendous impetus [3] to this subject came soon after the pioneering work of Russian scientists [4] because they showed that damage in ion implanted semiconductors is removed by pulsed or continuous wave laser irradiation. The term "laser annealing" was then coined, and it is used to day, conventionally, to indicate the overall effects produced during high power irradiation. In the silicon technology laser annealing has several attractive characteristics : it can heat only the region of interest without influencing the inside structure, the process is very fast : one can melt and solidify surface layers in times as short as nanoseconds.

In addition to the potential relevance in the investigation of new processing technologies for silicon integrated circuits, laser irradiation has offered a powerful method to study crystal growth and the impurity incorporation under conditions far from equilibrium.

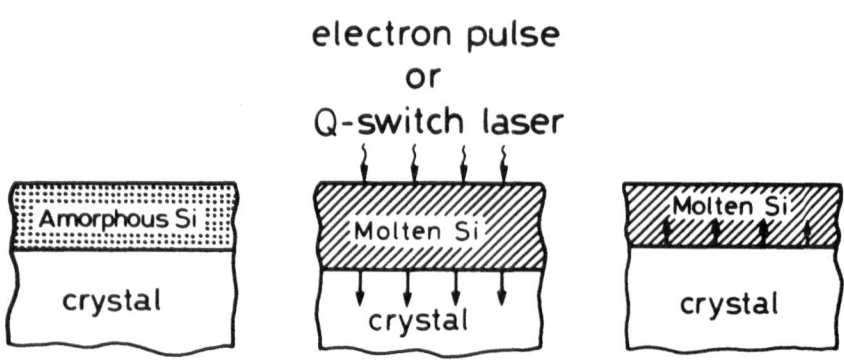

Fig. 1 Schematic of crystallization in amorphous Si on single crystal substrate by electron or Q-switched laser pulse irradiation.

New phases, metastable solutions, formation of amorphous silicon from the melt are only few of the possible new phenomena under investigation [5] .

In these lectures I will describe the effects of laser irradiation on ion implanted semiconductors, mainly silicon, for what concerns the structure change from amorphous to single crystal, or from amorphous to poly-layer. The basic process responsible for the structure change is reported schematically in Fig. 1. The shown case refers to a single crystal overlaid with an amorphous layer and irradiated with a laser or with an electron beam pulse of short duration. The light (or the electron) energy is absorbed by the sample and converted into heat. For a suitable energy density the near surface region can melt. If all the initial amorphous layer is liquid then the subsequent solidification occurs on a single crystal substrate and a liquid phase epitaxy will result. If the thickness of the maximum molten layer is lower than that of the amorphous layer a poly-layer will be produced. The previous considerations are valid below a characteristic value of the solid-liquid interface velocity during solidification.

The transient liquid formation is based on the assumption of an "instantaneous" ($\leq 10^{-11}$s) conversion of the photon energy into heat which then propagates inside the sample following the heat equation [6] . A quantitative thermal description of the process requires to account for the changes of optical and thermal parameters with temperature and structure of the irradiated material, for the absorption and release of latent heat during melting and solidification. The heat diffusion equation should be then solved numerically by computer [7] .

As a result of calculations one obtains the melt front penetration, the temperature distribution inside the sample and its time evolution, the velocity of the liquid-solid interface during melting or solidification. These considerations will be detailed in the second part of the paper.

The last section will be devoted to the impurity behaviour [8-9] . After irradiation there is usually a considerable redistribution of impurities introduced for instance by ion implantation. The shape of the final profile depends not only on the dopant species but also on the dynamics of the melting and solidification. One of the most striking manifestation of the rapidly moving liquid-solid interface is the incorporation, or trapping of dopants in the solid at concentrations in excess of the equilibrium solubilities. The interfacial segregation coefficient becomes a unique function of the regrowth velocity. There are several phenomena as constitutional supercooling and cell structure formation which also play a relevant role under these extreme conditions.

Laser irradiation has also stimulated a series of elegant "in situ" measurements to establish the validity of the thermal description over the non-thermal hypothesis [10] . These include time-dependent reflection [11-12] and transmission [13] during and following an excitation pulse. The abrupt increase in reflectivity and the drop in transmissivity agree with the optical response of molten silicon, being metallic in character. Other measurements as time-dependent Raman scattering [14] , time-of-flight of evaporated surface atoms [15] , time-dependent X-ray Bragg scattering [16] and thermionic-electron emission [17] , provide information on different mechanisms : a particular phonon mode, the surface temperature, the strain distribution, the electron kinetic energy and the lattice vibration respectively. Their relationship with the thermodynamic meaning of temperature has been discussed during several round tables at the school. We mention only the fact that in many of these "in situ" measurements a large number of shots is required. In the Raman and X-ray diffraction experiments several thousand of laser pulses are required to obtain a significant number of counts. It is then necessary for instance to guarantee against fluctuation in the energy density, beam uniformity etc, which is not at all a simple matter for lasers.

Measurements of the time-dependent electrical resistivity [18] on the other hand provide several useful informations as the melt front extension, the velocity of the solid-liquid interface both during melting and solidification. All these data are obtained in a single pulse and for their relevance in the laser-melting description we will report later some results.

74

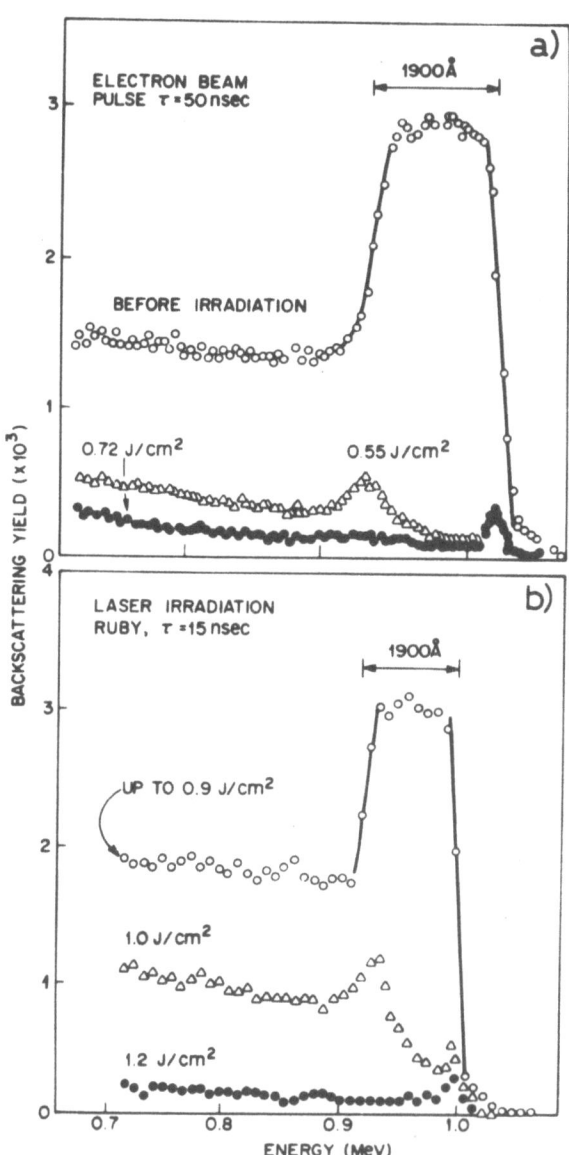

Fig. 2 Analysis with 2.0 MeV and 1.8 MeV He$^+$ beams in combination with channeling effect of recrystallization of a 1900 Å-thick amorphous Si on a (100) oriented substrate after irradiation with electron beam (a) and with ruby laser (b) pulse of different energy density.

2. CRYSTALLIZATION OF IRRADIATED LAYERS

Ion implantation offers a reproductible and clean method to produce amorphous layers few thousand angstroms thick. As a starting point we consider the amorphous to single crystal transition and its threshold dependence on the energy density value [19] . This is illustrated in Fig. 2a and b for pulsed electron beam and ruby laser irradiation, respectively. In both samples the amorphous layer created by ion implantation was 1900 Å thick. The sample structure is analyzed by MeV He$^+$ backscattering in combination with channeling effect techniques. Details of the methods were presented by P. Baeri at the school [20].

The aligned yield recorded after irradiation with a 0.55 J/cm^2 energy density of the electron beam decreases drastically and at 0.7 J/cm^2 coincides with that of unimplanted Si single crystals. The amorphous layer has become single crystal with the same orientation as the substrate. Similar results are found for ruby laser irradiation. Up to 0.9 J/cm^2 the aligned yield does not change. It decreases at 1.0 and at 1.2 J/cm^2 it reaches the unimplanted level. The high yield at 0.55 J/cm^2 for the electron beam and at 1.0 J/cm^2 for the ruby laser is caused by residual disorder located mainly at the original crystal - amorphous interface. The transition to single crystal occurs at a well defined energy density value and involves the entire disordered layer. Thicker amorphous layers require higher energy densities to become single crystals.

Changes in the structure of the irradiated layer can also be analyzed by diffraction for both transmission and reflection of high-energy electrons. As an example the patterns obtained by the reflection of electrons (RHEED) are shown in [21] Fig. 3. The electrons impinge at a glazing angle of few degrees from the surface, and they probe the first few hundred angstroms of the sample. The pattern of the as-implanted Si layer is a diffuse halo characteristic of the amorphous structure. Rings (b) appear in the pattern after ruby laser irradiation and with increasing energy density they change into spots (c).

Rings and spots indicate the presence of a polycrystalline and of a single crystal layer respectively. The poly-layer is produced when the thickness of the melted layer does not penetrate all the damaged layer. Solidification occurs before the liquid wets the single-crystal substrate. This statement is clearly demonstrated by the transmission electron microscopy of a cross section of a Si-sample with an initial amorphous layer 1900 Å thick [22] . After irradiation with 0.2 J/cm^2 ruby laser pulse a poly-layer extends from the surface to a depth of about 1000 Å, a residual thin amorphous layer is still present. At the end of the amorphous layer is present a region of damage due to the ion implantation process performed in this case at room temperature. The final aspect is

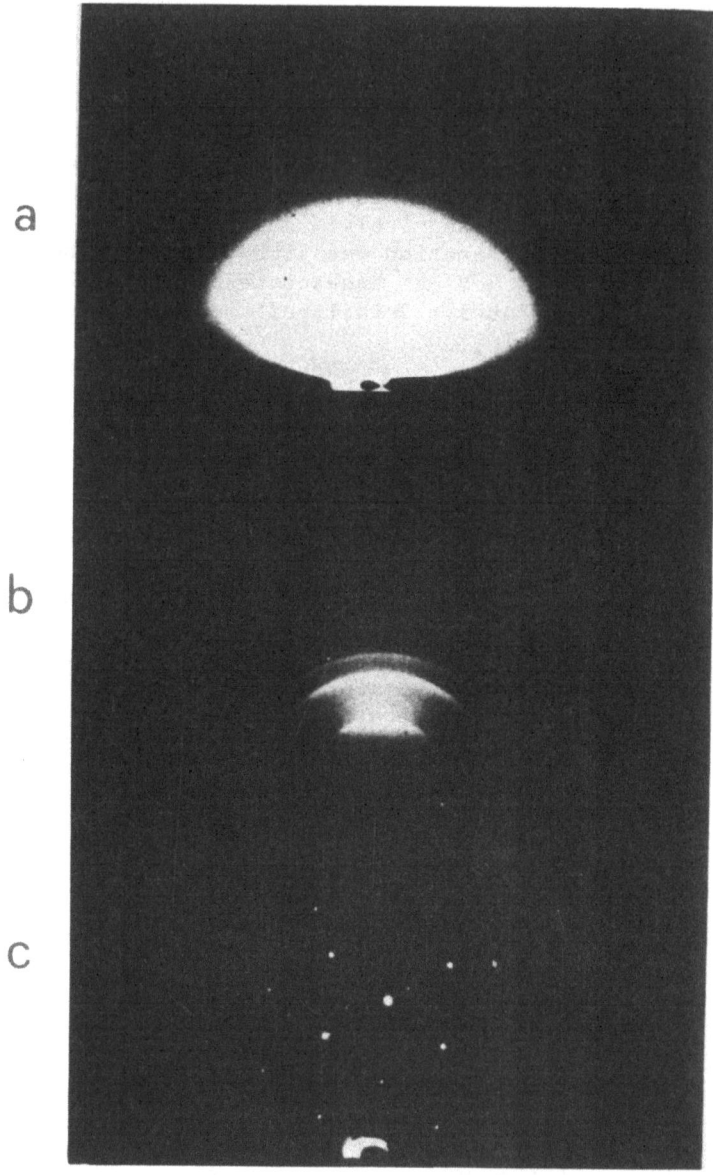

Fig. 3 Reflection high-energy electron diffraction (RHEED) patterns of an as-implanted amorphous layer (a), after 1.25 J/cm^2(b) and 5.0 J/cm^2 (c) ruby laser pulses, respectively. The azimuth direction is along the <110> .

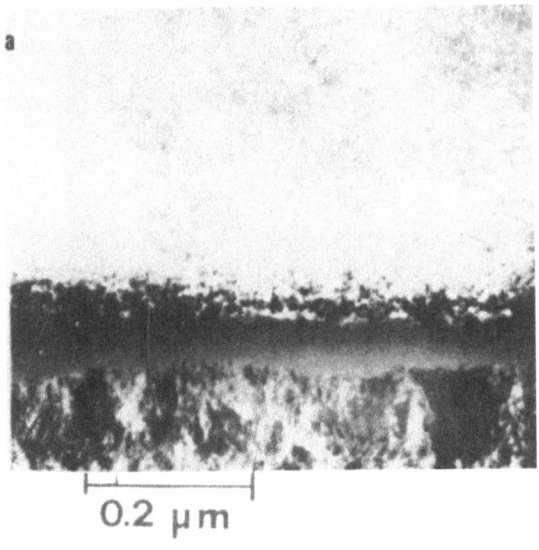

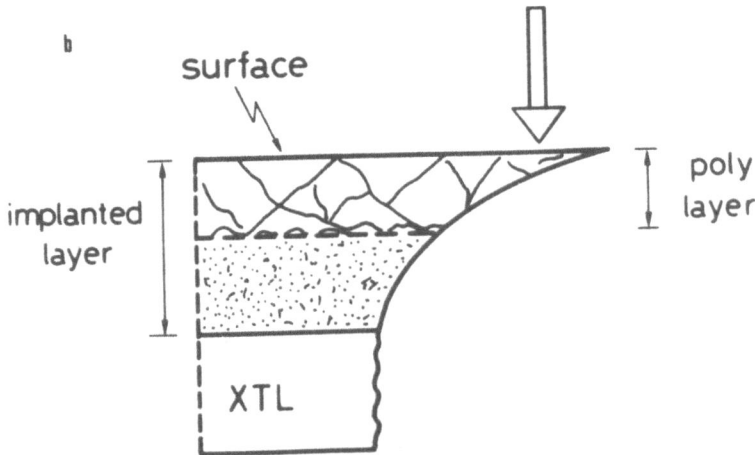

Fig. 4 Transmission electron micrograph of a cross section of a 1900 Å thick amorphous layer on Si <100> substrate after irradiation with 0.2 J/cm^2 ruby laser 15ns pulse duration (a); schematic of a back-thinned implanted silicon sample for TEM analysis (b).

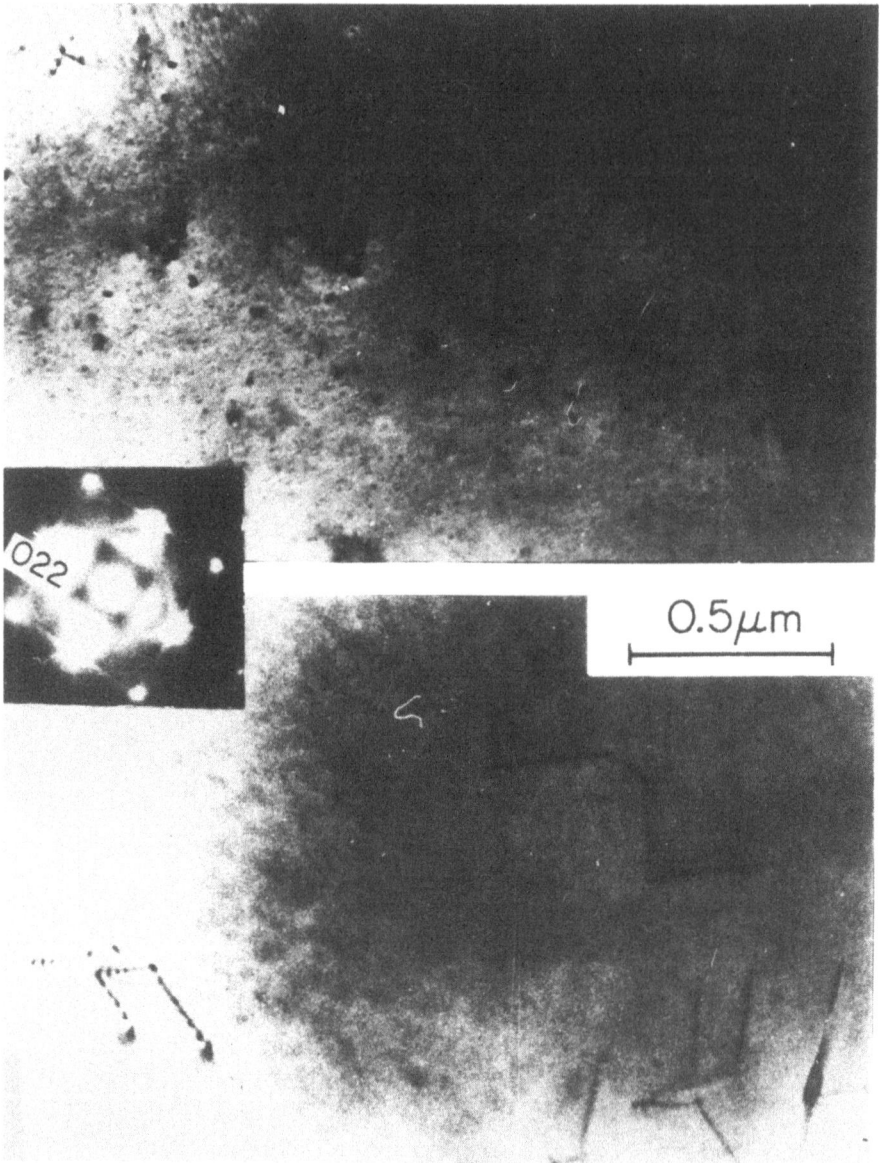

Fig. 5 Transmission electron micrograph of 10^3Å thick (a) and 4.5×10^3 Å thick (b) amorphous implanted Si layers after irradiation with 1.5 and 2.5 J/cm^2 ruby laser pulses respectively. The dislocations originate from small loops and their lengths scale with the thickness of the initial amorphous layer.

sketched in Fig. 4b.

Extended effects in the crystallized layers remain for energy density values just above the threshold for the amorphous-to-single crystal transition [23] . In <100> oriented substrate dislocation V - shaped pairs are distributed throughout the regrown layer. The dislocations are all of the same length and originate from a thin layer located just behind the initial amorphous region. This is illustrated in the two TEM reported in Fig. 5 for a 1000 Å and a 4500 Å thick Si amorphous layer on <100> Si substrate after laser irradition with a 1.5 and 2.5 J/cm^2 ruby laser 50ns duration [23], respectively. In the thinner annealed layer the residual defects are small dislocation loops about 1200 Å below the surface, in the thicker annealed layer the defect structure is similar, but the small loops are now located at about 4500 Å below the surface. The length of the V-shaped dislocations increases with the thickness of the amorphous layer, and they originate from the small loops.

It is then the nature of pre-existing defects in the transition region between the amorphous and the single crystal substrate free of defects, to determine the amount of residual disorder at the threshold value. In the case of <111> oriented substrate, stacking faults are the main type of extended defects at the threshold. They originate also at the pre-existing defects at the interface amorphous-single crystal. It has been shown [24] , for instance, that a reduction of black spots by thermal anneal at 400°C for 1/2hr decreases also the density of stacking faults. At this temperature no regrowth of the amorphous layer occurs by solid phase epitaxy.

The melt front penetration has been investigated [25] by dissolution of precipitates. High temperature diffusion of boron or phosphorus into silicon leads to the formation of precipitates and small loops. After Q-switched irradiation annealed defect-free-regions containing no phosphorus precipitates are found in layer of thickness increasing with the pulse energy density as shown in Fig. 6a and b respectively. Precipitates originate because the dopant concentration exceeds the solid solubility limit. They dissolve under conditions where impurity diffusion coefficients are extremely high as in liquids (~10^{-4} cm^2/sec) and for elevated quenching rates.

The fast deposition of energy associated with laser irradiation implies also a different description, or at least a more detailed one, of the amorphous recrystallization. The free-energy of the amorphous Si is higher than that of crystalline Si. Solid phase crystallization occurs because amorphous Si is thermodynamically less stable than crystalline Si. Crystal growth does not occur at room temperature because of kinetic barriers to atom motion. Fig.7 shows schematically the Gibbs free-energy diagram of amorphous, crystalline and liquid silicon [26] . The intersection of the free-

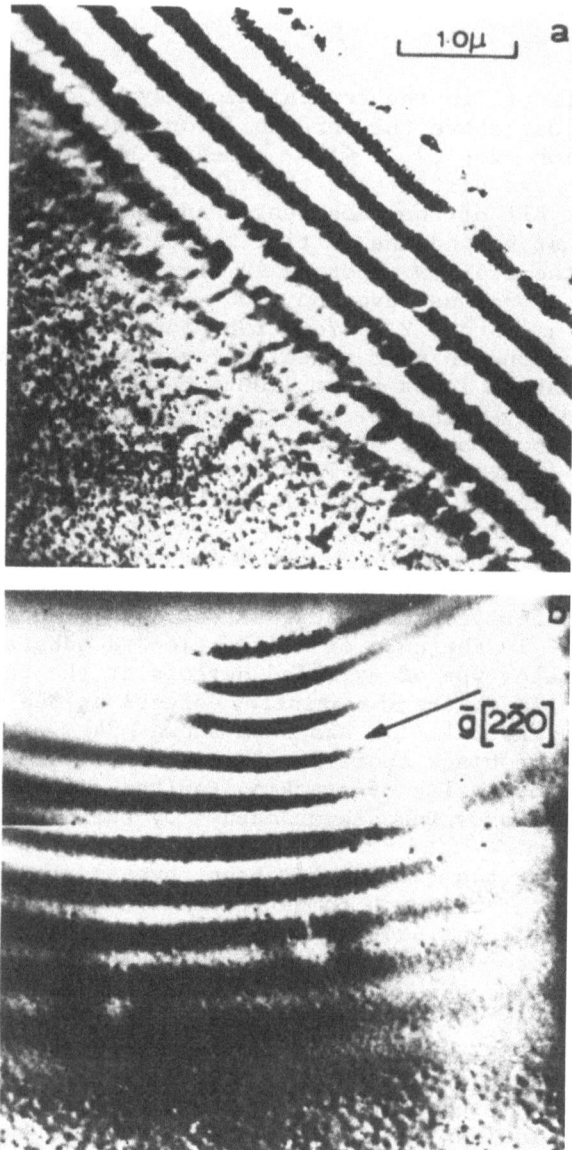

Fig. 6 Bright-field electron micrograph of a phosphorus-diffused silicon specimen shows dissolution of precipitates after ruby laser irradiation. The annealed depth increases with the energy density of the ruby laser single pulse. The thickness of the defect-free regions can be measured from thickness fringes. (a) 2.0 J/cm^2, (b) 3.0 J/cm^2.

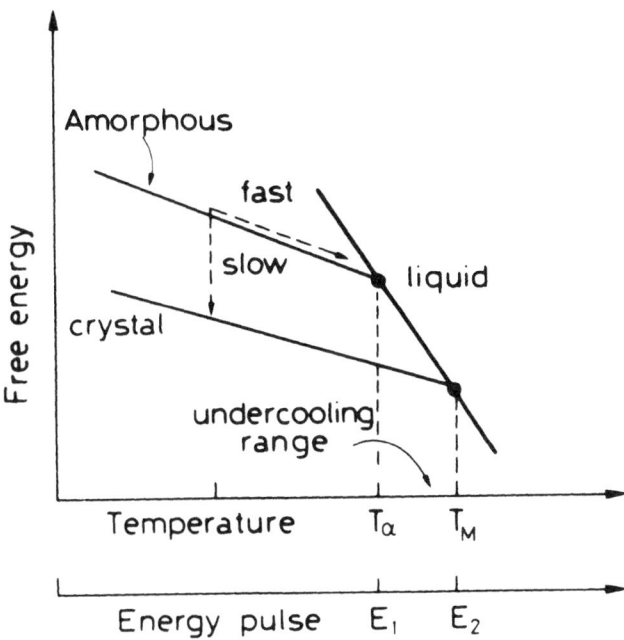

Fig. 7 Free-energy diagram of amorphous, crystalline and dense metallic liquid silicon. T_α and T_m are the melting points of the amorphous and of the crystalline phases,respectively. In the energy density E_1-E_2 the molten layer is undercooled.

energy curves of crystalline and liquid silicon is by definition the equilibrium melting temperature T_c. By a similar extension of the amorphous free-energy curve one obtains the melting temperature T_α, which is lower than that of the crystalline phase. This argument is based on the plausible assumption that the phase transition from the tetrahedral four-fold coordinated amorphous phase to the close-packed 11-12 fold coordinated metallic liquid is first order in nature [27].

The usual heating procedure does not allow the observation of the melting point depression because the amorphous silicon will crystallize in the solid phase before such high temperatures are reached. Only by pulsed irradiation one can hope to avoid crystallization. Heat of crystallization of amorphous Si obtained by ion implantation was obtained recently by differential scanning calorimeter measurements [28]. The obtained H_{ac} value was of 11.3 ± 0.8 and the estimated T_α is depressed of about 250K beneath T_c.

Pulsed electron heating experiments [29] gave some evidence for a substantial reduction in melting temperature of the amorphous material. So far experiments with laser irradiation have been unable to evidence such a depression.

The laser energy is absorbed exponentially with a coefficient depending on the light wavelength and on the target structure. Large thermal gradients are created during the heating so that usually the region at the interface with the substrate is just above T_α while the liquid at the surface is at a temperature above T_m. Undercooling effects are practically negligible.

Electron beam irradiation creates instead small temperature gradient and by a suitable value of the pulse energy density all the amorphous layer can be undercooled. Temperature distributions similar to these ones can be obtained by laser irradiation on the back side of the sample. The maximum of energy deposition occurs in the amorphous layer at the interface with the single crystal substrate. A heat sink is located on the same side of the heat source and the surface represents a barrier to the heat diffusion. No temperature gradient is practically present in the molten layer.

The upper left side of Fig. 8 reports [30] the calculated maximum temperature distributions during the heating for a 30 ns laser-pulse on a 1500 Å thick amorphous Si layer. For energy density in the $0.9-1.4$ J/cm^2 range the amorphous layer is molten at a temperature below the T_m melting point. In the upper right side of the same figure the channeling analysis is shown. Up to 0.7 J/cm^2 the yield coincides with that of the as-implanted sample. With increasing the energy density of the laser pulse a noticeable decrease of the aligned yield occurs in the energy part corresponding to the region at the interface with the single crystal substrate. About

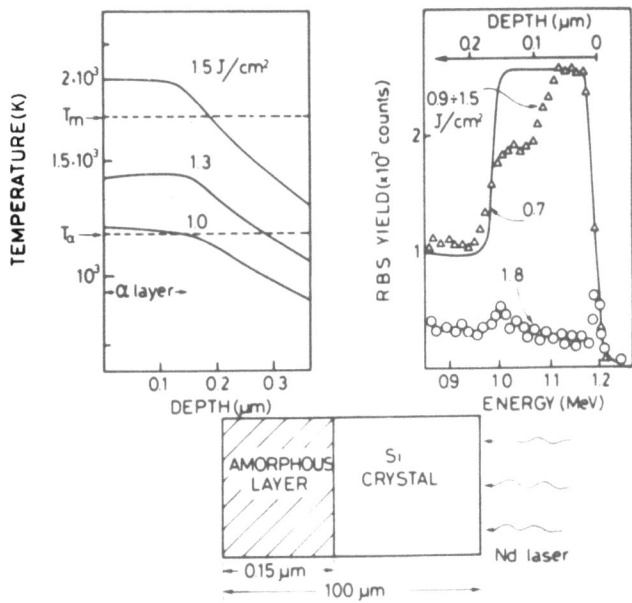

Fig. 8 Back-irradiation of a 1500 Å thick amorphous Si layer with 30 ns Nd glass laser pulse irradiation. The computed temperature profiles at several energy densities are shown in the left side. T_α and T_m represent the melting point of amorphous and crystalline Si respectively. Right hand side : experimental RBS spectra for aligned incidence of 2.0 MeV He$^+$ beam after irradiation at several energy density.

1/2 of the amorphous layer is regrown epitaxially with the under-lying substrate, the other 1/2 is a poly-layer as found by RHEED.

A double layered structure is formed for temperature of the molten layer below T_m, similar to the structure found after e-beam irradiation [29] . The experimental data agree closely with calcu-lations using T_α=1200K and $\Delta H_{\alpha-\ell}$=1250 J gr^{-1} for Si and T_m=1700K and $\Delta H_{c-\ell}$= 1780 J gr^{-1} for crystalline Si. The previous measure-ments with differential calorimetry give instead a difference of 400 J gr^{-1} instead of 530 J gr^{-1} as adopted in the calculations [30] .

The structure of an ion implanted semiconductor after front (a) or back laser (b) irradiation is reported schematically in Fig. 9. For front laser irradiation the sample surface reaches T_α

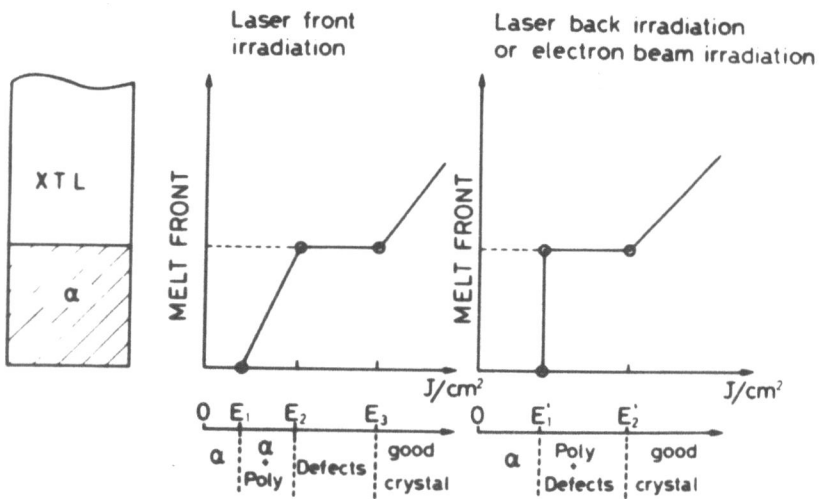

Fig. 9 Melt front penetration for front and back irradiation. The resulting structures are indicated in the lower part.

at an energy density E_1. With increasing the energy density the melt front at T_α penetrates inside and all the amorphous layer is molten at E_2. It becomes poly after solidification. In the energy range E_2-E_3 the absorbed energy increases the liquid temperature. Normal liquid at $T>T_m$ near the surface and undercooled liquid at $T<T_m$ near the interface coexist.

The E_2 value represents the energy density threshold for the amorphous-to-single crystal transition but the crystalline quality of the regrown layer is very poor. At energy values higher than E_3 all the initial amorphous layer is above T_m with part of the underlying single crystal substrate molten.

In back laser or electron beam irradiation the amorphous layer starts to melt at an energy density value E'_2. All the layer becomes liquid simultaneously for the low thermal gradient. Up to E'_2 value the underlying single crystal is still solid. In the $E'_1 - E'_2$ range the molten layer is undercooled, nucleation and fast crystallization occur easily. Above E'_2 a good crystalline epitaxial layer results.

According to the diagram of Fig. 7 one can start from a single crystal to obtain a liquid, if it is considerably undercooled during solidification an amorphous structure can result.

This condition if fulfilled when the interface moves at such high velocities that the liquid must be undercooled beneath T_m. Amorphization of crystalline Si has been obtained by ruby laser irradiation with 25ns pulse duration[31]. Few hundred Å thick amorphous layers were obtained for both <111> and <100> substrates. The critical parameter was the velocity of the solid-liquid interface [32].

3. THERMAL DESCRIPTION OF LASER IRRADIATION

The basic process of laser irradiation involves heat generation by the absorption of light and cooling by heat conductivity into the substrate. It has been clearly stated during this school that at least for nanosecond pulse duration the basic assumption of an instantaneous conversion of the absorbed light into heat is justified. Heating and cooling stages are then determined by numerical solutions of the heat equation including a source term due to the absorption of light at a certain depth, the changes of optical and thermal parameters with temperature and structure of the irradiated layer and the latent heat absorbed or generated during phase transitions.

For simplicity we assume a light beam travelling along the z axis normal to the specimen surface, which is uniform in the x-y plane. The target composition is homogeneous in this plane and structural changes occur only in the z direction. Edge effects are also neglected, i.e. the cross section of the laser beam is assumed to be much greater than the heated sample thickness. The heat equation becomes :

$$\frac{\partial T}{\partial t} = \frac{\alpha}{\rho C_p} \ I(z,t) + \frac{1}{\rho C_p} \frac{\partial}{\partial z} (K \frac{\partial T}{\partial z}) , \qquad (1)$$

where $I(z,t)$ is the power density of the laser light at depth z and time t. T is the temperature, ρ, C_p, K and α the density, specific heat, thermal conductivity and absorption coefficient of the sample.

In a homogeneous medium :

$$I(z,t) = I_o(t) \ (1-R) \ (\exp-\alpha z) \tag{2}$$

$I_o(t)$ being the temporal power output from the laser and R the target reflectivity. Before considering in detail the solution of Eq. 1, it is interesting [6] to illustrate some simple consequences for a rectangular laser pulse of duration τ_p and for constant values of ρ, α, C_p and K.

The heating process involves two characteristic lengths, the absorption α^{-1} and the heat diffusion length $\sqrt{2D\tau_p}$, with $D = \dfrac{K}{\rho C_p}$ the thermal diffusivity. If $\alpha^{-1} < (2D\tau_p)^{1/2}$ the heat source becomes a surface source (Fig. 10a) and the temperature rise is given by [33] :

$$\Delta T(z,t) = [2I_o (Dt)^{1/2}/K] \ \text{ierfc} \ [z/2(Dt)^{1/2}] \ (1-R) \tag{3a}.$$

For $t < \tau_p$, at the surface z=0 one obtains :

$$\Delta T(0,t) = (2I_o/K) \ (Dt/\Pi)^{1/2} \ (1-R). \tag{3b}$$

Approximately the heat required by a layer of thickness $\sqrt{2D\tau_p}$ to increase its temperature of ΔT should be equal to the absorbed energy, i.e. :

$$\Delta T \rho C_p (2D\tau_p)^{1/2} \sim I_o (1-R) \tau_p \Rightarrow T = \frac{I_o(1-R)}{R} \ (\frac{D\tau_p}{2})^{1/2}. \tag{4}$$

The energy density required to reach a given temperature, as for instance the melting point T_m, is given by :

$$E_t = \frac{(T_m - T_o)}{(1-R)} \ (\frac{\Pi}{4} \ K\rho C_p)^{1/2} \ (\tau_p)^{1/2}.$$

The threshold energy density is proportional to $(\tau_p)^{1/2}$, and is independent of the absorption coefficient. Heating and cooling rate are both characterized by τ_p. The heating rate is given by :

$$\frac{\Delta T}{\tau_p} = \frac{(I-R) I_o}{\rho C_p (2D\tau_p)^{1/2}}. \tag{5}$$

For instance for $\Delta T = 10^3$K and $\tau_p = 10$ns, the heating rate amounts to 10^{11}K/s. In the case of Silicon, the previous consideration requires an absorption coefficient larger 10^5 cm^{-1}, i.e. a wavelength of about 0.6 µm or lower. Thermal diffusivity ranges in between 0.1 and 10 cm^2/sec.

The other extreme case, $\alpha^{-1} \gg (2D\tau_p)^{1/2}$, typical of Nd:1.06 µm

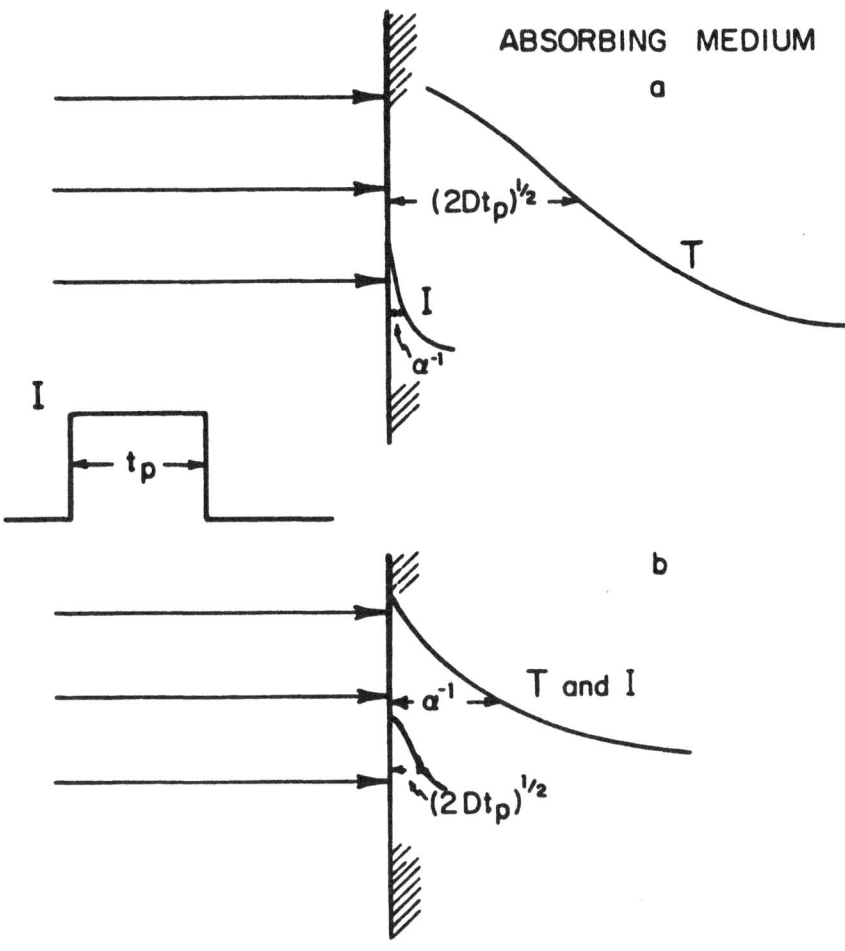

Fig. 10 Schematic laser pulse intensity and temperature profiles for a penetration depth, α^{-1}, of the light small (a) and large (b) compared to the thermal diffusion length.

wavelength in Si single crystal, implies the non-relevance of the heat diffusion and the temperature rise is given by :

$$\Delta T(z,t) = (1-R)I_o \; \alpha \; \exp \; (-\alpha z) \; t/\rho \; C_p \qquad \text{at } t < \tau p. \qquad (6a)$$

At the surface, z=0, and at the end of the pulse, $t = \tau_p$:

$$\Delta T(0,\tau_p) = (1-R) \; I_o \; \alpha\tau_p/\rho C_p . \qquad (6b)$$

The heating rate, given by $\Delta T(z)/\tau_p$ is then independent of the pulse duration and decreases exponentially with depth. The cooling rate can be estimated assuming that heat diffusion occur over a distance α^{-1}, then a time $\alpha^{-2}/2D$ is required so that :

$$\left.\frac{\Delta T}{\Delta t}\right|_{cooling} = \frac{(I-R)I_o\alpha^3\tau_p\,2D}{\rho C_p}, \tag{7}$$

of the order of $10^8-10^9 K/s$.

A detailed treatment requires the numerical solution of the heat eq. 1. To this aim the layer of interest is divided in several slices of thickness Δz, and for each of them we should know the following quantities :

1) Temperature of the i-th layer, T_{bi},

2) Structure of the i-th layer $\left\langle\begin{array}{l} \text{crystalline (0)} \\ \text{amorphous (1) ,} \\ \text{liquid (2)} \end{array}\right.$

3) Fraction of the i-th layer which is melted FF_i,

4) Thermal and optical parameters for each layer, function of

$(T_{bi}, \left\{\begin{array}{l}0\\1,FF_i\\2\end{array}\right.)$.

Let us compute the temperature change with time in the i-th slice. The incident power density is given by $I_i = I_{i-1}e^{-\alpha_{i-1}\Delta z}$ ($I_1 = I_o(t)(I-R)$), and the energy absorbed during Δt is :

$$\Delta Q_{abs} = I_i(1-e^{-\alpha_i\Delta z})\,t\ . \tag{8a}$$

The energy exchanged by thermal diffusion with the slice i-1 and i+1 is given by :

$$\Delta Q_{diff} = \left| K_-\,\frac{T_{bi-1}-T_{bi}}{\Delta z} + K_+\,\frac{T_{bi+1}-T_{bi}}{\Delta z}\right| ,$$

with $K_- = (K_{i-1}+K_i)/2$, $K_+ = (K_{i+1}+K_i)/2$. \qquad (8b)

The temperature rise is then given by :

$$T_{bi}(t+\Delta t) = T_{bi}(t) + \frac{\Delta Q_{abs} + \Delta Q_{diff}}{\rho C_p\Delta z}\ . \tag{9}$$

If $T_{bi}(t)<T_m\leq T_{bi}(t+\Delta t)$ melting should be considered. If instead $T_{bi}(t)>T_m\geq T_{bi}(t+\Delta t)$, solidification has occured. In both cases the latent heat should be included. The energy available for melting if $T_{bi}(t)\leq T_m$, or released from solidification if $T_{bi}(t)>T_m$, is

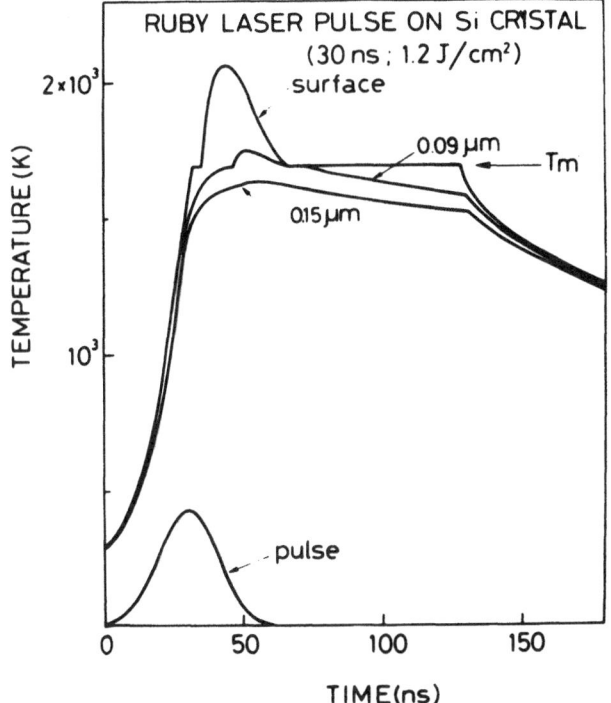

Fig. 11 Temperature-time dependence at different depths for Si single crystal irradiated with 30 ns ruby laser pulse of 1.2 J/cm^2 energy density.

obtained by:

$$\Delta Q'_i = \Delta Q_{abs} + \Delta Q_{diff} - (T_m - T_{bi})\rho C_p \Delta z. \qquad (10)$$

This energy determines the variation in the fraction of the molten layer, FF_i intact:

$$\Delta FF_i = \frac{\Delta Q'_i}{\rho \Delta z \Delta H_m}, \qquad (11)$$

with ΔH_m enthalpy of melting or solidification. The new value for FF_i at time, $t + \Delta t$, is:

$$FF_i(t + \Delta t) = FF_i(t) + \Delta FF_i. \qquad (12)$$

The new fraction can be lower, equal or larger than 1. If $0 \leq FF_i \leq 1$ the temperature $T_{bi}(t+\Delta t)$ is maintained at T_m; if $FF_i > 1$, the surplus of absorbed energy $\Delta Q''_i = (FF_i - 1)\rho \Delta z \Delta H_m$ is used to rise the temperature of the slice above T_m, i.e. $T_{bi}(t+\Delta t) = T_m + \Delta Q''_i / \rho C_p \Delta z$. Then the calculation proceeds to the next slice and so on. The numerical approach requires as stability condition the following inequality $K \Delta t / \rho C_p (\Delta z)^2 < 1/2$ between Δt and Δz, time interval and space interval respectively.

In addition one uses as boundary conditions, the temperature distribution in the target at $t=0$, no loss of heat occurs at any time from the irradiated surface, and for a bulk sample $\lim_{z \to \infty} T$ is constant at any time t.

As an illustration of the method we report the following examples which refer to ruby laser irradiation of Si single crystal. The pulse duration is 30 ns, i.e. a gaussian shape is assumed with a FWHM of 30 ns. The time dependence of temperature at three different depths inside the crystal is shown in Fig. 11 for 1.2 J/cm^2 energy density pulse. The surface layer reaches the melting temperature just after 30 ns from the switch-on of the laser pulse, then the temperature remains constant because the absorbed energy is used as latent heat for melting. The temperature increases later and reaches a maximum value of 2×10^3K at the end of the pulse. The layer at a distance of 90 nm from the surface reaches the melting point at a later time and so on. Cooling occurs with a rate of 10^9K/sec for all the considered layers, while the heating rate if of the order of 10^{11}K/sec.

Melt front penetration and its dependence on the energy density are shown in Fig. 12. Threshold for melting just the surface layer corresponds to 0.8 J / cm^2. Increasing the energy density increases also the thickness of the molten layer and the time interval during which the surface is liquid. In ruby laser irradiated Si

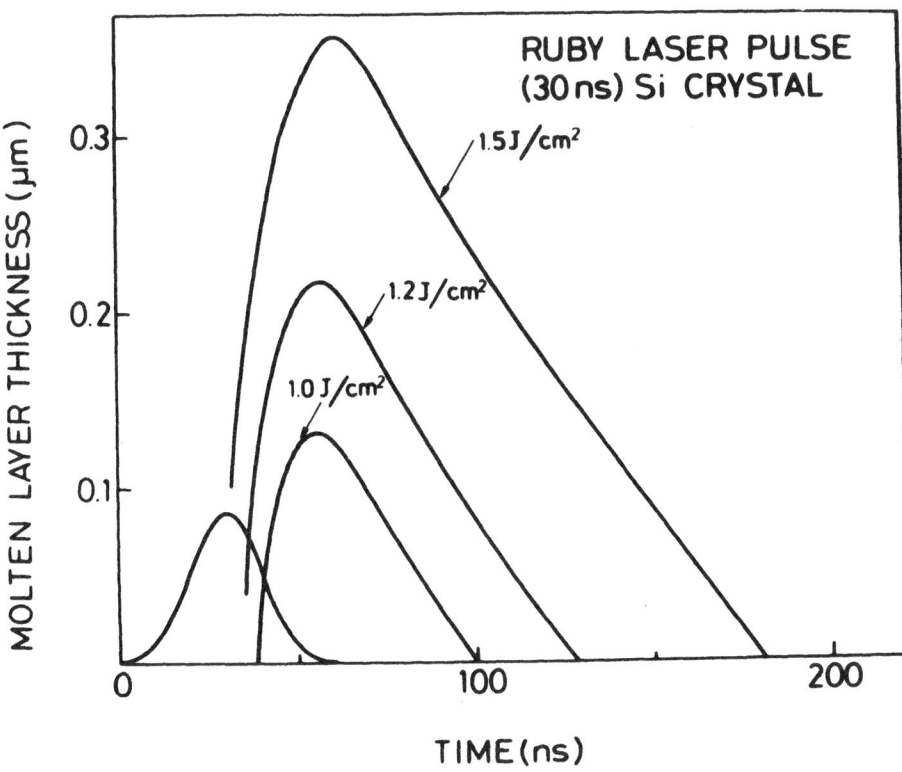

Fig. 12 Kinetics of melt front in Si single crystals irradiated with 30 ns ruby laser pulses of different energy densities.

a typical rate of 0.6 μm / J cm^{-2} is obtained. The melting pro-ceeds with a planar front at a velocity of about 10 m/s, solidifica-tion usually occurs in a time interval of about ten times the pulse duration.

During solidification the heat of melting liberated at the ad-vancing solid-liquid interface has to be transported by heat conduc-tion into the substrate, i.e. :

$$\Delta H_m \rho \cdot v = K_s \left. \frac{\partial T}{\partial z} \right|_s \qquad K_\ell \left. \frac{\partial T}{\partial z} \right|_\ell , \qquad (13)$$

where v is the velocity of the liquid-solid interface and the sub-scripts s and ℓ refer respectively to the solid and liquid phases. The equation does not include undercooling effects at the interface.

Usually the temperature gradient in the liquid is much smaller than that in the solid so that the velocity is determined mainly by the temperature gradient in the solid. For instance, if the energy density is just enough to melt the layer initially at room temperature and if $\alpha^{-1} \ll \sqrt{2D\tau_p}$, one can estimate:

$$\frac{\partial T}{\partial z} \simeq \frac{T_m - 300}{\sqrt{2D\tau_p}} \quad ; \text{ for } Si\,(D=0.1 \text{ cm}^2/s): v = \frac{1 \times 10^{-3}}{\sqrt{\tau_p}}(m/sec) . \tag{14}$$

With increasing the energy density, also the melt duration increases too and solidification starts after the end of the laser pulse. The time interval τ_p becomes then the melt duration t_m which is proportional to the square of the energy density, E^2. The velocity becomes inversely proportional to the energy density value. In the other case, $\alpha^{-1} > \sqrt{2D\tau_p}$, the regrowth is independent of the pulse duration, and it is given by $T_m \alpha/e$. The maximum velocity is reached in any case near the energy threshold value. Short pulses reduce in addition considerably the heat diffusion length.

This analysis is based on a simple description of the process, and it is necessary to support it by experiments. With this aim, measurements [18] of the electrical conductance of Si during the irradiation process were undertaken. They yield direct measurements of important dynamic parameters as the melt depth, the front velocity and the solidification. In these experiments one takes advantage of the large increase (~30 times) of conductance of Si upon melting. Samples are irradiated with ruby laser pulse and for each pulse the electrical conductivity is measured as function of time. Particular care must be paid to reduce substantially the photoconductive response, so that the entire melt conductance is not masked by it. Si samples doped with Au, to decrease the carrier lifetime, were used.

Simultaneous measurements of the surface reflectivity by a Ne-He laser were used to confirm that melting had occured, to determine the time during the laser pulse at which the melt started, and to measure the melt duration [34]. A typical set of measurements is shown in Fig. 13. The time dependent voltage V(t) is reported for a variety of incident laser energy densities in the lower part of the figure. The signal observed for energy densities below the melt threshold, $0.8 J/cm^2$, is due to the photoconductivity in Si. Above the threshold value the transient voltage curves show the conductance increasing in magnitude and duration after the photoresponse

The reflectance signal reported in the upper part of the same Fig. 13 indicates the onset of the high reflectivity phase and its duration. It is clear that at $0.57 J/cm^2$ no melting of the surface has occured, and that at $0.2 J/cm^2$ the surface remains liquid for a larger time than at $1.65 J/cm^2$ irradiation.

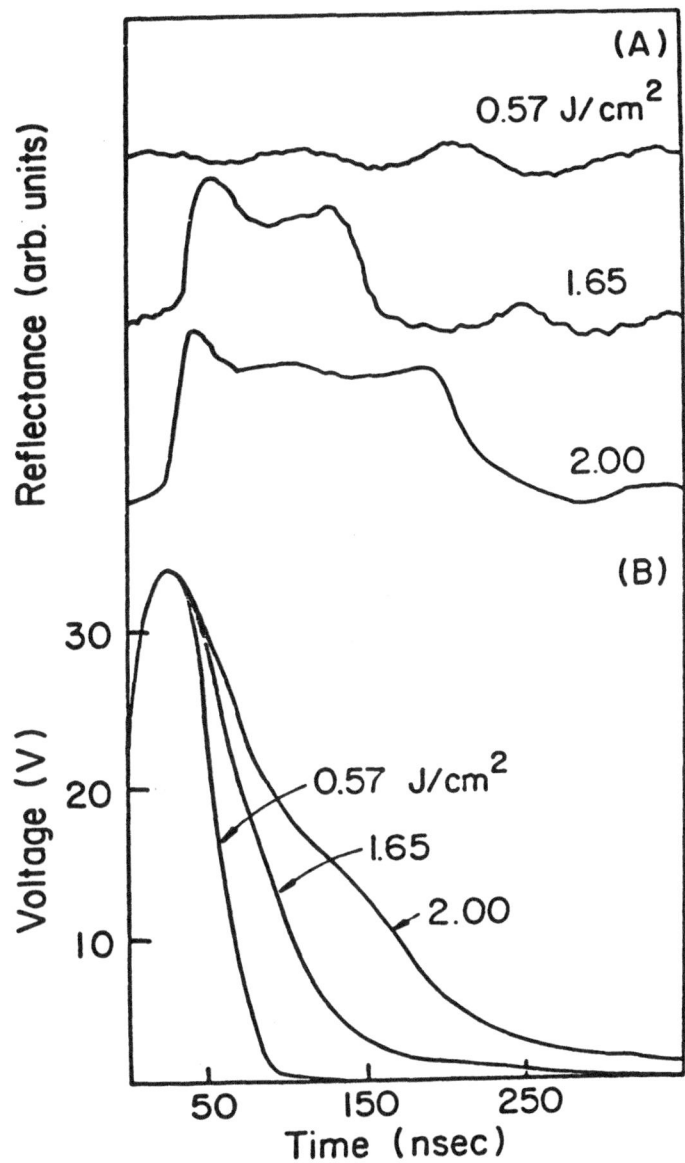

conductance, as measured by the voltage across the scope load re-
sistor, as a function of time.

94

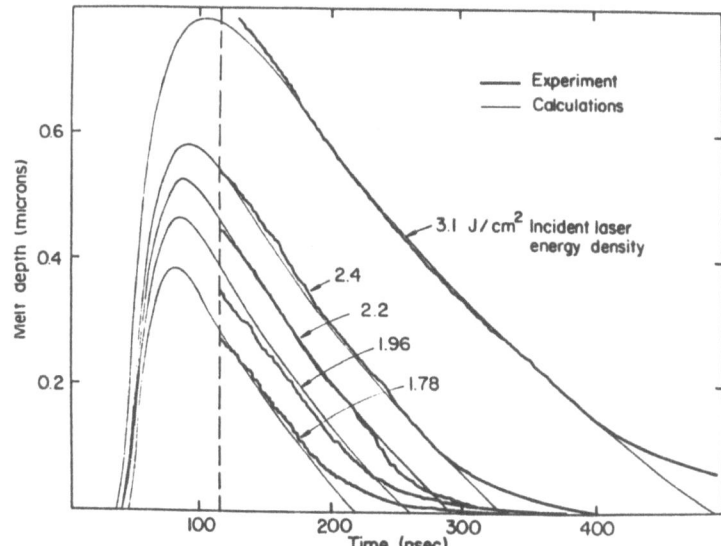

<u>Fig. 14</u> Experimental (heavy line) and computer calculated (light line) melt depths as a function of time for several incidence laser energy densities above the melt threshold.

Comparison of experimental data with calculation is reported in Fig. 14. The voltage dependence data shown in Fig. 13 are converted into melt depth through a detailed analysis in which the other contributions to the conductance in addition to that of molten Si alone are estimated. The good agreement between the experimental results and the calculations support the thermal description. The slopes of the curves in Fig. 14 provide the velocity of the liquid-solid interface during cooling. Also these values of few meters per second are in agreement with the predictions of the thermal model.

4. DOPANT INCORPORATION

The formation of a liquid layer modifies the profiles of the implanted impurities. Diffusion in molten Si occurs with a coefficient of 10^{-4} cm^2/sec and detectable migration takes place during the time interval, ~100ns, the near surface region remains liquid. Impurities after irradiation are then electrically active and redistribute in depth [35]. The sheet resistance of Si samples implanted with 40 keV - 10^{16} As/cm^2 is reported in Fig. 15 as a function of the energy density for Nd/YAG double frequency pulse. At an energy density of about 0.4 J/cm^2 the abrupt decrease of the sheet resistance (o) is related to the melting of the surface region. Dissolution of As complexes occurs also at the same energy value as shown for the samples irradiated after thermal annealing (o). The As distribution is shown in Fig. 6 before and after irradiation with 0.6 J/cm^2 energy density pulse.

Profile broadening is not the only observed effect. Several impurities show in addition a partial surface accumulation as reported in Fig. 17 for Sb [36] and Cu [37]. For Sb the profile depends on implantation and irradiation conditions whilst for Cu complete surface accumulation is always observed. The redistribution of the impurities during solidification is governed by the equilibrium distribution coefficient, K_0, defined as $K_0 = \frac{c_s}{c_\ell}\big|_{eq}$ where c_s and c_ℓ are the concentrations in the solid and liquid phase determined by the phase diagram at a fixed temperature close to the melting point. If $K_0<1$ impurities are rejected at the surface toward the liquid where they are allowed to diffuse.

Normal crystal growth occurs at a velocity of the solid-liquid interface of about 10^{-5}m/s, i.e. five orders of magnitude lower than that obtained by laser irradiation. In these conditions K_0 is meaningless and the redistribution is determined by the interfacial redistribution coefficient $K'= \frac{c_s}{c_\ell}\big|_i$. Experiments indicate that K' for a large number of impurities should be much larger than K_0, and in addition the dopant concentration retained in substitutional lattice sites can be several order of magnitude larger than the equilibrium solubility [36].

The dynamic of the dopant redistribution is reported in Fig.18 where the profiles both in the liquid and solid phase are shown at different stages of the solidification process. The continuous rejection of impurity at the interface will result in surface accumulation, as the last layer will freeze. These profiles are obtained following a procedure similar to that reported with some details in the previous section on the heat flux calculation.

The process described by the calculations [7] is the following: i) diffusion in the liquid phase of the initial implanted distribution, accounting for the 1-s interface kinetics, as determined by

96

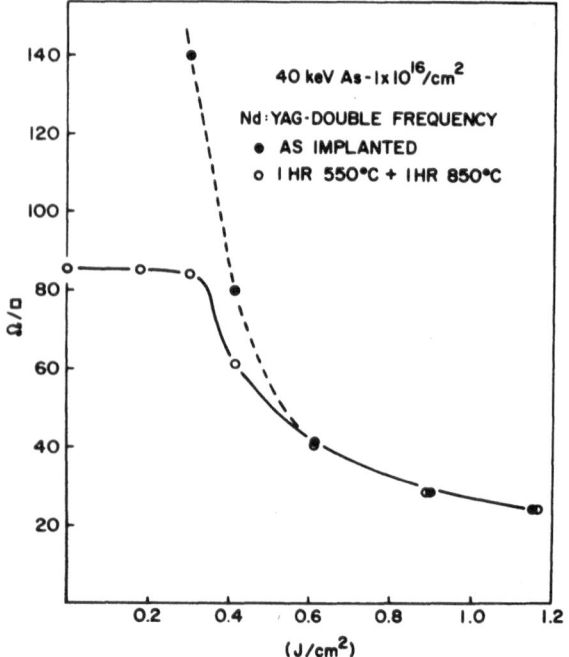

<u>Fig. 15</u> Sheet resistance of 40 keV-10^{16} As at/cm^2 implanted Si
samples vs. energy density of 20 ns pulse duration Nd:YAG double
frequency. Full circles refer to as-implanted samples, open- to
thermally annealed samples.

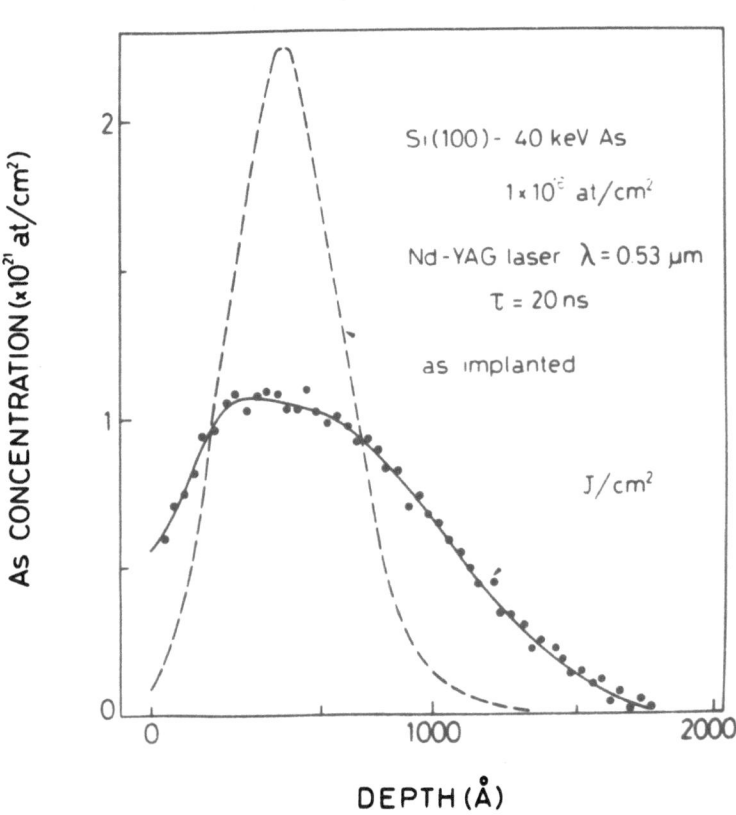

Fig. 16 Broadening of the As-impurity profile after irradiation with a 20ns pulse duration of Nd:YAG double frequency –0.6 J/cm².

98

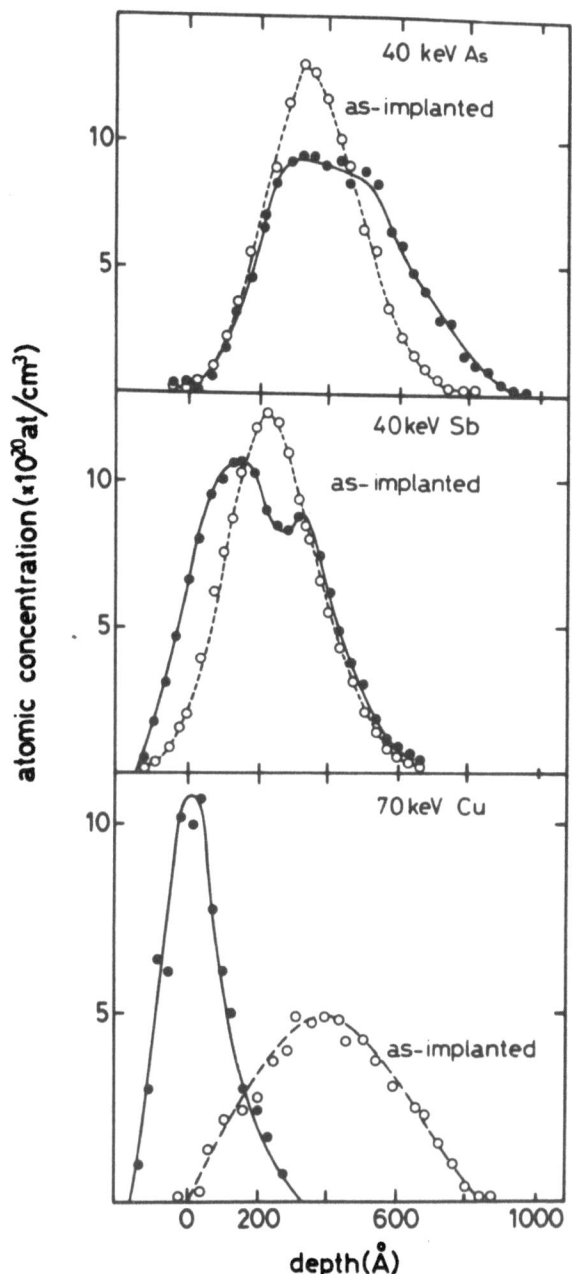

Fig. 17 Changes in the impurity profiles after laser irradiation.

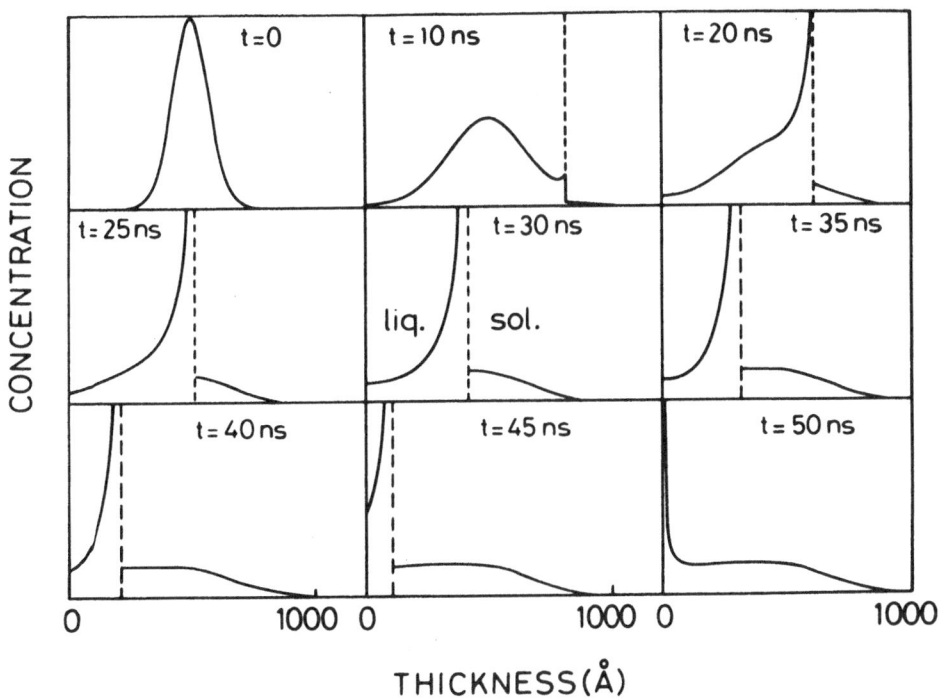

THICKNESS(Å)

Fig. 18 Calculated impurity profiles at different times for an initial gaussian concentration distribution. The solidification proceeds from right to the left with a solid-liquid interface velo-city of 2m/s and for an interfacial distribution coefficient K'=0.1.

heat flow calculations
ii) impurity rejection from the solid to the liquid phase, according to a K'fitting parameter;
iii) diffusion of the impurities rejected by the interface in the liquid.

This simple description combine both mass and heat transport, but no interference between these two processes is considered. From the fit of the experimental data with calculations it has been in-ferred that K' is velocity dependent [38] . The amount of surface segregation changes with the liquid-solid velocity during solidifi-cation. The velocity can be changed in several ways : by changing the pulse duration, or the coupling of the laser beam with the sub-strate, or the target temperature [39] .

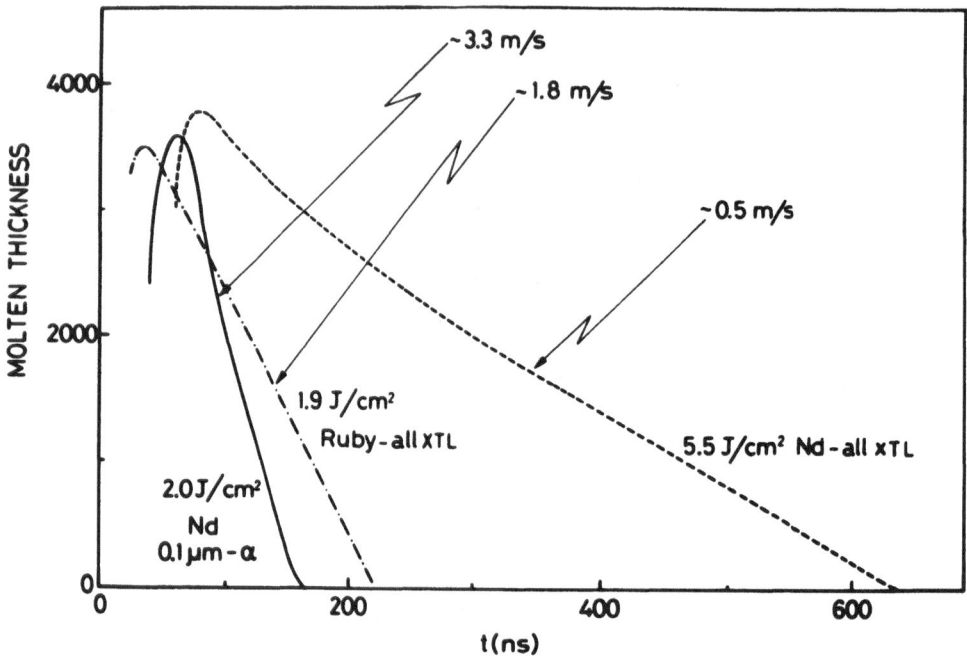

Fig. 19 Kinetics of the melt thickness for Nd and ruby laser pulse irradiation of Si single crystals and of Si samples with 0.1 µm thick amorphous layer.

Figure 19 reports as an illustration [40] the molten thickness vs time for a ruby (0.69 µm) and Nd (1.06 µm) laser irradiation of Si single crystals and 10^2 nm thick amorphous layer on single crystals. The solid-liquid interface velocity given by the slope of molten thickness-time and averaged over the last hundred nm changes from 0.5 to 3.3 m/s. In the presence of impurities the solidification leads to segregation. As an example the depth profiles of Te implanted in Si and irradiated with a Nd laser pulses are shown in Fig.20. The lower part reports the profile after irradiation with 2.0 J/cm^2 for aligned and random incidence of the analyzing 2.0 MeV He$^+$ beam.

The Te accumulation at the surface amounts to 20% and is easily obtained by the aligned yield. The large attenuation for the in-depth distribution indicates that the majority of the remaining Te atoms are substitutionally located. The calculated solidification velocity is ~3.0 m/s. Channeling effect in combination with MeV He$^+$ Rutherford backscattering provides a simple and a reliable method to determine the total amount of rejected impurities at the surface. The profile determination is limited by the depth resolution of the technique which is hardly pushed below the 5nm.

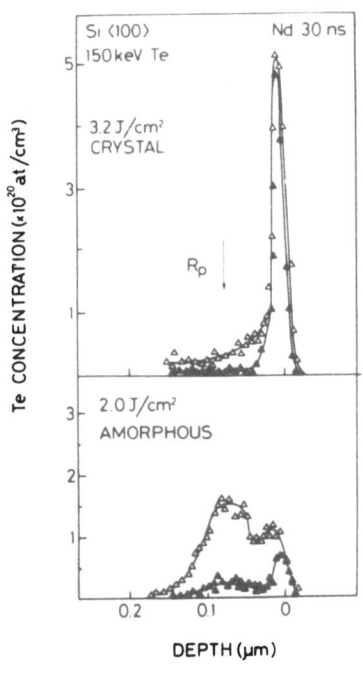

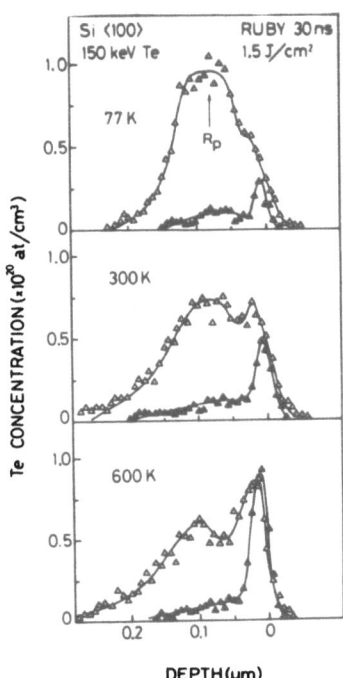

Fig. 20   Fig. 21

Fig. 20   Te depth profiles for Nd laser irradiation of Si single crystals (upper part) and of Si with 0.1 μm thick amorphous layer (lower part). The arrow indicates the position of the projected range for the as-implanted Te distribution. The data refer to random (Δ) and to aligned (Δ) incidence of the 2.0 MeV He[+] analyzing beam.

Fig. 21   Te depth profiles after irradiation with 30 ns ruby laser pulse of 1.5 J/cm^2 at different substrate temperatures.

If the implanted sample is annealed before irradiation the Te profile shows a much larger surface accumulation as shown in the upper part of Fig. 20. The small absorption coefficient of the Nd wavelength in Si single crystal produces small temperature gradient and the calculated solidification velocity is about 0.8 m/s. To reduce the required energy density for melting the sample was heated at about 300°C during irradiation. The K' fitting values are 0.5 and 0.03 for the higher and lower velocity respectively.

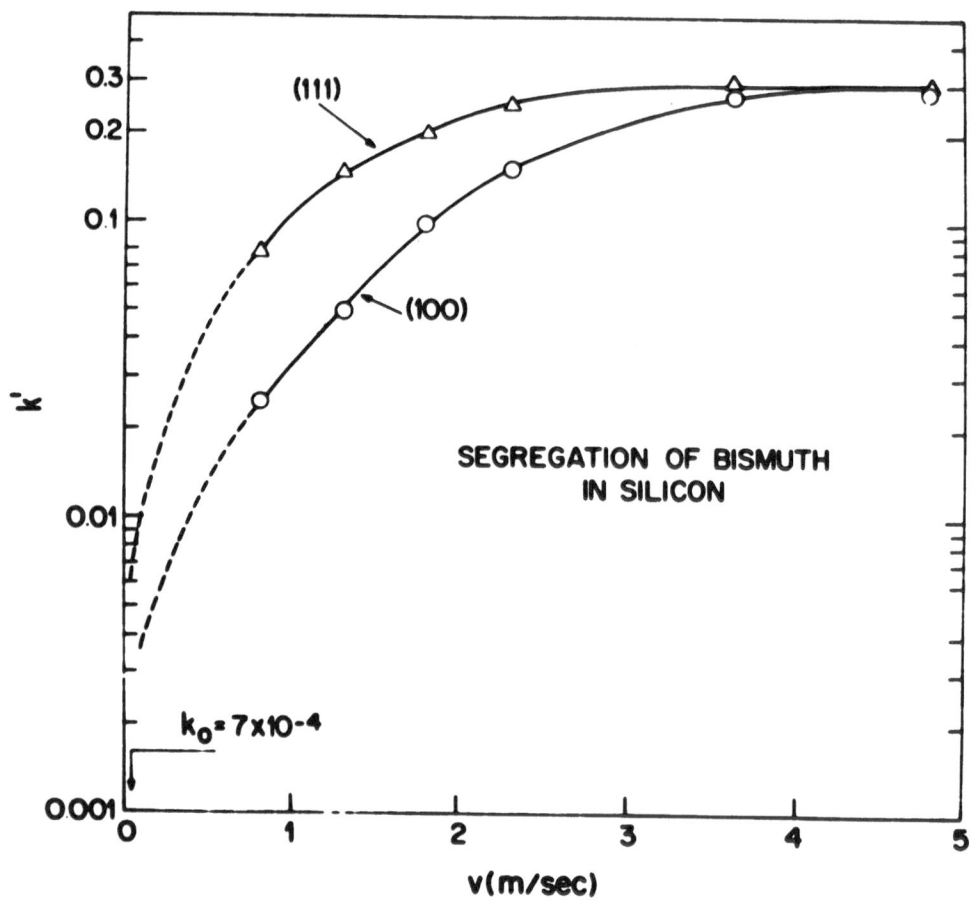

<u>Fig. 22</u> Bi segregation coefficient (K') in (100) and (111) Si
oriented substrates as a function of the liquid-solid interface
velocity.

The estimated K_0 equilibrium value is ~10^{-4}.

Similar results are obtained by changing the substrate temperature during irradiation [40] . As an illustration Fig. 21 reports
Te distribution profiles after irradiation with ruby laser pulse of
30ns duration and 1.5 J/cm^2 energy density at different substrate
temperatures. The higher surface accumulation is obtained at 600 K
substrate temperature. The solid-liquid interface velocity changes
are associated to the dependence of the thermal conductivity on
temperature.

The major result of these experiments is the unique relation between the K' value needed to fit the surface accumulation and the solid-liquid interface velocity independent of the way it was obtained. A typical K'-v trend is shown in Fig. 22 for Bi implanted Si samples [38] . The K' value seems to saturate at a value <1 . Although measurements at higher velocities would be of interest they are limited by the amorphous formation occuring at solidification velocities of ~20 m/s. Under these conditions the usual definition of K' loses its significance.

The data of Fig. 22 refer to different substrate orientations. At low velocities the (111) oriented Si substrates give rise to less segregation than the (100) crystal substrates [38] . At large velocities the two values almost coincide. The understanding of the orientation dependence requires a detailed description of the crystal-liquid interface morphology. So far two different possible mechanisms have been considered. The amount of undercooling required to grow a given liquid at a fixed speed is larger for (111) than for (100) substrate orientation. The increased undercooling give rise to an increased amount of trapped impurities [41] . The other mechanism [42] is based on the atomistic description of the different faces. The density of ledges in the (111) plane is smaller than that in the (100) plane thus requiring a greater lateral velocity to maintain the same solidification rate. This faster lateral growth will produce larger dopant trapping in the (111) oriented substrates.

The dependence of K' on velocity is related to the kinetic aspects of the segregation process. In a simple view two different times should be considered; the time required for the interface to move one interatomic distance during solidification, $\tau_g = \frac{\lambda}{v}$, with λ interatomic distance, and the residence time, $\tau_r = \frac{\lambda^2}{D_i}$, with D_i diffusion of coefficient of the impurity at the interface. Trapping of the impurity can occur if the dopant resides in the near interface region longer than the time required to regrow it, i.e., if $\tau_g < \tau_r$. In the opposite case the dopant is rejected into the high solubility phase, i.e., into the liquid. The critical velocity above which trapping K' increases steeply is then $v_{crit} = D_i/\lambda$.

The diffusion coefficient D_i is estimated [43] by the relationship $D_i = \sqrt{D_s D_\ell}$, where the subscripts s and ℓ refer to solid and liquid phases respectively. The interface is assumed to have properties intermediate between those of the two adjacent media. D_ℓ is of the order of $10^{-4} \mathrm{cm}^2/\mathrm{s}$ for any dopant in Si, while D_s ranges from $10^{-11} \mathrm{cm}^2$ for the slow (Bi,As,Sb...) to $10^{-4} \mathrm{cm}^2/\mathrm{s}$ for the fast (Cu,Ag,Au...) diffusers in Si at temperature close to the melting point.

These D_i estimates give v_{crit} values of the order of few cm/s and of few m/s for slow and fast diffusers in Si, respectively.

The experimental data so far obtained by laser irradiation agree with these estimates. In the case of fast diffusers the velocity obtained by irradiation is not sufficient to trap them into the lattice [44] .

The concentration of dopant trapped in substitutional positions exceeds the solid solubility value of several orders of magnitude. However the maximum solubility obtained by laser irradiation seems to be limited by three mechanisms [45] : interface stability, mechanical stress and thermodynamics limit.

The solid-liquid interface can proceed flat according to the classical work of Mullins and Sekerka [46] if:

$$G_\ell > \frac{vm\ c_s(i)}{D_\ell}\ (\frac{1-K'}{K'}).$$

The term on the left is the temperature gradient in the liquid close to the interface and that on the right is the gradient obtained from the phase diagram and taking into account the dopant concentration in the liquid close to the interface. In the case the disequality is not fulfilled, interfacial instability develops and cell formations occurs. The instability is caused by constitutional supercooling in the liquid at the interface. By increasing the thermal gradient in the liquid the onset of instability is delayed. This implies that the regrowth velocity must be increased.

An example of cell formation in laser irradiated In implanted Si samples is shown by the TEM micrographs reported in Fig. 23 [47] . The upper part and the lower part refer to a sample in which the regrowth velocity was 3 m/s and 1.5 m/s,respectively. The interface breakdown and then the cell size occur on the scale of D/v. At the high regrowth velocities encountered in laser irradiated semiconductors the cell size ranges between few hundreds and few thousands angstroms.

Constitutional supercooling is the most common limit to supersaturation but other mechanisms can be present. In the case of boron the large difference in covalent radius between the dopant and the Si host results in large uniaxial strain [45] . The limit to the maximum substitutional concentration is due to the limit of elastic properties of Si and extended defects results at high concentrations.

An absolute thermodynamic limit [48] to solute trapping is given at the intersection between the solidus and the liquidus free energy curves. This is a very general limiting condition and it is not yet clear if it can be reached.

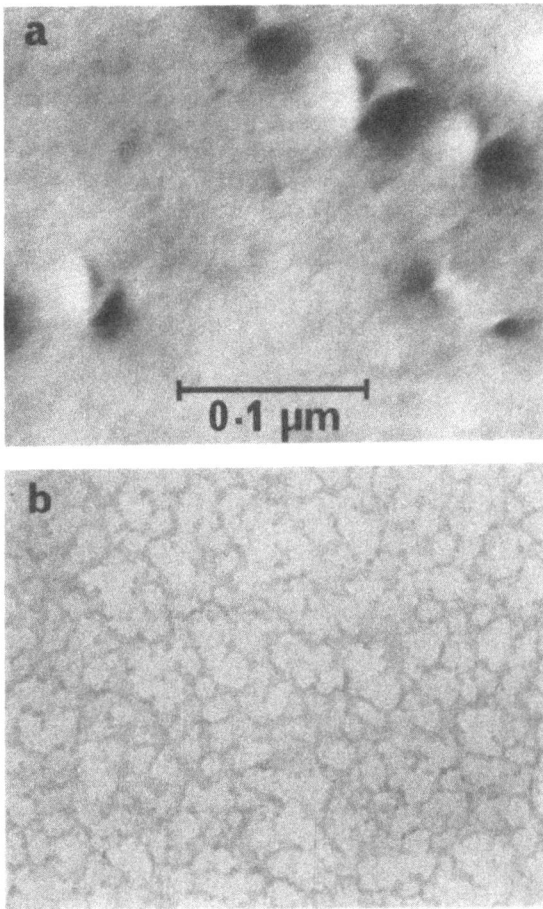

Fig. 23 Transmission electron micrographs showing the cell formation and its dependence on the solid-liquid velocity for In implanted Si samples, ruby laser irradiated.

5. CONCLUSIONS

The aim of these lectures was to present the laser irradiation effects in ion implanted semiconductors and to describe them in terms of a melting model. Structure changes, residual defects and dopant incorporation are all explained by this simple thermal description. In addition several predictions of the model have been and can be used as a guide for further investigations.

REFERENCES

1. Poate, J.M. and Foti, G. editors "Surface Modification and Alloying" Nato Advanced Institute Series (New York, Plenum Press, 1983).
2. Baeri, P., Campisano, S.U., Foti, G. and Rimini, E., J. Appl. Phys. 50 (1978) 788.
3. See for instance the proceedings of the Material Research Society Conferences (Boston Meeting) on this subject.
   (i) Laser-Solid Interactions and Laser Processing (Edited by S.D. Ferris, H.J. Leamy and J.M. Poate). A.I.P. Conf. Proc. 50 New York, 1979;
   (ii) Laser and Electron Beam Processing of Materials (Edited by C.W. White and P.S. Peercy). (New York, Acadamic Press, 1980);
   (iii) Laser and Electron-Beam Solid Interaction and Materials Processing (Edited by J. Gibbons, L.D. Hess and T.W. Sigmon). (New York, North Holland, 1981).
   (iv) Laser and Electron Beam Interaction with Solids (Edited B.R. Appleton and G.K. Celler), (New York, North Holland, 1982).
4. Khaibullin, I.B. , Shtyrkov, E.J., Zaripov, M.M., Galyautdinov, M.F. and Zakirov, G.G., Sov. Phys. Semicond. 11 (1977) 190; Antonenko, A.Kh., Gerasimenko, N.N., Dvurechenskii, A.V., Smirnov, L.S. and Tseiitlin, Sov. Phys. Semicond. 10 (1976) 81.
5. Rimini, E., Physica Scripta-October (1982); Poate, J.M. and Brown, W.L., Physics Today, 35 (1982) 6-24.
6. Bloembergen, N. in "Laser-Solid Interactions and Laser Processing" (S.D. Ferris, H.J. Leamy and J.M. Poate eds) p.1 A.I.P. n.50, New York, 1979.
7. Baeri, P. and Campisano, S.U. in "Laser and Electron Beam Processing of Semiconductor Structures" J.W. Mayer and J.M. Poate editors (New York, Academic Press, 1982), chap. 4
8. White, C.W., Zehener, D.M., Campisano, S.U. and Cullis, A.G. in "Surface Modification and Alloying" J.M. Poate and G. Foti Editors (New York, Plenum Press, 1983); Chap. 4.
9. Campisano, S.U. "Non equilibrium dopants incorporation in silicon melted by laser pulses" Appl. Phys. (in press).

10. Van Vechten, J.A. "Plasma annealing and laser sputtering : role of the Frenkel excitons" - this volume.

11. Auston, D.H., Surko, C.M., Venkatesan, T.N.C., Slusher, R.E. and Golovchenko, J.A. Appl. Phys. Lett. 33 (1978) 437.

12. Auston, D.H., Surko, C.M., Venkatesan, T.N.C., Slusher, R.E. and Golovchenko, J.A., Appl. Phys. Lett. 35 (1979) 635.

13. Lowndes, D.H., Phys. Rev. Lett. 48 (1982) 267.

14. Lo, H.W. and Compaan, A., Phys. Rev. Lett. 44 (1980) 1604.

15. Stritzker, S., Pospieszcyk, P. and Tagle, J.A., Phys. Rev. Lett. 47 (1981) 356.

16. Larson, C., White, C.W., Noggle, T.S. and Mills, D., Phys. Rev. Lett. 48 (1982) 337.

17. Yen, R., Liu, J.M., Kurz, H. and Bloembergen, N., Appl. Phys. A 27 (1982) 153.

18. Galvin, G.J., Thompson, M.O., Mayer, J.W., Hammond, R.M., Paulter, N. and Peercy, P.S., Phys. Rev. Lett. 48 (1982) 33.

19. Foti, G. and Rimini, E. in "Laser and Electron Beam Processing of Semiconductor Structures" J.W. Mayer and J.M. Poate editors (New York, Academic Press, 1982), Chap. 2.

20. Baeri, P. "Rutherford Backscattering : from principles to application in laser annealing", unpublished.

21. Vitali, G., Bertolotti, M., Foti, G. and Rimini, E., Phys. Lett. A63 (1977) 351.

22. Cullis, A.G., Webber, H.C., Mc Caugham, D.V. and Chew, N.G., in "Laser and Electron Beam Processing of Materials" (C.W. White and P.S. Peercy eds), p. 183 (New York, Academic Press, 1980).

23. Foti, G., Rimini, E., Tseng, W.F. and Mayer, J.W., Appl. Phys. 15 (1978) 365.

24. Campisano, S.U., Foti, G. and Servidori, M., Appl. Phys. Lett. 36 (1980) 279.

25. Narajan, J., Appl. Phys. Lett. 34 (1979) 312.

26. Spaepen, F. and Turnbull, D. in 3(i), p. 73.

27. Bagley, B.G. and Chen, H.S. in 3(i), p. 97.

28. Poate, J.M. "Some Thermodynamics Properties of Amorphous Si" IBMM- Grenoble Sept. 5-10 1982 (Nucl. Instr. Meth., to be published).

29. Baeri, P., Foti, G., Poate, J.M. and Cullis, A.G., Phys. Rev. Lett. 45 (1980) 2036.

30. Baeri, P., Campisano, S.U., Grimaldi, M.G. and Rimini, E., J. Appl. Phys. (Nov. 1982).

31. Cullis, A.G., Webber, H.C. and Chew, N.G. in 3 (iv) p. 131.

32. Cullis, A.G., Webber, H.C., Chew, N.G., Poate, J.M. and Baeri, P. Phys. Rev. Lett. 49 (1982) 219.

33. Ready, J.F. "Effects of higher power laser irradiation" (New York, Academic Press, 1971), Chap. 4.

34. Galvin, G.J., Thompson, M.O., Mayer, J.W., Peercy, P.S., Hammond, R.B. and Paulter, N. "Time-Resolved Conductance and Reflectance Measurements of Silicon Solidification from the Melt During Pulsed Laser Annealing" Phys. Rev. to be published.

35. Rimini, E., Chu, W.K. and Mader, S.R., J. Appl. Phys. 52 (1981) 3696.

36. White, C.W., Wilson, S.R., Appleton, B.R. and Young, F.W. Jr.,
     J. Appl. Phys. 51 (1980) 738.
37. Baeri, P., Campisano, S.U., Foti, G. and Rimini, E., Phys. Rev.
     Lett. 41 (1978) 1246.
38. Baeri, P., Foti, G., Poate, J.M., Campisano, S.U. and Cullis,
     A.G., Appl. Phys. Lett. 38 (1981) 800.
39. Cullis, A.G., Webber, H.C., Poate, J.M. and Simons, A.L., Appl.
     Phys. Lett. 36 (1980) 320.
40. Foti, G., to be published.
41. Jackson, K.A. and Gilmer, G.H., to be published.
42. Turnbull, D. and Spaepen, F. Chapter 2 in "Laser Annealing of
     Semiconductors" eds J.M. Poate and J.M. Mayer, (New York,
     Academic Press, 1982).
43. Jindall, B.K. and Tiller, W.A., J. Chem. Phys. 49 (1968) 4632.
44. Aziz, M.J., J. Appl. Phys. 53 (1982) 1158.
45. White, C.W., in ref. 3(iv) p. 109.
46. Mullins, W.W. and Sekerka, R.F., J. Appl. Phys. 35 (1964) 444.
47. Cullis, A.G., Series, R., Webber, H.C. and Chew, N.G. in "Semi-
     conductor Silicon 1981" Electrochemical Society, New York, 1981,
     p. 518.
48. Baker, J.C. and Cahn, J.W., Acta Met. 17 (1969) 575.

TRANSPORT THEORY

G. Dewel

Chimie-Physique II, Université Libre de Bruxelles
Campus Plaine C.P. 231   1050   Bruxelles, Belgium.

1. INTRODUCTION

The evaluation of transport coefficients usually involves two steps. First, one derives irreversible kinetic equations for the carriers distribution functions. Then, in the limit of long times and large distances, these kinetic equations lead to transport equations for the (quasi-) conserved quantities. This last step provides explicit expressions for the transport coefficients in terms of the microscopic properties of the particles and the interactions between them. Indeed, even in the simplest cases (Boltzmann equation, Landau equation) the kinetic equations are extremely difficult to solve. However when all quantities vary slowly in space and time a great simplification occurs because under these conditions the evolution of the system is completely described by local values of the thermodynamic variables ("Hydrodynamical contraction").

The first part of this programme contains the difficult question of the microscopic definition of irreversibility : how is it possible to get irreversible kinetic equations from reversible equations of motion ? Important progress have been made recently in this field [1] but we shall consider here a more phenomenological approach. The main problem is then to generalize the Boltzmann equation to more realistic situations. In the case of semiconductors the need to generalize the Boltzmann equation has appeared only rather recently with the trend towards smaller and faster devices and the interest on the dynamics of hot electrons [2] .
On the contrary in the case of dense gases, this work began just after the second world war and culminated in the sixties with the derivation of various generalized master equations [3-4-5] .

Despite this effort we must recognize that even now we do not have a completely satisfactory theory of transport phenomena in dense systems. Although it is rather easy to obtain formal general kinetic equations, many difficulties appeared when explicit calculations were performed [6].

On the other hand, the linear response theory provides general expressions for the transport coefficients. For instance in the case of the electrical conductivity, the Green-Kubo formula may be written :

$$\sigma^{\alpha\beta} = \frac{1}{\Omega} \int_o^\infty dt \int_o^\beta d\lambda < J^\alpha(-i\lambda) \; J^\beta(t) >,$$

where :

$$< A > = \frac{1}{Z} \, \text{Tr} \, \{ \, e^{-\beta H} \, A \, \} \tag{1}$$

and $J^\alpha(t) = e^{iHt} \, J^\alpha \, e^{-iHt}$

(Ω = volume; Z = the partition function).

At first sight tne autocorrelation method seems very different from the standard kinetic equation-approach. However, these expressions (1) are purely formal. In order to get explicit expressions, one still needs a method to analyze the many body dynamics contained in the evolution operators exp (iHt). Such a method is provided by the kinetic theory. In fact it can be shown that the evaluation of the Green-Kubo formulae can be reduced to the solution of a typical kinetic equation [7]. Nevertheless these formulae provide a good starting point in the cases where the traditional approach fails (e.g. spin diffusion in localized spin systems). In this note we give a simple introduction to kinetic theory. It has indeed been stressed that the simple relaxation time approximation to the classical Boltzmann equation is not sufficient to account for the dynamical properties of photoexcited carriers at high densities [8]. The paper is organized as follows. Section 2 is devoted to a short introduction to the generalized kinetic equations. In Section 3 the derivation of the transport equations is illustrated on the simple case of a weakly coupled neutral Fermi system. In section 4 we discuss briefly the problem of the dynamics of a system of charged particles.

2. THE GENERALIZED KINETIC EQUATIONS (LINEAR CASE)

A basic quantity in the description of quantum systems is the one-particle Wigner distribution function $f(\underline{r},\underline{k},t)$ which is the natural generalization of the classical distribution function to the quantum case :

$$f(\underline{r}, \underline{k}, t) = \frac{1}{(2\pi)^3} \int d\underline{r}' \, e^{i \underline{k}.\underline{r}'} \, \mathrm{Tr}\{ \rho(t) \, \psi^+ (\underline{r} - \frac{\underline{r}'}{2}) \, \psi(\underline{r} + \frac{\underline{r}'}{2}) \}$$

$$(2).$$

$\psi^+(r)$ and $\psi(r)$ are the creation and destruction operator at the point $\underline{r}$, they satisfy the usual equal time commutation relations for fermions or bosons; $\rho(t)$ is the density matrix.

For linear deviations from thermal equilibrium, the Fourier transform with respect to $\underline{r}$ of the Wigner function , $f_q(k ; t)$, obeys the following generalized kinetic equation [4]:

$$\partial_t f_q(k; t) + i [\epsilon_{k + q/2} - \epsilon_{k - q/2}] \, f_q(k; t)$$

$$+ \sum_{k'} M_{kk'} \, f_q(k'; t) = \sum_{k'} \int_0^t d\tau \, G_q(t - |k,k') \, f_q(k'; \tau) + D_q(k; \tau)$$

$$(3).$$

The second term on the left hand side (l.h.s.) of (3) represents the free motion of the particle; in the long wavelength limit it takes the more familiar form $i \, \underline{q} . \, \underline{v}_k \, f_q(k,t)$ ($\underline{v}_k$ being the velocity of the particle).
The third term is a non-dissipative contribution which describes the mean field on one particle created by all the others. The kernel $G_q(t|k,k')$ is a non-Markovian nonlocal collision operator; it accounts for the finite duration and the finite extension of the collision process. $D_q(k|\tau)$ describes the decay of the correlations that existed between the particles in the initial state. Similarly the stationary state which is established in presence of an external field $E_{q,\omega}$ can be determined from the following equation for the Laplace transform $f_{q,\omega}(k)$ of the Wigner function[9] :

$$i [\epsilon_{k + q/2} - \epsilon_{k - q/2}] \, f_{q,\omega}(k) + \sum_{k'} M_{kk'} \, f_{q,\omega}(k') -$$

$$E_{q,\omega} G_{q,\omega} = \sum_{k'} \psi_{q,\omega}(k,k') \, f_{q,\omega}(k')$$

$$(4).$$

$\psi_{q,\omega}(k,k')$ is the Laplace transform of the kernel appearing in (3). The last term on the (l.h.s.) of (4) describes the acceleration of the particles by the external field; it also involves interference effects between the collision processes and the field which are usually neglected in the traditional approach.
For instance the so-called relaxation effect, which is a consequence of the deformation of the screening cloud by the field, reduces the mobility of charged carriers with respect to its value obtained from the Boltzmann equation [9] .

It is clear that the complexity of the many body problem has been hidden in the definitions of the various operators appearing in Eqs. 3 and 4. Well defined rules have been derived to obtain explicit expressions for these quantities [4]. Since these rules are rather technical we shall not give them here but merely discuss the physical meaning of the approximation schemes.

Any simplifications of the kinetic equations 3 or 4 must be based on an analysis of the characteristic time and length scales of the problem. In simple systems one usually defines :
- the collision time τ_c which is the duration of the collision.
In classical systems, it is related to the range of the potential r_o : $\tau_c \approx r_o/\langle v \rangle$ ($\langle v \rangle$ being the mean volicity). In quantum systems, it can be estimated from the time energy uncertainty relation. In general, it can be obtained from the non-universal time dependence of the transition probabilities [10].
- the relaxation time τ_R, or average time between two collisions; the mean free path l_o is the mean distance travelled by a particle between two collisions. This time can become very long; for instance, in the case of a degenerate Fermi liquid at low temperatures we have :

$$\tau_R \approx \frac{l}{kT^2}$$

- the characteristic frequency ω of an external perturbation or the hydrodynamic time τ_h which is the time necessary for a particle to travel through a hydrodynamic length L_h which is a measure of the spatial extension of the inhomogeneities in the system.

The dynamical behavior of simple many body systems may then be decomposed into various regimes :

a) $\tau_c < \tau_R < \tau_h$: Hydrodynamic or collision dominated regime.
The traditional transport theory is applicable in this case. After a short time of the order of τ_R the system approaches a state of *local equilibrium*; the parameters characterizing the distribution function (temperature, chemical potential ...) vary slowly from one region of the system to another. Then on the time scale τ_h, these local variations of the conserved variables are smoothed out to constant values over all the system.

b) $\tau_c < \tau_h < \tau_R$: Collisionless regime.
In this case, one is no more interested in transport coefficients which are not well defined because the relaxation time is so long that one does not reach local equilibrium during the characteristic period of the disturbance. This regime is meaningful in systems where the relaxation time is very long (Fermi liquids). The self consistent field may induce collisionless collective modes involving the cooperative motion of all the system. Typical examples are the plasmons or the zero sound in neutral Fermi liquids (24).

Contrary to the case of the hydrodynamic regime, the collisions which act to disrupt the mean field and hence damp these modes, may be treated by perturbation theory since few collisions only take place on the time scale τ_h.

c) $\tau_h < \tau_c < \tau_R$: In this extreme case the non-Markovian character of the collision operator must be taken into account in the evaluation of the damping of the collision modes.

Of course the main difficulties appear in the intermediate region where $\tau_R \approx \tau_h$. However the concepts introduced in the hydrodynamic regime can be extended with some success to larger wavelengths [11] .

3. FROM KINETIC THEORY TO HYDRODYNAMICS

a) The collision operator.

In this chapter we will be interested in the dynamical behavior for long times and large distances. In this limit we expect the familiar hydrodynamical contraction of the description to be valid. How this comes about will be discussed below using a prospection operator method. For the sake of simplicity we consider here the weakly coupled Fermi systems. Linearizing around absolute equilibrium, the corresponding kinetic equation may be written as a weakly coupled Uehling-Uhlenbeck equation [12] :

$$\partial_t \phi_q(k;t) + i \, \underline{q} \cdot \underline{v}_k \, \phi_q(k;t) =$$

$$n \sum_\ell \sum_{k_1} | V(1) - V(k-1-k_1) |^2 \, \delta(\varepsilon_k + \varepsilon_{k_1} - \varepsilon_{k-1} - \varepsilon_{k_1+1})$$

$$x \quad f^\circ_k \, f^\circ_{k_1} \, (1-f^\circ_{k-1}) \, (1-f^\circ_{k_1+1}) \, [f^\circ_k(1-f^\circ_k)]^{-1}$$

$$x \quad [\phi_q(k_1+1;t) + \phi_q(k-1;t) - \phi_q(k_1;t) - \phi_q(k;t)] \qquad (5),$$

where $\sum_k = \dfrac{\Omega}{8\Pi^3} \int d\underline{k}$; n = density.

V(q) is the Fourier transform of the interacting potential. The deviation $\phi_q(k;t)$ is defined by the relation :

$$f_q(k;t) = f^\circ_k + f^\circ_k(1-f^\circ_k) \, \phi_q(k;t) , \qquad (6)$$

where f°_k is the equilibrium Fermi-Dirac distribution :

$$f^\circ_k = \frac{1}{e^{\beta[\varepsilon_k - \mu]} + 1}$$

(μ = chemical potential).

In Eq. 5 we have neglected the mean field term because in this case it only plays an important role in the collisionless regime at very low temperatures.

In the following it will be convenient to use a bracket notation for the integration on $\underline{k}$. We define the scalar product of two functions of k by :

$$<\phi \mid \chi> = \sum_k \frac{1}{s^2(k)} \phi(k) \chi(k) , \qquad (7)$$

where the weighting factor $s^{-2}(k)$ is :

$$\frac{1}{s^2(k)} = f^\circ_k (1-f^\circ_k) .$$

We introduce also the Laplace transform :

$$\phi_q(k;w) = \int_0^\infty dt\, e_{i\omega t}\, \phi_q(k;t) . \qquad (8)$$

Eq. 5 takes now the form :

$$[-i\omega + i\underline{q}.\underline{v}_k - C]\, \phi_q(k;\omega) = \phi_q(k;t=o). \qquad (9)$$

The collision operator then becomes :

$$C\phi_q(k;\omega) = n \sum_1 \sum_{k_1} \mid V(1) - V(k-1-k_1)\mid^2$$

$$x \quad \delta(\varepsilon_k + \varepsilon_{k_1} - \varepsilon_{k-1} - \varepsilon_{k_1+1}) \frac{S(k)}{S(k_1)\, S(k-1)\, S(k_1+1)}$$

$$x[\phi_q(k_1+1;\omega) + \phi_q(k-1;\omega) - \phi_q(k;\omega) - \phi_q(k_1;\omega)] \qquad (10).$$

With the scalar product defined above we have:

$$<\chi \mid C \mid \phi> = <\phi \mid C \mid \chi> ,$$

$$<\phi \mid C \mid \phi> \leqslant o . \qquad (11)$$

C is symmetric and non positive. The corresponding eigenfunctions and eigenvalues are defined in the usual way :

$$C\, \Phi_i = \lambda_i\, \Phi_i . \qquad (12)$$

From (11) it follows :

$$\lambda_i = \frac{<\Phi_i \mid C \mid \Phi_i>}{<\Phi_i \mid \Phi_i>} \leqslant 0 . \qquad (13)$$

The eigenvalues ($\lambda_i \neq 0$) can be interpreted as reciprocal relaxation times for the approach to equilibrium. Indeed assuming that the eigenfunctions form a complete set, the solutions of (5) for spatially homogeneous disturbance become trivially :

$$\phi_o (t) = \sum_i \alpha_i \Phi_i \exp(\lambda_i t) . \qquad (14)$$

The five eigenfunctions with zero eigenvalues correspond to the summational invariants : 1, k, k^2. With the orthogonality condition : $<\Phi_i \mid \Phi_j> = \delta_{ij}^{kl}$, we readily obtain their explicit form :

$$a_1 = \frac{1}{\alpha_1} ,$$

$$a_i = \frac{k_i}{\alpha_2} \qquad (i \equiv 2,3,4 \equiv x,y,z)$$

$$a_5 = \frac{1}{\alpha_5} [k^2 - \frac{<1 \mid k^2>}{\alpha_1}] , \qquad (15)$$

where the α_i are normalization constants; they can be expressed in terms of thermodynamic quantities. In our model we get :

$$\alpha_1^2 = \frac{1}{\beta} (\frac{\partial n}{\partial \mu})_{T,V} \quad ; \quad \alpha_2^2 = \frac{m \, n}{\hbar^2 \beta} ; \quad \alpha_5^2 = (\frac{2m}{\hbar^2})^2 k \, T_o^2 \, C_V n \qquad (16);$$

$$(16)$$

n is the number density, T_o the equilibrium temperature $\beta = \frac{1}{kT_o}$

and C_V the specific heat per particle at constant volume.

b) The transport equations.
Here the hydrodynamical variables correspond to the $1, k, k^2$ "moments" of the distribution function. Indeed the deviations from their equilibrium values of the temperature and the densities of particle and momentum (respectively $\delta T_q(\omega)$, $\delta n_q(\omega)$, $\delta u_q^i(\omega)$) are given by :

$$\delta n_q(\omega) = \alpha_1 A_{1,q}(\omega) ,$$

$$\delta u_q^i(\omega) = \frac{\alpha_2 \hbar}{m} A_{i,q}(\omega) \qquad (i=x,y,z) ,$$

$$\frac{\delta T_q(\omega)}{k T_o^2} = \frac{2m}{\hbar^2 \alpha_5} A_{5,q}(\omega),$$ (17)

where $A_{i,q(\omega)} = <a_i \mid \phi_q(\omega)>$.

Despite the importance of the eigenvalue problem (12) in transport theory, the $\{\phi_i\}$ and the $\{\lambda_i\}$ remain largely unknown in many cases. For most systems of particles interacting through purely repulsive potentials $[V(r) \approx 1/r^n ; n > 2]$, it is expected (and sometimes proved rigourously [13])that there is a finite gap in the spectrum of the collision operator between the zero (degenerate) eigenvalue and the smallest nonvanishing eigenvalue λ_o . On the other hand in the case of systems which have undergone a phase transition with the breakdown of a continuous group of the hamiltonian, it has been shown that the continuous spectrum starts immediately at zero (Goldstone modes) [14] . Because of the intricacy of the collision term, it is often useful to consider model equations with the same basic features as the kinetic equations but simpler to solve. One of these models is the B.G.K. model [15] :

$$C_{B.G.K.} \phi_q = \frac{1}{\tau R} \{ \sum_{j=1}^{5} a_j <a_j \mid \phi_q> - \phi_q \}.$$ (18)

This operator satisfies the conservation laws; it has indeed a fivefold zero eigenvalue,all others being equal to the collision frequency τ_R^{-1}.
More elaborated models taking into account the energy dependence of the relaxation time have also been introduced [16] .

To proceed we introduce a projection operator P which projects any suitably integrable functions onto the hydrodynamical subspace spanned by the a_i defined in (15) :

$$P h(k) = \sum_{i=1}^{5} a_i(k) <a_i \mid h>,$$

$$Q = I - P.$$ (19)

Q is the complement .
In particular, we have :

$$P \phi_q(k;\omega) = \sum_{i=1}^{5} a_i(k) A_{i,q}(\omega),$$

which is precisely the linearized local equilibrium distribution function.
We now project Eq. 9 onto P and Q :

$$- i\omega P \phi_q + PL'(P+Q) \ \phi_q = P \ \phi_q(t=o) , \qquad (20a)$$

$$- i\omega Q \phi_q + QL'(P+Q) \ \phi_q = Q \ \phi_q(t=o) , \qquad (20b)$$

where $L' = i\underline{q}.\underline{v}_k - C$.

Solving for $Q \ \phi_q$ and substituting in (20a) yields an equation for $P\phi_q$ which we project on the various basis vectors a_i to obtain :

$$\sum_{j=1}^{5} [- i\omega \delta_{ij}^{kr} + \langle a_i \ | \ L' \ | \ a_j \rangle + \langle a_i \ | \ \psi \ | \ a_j \rangle] \ A_{j,q}(\omega)$$

$$= A_{i,q}(t=o) + I_i(q,\omega) , \qquad (21)$$

where :

$$\langle a_i \ | \ \psi \ | \ a_j \rangle = \langle a_i \ | \ PL'Q \ [i\omega - QL'Q]^{-1} QL'P \ | \ a_j \rangle \qquad (22)$$

and the second term of the (r.h.s.) of (21), which is a correction to the initial condition, is given by :

$$I_i(q,\omega) = \langle a_i \ | \ PL'Q \ [i\omega - QL'Q]^{-1} \ Q\phi_q \ (k;t=o) \rangle . \qquad (23)$$

On account of the fact $C|a_i\rangle = o$, the various terms in can be simplified :

$$\langle a_i \ | \ L' \ | \ a_j \rangle = \langle a_i | \ i\underline{q}.\underline{v}_k | \ a_j \rangle . \qquad (24)$$

Using (16) these matrix elements can be expressed in terms of thermodynamic expressions :

$$\langle a_i \ | \ \psi \ | \ a_j \rangle = \langle \ _{i,q}(k) | \ [\ i\omega - QL'Q \ ]^{-1} | \ \hat{J}_{j,q}(k) \rangle , \qquad (25)$$

where :

$$\hat{J}_{i,q}(k) = Q(i\underline{q}.\underline{v}_k \ a_i(k))$$

$$= i\underline{q}.\underline{J}_i(k) . \qquad (26)$$

$J_i(k)$ is the part of the current associated with to the slow variable a_i which is orthogonal to the hydrodynamic subspace.

Finally :

$$I_i(q,t) = \langle \hat{J}_{i,q}(k) \mid [i\omega - QL'Q]^{-1} Q \mid \phi_q(k;t=o)) \rangle. \tag{27}$$

Equations 21 are the generalized hydrodynamic equations. In order to make these more transparent we insert the Laplace transforms and use the ordinary hydrodynamic variables $\delta n_q(t)$, $\delta u_q^i(t)$ and $\delta T_q(t)$ defined in (17).

Inserting the explicit expressions for the matrix elements (24) we get, assuming that q is oriented along 0_x :

$$\partial_t \, \delta n_q(t) + i \, q_x \frac{n}{m} \, \delta u_q^x(t) = o \,, \tag{27a}$$

$$\partial_t \, u_q^y(t) - \int_o^t dt' \, K_{33}(q;t - t') \, \delta u_q^y(t') = I_3(q;t) \,,$$

$$\partial_t \, u_q^z(t) - \int_o^t dt' \, K_{44}(q;t - t') \, \delta u_q^z(t') = I_4(q;t) \tag{27b},$$

$$\partial_t \, u_q^x(t) + iq_x \frac{1}{n} (\frac{\partial P}{\partial n})_T \, \delta n_q(t) + iq_x \frac{1}{n} (\frac{\partial P}{\partial T})_n \, \delta T_q(t)$$

$$- \int_o^t dt' \, K_{22}(q;t - t') \, \delta u_q^x(t') = I_2(q;t) \,,$$

$$\partial_t \, \delta T_q(t) + iq_x \frac{T}{\rho C_V} (\frac{\partial P}{\partial T})_n \, \delta u_q^x(t)$$

$$- \int_o^t dt' \, K_{55}(q;t-t') \, \delta T_q(t') = I_5(q;t) \,, \tag{27d}$$

where P is the pressure the dissipative kernels $K_{ii}(q;\tau)$ and the initial condition terms $I_i(q;t)$ are obtained directly by inverting expressions (25) and (27) respectively :

$$K_{ii}(q;t) = \langle \hat{J}_i \mid \exp [\{ C - Q \, iq_x \, v_k^x \, Q \} t] \mid \hat{J}_i \rangle$$

$$= \langle \hat{J}_i \mid \exp [\hat{C}t] \mid \hat{J}_i \rangle. \tag{28}$$

We have indeed trivially QCQ = C, using (26); $K_{ii}(q;t)$ may also be written as :

$$K_{ii}(q;t) = - q_x^2 \langle J_i(k) \mid \exp [\hat{C}t] \mid J_i(k) \rangle \tag{29}$$

and finally :

$$I_i(q;t) = -\langle \hat{J}_{i,q}(k) \mid \exp [\hat{C}t] \mid Q \, \phi_q(k;t=o) \rangle. \tag{30}$$

On comparing equations 27 with the usual (linearized) balance equations, one can make the following remarks :
a) The thermodynamic coefficients are the same ;
b) The transport equations (27) are noninstantaneous; they depend on the past history of the hydrodynamic variables ;
c) There is an explicit dependence on the initial value of nonhydrodynamical variables through the terms $I_i(q;t)$ (cf. Hibert's paradoxe) ;
d) The dissipative kernels are non-local in space.

Up to present, we have made no assumptions about time or length scales in the derivation of equations 27 from the kinetic equation 9, we now study the hydrodynamic limit ($t > \tau_R$; $L_h > 1$). To simplify the discussion we denote the hydrodynamical variables as a vector $\underline{H}_q(t)$ and we introduce two matrices : one with components $R_{ij}(q) = <a_i \mid iq_x v_k^x \mid a_j>$ corresponding to the thermodynamic coefficients and one with components $K_{ij}(q;\tau)$, cf (28). Equations 27 may then be written in compact form as :

$$\partial_t H_q(t) + R(q) H_q(t) - \int_o^t d\tau K_q(\tau) H_q(t-\tau) = I(q,t) .$$

(31)

The operator $\hat{C}$:

$$\hat{C} = C - Q (i q_x v_k^x) Q ,$$

(32)

which determines the time evolution of $K_q(\tau)$ and $I(q;t)$, is quite awkward to compute. Fortunately in the hydrodynamic limit: $L_h > 1$ or $q < \lambda_o/<v>$, the second term in (32) is a small correction to C. As a result the operator $\exp [\hat{C}t]$ possesses a perturbation expansion in powers of q around $\exp [Ct]$. This expansion can be obtained by iterating the identity :

$$\exp [\hat{C}t] = \exp [Ct] + \int_o^t d\tau \exp [\hat{C}\tau] Q(iq_x v_k^x) Q$$

$$x \exp [C\tau] .$$

(33)

Since in $K_q(\tau)$ and $I(q;\tau)$ the time evolution operator acts only on functions which have no components in the hydrodynamic subspace, the long time behavior of these terms is dominated by $\exp [\lambda_o t]$, where λ_o ; ($\lambda_o < o$) is the smallest nonvanishing eigenvalue of the collision operator.
The operators $K_q(\tau)$ and $I(q,\tau)$ thus decay to zero with a characteristic time λ_o^{-1} :

$$\left. \begin{matrix} K_q(\tau) \\ \\ I(q,\tau) \end{matrix} \right\} \rightarrow 0 \text{ for } \tau > 1/\lambda_o = \tau_R .$$

(34)

As a result, for times $t > \lambda_o^{-1}$, the inhomogeneous terms die out and the upper limit of the convolution product in the (l.h.s.) of (31) can be extended to infinity to give :

$$\partial_t H_q(t) + R_q H_q(t) - \int_o^\infty K_q(\tau) H_q(t-\tau) \, d\tau = 0. \qquad (35)$$

Moreover, we can expand $H_q(t - \tau)$ as follows :

$$H_q(t-\tau) = H_q(t) - \tau \frac{\partial H_q(t)}{\partial t} + \ldots$$

$$\approx H_q(t) [1 - O(q)]. \qquad (36)$$

When the gradients are small, the temporal variation of $H_q(t)$ appearing in (36) is also small and of the order q as it can be tested from Eq. 27. This series is in fact an expansion in powers of q (Chapman-Enskog expansion) [17]. Inserting (36) and expanding Equ. 35 up to order q^2 we can localize in time the convolution product as follows :

$$\int_o^\infty d\tau \, K_q(t-\tau) \, H_q(\tau) \xrightarrow{q \to o} - q^2 \, [\int_o^\infty K^{(2)}(\tau) \, d\tau] \, H_q(t) + O(q^3). \qquad (37)$$

Indeed from (29), the matrix K is at most of order q^2. As a result Equ. 35 reduces to the traditional hydrodynamic equation which can be written in compact form as :

$$\partial_t H_q(t) + R(q) H_q(t) + q^2 M H_q(t) = o,$$

where :

$$M = - < J_i(k) \mid \frac{1}{C} \mid J_i(k) >. \qquad (38)$$

This derivation of the hydrodynamic equations heavily relies on the existence of a gap in the spectrum of the collision operator. In the cases where the continuous spectrum starts at zero and using similar arguments, it can be shown that the long wavelength Goldstone modes give a negligible contribution to hydrodynamics [18]. This calculation also provides microscopic expressions for the transport coefficients in terms of the one particle distribution function. From (38), the shear viscosity is given by :

$$\eta = - \rho < J_3(k) \mid \frac{1}{C} \mid J_3(k) >. \qquad (39)$$

sing (26) we get :

$$\eta = - \frac{m^2}{kT} < v_k^{\ x} \ v_k^{\ y} \ | \ \frac{1}{C} \ | \ v_k^{\ x} \ v_k^{\ y} > .$$ (40)

Similarly for the thermal conductivity:

$$K = - \frac{1}{kT^2} < [\epsilon_k - \frac{5\,P}{2n}] \ v_k^{\ x} \ | \ \frac{1}{C} \ | \ [\epsilon_k - \frac{5\,P}{2n}] \ v_k^{\ x} > .$$ (41)

Equ. 40 can also be written as :

$$\eta = - \frac{m^2}{kT} \int \ d\underline{k} \ v_k^{\ x} \ v_k^{\ y} \ \chi^{xy}(k),$$ (42)

where χ^{xy} is the solution of the following integral equation :

$$C \ \chi^{xy} = v_k^{\ x} \ v_k^{\ y} \ f^\circ_{\ k} \ (1-f^\circ_{\ k}).$$ (43)

Useful variational methods have been developed to study these integral equations [19] . In this form the equivalence with the theory of Chapman-Enskog is transparent. However the method presented here allows to analyze the relaxation of the exact solution toward the so-called normal solution.
With minor changes the projection method can also be applied to collision operators including non-Markovian and nonlocal corrections. For instance in the case of a nonlocal collision operator the kinetic equation becomes :

$$[- i\omega + iq_x v_k^{\ x} - C_q] \ \phi_q(k;\omega) = \phi_q(k;t=o) .$$ (44)

We can go through all the steps described above. For the sake of brevity we limit ourselves to the transverse excitations, the corresponding current now becomes (cf. 25) :

$$J_{3,q} = C \ Q \ \{ \ [\ i \ q_x \ v_k^{\ x} - C_q] \ v_k^{\ y} \ \} ,$$

(C is a constant). (45)

Expanding the collision operator in powers of q:

$$C_q = C_o + i \ q_x \ T_o^{\ x} + q^2/2 \ C''.$$

We get at the lowest order :

$$J_{3,q} \approx C \ i \ q_x \ Q \ \{ \ [v_k^{\ x} + T_o^{\ x}] \ v_k^{\ y} \ \} .$$ (46)

T_o^x is a potential contribution to the transverse current ("colli-sional transfer"). It is directly related to the non-local charac-ter of the collision operator. Similarly we obtain for the shear viscosity :

$$\eta = C_1 < [v_k^y \, (v_k^x + T(o)^x)] \quad Q \mid \frac{1}{C} \mid Q \, [\, (v_k^x + T_o^x)v_k^y] >$$

$$+ \, C_2 < v_k^y \mid C'' \mid v_k^y >,$$

or $\eta = \eta' + \eta''$,   (47)

where η'' is the direct term which has no counterpart in a purely local theory. For instance the Landau theory transport coefficients of the Fermi liquid do not contain any direct term. Similarly non-Markovian contributions play a crucial role in the evaluation of the bulk viscosity. Thus even when we are only interested in trans-port coefficients it is premature to perform the limit $q \to 0$; $\omega \to 0$ in the collision operator of the kinetic equations. In conclusion microscopic expressions for the transport coefficients *including their potential part* can be obtained from the kinetic equation for the *one particle* Wigner distribution function.

On the other hand one can also obtain microscopic expressions for the hydrodynamic modes. These collective modes correspond to the eigenvectors $\hat{a}_{i,q}$ of the linear problem defined by equation 38. In a normal Fermi system (one component) one has five hydrodynamic modes the frequencies of which are :

$$\lambda_{1,2} = \pm \, i \, C \, q - \Gamma q^2,$$

$$\lambda_{3,4} = - \, \eta q^2/\rho,$$

$$\lambda_5 = - \, K \, q^2/n \, c_p,$$ (48)

where C is the sound velocity :

$$C^2 = \frac{C_p}{C_v} \, (\frac{\partial P}{\partial n})_T$$

and

$$\Gamma = \frac{1}{2} \, [\, \frac{4\eta}{3\rho} + (\frac{1}{C_v} - \frac{1}{C_p}) \, \frac{K}{n} \,] \, .$$

$\lambda_{1,2}$ describe damped sound-wave propagations; $\lambda_{3,4}$ express the vis-cous damping of the transverse velocity and λ_5 corresponds to the damping of the thermal mode.
For an m-component normal system there are (4 + m) hydrodynamic modes.

They play an important role in statistical physics. Any long wave-length disturbance may be described as a superposition of these modes. They are also useful in problems where the *nonlinear* terms are important. Using a projection operator method it is possible to derive kinetic equations which describe the nonlinear dynamics of the hydrodynamic modes $\{\hat{a}_{i,q}\}$ [20] :

$$\partial_t \, \hat{a}_{i,q} = \lambda_i \, \hat{a}_{i,q} + \sum_{q'} \sum_{k,l} \, V^{kl}_{q,q'} \, \hat{a}_{q-q',l} \, \hat{a}_{q',k} \, . \qquad (49)$$

λ_i is given by (48).
These mode—mode coupling equation have been used to study the anomalous behavior of the transport coefficients near a critical point [20] or the long time decay of the Green—Kubo integrand (21). Similar equations can also be applied to describe the dynamics near an instability in system driven far from equilibrium (22).

4. REMARKS ABOUT THE DYNAMICAL BEHAVIOR OF A SYSTEM OF CHARGED PARTICLES.

a) Collisionless regime
Since it has been suggested that the dense electron-hole (e-h) plasma created by the pulse might play an important role in the problem of laser annealing [23] it seems appropriate to make a few remarks about the dynamics of a system of particles interacting through Coulomb forces.
The long range of these forces calls indeed for a special treatment. In this section, we consider only a one-component plasma (O.C.P.) moving in an inert neutralizing background. Because in a plasma each particle interacts simultaneously with a large number of other ones, the mean field term plays now an important role. On the short time scale (collisionless regime), the plasma may be described by the quantum Vlassov equation which becomes in the linear case [10] :

$$\partial_t \, f_q(k;t) + i \, [\, \varepsilon_{k+q/2} - \varepsilon_{k-q/2} \,] \, f_q(k;t)$$

$$- \, i \, V_q \, [\, f^\circ_{k+q/2} - f^\circ_{k-q/2}] \, \sum_{k'} f_q(k';t) = o \, , \qquad (50)$$

where $V_q = \dfrac{4 \Pi e^2}{q^2}$.

In the long wavelength limit, it takes the more familiar form :

$$\partial_t \, f_q(k;t) + i \, \underline{q}.\underline{v}_k \, f_q(k;t) - i \, V_q \underline{q}.\underline{v}_k (\frac{\partial f^\circ_k}{\partial \varepsilon_k}) \sum_{k'} f_q(k';t) = 0 \, .$$

$$\qquad (51)$$

In the degenerate case : $\partial f^\circ_k / \partial \varepsilon_k \approx -\delta(\varepsilon_k - \mu)$; one recovers the well known Landau-Silin equation which has been used to describe the dynamical properties of Fermi liquids in the collisionless regime [24] . The physical meaning of these equations is simple; the charge fluctuations induce a space charge field :

$$\underline{E}_{-p} = i\underline{q} \, 4\pi^2 \, \sum_{k'} f_q(k';t) \, / \, q^2 ,$$

which in turn acts to drive the electrons. Equations 50, 51 admit collective modes (plasmons); the corresponding frequency is given in the degenerate case :

$$\omega_{q \to o} \approx \omega^\circ_p \, [\, 1 + \frac{3}{10} \left(\frac{qv_F}{\omega^\circ_p}\right)^2 + \ldots \,] ,$$

where $\omega^\circ_p = \left(\frac{4\pi n \, e^2}{m}\right)^{1/2}$. (52)

Even in the collisionless regime these plasma oscillations are damped when q is greater than some critical wave vector. This Landau damping can be pictured as the decay of the plasmon in a single-pair excitation.

The plasma oscillations therefore appear as the collisionless mode of a charged system. The existence of these high frequency oscillations throw considerable doubt on the applicability of the standard theory of hydrodynamic modes (see Chap. 3) to the case of charged particles. Another important consequence of the collective behavior of a system of charged particles is the screening which is described by the frequency and wave vector dependent dielectric function $\varepsilon(q, \omega)$. The dielectric function derived from Equ. 50 is given by [24]:

$$\varepsilon(q, \omega) = 1 - \frac{4\pi e^2}{q^2} \, \sum_k \frac{[f^\circ_{k+q/2} - f^\circ_{k-q/2}]}{(\varepsilon_{k+q/2} - \varepsilon_{k-q/2} - \omega - i\varepsilon)} . \quad (53)$$

Formula 52 is equivalent to the well-known Random Phase Approximation (R.P.A.).

An important question arises in connection with the Vlassov equation : *the stability of the plasma.*

In the case of the Fermi-Dirac (Maxwellian) distribution, the Landau damping guarantees the stability of the equilibrium plasma. However for certain nonequilibrium distribution $f_q(k)$, the collective mode (which can be calculated from (51) f°_k, $\overset{q}{\to} f_q(k)$ grows with time (unstable plasma). The instability may show up at some finite value of q (microscopic instability). This often signals a transition to an ordered state (e.g. permanent spin density waves [25]).

After a short time, the linear Vlassov equation becomes invalid to describe this unstable mode and the full non-linear Vlassov equation as well as other nonlinear couplings must be taken into account in order to provide a stabilizing mechanism. Mode-mode coupling equations similar to (49) but involving now the collisionless modes can also be derived [26] . These plasma instabilities are to be contrasted with the instabilities in open systems driven far from equilibrium discussed in Binder's lecture [27] . In the latter case one makes the hypothesis that the system remains in a state of local equilibrium throughout the transition [28] ; the basic equations are here the hydrodynamic equations (cf. Rayleigh-Bénard instability). On the contrary the plasma instabilities discussed above depend upon the detailed form of the distribution function (e.g. Bump-in-tail instability). The integration over the velocity to obtain the conserved variables would remove at the same time the mechanism of these plasma instabilities. By the way it must be stressed here that strong deviations from the equilibrium distribution of the photoexcited carriers have been characterized at high excitation levels [8] .

b) Hydrodynamic regime

At lower frequencies the collisions play an important role and the kinetic equation for $\phi_q(k,\omega)$ defined in (6) then takes the form :

$$[i\omega + i\underline{q}\cdot\underline{v}_k + i V_q - C_q(\omega)] \quad \phi_q(k,\omega) = \phi_q(k|t = o), \quad (54)$$

where V_q is the Vlassov operator obtained from (51):

$$V_q = \underline{q}\cdot\underline{v}_k V_q \sum_{k'} f^\circ_{k'}(1-f^\circ_{k'}) \chi \qquad (55)$$

and $C_q(\omega)$ [10] is the wavevector and frequency-dependent collision operator. In the weak coupling limit (e.g. degenerate Fermi systems) it reduces to the Balescu-Lenard quantum operator $C_q^{LB}(\omega)$ [10] which describes binary collisions between charged particles by taking into account the dynamic screening through the wavevector and frequency dependent dielectric constant $\varepsilon(q,\omega)$. This collision operator includes contributions of many particle collisions contained in $\varepsilon(q,\omega)$. Because we shall not need it, we do not give here the explicit form of $C_q^{LB}(\omega)$. We can now proceed as in the previous section and project equation (54) onto the hydrodynamic subspace to obtain equation analogous to (21),(22) and (23) but in the plasma case we must substitute now in these relations the definition :

$$L' = i \underline{q}\cdot\underline{v}_k + i V_q - C_q(\omega). \qquad (56)$$

The main difference with the neutral case lies in the presence of the *Vlassov operator* which from (55) is of order q^{-1} and thus *non-*

analytic in q. Following the method presented in section 2 , we obtain then the equation for the hydrodynamic vector in compact form (cf.35) :

$$\partial_t H_q(t) + R_q H_q(t) - \int_o^\infty K_q(\tau) H_q(t-\tau) \, d\tau = 0, \tag{57}$$

where now according to (56) the Vlassov term has to be included in the definition of the matrix R_q :

$$R_{ij} = \; < a_i \mid i \, \underline{q} \cdot \underline{v}_k + iV_q \mid a_j >. \tag{58}$$

In the plasma case, the Taylor expansion :

$$H_q(t-\tau) = H_q(t) - \tau \frac{\partial H_q}{\partial t} + \ldots \tag{59}$$

can no longer be used to extract from (57) a local description in time. Indeed because of the Vlassow term R_q contains a term of order q^{-1}. The iteration of this term when computing the time derivatives appearing in (59) introduces higher and higher powers of q^{-1} and the correction terms coming from the nonlocality in time cannot be neglected. A different procedure for reducing Equ. 59 to a local form is needed. This problem has been extensively studied by M. Baus [29] . As already pointed out the most striking features of the plasma is that the sound modes of the neutral fluids are shifted into high frequency oscillations (52). Because of the presence of these high frequency modes the hydrodynamic frequencies cannot be obtained from the usual hydrodynamic equations [29] but they are expressed in terms of the frequency dependent collision operator evaluated at the plasma frequency ω_p . This work has also been generalized to two component plasmas where the fluctuations in charge and mass density are no longer proportional to each other [30] . It would be highly interesting to generalize this theory to solid state plasmas by including the coupling with the phonons in order to clarify the discrepancies between various predictions about the hydrodynamics of the photoexcited (e-h) plasma. (expansion [31] or confinement [32]).

REFERENCES

1. Misra,B. and Prigogine,I. Proceedings of the Workshop on long time Prediction in Nonlinear Conservative Systems. Austin(1981) to be published.
2. For a recent review see : J. de Physique, Paris, C7-1981. Proceedings of the Third International Conference on Hot Carriers in Semiconductors (Montpellier 1981).

3. Van Hove, L. Physica 23 (1957) 441.
4. Prigogine, I. and Résibois, P. Physica 27 (1961) 629.
5. Zwanzig, R. Lectures in Theoretical Physics III (Boulder) (New York, Interscience Publishers, 1961), p. 106.
6. See for instance : Dorfman, J.R., in J. Stephenson, ed. Statphys. 14, Proceedings of the XIVth International Conference on Thermodynamics and Statistical Mechanics (Amsterdam, North Holland, 1981).
7. Résibois, P., J.Chem. Phys. 41 (1964) 2979.
8. Smirl, A.L. these proceedings.
9. Résibois, P. Phys. Rev. B 138 (1965) 281.
10. Balescu, R. Statistical Mechanics of Charged Particles (New York, Interscience, 1963).
11. de Schepper, I.M. and Cohen, E.G.D. J. Stat. Phys. 27 (1982) 223.
12. Uehling, E.A. and Uhlenbeck, J.E. Phys. Rev. 43 (1933) 552.
13. Pao, Y. Commun. Pure Appl. Math. 27 (1974) 407.
14. Dewel, G. Physica 89 A (1977) 245.
15. Williams, M.M.R. Mathematical Methods in Particle Transport Theory (London, Butterworth, 1971).
16. Cercignani, C. Mathematical Methods in Kinetic Theory (New York, Plenum Press, 1969).
17. See for instance : Résibois, P. and de Leener, M. Classical Kinetic Theory of Fluids (New York, Wiley, 1977).
18. Beck, H. Phys. Kondens. Materie 12 (1971) 330.
19. Jensen, M., Smith, H. and Wilkins, J.W. Phys. Rev. 185 (1969) 323.
20. Kawasaki, K. Ann. Phys. 61 (1970) 1.
21. Zwanzig, R. in S. Rice, K. Freed and J.C. Light, eds., Statistical Mechanics : New Concepts, New Problems, New Applications (Chicago, University of Chicago Press, 1972).
22. Graham, R. and Pleiner, H. Phys. of Fluids 18 (1975) 130.
23. Van Vechten, J.A. J. Physique 41 (1980) C4-15.
24. Pines, D. and Nozières, P. The Theory of Quantum Liquids (New York, Benjamin, 1966).
25. Overhauser, A.W. Phys. Rev. Letters 4 (1960) 462.
26. Davidson, R.C. Methods in Nonlinear Plasma Theory (New York, Academic Press, 1972).
27. Binder, K. these proceedings.
28. Glansdorff, P. and Prigogine, I. Thermodynamic Theory of Structure Stability and Fluctuations (New York, Interscience, 1971).
29. Baus, M. Physica 79 A (1975) 377.
30. Baus, M. Physica 88 A (1977) 319, 336.
31. Combescot, M. Phys. Lett. 85A (1981) 308.
32. Van Vechten, J.A. and Wautelet, M. Phys. Rev. 23 (1981) 5543.

128

PHASE DIAGRAMS AND SEGREGATION

Franz Rosenberger[+]

Department of Physics, University of Utah, Salt Lake City,
Utah 84112, U.S.A.   + Support by NSF (Grant DMR79-13183)
and NASA (Grant NSG 1534) during the writing of this paper
is gratefully acknowledged.

1. INTRODUCTION

Phase diagrams (or constitution diagrams) are graphical representations of the *equilibrium* relationships between phases of elements, compounds, solutions and their mixtures. Segregation pertains
to the fact that most materials systems (of technological interest)
undergo compositional changes on phase change. The equilibrium segregation behavior of many systems can, in principle, be deduced
from their phase diagrams. Hence, the interpretation of phase diagrams is at the heart of designing materials preparation processes.

In this lecture we present a concise overview of the various
types of liquid-solid phase equilibria. The corresponding phase
diagrams are developed purely phenomenologically, heavily relying
on (thermodynamically founded) graphical arguments. Diagrams for
vapor-condensed phase equilibria are only briefly touched upon.
There is not enough room for a full coverage of the complex situations that can arise in interaction with vapors. However, it is
suggested that mass transfer between vapors and condensed phases
plays a more important role in laser annealing than has commonly
been assumed to-date.

The second part of the presentation focuses on segregation
coefficients, thermodynamic models for their estimation and on the
dynamics of segregation. A brief comparison is made with segregation trends observed in laser annealing. Finally it is shown that
convective transport may play a significant role even during the
short duration of typical laser melting processes.

Parts of this presentation are a condensed version of corresponding sections of the author's monograph on Fundamentals of Crystal Growth [1] . Ample references are given for reading beyond the
introductory level of this lecture.

2. PHASE EQUILIBRIA [1-3]

Two phases can stably coexist only if the chemical potential
for each of their components, $i\mu = 1,2,3 \ldots$, is equal in both phases,
i.e. :

$$\mu_i^{(1)} = \mu_i^{(2)} . \tag{1}$$

Hence, if the state variables of a system are changed, phase coexistence will only be preserved if the resulting changes in the components' chemical potentials are the same for both phases :

$$\Delta\mu_i^{(1)} = \Delta\mu_i^{(2)} . \tag{2}$$

Chemical potential changes and changes in intensive state variables
such as temperature T, total pressure p and electric field strength
F etc. are related through the Gibbs-Duhem equation :

$$d\mu_i = - s_i dT + v_i dp + p_i dF + \ldots \tag{3}$$

where s_i, v_i and p_i are respectively the molar entropy, the molar
volume and molar polarizability of component i in the phase under
consideration. Note that s_i, v_i and p_i may be concentration dependent, i.e., take on different values depending on the kind and concentration of other components in the phase. Thus, from Eqs. 2 and
3 it is obvious that, once a phase equilibrium exists, a change in
a state variable must in general be accompanied by a specific change in another state variable if equilibrium is to be regained. For
combinations of pressure and temperature changes this restriction
leads for *monocomponent* systems to the Clausius-Clapeyron relation
which can be written as :

$$\frac{dp}{dT} = \frac{\Delta s}{\Delta v} \tag{4}$$

where $\Delta s = s^{(2)} - s^{(1)}$ and $\Delta v = v^{(2)} - v^{(1)}$. Eq. 4 can be employed
to compute the coexistence curves in Fig. 1, i.e., the melting
"point" curve $T_m(p)$ and the vapor pressure curves $p(T)$ of the liquid
and solid phase, respectively, provided that the $s(T,p)$ and $v(T,p)$
are known for the coexisting phases.

The shaded areas in Fig. 1, that extend the phase stability ranges beyond the coexistence lines, are to indicate metastable states.

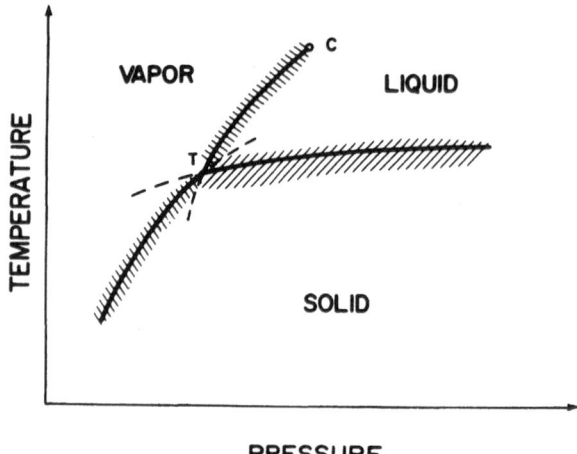

<u>Fig. 1</u> Schematic T-p phase diagram for a monocomponent material that expands on melting.

How much, for instance, a liquid can be undercooled before nucleation and growth of the solid sets in, that depends on the thermal history of the sytem and the underlying kinetics. These considerations, however, go beyond the scope of this lecture.

For systems with more than one component (independent chemical constituents, see [2]), Eqs. 3 and 4 afford already a glimpse at the complexity of phase equilibria. If p and T are uniform throughout a system, a temperature change can only be compensated by a pressure change if the $\Delta s_i / \Delta v_i$ are the same for all components. In most real systems this can only be fulfilled if the variation in T and p is accompanied by changes in the composition of the phases and, hence, in the values of s_i and v_i. Consequently, in general, coexisting multicomponent phases differ in their composition, i.e., in their ratio of component concentrations.

As indicated above, the mass transfer equilibrium conditions (1) restricts the number of intensive state variables that can be chosen independently to uniquely specify an equilibrium state. For multicomponent, multiphase systems Gibbs has formulated this restriction of the "number of degrees of freedom" f in the famous phase rule :

$$f = C - P + 2 \tag{5}$$

where C and P are the number of components and phases, respectively. Note that in the above form only two intensive state variables are accomodated (such as p and T). If more than two terms must be accounted for in Eq. 3, f increases correspondingly. Application of the phase rule, for instance, to a monocomponent three-phase system, where $f = 1 - 3 + 2 = 0$, shows that at this triple point (T in Fig. 1) the state of the system is uniquely specified. Only a certain combination of p and T that is specific for the material allows for the coexistence of three phases. For a binary three-phase system, however, where $f = 2 - 3 + 2 = 1$, there is no triple "point". In order to specify a specific state of three coexisting phases one must state one more variable. For instance the specification of the composition of one phase fixes the (different) composition of the other phases and, thus, uniquely determines the system (Section 5).

3. BINARY LIQUID-SOLID PHASE DIAGRAMS [1,3-5]

Depending on the underlying length scale, mixing of components can occur in three different forms :
1) Mechanical mixtures are intimate associations of the components in a state of subdivision large enough that only a small fraction of the atoms interact directly with those of the other component. Thus, to represent a mechanical mixture, component particles must not be smaller than a few hundred Angstroms ;
2) Solutions are mixtures on an atomic scale;
3) Mechanical mixtures of solutions have particles of the above minimum dimensions that consist of solutions.
The molar Gibbs function of a phase formed by a binary mechanical mixture is simply additive with respect to the pure components :

$$g^m = X_A h_A + X_B h_B - T(X_A s_A + X_B s_B) \tag{6}$$

where the X_i are the mole fractions and the h_i and s_i the molar enthalpies and entropies, respectively.

In solutions the Gibbs function is not additively composed of the pure component contributions. The changed neighbor interaction causes changes in the internal energy and entropy (motional randomness). Consequently, in solutions one must replace the molar parameters h_A, h_B, s_A and s_B by the partial molar quantities $\bar{h}_A$, $\bar{h}_B$, $\bar{s}_A$ and $\bar{s}_B$ where the bar indicates the dependence on the other component, i.e., the non-ideality. In addition to these changes, there is a configurational entropy of mixing connected with the formation of a solution. Even if atoms A and B would interact virtually identically, so that the motional entropy of the atoms in the pure components were approximately the same, the mere bringing of the components into states of lower order causes an increase of the entropy by the configurational entropy of mixing:

$$\Delta s_m = - R[X_A \ln X_A + X_B \ln X_B].\tag{7}$$

Therefore, the molar Gibbs function of a real solution of a phase can be expressed as:

$$g^s = S_A \bar{h}_A + X_B \bar{h}_B - T(X_A \bar{s}_A + X_B \bar{s}_B) + RT[X_A \ln X_A + X_B \ln X_B].\tag{8}$$

Depending on whether $g^m \gtrless g^s$ we can decide whether a solution or a mechanical mixture will form under given conditions. In the following we will discuss (solid) solution formation in a sequence of increasing deviations from ideal behavior, i.e., in terms of increasing differences between the unbarred and barred parameters in Eqs. 6 and 8.

3.1 Ideal solutions

If the A and B interactions are essentially the same as in the pure components, although A and B atoms can be distinguished, the solution is called ideal. With $\bar{h}_i = h_i$ and $\bar{s}_i = s_i$, using Eqs. 6-8, the Gibbs function of an ideal solution is:

$$g^{ids} = g^m - T\Delta s_m = g^m + RT[X_A \ln X_A + X_B \ln X_B].\tag{9}$$

Based on the chemical potentials:

$$\mu_i = \mu_i^* + RT \ln X_i,\tag{10}$$

with μ_i^* the chemical potential of the pure component at the pressure and temperature of the solution, Eq. 9 can be rewritten to:

$$g^{ids} = X_A \mu_A + X_B \mu_B.\tag{11}$$

Since Δs_m is always positive for all compositions and finite temperatures, $g^{ids} < g^m$ and, hence, an ideal single solution is always

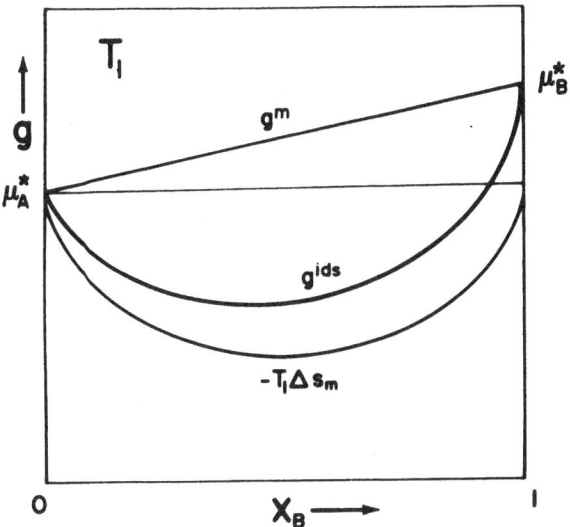

Fig. 2 Gibbs function for a mechanical mixture, entropy of mixing and Gibbs function for an ideal solution as function of composition (schematic).

more stable than a mechanical mixture. Similarly, it can be shown that under these assumptions the single solution is always more stable than any mechanical mixture of solutions.

For the construction of phase diagrams it is useful to visualize the concentration dependence of g^m and g^{ids}. Schematic plots of these functions for a fixed temperature and pressure are given in Fig. 2. As for the temperature dependence, one can see from the Gibbs-Duhem equation 3 and Eqs. 9-11 that g decreases with increasing T at fixed composition and pressure. Furthermore, due to the increasing weight of the term $T s_m$, the g-curves become more convex with increasing T. Thus, we can schematically illustrate the

134

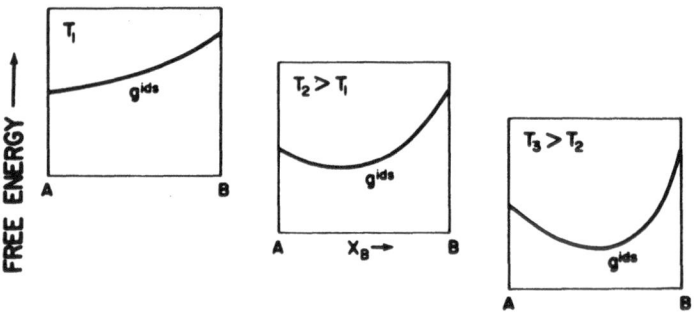

Fig. 3 Constant pressure Gibbs function (free energy) curves for ideal solution at various temperatures.

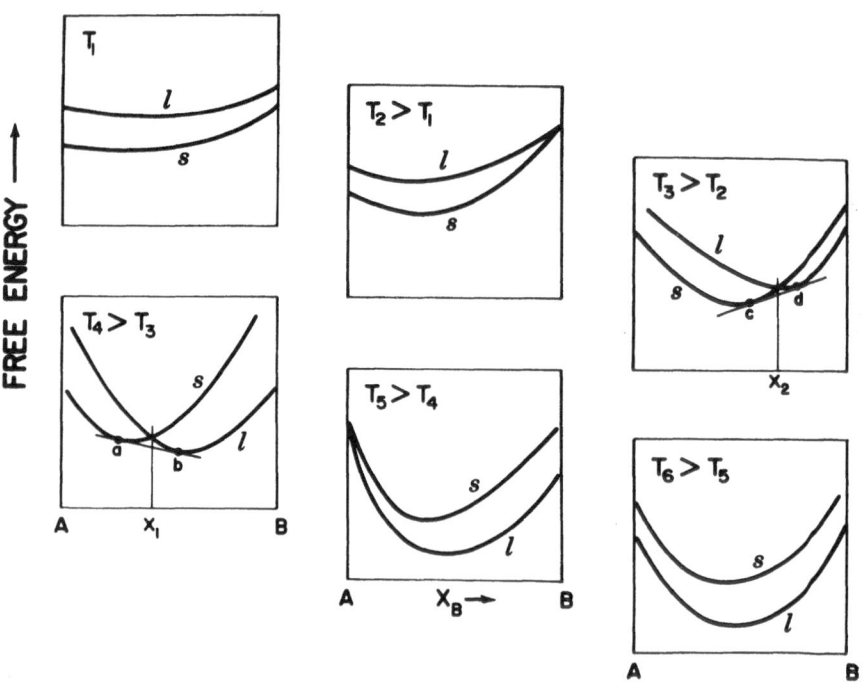

Fig. 4 Free energy curves for the liquid and the solid phase in an ideal system with the temperature increasing from $T_1 \to T_6$.

temperature dependence of the free energy of an ideal solution as shown in Fig. 3.

For a system in which the liquid as well as the solid phase form ideal solutions, both states have free energy curves of the shown form. At a high temperature (T_6 in Fig. 4) the free energy for the liquid solution is lower over the whole composition range. Thus only the liquid solution is stable. As the temperature is lowered, g increases for both liquid and solid solution. In general, at a given temperature, the liquid form of a substance has a higher entropy than the solid. Consequently, the free-energy function for the liquid will be raised more than that for the solid. Eventually the two functions begin to intersect upon lowering of T. At the melting point of pure A, T_5, liquid A and solid A are in equilibrium. For all other compositions the stable state at T_5 is a single liquid solution.

As we lower the temperature further, for instance to T_4, $g^{(s)}$ and $g^{(l)}$ intersect at some intermediate composition X_1. What is the stable state of an alloy of *overall* composition X_1 ? As can readily be shown (see e.g. [1]) the state with the lowest g possible is a mixture of solid and liquid solutions with the compositions of points a and b which are given by the common tangent to $g^{(s)}$ and $g^{(l)}$. This applies to any overall composition between X_a and X_b. Within these boundaries the liquid phase always has the composition X_b and the solid phase X_a at the given T and p for which we have plotted $g^{(l)}$ (X) and $g^{(s)}$ (X). The relative amounts of solid and liquid solution are governed by the lever rule :

$$\frac{n^{(s)}}{n^{(l)}} = \frac{X_b - X_1}{X_1 - X_a} \tag{12}$$

where $n^{(s)}$ and $n^{(l)}$ are the total number of moles in the solid and liquid phase, respectively. Systems with overall composition to the left of X_a form at T_4 a single solid solution. Systems with overall composition to the right of X_b form a single liquid solution at T_4. This "thought-experiment" can be continued through temperatures T_3 and T_2 until at T_1 the whole system consists of a single solid solution. Based upon the various steps in Fig. 4 one can construct an equilibrium phase diagram for the system A-B as depicted in Fig. 5a. Each temperature determines uniquely the composition of a solid and a liquid solution which can be in equilibrium with each other. All liquid solution freezing points form the *liquidus line*; the solid solution melting points form the *solidus line*. Constant temperature lines between solidus and liquidus are descriptively called *tie-lines*. Liquidus and solidus separate the phase diagram into three regions :
  - a one-phase liquid region,

136

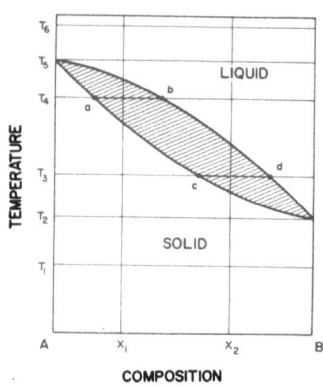

Fig. 5a

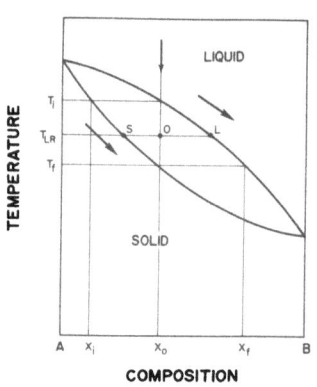

Fig. 5b

Fig. 5 Constant pressure equilibrium phase diagram for ideal binary system
(a) Phase diagram resulting from the free-energy curves in Fig. 4.
(b) Composition changes on freezing of solution X_0.

  — a one-phase solid region,
  — a two-phase liquid-solid region.
Only the first two have physical significance. The third region, here lens-shaped, is merely an "information storage space". Equilibria between solid and liquid solutions of the same composition, say X_1, are excluded by this area.

Let us follow the compositional changes that occur upon progressive *equilibrium solidification* of a melt in a system corresponding to Fig. 5b. As we lower the temperature of a melt of, say, initial composition X_0 the first minute crystal appearing at T_i has the composition X_i. Further withdrawal of heat and lowering of T, in order to solidify a larger portion of the system, results in changes of the composition of the diminishing melt toward X_f. Concurrently, the composition of the whole growing solid is shifted toward X_0. Note that this requires, superimposed to the solidification flux from the liquid to the solid phase, a diffusion flux of

B into the already solidified portion of the system. Close to T_f, the remaining melt (now a small droplet) contains a bit less B than X_f and the solid counterpart a bit less than X_0. Finally *at* T_f the whole melt is solidified and the system has again a uniform composition.

This consideration is based on the assumption that diffusion can establish equilibrium between the bulk phases throughout the whole process. Since diffusion is a slow process, particularly in solids, very low crystallization rates are required to avoid concentration gradients in the forming solid. On the other hand, one can conduct such a freezing process intentionally rapidly enough to override the equalizing diffusion in order to achieve separation and purification, as will be briefly discussed in Section 7.

There are some real binary system which behave close to ideally in the solid and liquid states. Such systems present the possibilit of relatively simple quantitative calculations of phase diagrams. An example is given in Fig. 6 in form of the germanium-silicon system. Note that Ge and Si are isomorphous, i.e., they possess the same (diamond) crystal structure in the pure solid state. In the following we will see that even in some systems with significant deviations from ideality, complete intersolubility can occur when the components have the same crystal structure. This has lead to the custom of calling lens-shaped phase diagrams in general isomorphous.

3.2 Real solutions

For a discussion of real solution behavior it is advantageous to rewrite the Gibbs function of Eq. 8 in the form :

$$g^s = g^{ids} + \text{correction term}$$

$$= g^m + \Delta g^{xs} - T\Delta s_m \tag{13}$$

where the excess free energy of dissolution :

$$\Delta g^{xs} = X_A(\bar{h}_A - h_A) + X_B(\bar{h}_B - h_B) - T[X_A(\bar{s}_A - s_A) + X_B(\bar{s}_B - s_B)]$$

$$= \Delta h^{xs} - T\Delta s^{xs} \tag{14}$$

characterizes the difference between the actual Gibbs function and the value of g if the solution were ideal.

There are several theoretical approaches to calculate Δg^{xs} under idealized assumptions. In the widely used *regular solution* concept one assumes that $\Delta s^{xs} = 0$ and, hence, $\Delta g^{xs} = \Delta h^{xs}$. The heat of dissolution Δh^{xs} is experimentally as well as theoretically

138

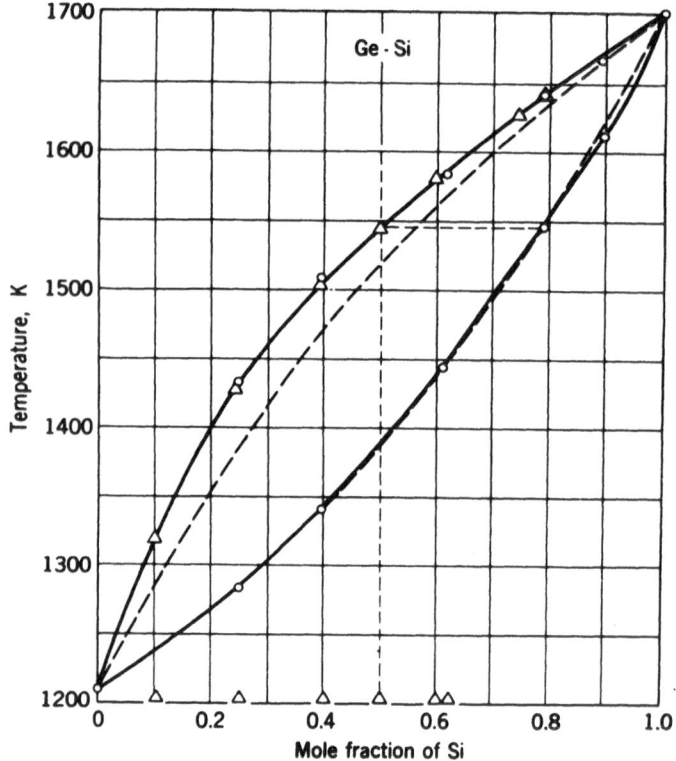

Fig. 6 Phase diagram for germanium-silicon. Full lines : experimental data; dashed lines : calculated. After [2] .

more readily accessible than Δg^{xs} (see also Section 7). Besides this utilitarian justification of setting $\Delta s^{xs} = 0$, this regular solution concept in which only the configurational entropy change is accounted for holds rather well for numerous real solutions, in particular for liquid solutions.

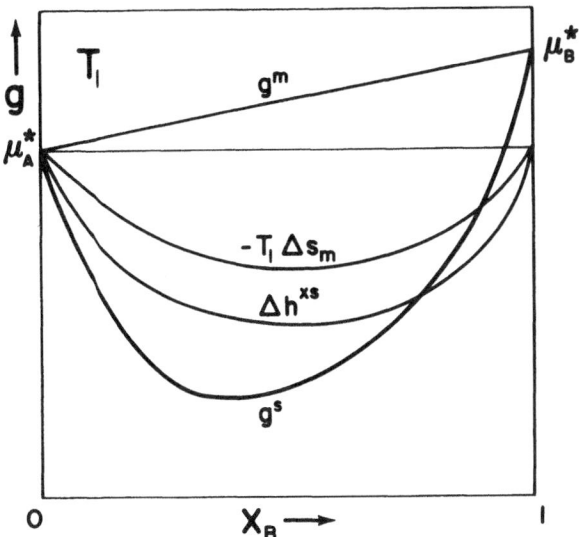

<u>Fig. 7</u> Free-energy vs. composition curves for negative Δh^{xs}.

There are many different ways the free energy of a real solution may deviate from ideality. Yet there is only a small number of *distinctly* different types of binary phase equilibria. In the following we will base the classification of phase diagrams on the sign and magnitude of Δh^{xs} (and Δg^{xs}).

3.2.1 Regular solutions with $\Delta h^{xs} < 0$

The theory of regular solutions shows that $\Delta h^{xs}(X)$ is a symmetric parabolic function. Consequently, the Gibbs function curves, at constant temperature, have the same convex downward shape as those for ideal solutions, as schematically indicated in Fig. 7

140

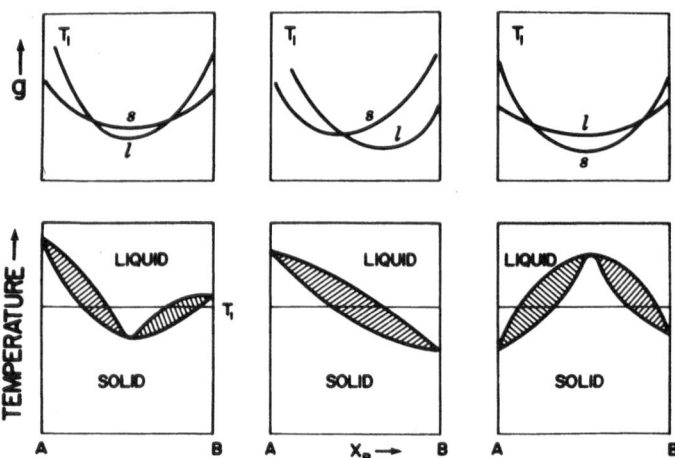

<u>Fig . 8</u>   Schematic phase diagrams for negative Δg^{xs} : (a) $\Delta g^{xs}(1)$
more negative than $\Delta g^{xs}(s)$; (b) Comparable excess Gibbs function
for solid and liquid solutions ; (c) $\Delta g^{xs}(1)$ less negative than
$\Delta g^{xs}(s)$.

Now, going beyond regular solutions, for all systems in which
both liquid and solid phases have a moderate negative Δh^{xs}, we can
thus expect Gibbs function curves that are convex downward at all
temperatures. However, depending on whether :

$$|\Delta g^{xs}(s)| \gtrless |\Delta g^{xs}(1)| ,$$ (15)

one can obtain three different types of phase diagrams as depicted
in Fig. 8. Solidus and liquidus do not necessarily simply descend
anymore from T_m high to T_m low as in ideal systems. Extrema can
appear at intermediate compositions. At these extremum points where
solidus and liquidus become identical solid and liquid solutions of
the same composition can coexist in equilibrium. Such congruent
melting or freezing situations are very desirable in materials preparation since the process is not plagued by the partial rejection
of a component at the interface. The interfacial accumulation of a
component can act morphologically destabilizing [6] , thus limiting
the maximum feasible solidification rate that results in a

macroscopically smooth interface. Note also that even "very real"
solutions can have a lens-shaped phase diagram if the excess Gibbs
function of dissolution is comparable for the solid and liquid phase
A lens shape is thus no criterion for the ideality of a solution.

An important point to be made here is that real systems with
negative excess heat of dissolution Δh^{xs} have no miscibility gap,
i.e., they form solutions over the whole composition range. Beyond
that, the attractive interaction between components A and B (a negative Δh^{xs} implies that heat is given off during solution formation)
leads often to intermediate phase formation and disorder-order transitions on freezing and further cooling, respectively. A classical
example for ordering is the Cu-Au system that forms ordered phases
about the stoichiometric ratios CuAu and Cu_3Au [7] .

Ordered solutions are only likely with components that have
very similar structures, i.e., that are isomorphous. Components
that differ strongly in properties such as atomic size, electronic
and crystal structure are unlikely to form ordered solutions. However, because of these differences, stable intermediate phase formation may be favored. There are numerous conditions that result in
intermediate phases. However, somewhat simplifying one can expect
that the more dissimilar the electronic structure of the components,
the less metallic and the more ionic or covalent will be the bonds.
Because of the directed nature of the bonds and the high order in
intermediate phases, their entropy (vibrational and configurational)
as well as enthalpy and, consequently, the free energy are typically
low. Hence, these *compound-like intermediate phases* show high stability.

Figure 9 depicts schematically a set of g(x)-curves for the
stable solid phases in such a compound-forming binary system. Due
to the different structures (crystal structure and/or lattice parameter) of the components, the g(α) and g(β) intersect sharply in the
central portion rather than joining smoothly as the g(X) for isomorphous components do (Figs. 4 and 8). From bond arguments one may
expect that stable, compound-like intermediate phases form with sharp
g(γ) minima at small stoichiometric ratios. However, as is typical
for solid solutions, the free energy does not vary as rapidly with
composition as one would anticipate from, for instance, gaseous reactions. Hence, g(γ) has a comparably broad minimum, though much
narrower than g(α) and g(β). Also, intermediate phase formation has
a drastic effect on the intersolubility of the two components, as
indicated in Fig. 9. Without γ-formation the miscibility gap would
extend from X_a to X_b. The intermediate phase formation shifts these
limits to X_a' and X_b'. Hence, a broader miscibility gap arises than
is typical for systems with $g^{xs} > 0$ (see below). Note also the relatively narrow stability range for the compound-like phase itself.
Some of the technologically most important compound semi-conductors,
such as the III-V systems, fall into this category.

142

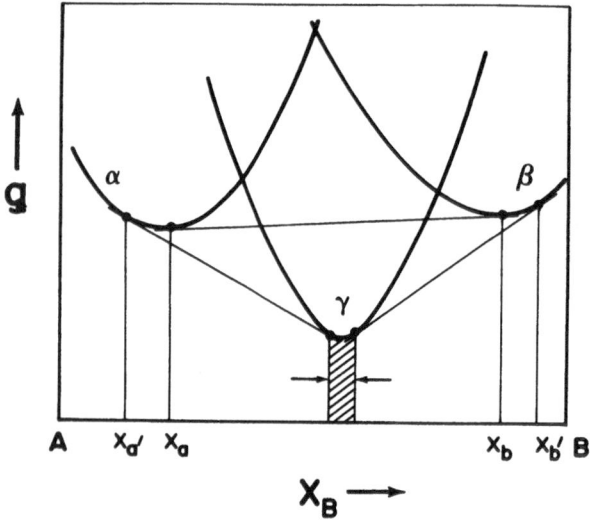

Fig. 9 Typical free-energy curves for A- or B-rich phases and an intermediate phase in binary system in which the components possess a large electrochemical dissimilarity.

Intermediate phases may also form when the interatomic forces between like-atoms are equal or even stronger than the A-B interactions. These intermediate phases tend to be random solid solutions that are energetically favored by the entropy of mixing term. Typically, an exact stoichiometric ratio is of minor significance in these *solid solution type intermediate phases* and, hence, they have a minimum in g(α) that is broader than in Fig. 9. Consequently the γ stability range is also wider and the intersolubility restrictions less drastic.

The above discussion is based on $\Delta h^{xs} < 0$. It must be pointed out, however, that intermediate phase formation has also been

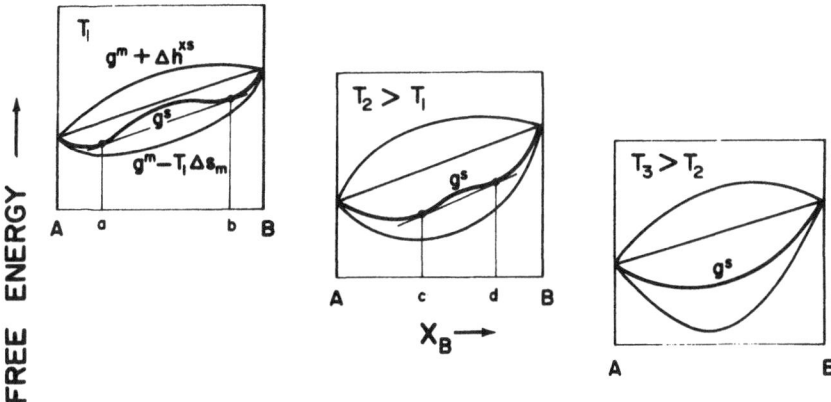

<u>Fig. 10</u> Free-energy composition curves for positive Δh^{xs} at several temperatures.

observed for some systems with $\Delta h^{xs} > 0$.

3.2.2 Regular solutions with $\Delta h^{xs} > 0$

   With solutions possessing a positive Δh^{xs} the situation is more complex. As illustrated in Fig. 10 for one phase, the enthalpy terms now compete with the entropy of mixing term. At high temperatures, where the $T\Delta s_m$ term is large, g^s will still be convex downward. Upon lowering of T, however, $T\Delta s_m$ will eventually become comparable with Δg^{xs}, and $g^s(X)$ will eventually become inflected (see points c and d in Fig. 10). It can be shown that ordinarily the slope of g^s must be negative at the pure component ends of the composition coordinate. Hence, a g^s-curve remains inflected upon further lowering of T but the inflected part broadens. When an inflection is present, then, using again the common slope argument, systems with overall composition X between points c and d will exist in the form of a mixture of two solutions of composition X_c and X_d, respectively.

   By scanning in this manner through the entire temperature range in which the phase exists, one can construct the *solvus* line in phase diagrams, as schematically shown in Fig. 11. The solvus separates the equilibrium existence range of one solution from the temperature-composition combinations for which two coexisting solutions of different composition are more stable. Hence, the area between solvus and composition coordinate can be thought of as an instability range for single solutions. This miscibility gap is similar to the areas between solidus and liquidus encountered earlier. Note, however,

144

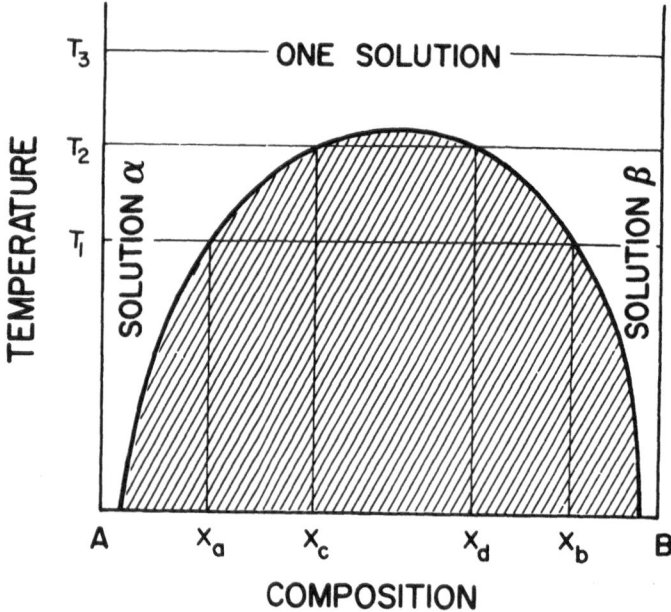

<u>Fig. 11</u> Miscibility gap for a phase corresponding to the g^s-curves in Fig. 10.

that there the two different solutions in equilibrium belong to different phases, whereas here we have two different solutions of the same phase. The relative amounts of the solutions can again be obtained with the lever rule (12) from the corresponding sections of the tie-line.

Now proceeding to equilibria between two phases, the relative magnitudes of $\Delta g^{xs}(1)$ and $\Delta g^{xs}(s)$ determine again the nature of a phase diagram for systems with positive Δg^{xs}. Fig. 12 depicts schematically the three distinctly different types one can obtain. Each type shows a miscibility gap at low temperatures.

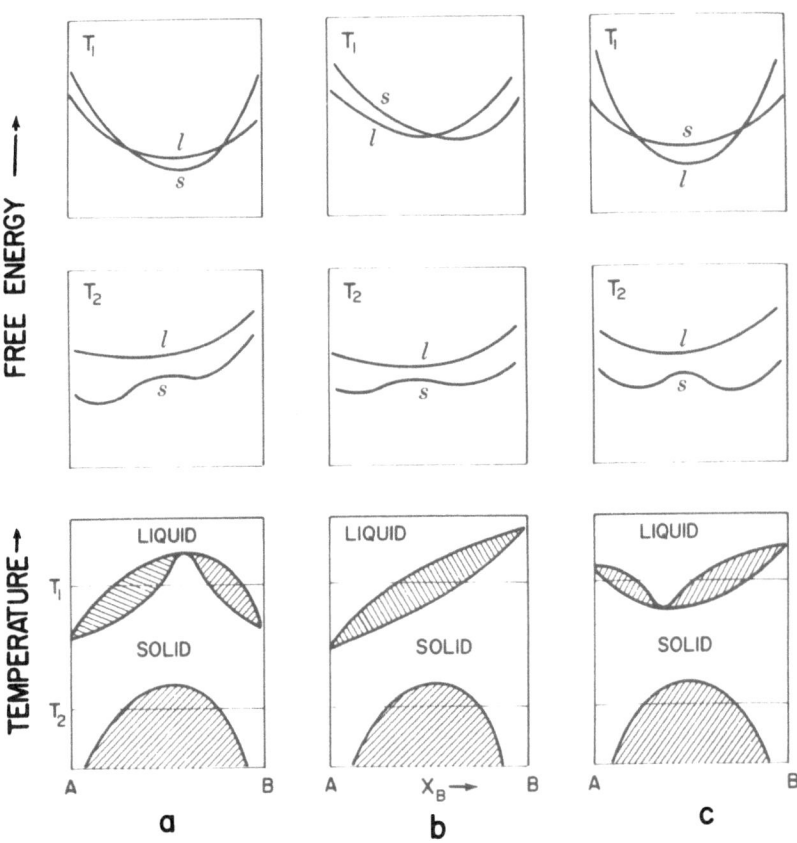

<u>Fig. 12</u> Schematic phase diagrams for positive Δg^{xs}: (a) liquid more
positive than solid; (b) comparable; (c) liquid less positive than
solid.

    Let us outline the compositional changes that occur upon cooling
in such systems with miscibility gap in the solid phase. We pick
for the discussion Fig. 12c which is shown on a larger scale in
Fig. 13. When a liquid solution of composition X_0 is cooled down
at T_1 ,it becomes saturated with respect to a solid solution of a
composition given by the intersection of the tie-line with the solidus. Upon quasistatic lowering of T, the whole system will be
solidified at T_2 where it regains its homogeneous composition X_0.
Now, if we cool the system further down at T_3, at the solvus line,
the solid solution X_0 becomes saturated with respect to a second

146

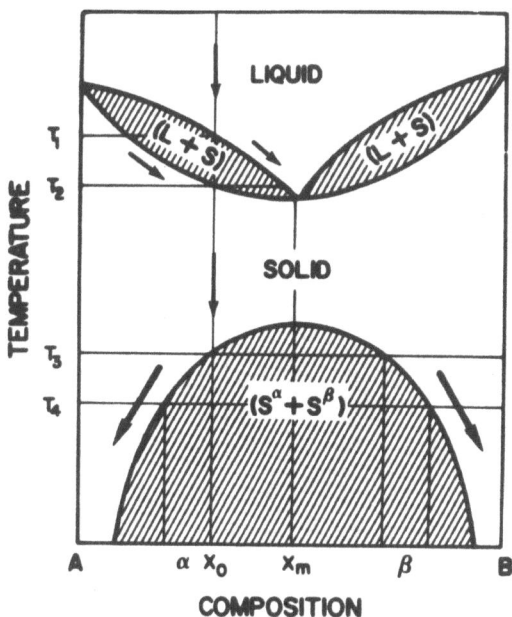

Fig. 13 Composition changes on cooling in a binary system with solid miscibility gap.

solid solution of composition X_β. The composition is given again by the tie-line intersection with the solvus. As we cool down even further to T_4, slowly enough to allow for maintenance of equilibrium conditions, the mother phase α will continuously decrease its B content, thereby remaining saturated with respect to a β phase of continuously increasing B content. Their relative amounts are again proportional to the inverse distances on the tie-line as divided by the unchanged overall composition X_0 (lever rule). It should be emphasized here again that this discussion is restricted to equilibrium conditions. With practical solidification rates, however, solution separation of this type is often not found due to the slow diffusion rates in solids.

3.2.3 Other solution models

In recent years there has been considerable progress in the quantitative calculations of (binary) phase diagrams, in particular for III-V compound systems. Most of the models used are variations of the *simple solution* model [8,9] in which some simple temperature dependence is assumed for Δg^{xs}. The earlier efforts in this

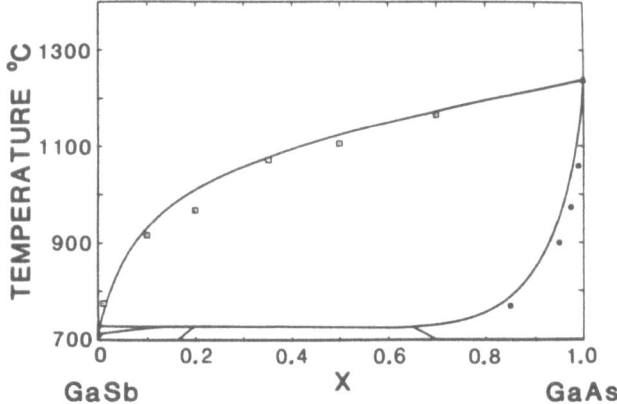

TEMPERATURE °C

1300

1100

900

700

0   0.2   0.4   0.6   0.8   1.0

GaSb   X   GaAs

Fig. 14   Pseudo-binary phase diagram for GaSb-GaAs. Solid lines :
calculated; symbols : experimental data. After [10] .

direction depend on the existence of some experimental phase diagram
data from which some fitting parameters can be determined. More
recently, however, these liquid phase and solid phase interaction
parameters have been estimated without adjustable parameters from
electronegativities, energies of vaporization and entropies of fusion of the components and certain bonding models which are based
on band gap data. Thus, very successful predictions of liquidus and
solidus data have been made. For a review of this area see [9-11].
A particular impressive example is given in Fig. 14. Note the
large compositional gap between the liquidus and solidus for this
GaSb-GaAs pseudo-binary, indicating a large Δh^{xs} in the solid.

3.3 Invariant reactions

In most binary systems encountered so far the composition changes continuously during a liquid-solid transition. Consequently,
the temperature had to be lowered continuously to complete solidification. Now, from the phase rule, $f = C - P + 2$, one would expect
that even this last degree of freedom (here T, after P is fixed) is
lost when :
1) the effective number of components is reduced by one. For instance, when solid and liquid have the same composition and hence
become thermodynamically indistinguishable. Then $C = 1$ even in a

148

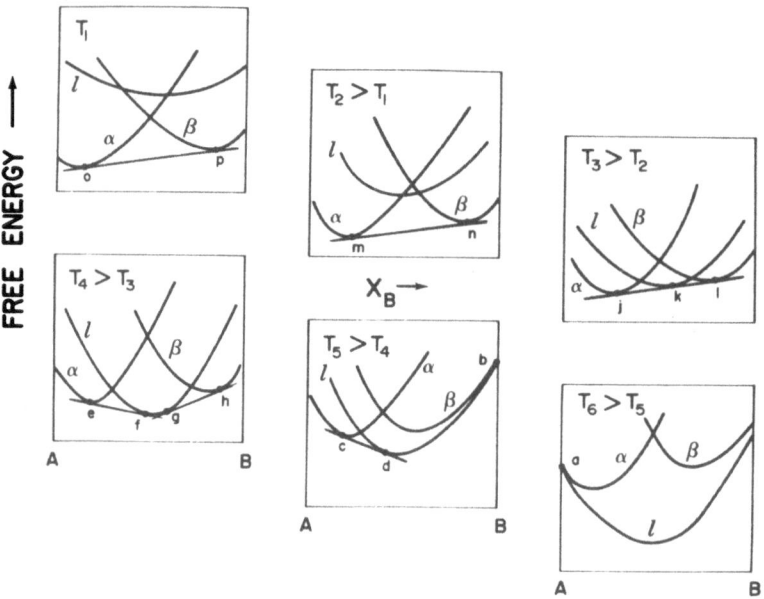

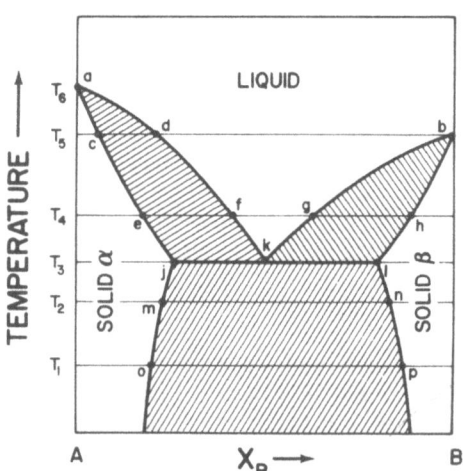

Fig. 15 Free-energy composition curves (a) and the resulting temperature-composition equilibrium diagram (b) for a eutectic system.

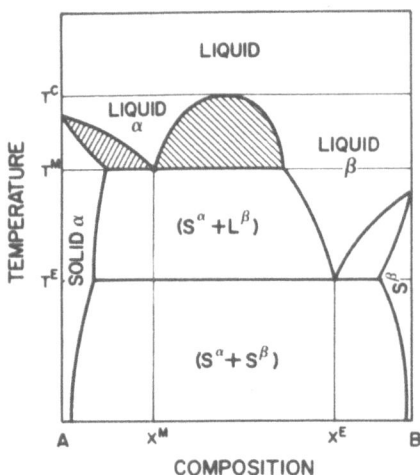

Fig. 16   Phase diagram incorporating a monotectic reaction.

binary system with two chemically distinguishable species in two
phases.   Such congruently melting systems, were encountered in
Figs. 8 and 12.
2) the number of phases of different composition is increased by one.
This can occur when two phases react to form a new, third one or
when one mother phase decomposes into two other new phases.

     In an invariant equilibrium situation (i.e., for f = 0) the
relative amount of the phases is only determined by the heat flow
to or from the system since the temperature, here invariant, has
lost the regulatory function that it had concerning composition.

     The most commonly occuring invariant reaction in binary solu-
tion systems, the *eutectic reaction*, consists in the separation of
two different solid solutions from a liquid phase or, more generally,
in :

$$L^{(1)} \xrightarrow[\text{heating}]{\text{cooling}} S^{(1)} + S^{(2)} .$$

To see when eutectic reactions can occur, let us look at the g(X)
scheme of a system that contains only two final solid solution pha-
ses and one liquid phase.   Positive Δg^{xs}'s are assumed.   If $\Delta g^{xs(s)}$
is sufficiently larger than $\Delta g^{xs(1)}$ so that at a certain temperature there is a common tangent to $g^{(1)}$ *and both* g^{α} and g^{β}, a eutectic reaction arises (T$_3$ in Fig. 15).   The eutectic composition

determines the lowest possible solidification and melting temperature for the system.

As a liquid solution of eutectic composition X_k is cooled down,
at the eutectic temperature T_3 the solid solutions α and β form.
Further extraction of heat does not lower T until the liquid is
quantitatively transformed to α and β. A detailed discussion of
other possible compositional changes in such systems can be found,
e.g., in [1]. Eutectic transitions are at the heart of numerous
metallurgical processes.

Systems with a miscibility gap in the liquid phase show a *monotectic* invariant transition characterized by:

$$L^{(1)} \xrightleftharpoons[\text{heating}]{\text{cooling}} L^{(2)} + S^{(1)}.$$

Whereas eutectics appear frequently without other invariant proces
ses in the same system, this is rarely the case with monotectics.
Typically they form combinations with other transitions, as shown
in Fig. 16. Usually there is enough difference in density between
$L^{(\alpha)}$ and $L^{(\beta)}$ that, if the liquids are kept above T^M and below T^C,
they form separate layers in a container. At $T > T^C$, however, the
compositional difference between the two liquids disappears and
they begin to form one liquid solution.

As one cools a liquid alloy of the monotectic composition X^M,
at T^M $L^{(\alpha)}$ begins to decompose into $L^{(\beta)}$ and $S^{(\alpha)}$. This invariant
reaction continues upon extraction of heat until $L^{(\alpha)}$ has disappeared.
Further lowering of T will shift the composition of α towards A and
of L towards B. When the eutectic temperature T^E is reached the
composition of L is X^E.

The third type of the three-phase invariant reactions is the
peritectic reaction:

$$L^{(1)} + S^{(1)} \xrightleftharpoons[\text{heating}]{\text{cooling}} S^{(2)}.$$

The g(X) curves that lead to the peritectic reaction are rather similar to those of eutectic systems. In eutectic systems (Fig. 15),
before reaching the common tangent situation for all three phases
(T_3) upon lowering of T, a two-tangent situation (T_4) is encountered.
In peritectic systems this sequence is reversed. This is due to the
relative position of the g(X) minima along the composition axis.
In a eutectic the "liquid minimum" lies between those of the two
solid phases. In the peritectic case, however, it lies outside the
"solid minima". This situation is favored by large differences in
the melting points of the two components. (Example : FeO and MnO
with T_m = 1365° and 1785° C, respectively, T_p : 1430°C.)

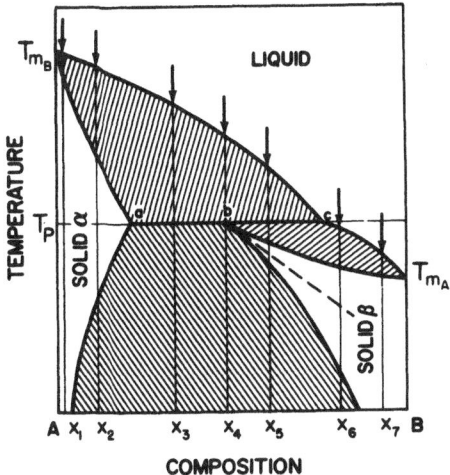

Fig. 17 Peritectic phase diagram with alloys of various overall compositions.

Referring to Fig. 17, we will now discuss the result of equilibrium cooling of a liquid solution of peritectic composition X_4. When the cooling path crosses the upper liquidus, solid α forms until the composition of the remaining liquid corresponds to c. Then, at T_p, L and α react completely to β. The slightest further withdrawal of heat causes some α to separate again from β according to the solvus line. For a discussion of the compositional changes on cooling for the solutions of non-peritectic composition (X_1-X_3, X_5-X_7) see [1].

In concluding this section, a significant difference in mass transfer between eutectic and peritectic reactions should be pointed out. In a eutectic solidification the compositional readjustment in the solid adjacent to the interface with the liquid occurs mostly by diffusion through the liquid phase. In a peritectic solidification however, diffusion in the solid phase is the rate limiting parameter. The peritectic reaction, i.e., $L_p + \alpha_p \rightarrow \beta_p$, takes place at the interfaces between the α-phase particles and the surrounding melt. After formation of the first layer of β_p on the α_p particles, the reaction rate decreases drastically. So, under any practical circumstances α_p will more or less stay as is while L will follow the lower liquidus (Fig. 17) and continue to form layers of β that are increasingly more rich in B than β_p. The product of the reaction is thus a mixture of α_p and "cored" β.

4. TERNARY AND MANY COMPONENT LIQUID-SOLID SYSTEMS

An increase in the number of components in a system obviously increases the complexity of phase data presentation (and experimental determination !). Even for the simplest multi-component system, a ternary phase, we see from the phase rule (f=3 - 1 + 2 = 4) that four intensive state variables (p, T and two concentrations) must be specified in order to uniquely determine the state of that phase. Hence, if we fix the pressure, i.e. for f = 3, three dimensions are required to represent an isobaric, ternary phase diagram. There is not enough room for a detailed discussion of phase equilibria involving more than two components. We will only illustrate some conceptual difficulties with the simplest ternary system, an isomorphic system, and otherwise give references for more, in-depth reading.

In three-dimensional representations of ternary equilibria the direction normal to the composition plane is typically used as temperature axis. Such an isobaric presentation for a liquid-solid system without miscibility gaps is given in Fig. 18. A liquidus and solidus surface are suspended between the components' melting points. Intersection of these surfaces with isothermal planes (e.g. T_1 in Fig. 18) results in solidus and liquidus isotherms. Conventional two-dimensional ternary phase diagrams consist in projections of these isotherms into the composition plane (triangle). The fundamental difference to isobaric binary systems is that solidus and liquidus composition are not uniquely specified by the temperature (in contrast to, e.g., the binary tie-line that results from the intersection of the T_1 plane with the A-C binary in Fig.18). Furthermore, the orientation of ternary tie-lines is *a priori* not defined either. Without specific experimental data one can only predict, as schematically indicated in Fig. 19, that the direction of the tie-lines will gradually change from the direction of one bounding, binary tie-line to that of the other. Only if during the transition from a liquid solution L_1 to the solid solution α_1 the ratio of component C and B does not change, then will the tie-line aim at the corner representing pure A.

Simple ternary eutecties are discussed in [1] . Detailed analyses of ternary eutectic, peritectic and other invariant transformations with emphasis on theory can be found in [4]] and with extensive reference to experimental data in [13] . Calculations of ternary and quaternary III-V phase diagrams have been reviewed in [9].

Efficient discussions of quaternary and many-component phase diagrams require considerable insight into topological schemes, combinational methods, multidimensional vector spaces and matrix algebra. For an in-depth discussion of phase equilibria in multicomponent systems see [14] .

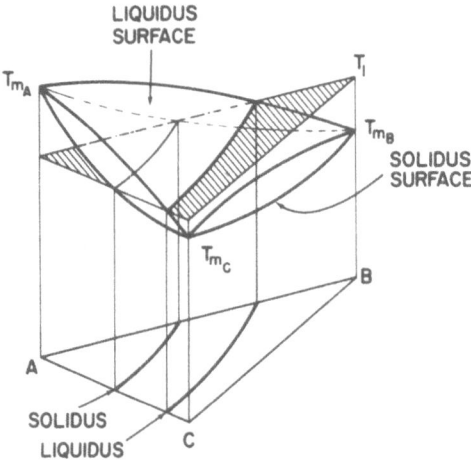

Fig. 18   Three-dimensional phase diagram of a hypothetical isomorphous ternary system A-B-C with $T_{mA} > T_1 > T_{mB} > T_{mC}$.

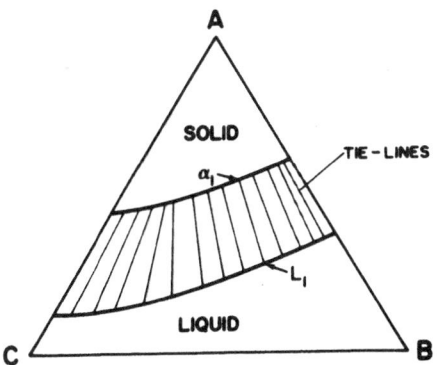

Fig. 19   Isothermal section at T_1 of the spatial phase diagram of
Fig. 18.

5. VAPOR-LIQUID-SOLID PHASE DIAGRAMS

In the interest of simplicity, we have focused the foregoing presentation upon solid-liquid equilibria. Yet, for modern materials preparation, vapor-condensed phase equilibria are equally important. Vapors may constitute the nutrient (or part of it) in a crystal growth system. Or, less desirably, evaporation from condensed phases may result in (intolerable) materials losses or compositional changes. In both cases one must understand the coexistence conditions over a considerable pressure range in order to choose advantageous conditions for the process. Hence, in materials preparation one is frequently faced with complex (isothermal, isobaric) sections through or projections of the V-L-S stability ranges of a T-p-X phase diagram. It cannot be the goal of this introductory lecture to cover V-L-S equilibria in any depth. Therefore we will here only whet the reader's appetite with a schematic example for a three-dimensional (T-p-X) phase space figure and a corresponding real system in which a solid can be obtained from a liquid, either by the traditional lowering of the temperature or by a temperature increase.

5.1 Binary systems

Figure 20 represents the T-p-X phase diagram for a binary system in which a (slightly) dissociating compound AB forms which may be congruently molten or sublimed, depending on the specific T-p combination. One sees that the AB-center plane (existence range) is braced on both sides by eutectica of liquid phases and (at lower p) gaseous phases. Details of this figure, which includes two quadruple states and four three-phase equilibria blades are difficult to grasp (and present) without a systematic introduction into this topic; see e.g [1] . Here we want only to emphasize the usefulness of considering the whole phase space diagram with a striking example (Fig. 21) of a solidification process that cannot be understood without an account for the vapor participation.

Figure 21 depicts part of the T-p projection of the Cd-Te phase space. The sickle-shaped CdTe V-L-S equilibrium curve forms together with the joining vapor pressure curve for pure cadmium a reentrant angle in which Cd-rich liquid solutions of Cd-Te are stable. As a consequence of this particular coexistence line shape (which can be deduced from Fig. 20) there are for each fixed Cd-pressure (between about 2.5 and 6.5 atmospheres) two possibilities of obtaining the compound CdTe. Either by lowering of T and crossing the phase line from a nearly stoichiometric liquid solution (Path A→C). Or by raising of the temperature (thus vaporizing the excess Cd from the "melt") from the Cd-rich side (Path F→D).

These fundamentally different paths lead to somewhat different stoichiometries in the resulting CdTe. Hence, the specific choice

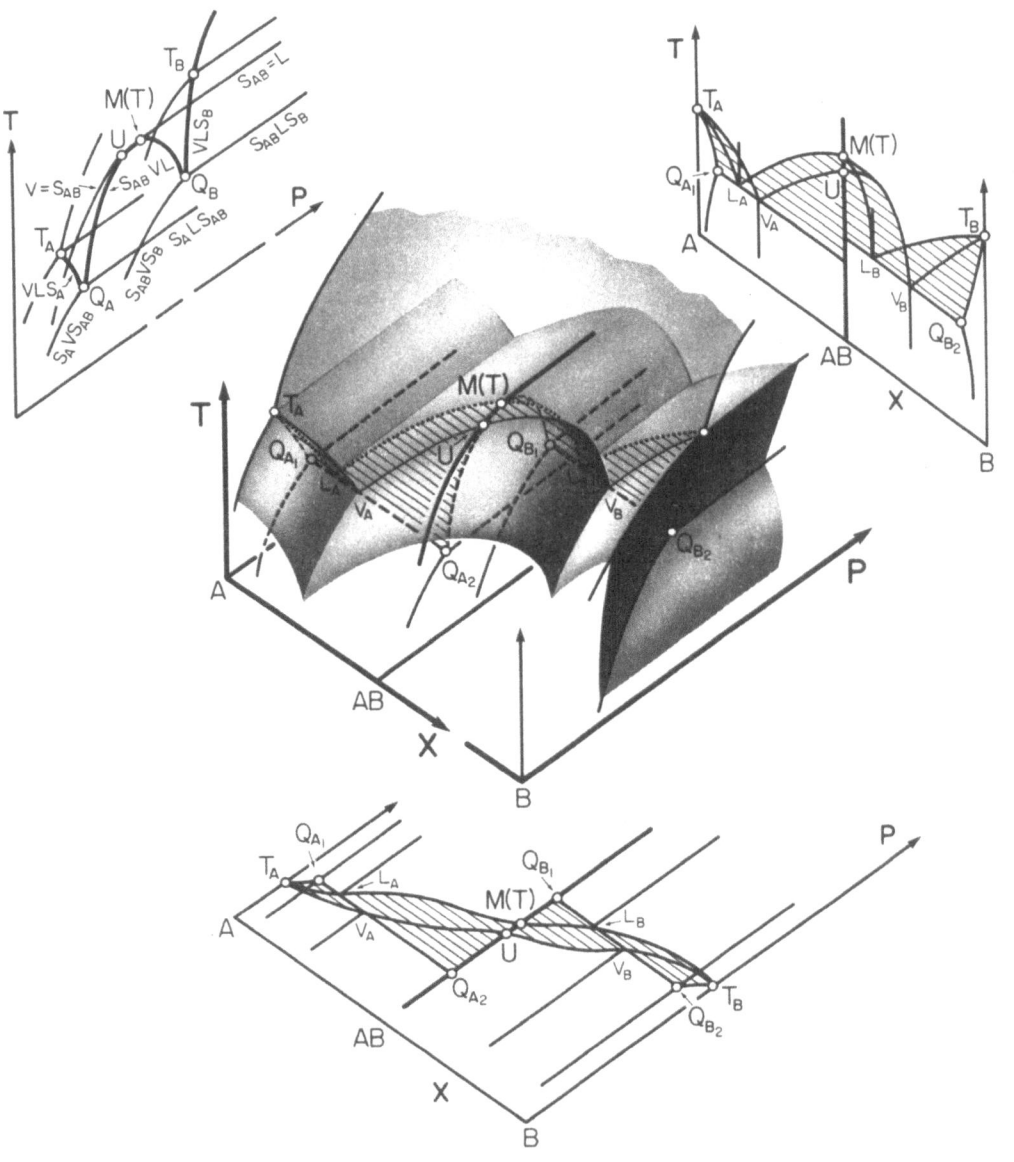

Fig. 20 T-p-X phase diagram for a binary system in which a slightly dissociating compound forms which may be congruently molten or sublimed. Three-dimensional phase figure and corresponding T-p, T-X and p-X projections.

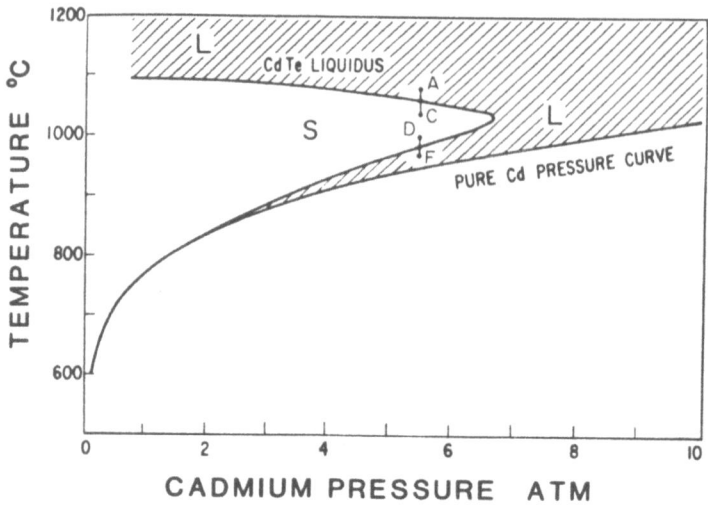

Fig. 21 T-p projection of the "Cd-rich half" of the Cd-Te system. After [41] .

will have to be based on the specific device application for the material.

5.2 Possible vapor transport in laser melting

In the few years that laser annealing has been investigated, mass transfer in this process has almost exclusively been discussed in terms of liquid-solid interaction. However, as the following simple model consideration shows, species exchange with the ambient vapor space can play a significant role even in laser melting processes of rather short duration. Solutions to the time-dependent diffusion equation reveal that after an abrupt change in the concentration boundary conditions, steady state diffusion conditions are reestablished over a distance d within the characteristic diffusion time :

$$\tau \approx \frac{d^2}{D_{AB}} \tag{16}$$

where D_{AB} is the binary diffusivity of the system under consideration. Thus, the distance over which significant diffusion occurs within a time span τ is:

$$d \approx (\tau D_{AB})^{1/2}. \tag{17}$$

Taking typical diffusion coefficients for vapors, liquids and solids, respectively as 10^{-1}, 10^{-4} and 10^{-8} cm^2/sec one sees that for:

$$\tau = 10^{-3} \text{ sec}, \quad \left\{ \begin{array}{l} d_{vapor} \approx 10^{-2} cm \\[2ex] d_{melt} \approx 3 \times 10^{-4} cm = 30,000 \text{ Å} \\[2ex] d_{solid} \approx 300 \text{ Å}, \end{array} \right.$$

whereas for:

$$\tau = 10^{-9} \text{ sec}, \quad \left\{ \begin{array}{l} d_{vapor} \approx 1,000 \text{ Å} \\[2ex] d_{melt} \approx 30 \text{ Å} \\[2ex] d_{solid} \to 0. \end{array} \right.$$

Compare these characteristic (vapor) diffusion distances with a typical molten layer depth of, say, 5,000 Å. Then it becomes evident that significant diffusive mass transfer between vapor and melt can occur during laser melting of not too short duration, even taking into account the typically three orders of magnitude lower mass density of vapors.

6. EXPERIMENTAL DETERMINATIONS OF PHASE DIAGRAMS

One of the big impediments in materials research is the lack of phase diagrams for many systems of technological interest. Experimental phase diagram studies are widely considered to be sufficiently tedious so that preference is given to "extrapolations from similar materials". On many occasions, this attitude has led to unwarranted research efforts that by far exceed the involvement and funds required for definitive (though, admittedly, less glamours) experiments on phase equilibria. For an introduction to experimental techniques in this area the reader is referred to [4,20,21].

158

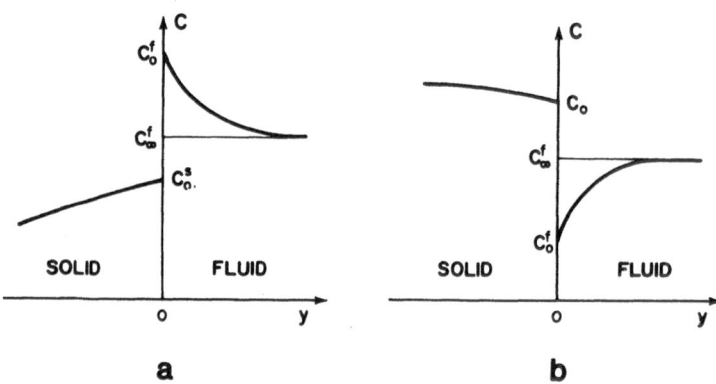

a   b

<u>Fig. 22</u> Typical dynamic concentration profiles near solid-fluid
interfaces at finite growth rates. The considered component is
(a) partially rejected ($k_0 < 1$), (b) preferentially incorporated
into the solid ($k_0 > 1$).

7. SEGREGATION COEFFICIENTS

In practice, crystallization is rarely conducted so slowly
that a system can readjust its composition homogeneously throughout the *bulk* of its phases according to the phase diagram. Hence,
at progressing interfaces one will typically encounter dynamic concentration gradients, as indicated in Fig. 22. At the same time
(local) equilibrium may still persist between fluid and solid directly on the interface.

In order to quantitatively describe the changes in composition
occuring in phase transitions one defines segregation coefficients.
Three different definitions are advantageously used depending on
the particular situation. If, for instance, there is no clear evidence that local equilibrium exists between solid and fluid at the
interface, one will cautiously base a discussion of the interfacial
transfer of a component on the *interfacial segregation coefficient*
k_0, which is defined as:

$$k_o = \frac{X_0^s}{X_0^f} = \frac{C_0^s}{C_0^f} \times \frac{C_0^f + [\text{host concentr. in fluid}]}{C_0^s + [\text{host concentr. in solid}]} \qquad (18)$$

where the (molar) concentrations in the solid and fluid phases (super-
scripts s and f, respectively) are taken directly at the interface
(subscript zero) as indicated in Fig. 22.

At low solute concentrations (where $C_0 \ll$ [host concentr.])
and comparable densities of fluid and solid,(18) can be approximated
by :

$$k_o \approx \frac{c_0^s}{c_0^f} \,. \tag{19}$$

Most segregation work has been concerned with impurity-in-melt systems. Hence, it has become customary to use (19) as definition of
k_0, then often called distribution or partition coefficient. We
will follow this custom. However, in employing (19) to solution or,
in particular, to incongruent vapor growth considerable errors can
result in comparison with (18).

For a description of the efficiency of, say, a purification
process k_0 is not too informative. Whenever segregation between
bulk phases is of interest one will refer to the composition of
the bulk fluid rather than to the interfacial c_0^f and use the *effective segregation coefficient* k that is defined for the impurity-in-
melt case as :

$$k = \frac{c_0^s}{c_\infty^f} \tag{20}$$

where the subscript ∞ implies "far away from the interface". With
increasing growth rate V^s, the partial rejection ($k_0 < 1$) or prefe-
rential incorporation ($k_0 > 1$) of the solute causes an increasing
deviation of c_0^f from $c_\infty{}^f$. Eventually, at high enough V^s, k approaches unity and macroscopic segregation ceases, as quantitatively
described in Section 8. On the other hand, if the experimental conditions are such that the concentration gradient in the liquid becomes negligible ("ideal mixing case") and $k \approx k_0$, maximum efficiency
is obtained in a purification process.

If solid and fluid are in mass transfer equilibrium at the interface, then k_0 becomes equal to the *equilibrium segregation coefficient* k* which depends only on the thermodynamic properties of the
system rather than the mass transfer kinetics at the interface.
This case of :

$$k_0 = k^* \tag{21}$$

is frequently approached in practice. For solute concentrations
above 1% or so, equilibrium segregation coefficients can easily be

160

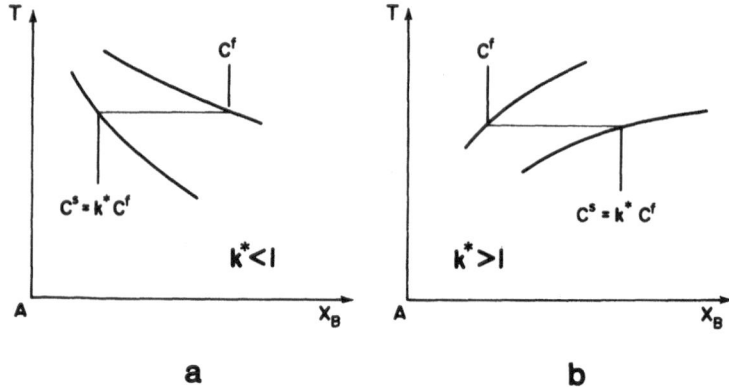

Fig. 23 Portions of phase diagrams in which the solidification temperature is (a) lowered, (b) raised by the solute, and corresponding segregation coefficients k* .

deduced from phase diagrams. Figure 23a shows that k*'s for components which lower the solidification temperature of a system are smaller than unity. In systems with upward sloping coexistence lines, as in Fig. 23b, k* > 1.

Equilibrium segregation coefficients depend in general on the temperature, the pressure and on the concentration of the component of interest, say B, as well as on the presence of other components. Therefore, it is not meaningful to call $k^*(T, P, X_B, X_i)$ a segregation "constant" as is customary. At low concentrations, segregation coefficients are often found to be relatively concentration independent. That "constant" range, however, depends on the individual system. For example, in germanium k_0-values for most impurities are found to be concentration independent in a range as wide as $10^{-2} > X_B > 10^{-10}$. However, as a counter example, $k_{Ca/NaCl}$ is 0.3 at 10^{-2} mole fractions in the melt and becomes 1.0 at 5×10^{-5}, i.e., no segregation occurs at this level. Hence, extrapolation of k*- values from higher concentrations (phase diagrams !) to impurity levels below 10^{-2} can be misleading.

7.1 Thermodynamic models

Equilibrium segregation coefficients k* can be predicted theoretically if one knows certain experimentally accessible parameters,

such as the heat and entropy of mixing in the liquid and solid solution. Even if some of these parameters are not known and need to
be estimated based on idealized models, the thermodynamic approach
can yield valuable information.

Assuming low solute concentrations, regular solution behavior
in melt and solid and heat capacities of solid and liquid to be
those of the pure host at its melting point T_2, one obtains [17,18]:

$$\ln(k^*) = \frac{1}{R} \left[\frac{\Delta h_{21}^d}{T} - \frac{\Delta h_{2s}^d}{T} + \frac{\Delta h_2(T_2)}{T} - \frac{\Delta h_2(T_2)}{T_2} \right]. \quad (22)$$

The first two terms on the rhs contain the heat of dissolution or
the energy required to transfer pure molten and solid component 2
into the host lattice and host melt, respectively. Depending on
whether $\Delta h_{21}^d \gtrless \Delta h_{2s}^d$, the segregation coefficient decreases or increases. In other words, the impurity will accumulate in the phase
in which it causes less free energy increase. The last two terms
contain the heat of fusion of the "pure impurity". Depending on
whether the melting point of $2 : T_2 \gtrless T$, the segregation coefficient
is shifted to lower or higher values as we have anticipated already from mere graphical arguments in Fig. 23. Note that all terms
contain T or T_2 in the denominator. From this, one can expect that,
for instance, crystallization from a solution around room temperature leads, in general, to stronger segregation than melt crystallization. Certainly the heat of dissolution term, particularly Δh_{21}^d,
will be different from the melt case but the 1/T dependence forms
in most systems an overriding effect, as is indeed observed.

Table 1 Values of segregation coefficients k^* and k_0 in germanium
and silicon. After [19].

Solute	Germanium		Silicon	
	Calculated	Experimental	Calculated	Experimental
Sn	2×10^{-2}	2×10^{-2}	8×10^{-3}	1.6×10^{-2}
As	1×10^{-1}	2×10^{-2}	1×10^{-3}	3×10^{-1}
Al	1×10^{-1}	7.3×10^{-2}	4×10^{-2}	2×10^{-3}
In	8×10^{-4}	1×10^{-3}	3×10^{-4}	4×10^{-4}

For covalent, elemental semiconductors segregation coefficients calculated from (22) agree well with experimental results [19], see
Table 1. The heats of dissolution were calculated based on simple
models and it was shown that the vibrational change of entropy
term for most impurities in Ge and Si is small as compared to the
overriding Δh_{21}^d term; hence, the regular solution approach is well
justified.

In ionic systems, the estimation of heats of dissolution is
considerably more complex than in covalent lattices. Satisfactory
results have yet been obtained only for monovalent impurities. For
segregation from alkali halide melts the activity coefficients were
calculated, assuming regular solutions, from Δh_{2s}^d data computed for
monovalent alkali and halogen impurities [20-22]. Very little,
however, is known about the impurity-host interaction in melts.
Some workers [20] drew qualitative conclusions for k^* by comparison
of experimental and theoretical results. Others [21] , calculating
activity coefficients based on a simple melt structure model [23],
obtained quite different results. Thus, the uncertainty in calculated segregation coefficients for monovalent impurities in solidification from alkali halide melts is typically ± 0.1. The above
approach has also been successfully extended to segregation from
aqueous alkali halide solutions [24] .

More recently considerable progress has been made in the prediction of k^* for electronically active impurities (donors) in III-V
compounds. The distribution coefficients are calculated with no
adjustable parameters, in an approach similar to the one mentioned
under "other solution models" in Section 3.2. For a review of these
efforts see [25].

7.2 Experimental observations and their interpretations

Under laboratory conditions, often some of the idealizing
assumptions employed in the above models, in particular the equilibrium conditions and absence of interaction with other solutes
(impurities), are not fulfilled. Hence, a large variety of segregation phenomena is observed, that cannot be analized with these
simple tools. In the following we will list only a few examples.
For a more detailed discussion and additional references see [1] .

Numerous elemental and compound semiconductors show a strong
anisotropy in segregation. For instance in the solidification of
germanium the ratio of the segregation coefficient for phosphorus
on faceted and non-faceted parts of the interface is 2.5 [26] .
Anisotropy in segregation is a non-equilibrium phenomenon. A dependence of equilibrium segregation on the crystallographic orientation of the interface would violate the thermodynamic state function
concept. Anisotropic k_0 values show often a pronounced growth rate
dependence.

Numerous mechanisms have been suggested to account for k_0 anisotropies. Most models involve adsorption of the solute on the interface and differences in the growth kinetics on faceted (atomically smooth) and non-faceted (atomically rough) parts of the interface [28-31] ; see also Prof. Müller-Krumbhaar's lecture. The adsorption models are also supported by the observation that the polarity of an interface can affect segregation. For instance for III-V compounds one has found sizeable differences in some k_0's between {111} surfaces bound by group III or group V atoms. A particularly original model invokes the Schottky-barrier formed by the melt-solid semiconductor contact (interface) [32,33]. A theoretical treatment of anisotropic k_0-phenomena in terms of interfacial electric fields has been presented recently [34] . Electrical phenomena in segregation have also been reported for the freezing of dilute aqueous solutions [35] .

In ionic systems one observes frequently a strong *concentration dependence* in the segregation of solutes that differ in their valence state from the (solid) solvent. This can be understood in terms of the vacancy concentration that is necessary for charge compensation of these solutes. If this extrinsically required vacancy concentration is comparable to the intrinsic Schottky defect concentration, incorporation of the solute is energetically favored. At high solute concentrations, however, requiring vacancy concentrations far in excess of the Schottky concentrations, segregation becomes pronounced. This concentration dependence has been analyzed in terms of the mass action law [36] . Furthermore, *solute-solute interactions* can significantly alter segregation coefficients. For instance, in the solidification of boron-doped but otherwise highly pure germanium melts one finds $k_0 = 7$. However, in the presence of rather low oxygen concentration this value approaches 1 [37] .

8. DYNAMICS OF SEGREGATION

Dynamic concentration profiles, as depicted in Fig. 22, and the resulting effective segregation coefficients have been calculated. For a rigorous treatment one needs to know the diffusive-convective fluxes in both phases adjacent to the interface. With the relative slowness of diffusion in solids, it is often well justified to consider the concentration distribution in the solid phase to be "as received" on segregation, i.e., to be frozen-in after the interfacial growth kinetics has lead to a specific value. For the melt, however, one is then still left with the complex problem of diffusion within the convecting fluid. In general this requires solutions to the coupled, time-dependent conservation equations for momentum, mass and components. Only with the recent availability of high speed computers has the (numerical) evaluation of such highly non-linear boundary value problems become feasible. Thus, strongly simplifying models have been resorted to in the past.

164

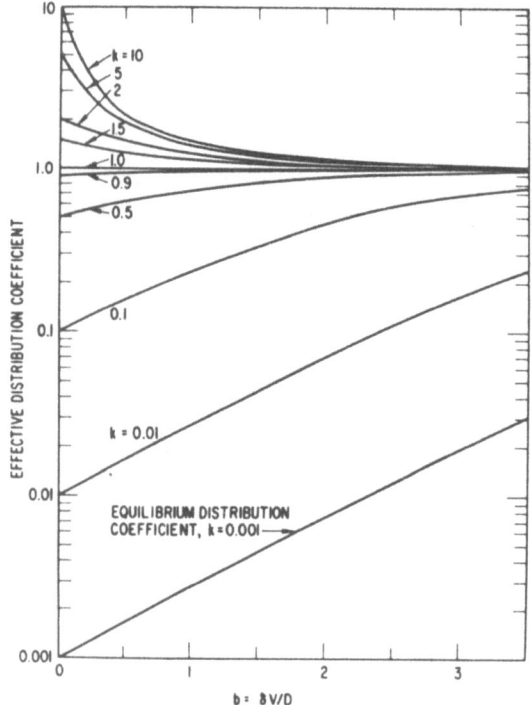

Fig. 24 Effect of normalized growth rate $b = \delta V^s/D$ on effective segregation coefficient.

The most widely used dynamic segregation model is that of Burton, Prim and Slichter (BPS) [38] . Steady state is assumed, i.e., c_o^f = constant (Fig. 22). The model is one-dimensional, isothermal and assumes constant density throughout the melt. Also it is assumed that diffusive-convective mixing keeps the solute concentration in the melt uniform to within a distance δ from the interface. Furthermore, considering the interface as stationary, most restrictively it is assumed that the convection velocity u within this layer of width δ results solely from the advective contribution of the crystallization flow into the interface, i.e., u = V^s. Under these simplifying conditions, the well-known BPS-relation :

$$k = \frac{k_0}{k_0 + (1 - k_0) \exp (-\delta V^s/D_{AB})} \tag{23}$$

is obtained. Plots of (23) for various k_0-values are given in Fig. 24, correctly showing the trend of k → 1 with increasing V^s that we have qualitatively anticipated.

The parameter δ in (23) is often interpreted, based on a classical (yet erroneous) paper by Nernst, as the distance from the interface over which mass transfer occurs by "diffusion only". This, however, is only justified as long as the characteristic diffusion distance $y \equiv D_{AB}/V^s$ is small as compared to δ. If $y \geqslant \delta$, interpretation of δ as "diffusion boundary layer width" becomes unrealistic. For details see [1] and for a recent, fluid dynamically somewhat more realistic model treatment see [39,40]. Note that in laser melting, with a representative $V^s \approx 10^2$ cm/sec and $D_{melt} \approx 10^{-4} cm^2/$ sec, $y \approx 100$ Å.

8.1 Observations in laser melting

There are several lectures and papers at this conference on segregation during the rapid solidification following laser melting; see in particular the lecture of Prof. Rimini. Hence, we will here only briefly refer to a few results that can readily be discussed in terms of the foregoing material.

Two distinctly different segregation behaviors have been observed :
a) Solutes with equilibrium segregation coefficients $0.3 \leqslant k^* \leqslant 1$ (e.g., B, P and As in silicon) show practically no segregation. For instance, as schematically depicted in Fig. 25a, a concentration distribution obtained from ion-implantation is simply broadened. Thus it appears that the effective segregation coefficient k has reached unity, and redistribution occurs only via diffusion in the melt. Calculated diffusion profiles agree astonishingly well with observed ones.
b) Solutes with $k^* \ll 1$ (e.g. Cu, Fe and Sb in silicon) show very strong segregation behavior, as illustrated in Fig. 25 b. Hence, for these solutes, the k-values apparently do not reach unity inspite of the V^s values around 10^2 cm/sec !! With these solutes, concentration values have been obtained that exceed equilibrium solubilities by up to one order of magnitude.

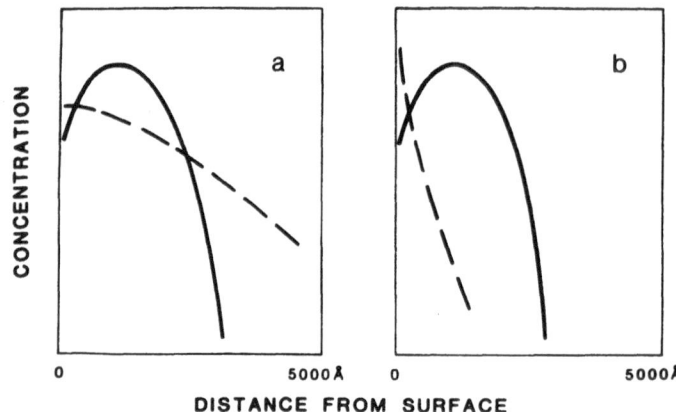

Fig. 25 Ion-implanted solute profiles (solid curves) in silicon and their redistribution on laser annealing (dashed profiles):
(a) No segregation, apparently diffusive redistribution only;
(b) Strong segregation. Based on [42,43] .

8.2 Convective transport during laser annealing

The good agreement between solute distribution curves, obtained from mere diffusion models, and experimentally determined profiles is astonishing to this author. Arguments have been made that convective flow cannot be established during the short times involved in laser melting. Such claims, however, cannot be supported by fluid dynamic arguments. Convection (i.e., momentum diffusion) is governed by the kinematic viscosity ν in analogy to the (binary) diffusivity in solute diffusion. From the similarity of the transport equations for momentum (Navier–Stokes equation) and solute (diffusion equation) one can expect that the ratio of the transient times for the development of convective and diffusive fluxes is $(\nu/D_{AB})^n$, where n is a positive number of order unity. Since ν-values in liquids are typically much greater than diffusivities and diffusive mass distribution is observed, convection flows should also develop during these experiments, given an appropriate driving mechanism.

Let us estimate the magnitude of the characteristic dimensionless groups that govern the various convection mechanisms.

Buoyancy-driven convection is controlled by the Rayleigh number[1]:

$$Ra = \frac{g \; \Delta T \; \beta \; L^3}{\nu \kappa} \qquad (24)$$

where g is the gravitational acceleration, ΔT the temperature difference across the characteristic length L, β the thermal expansion coefficient, ν the kinematic viscosity and κ the thermal diffusivity. For a typical laser molten film of width $w = 100\mu$ and depth $d = 0.5\mu$, with a $\Delta T = 500°C$, using the materials constants for Si, one obtains $Ra_w \approx 0.5$ and $Ra_d \approx 10^{-7}$ for L = w or L = d, respectively. These thermal Rayleigh numbers are way too small to drive significant convection through density gradients from thermal expansion. Even if one assumes sizeable density gradients from solute gradients, the corresponding solutal Rayleigh numbers are still too small, due to the smallness of the L's involved.

For an estimate of the likelihood of *surface-tension-driven* convection (σ- convection) we use here the Bond number [1]:

$$Bo = \frac{\rho \; \beta g \; L^2}{\partial \sigma / \partial T} \qquad (25)$$

which represents the relative strength of buoyancy- and σ- driven convection. The ρ in (25) is the mass density of the melt and $\partial \sigma / \partial T$ the temperature coefficient of the surface tension (of order 10^{-1} g/(sec^2 deg) for liquid Si). Evaluating (25) for our model case yields $Bo_w \approx 6 \times 10^{-4}$. In connection with the above Ra_w, this result clearly implies that σ-convection can play a significant role in laser melting. For an estimate of the possibly resulting convection velocities see Prof. Müller-Krumbhaar's lecture. With the high power densities used in laser annealing, another mechanism for convection must be considered : non-uniformities in the *radiation pressure* during melting [44]. In the center of the laser beam, the melt receives more (downward) momentum than in the "wings" of the intensity distribution. This has been estimated [44] to cause significant convection velocities. This author, however, would like to point out that in this geometry σ-convection tends to counteract the forced convection from (photon) momentum transfer. The surface tension of liquids decreases with increasing temperature. Hence, σ- convection would establish a flow pattern (viewed on the melt surface) from the center of the laser beam outwards; whereas the radiation pressure mechanism would drive flow towards the center.

9. CONCLUSIONS

Equilibrium phase diagrams and segregation coefficients are not only of importance for standard (slow) materials processing but can also give guidance for the rapid solidification associated with short duration laser melting. This is not surprising since the period of molecular vibration frequencies in condensed phases ($\approx 10^{-13}$ sec) are short as compared to the local heating times ($t \geqslant 10^{-9}$ sec) of most laser melting experiments. Thus quasi-equilibrium conditions can still be expected, unless significant activation barriers exist, such as those involved in solute adsorption and diffusion in the solid.

Though practically ignored to-date, it is suggested that mass transfer between the condensed phases and the ambient vapor may play a non-negligible role in laser annealing.

Mass, and heat transfer within melt films in laser annealing need considerable more (fluid dynamic) attention. In particular the question of diffusive versus convective solute transport appears not too well understood at this point.

REFERENCES

1. Rosenberger, F., Fundamentals of Crystal Growth I, Macroscopic and Transport Concepts (New York, Springer, 1979).
2. Moore, W.J., Physical Chemistry, 3rd ed. (Englewood Cliffs, Prentice Hall, 1962).
3. Gordon, P., Principles of Phase Diagrams in Materials Systems (New York, McGraw-Hill, 1968).
4. Reisman, A., Phase Equilibria (New York, Academic Press, 1970).
5. Kröger, F.A., The Chemistry of Imperfect Crystals (Amsterdam, North Holland, 1973) Vol. 1.
6. Chernov, A.A., J. Crystal Growth 24/25 (1974) 11;
    Chernov, A.A., Sov. Phys. Crystal. 16 (1972) 734;
    Delves, R.T., in Crystal Growth, B.R. Pamplin, ed. (Oxford, Pergamon, 1974) Vol. 1, pp. 40-103;
    Langer, J.S., Rev. Mod. Phys. 52 (1980) 1;
    Sekerka, R.F., in Crystal Growth, P. Hartman, ed., (Amsterdam, North Holland, 1973) pp. 402-443.
    Wollkind, D.J., in Preparation and Properties of Solid State Materials, W.R. Wilcox, ed. (New York, Marcel Dekker, 1979), Vol. 4, pp. 111-191.
7. Hansen, M., Constitution of Binary Alloys (New York, McGraw-Hill, 1958).
8. Panish, M.B. and Ilegems, M., in Progress in Solid State Chemistry, Vol. 7, H. Reiss and J.O. McCaldin, eds (New York,

Pergamon, 1972) pp. 39–83.

9. Stringfellow, G.B., J. Crystal Growth 27 (1974) 21.
10. Osamura, K. and Murakami, Y., J. Phys. Chem. Solids 36 (1975) 931.
11. Peuschel, G.-P., et al., Kristall Technik 14 (1979) 409.
12. Stringfellow, G.B., J. Phys. Chem. Solids 33 (1972) 665.
13. Haase, R., Schönert, H., Intern. Encyclopedia of Physical Chemistry and Chemical Physics, Topic 13, Vol. 1 (New York, Pergamon, 1969).
14. Palatnik, L.S., Landau, A.I., Phase Equilibria in Multicomponent Systems (New York, Holt, Rinehart and Winston, 1964).
15. Levin, E.M. et al., Phase Diagrams for Ceramicists (Am. Ceramic Soc., Columbus, 1964).
16. Macchesney, J.B. and Rosenberg, P.E., in Phase Diagrams, A.M. Apler, ed., (New York, Academic Press, 1970) Vol. 1, pp. 113–165.
17. Thurmoud, C.P. and Struthers, J.D., J. Phys. Chem. 57 (1953) 831.
18. Lehovec, K., J. Phys. Chem. Solids 23 (1962) 695.
19. Weiser, K., J. Phys. Chem. Solids 7 (1958) 118.
20. Ikeya, M., Itoh, N., Suita, T., Japan. J. Appl. Phys. 7 (1968) 837.
21. Gross, U., Thesis, University of Stuttgart (1970).
22. Douglas, T.B., J. Chem. Phys. 45 (1966) 4571.
23. Blander, M., J. Chem. Phys. 34 (1961) 697.
24. Rosenberger, F., Riveros, H.G., J. Chem. Phys. 60 (1974) 668.
25. Stringfellow, G.B., J. Phys. Chem. Solids 35 (1974) 775.
26. Dikhoff, J.A.M., Solid State Electron. 1 (1960) 202.
27. Hall, R.N., J. Phys. Chem. 57 (1953) 836.
28. Trainor, A., Bartlett, B.E., Solid State Electron. 2 (1961) 106.
29. Holmes, P.J., J. Phys. Chem. Solids 24 (1963) 1239.
30. Chernov, A.A., in Growth of Crystals (Rost Kristallov), A.V. Shubnikov, N.N. Sheftal, eds. (New York, Consultants Bureau, 1962) Vol. 3 p. 35.
31. Chernov, A.A., Sov. Phys. Uspec. 13 (1970) 101.
32. Zschauer, K.-H. and Vogel, A., Inst. Physics Conf. Series No.9 GaAs, p. 100 (1971).
33. Zschauer, K.-H., Festkoerperprobl. 15 (1975) 1.
34. Tiller, W.A. and Ahn, K.S., J. Crystal Growth 49 (1980) 483.
35. Anantha, N.G. and Chalmers, B., J. Appl. Phys. 38 (1967) 4416.
36. Wagner, C., J. Phys. Chem. 57 (1953) 738.
37. Edwards, W.D., J. Appl. Phys. 39 (1968) 1784.
38. Burton, J.A., Prim, R.C., Slichter, W.P., J. Chem. Phys. 21 (1953) 1987.
39. Wilson, L.O., J. Crystal Growth, 44 (1978) 247.
40. Wilson, L.O., J. Crystal Growth, 44 (1978) 371.
41. Lorenz, M.R., J. Appl. Phys. 33 (1962) 3304.
42. White, C.W., Narayan, J. and Young, R.T., Science 204 (1979) 461.

43. White, C.W. et al., J. Appl. Phys. 50 (1979) 2967.
44. Aziz, M.J., J. Appl. Phys. 53 (1982) 1158.

STATICS AND DYNAMICS OF PHASE TRANSITIONS : A BRIEF INTRODUCTION

K. Binder

Institut für Festkörperforschung, Kernforschungsanlage
5170 Jülich, Postfach 1913, W-Germany.

The first part of these lectures will be devoted to the *"order parameter"*-concept : "First order" and *"second-order"* transitions will be distinguished, the main ideas about *"critical phenomena"* mentioned, and some theoretical aspects of *phase diagrams* outlined. Then *structural transitions* and the concept of *"soft phonon modes"* are reviewed. The second part concerns the *kinetics of first order transitions : decay of metastable phases* by *nucleation and growth*; significance of *stability limits*; decay of unstable phases by *long-wavelength* instabilities. The third part discusses *transitions in systems driven far from equilibrium*.

Table 1 mentions some *condensed-matter systems* which can exist in several phases, and identifies an *extensive variable* which distinguishes them : the *"order parameter"* [1] . It is identically zero in the "disordered phase", nonzero in the ordered one. We then use the *thermodynamic potential* F which has the *"field"* H *conjugate* to the order parameter Φ as a "natural variable" (in addition to temperature T) :

$$\Phi = - (\partial F/\partial H)_T \ [\text{e.g. ferroelectric} : \vec{P} = -(\partial F/\partial \vec{E})_T] , \quad (1)$$

the other derivative of F involves the entropy S, $S = - (\partial F/\partial T)_H$. When we study the change of order with an independent variable, Φ may disappear at the transition continuously ($\Phi \propto (1-T/T_c)^\beta$, "second order transition", Fig. 1a) or discontinuously ("first order transition", Fig. 1b). At the second-order transition, "susceptibility" χ and specific heat C_H typically have powerlaw singularities :

Table I: Order Parameters for Phase Transitions in Various Systems

System	transition	order parameter
liquid-gas	condensation/evaporation	density difference $\Delta\rho = \rho_\ell - \rho_g$
binary liquid mixture	unmixing	composition -"- $\Delta c = c^{(2)}_{coex} - c^{(1)}_{coex}$
nematic liquid	orientational ordering	$\frac{1}{2} \langle 3\cos^2\theta - 1 \rangle$
quantum liquid	normal-fluid→suprafluid	$\langle\psi\rangle$, ψ = wave-function
liquid-solid	melting/crystallization	$\rho_{\vec{G}}$, $\vec{G}$ = reciprocal lattice vector
magnetic solid	ferromagnetic (T_C)	spontaneous magnetization $\vec{M}$
	antiferrromagnetic (T_N)	sublattice magnetization $\vec{M}_s$
solid binary mixture	unmixing	$\Delta c = c^{(2)}_{coex} - c^{(1)}_{coex}$
AB	sublattice ordering	$\psi = (\Delta I_c^{II} - \Delta I_c^{I})/2$
dielectric solid	ferroelectric (T_C)	spontaneous polarization $\vec{P}$
	antiferroelectric (T_N)	sublattice polarization $\vec{P}_s$
molecular crystal	oriental ordering	$Y_{\ell,m}(\vartheta,\varphi)$
...		

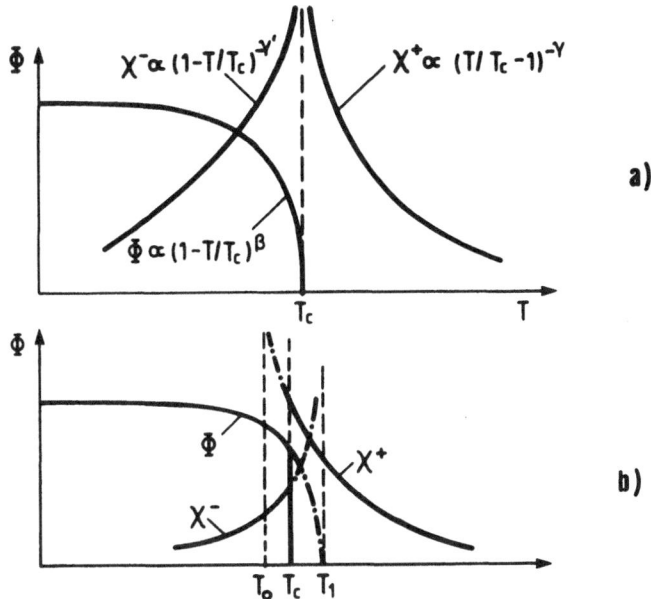

Fig. 1 Order parameter and associate "susceptibility" χ at a second-order transition (a) and at a first-order transition (b).

$$\chi = -(\frac{\partial^2 F}{\partial H^2})_T \propto \left| 1 - T/T_C \right|^{-\gamma},$$

$$C_H = - T (\frac{\partial^2 F}{\partial T^2})_H \propto \left| 1 - T/T_c \right|^{-\alpha}. \tag{2}$$

α, β, γ, ... are the so-called "critical exponents". At the first-
order transition, typically there is also a jump discontinuity ΔS
of entropy, and hence a latent heat $\Delta Q = T\Delta S$; second derivatives
of F often are finite at T_c and rather seem to diverge at the "stability limits" T_o and T_1 of the disordered and ordered phases,
respectively.

It is then natural to classify the transitions according to whether Φ is a *scalar quantity* or has *vector* or *tensor character* : for fluids or mixtures, the density or concentration differences obviously are scalar; for the λ- transition of He^4, the complex wave function has two components (real and imaginary part) and hence Φ is a vector, as in magnetic or dielectric systems; for orientational order both in liquid crystals and in molecular crystals, Φ is a tensor of second rank.

In Landau's phenomenological theory, F is expanded in powers of the order parameter density $\phi(\vec{x})$. In the scalar case :

$$\mathcal{F}\{\phi\} = F_0 + \int d\vec{x} \left\{ \frac{1}{2} r\phi^2(\vec{x}) + \frac{1}{4} u\phi^4(\vec{r}) - H\phi(\vec{x}) + \frac{1}{2} C[\nabla\phi(\vec{x})]^2 \right\} . \tag{3}$$

Here $\phi(\vec{x})$ is assumed small (no terms ϕ^6, etc.), slowly varying in space (no terms $[\nabla^2\phi(\vec{x})]^2$), the coefficients $u{>}0$, $C{>}0$, and for H=0 a symmetry against $\phi{\rightarrow}{-}\phi$ is required (no term $\phi^3(\vec{x})$, etc.) .

$\mathcal{F}$ is minimal for the homogeneous case , $\nabla\phi(\vec{x}){\equiv}0$, and hence (V is the system's volume; assume also $r=r'(T-T_c)$, $r'{>}0$, and H=0) :

$$\frac{1}{V}\frac{\partial \mathcal{F}}{\partial \phi} = r\phi_0 + u\phi_0^3 = 0, \begin{cases} \phi_0 = 0, & T{>}T_c \tag{4a} \\ \phi_0 = \pm\sqrt{-r/u} = \pm\sqrt{-r'/u}(1-T/T_c)^{1/2}, & T{<}T_c . \end{cases}$$

$$\tag{4b}$$

Indeed, these assumptions yield a 2nd-order transition, with $\beta=1/2$.

We next consider the response $\Delta\phi(\vec{x})$ to a wave-vector-dependent field $\Delta H(\vec{x})=\Delta H_{\vec{k}} e^{i\vec{k}\cdot\vec{x}}$. Then the functional derivative $\delta F/\delta\phi(\vec{x})$ becomes, linearizing $\phi(x)$ around ϕ_0 ($\phi(x)=\phi_0+\Delta\phi(x)$) :

$$\frac{\delta\mathcal{F}}{\delta\phi(\vec{x})} = r[\phi_0+\Delta\phi(\vec{x})] + u[\phi_0^3+3\phi_0^2\Delta\phi(\vec{x})] - \Delta H_{\vec{k}}e^{i\vec{k}\cdot\vec{x}} - C\nabla^2[\Delta\phi(\vec{x})] = 0 . \tag{5}$$

Using $\Delta\phi(\vec{x})=\Delta\phi_{\vec{k}}e^{i\vec{k}\cdot\vec{x}}$, Eq. (5) is solved by:

$$\frac{\Delta\phi_{\vec{k}}}{\Delta H_{\vec{k}}}{\equiv}\chi(\vec{k}) = [r+3u\phi_0^2 + Ck^2]^{-1} = \chi_T/[1+k^2\xi^2] \tag{6}$$

where suceptibility χ and correlation length ξ of order-parameter fluctuations diverge at T_c ($\xi\propto|1-T/T_c|^{-\nu}$) :

$$\chi_T = \begin{cases} [r'(T-T_c)]^{-1} \\ [2r'(T_c-T)]^{-1} \end{cases}; \xi = \begin{cases} C^{1/2}[r'(T-T_c)]^{-1/2}, & T > T_c' \\ C^{1/2}[2r'(T_c-T)]^{-1/2}, & T < T_c \end{cases} . \tag{7}$$

Hence we get exponents $\gamma=1$, $\nu=1/2$; note that the fluctuation-dissipation theorem connects the wave-vector dependent susceptibility $\chi(\vec{k})$ to the Fourier transform $S_T(\vec{k})$ of order-parameter correlations (="structure factor") :

$$k_B T \chi(\vec{x}) = S_T(\vec{k}) = \int d\vec{x} e^{i\vec{k}\cdot\vec{x}} [<\phi(0)\phi(\vec{x})>_T - \phi_0^2]. \qquad (8)$$

This description of 2nd-order transitions is at best qualitatively correct, since statistical fluctuations are neglected. To account for the latter, Eq.(3) is considered as effective hamiltonian from which one gets the free energy F via :

$$F = -k_B T \ln Z = -k_B T \ln \int d\{\psi\}[-\mathcal{F}\{\psi\}/k_B T] . \qquad (9)$$

A treatment of Eq. 9 with the renormalization group approach [2] shows : (i) Landau theory describes critical behavior correctly for spatial dimensionalities $d > d^* = 4$;(ii) for $d < d_c$ fluctuations destroy the ordered state, with $d_c = 2$ for order parameter dimensionality $n>2$ (ϕ=vector), $d_c=1$ for $n=1$ (ϕ=scalar): (iii) for $d_c <d<d^*$ the critical exponents depend on both n and d, $\chi(\vec{k})$ is a scaled function of ξ, $(\vec{k}) = \epsilon^{\gamma/\nu}\tilde{\chi}(k\xi)$, $\xi \propto |1-T/T|^{-\nu}$, F is a scaled function of H and $\tau = 1-T/T_c$, i.e. $F(\tau, H) = F^{regular}(\tau,H)+ |\tau|^{\gamma+2}\tilde{F}(H, |\tau|^{-\gamma-\beta})$. The exponents, α, β, γ, ν, ... are material-independent universal constants for each "universality class" (as specified by n,d).

A 1st-order transition in Landau theory arises when $u<0$ in Eq. 3, and a term $\frac{1}{6}v\phi^6(\vec{x})$, ($v>0$) must be added. While in the 2nd order case $\mathcal{F}(\phi)$ has two minima for $T<T_c$, which merge at T_c (Fig.2a). $\mathcal{F}(\phi)$ has three minima for $T_c<T<T_1$, and T_c is reached when they are equally deep, (Fig. 2b). One finds for $r=r'$ $(T-T_0)$ that : $T_c = T_0+3u^2/(32r'v)$, $T_1=T_0+u^2/(8r'v)$ and the order parameter at T_c jumps from $\phi_0=\pm\sqrt{-3u/4v}$ to zero.

When external parameters p (other than T,H) are varied, it may happen that u changes sign, and hence a line $T_c(p)$ of 2nd order transitions ends there (at a so-called tricritical point) and continues at a 1st-order line.

Another case where Landau theory predicts 1st-order transitions arises in the absence of a symmetry between Φ and $-\Phi$: then a term $\frac{1}{3}w\phi^3$ may exist in Eq. 3, and for $u>0$, $\mathcal{F}(\psi)$ may have two minima (Fig. 3a). The transition occurs when they are equally deep.

The simplest 1st-order transition occurs when considering the variation of the conjugate field at $T<T_c$(Fig. 3b). From Eq. 4 one finds the stability limit (spinodal) $\phi_s=\pm\sqrt{-r/3u}=\phi_0/\sqrt{3}$, $H_c=\pm(-2r/3)^{3/2}/\sqrt{u}$. It must be emphasized, that some of these predictions are again

176

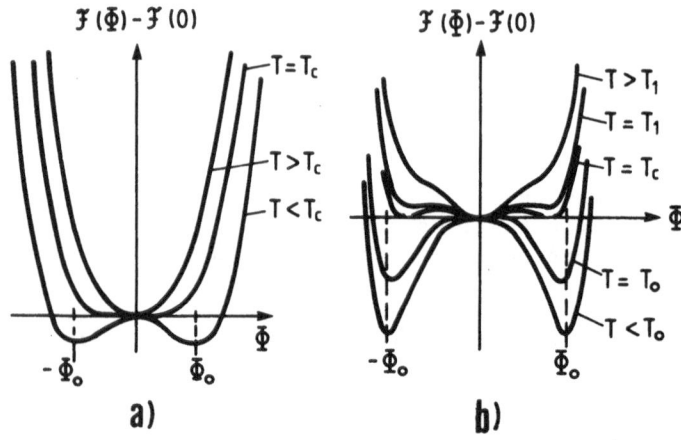

a) **b)**

<u>Fig. 2</u> Variation of the Landau free energy at transitions of 2nd
order (a) and 1st order (b).

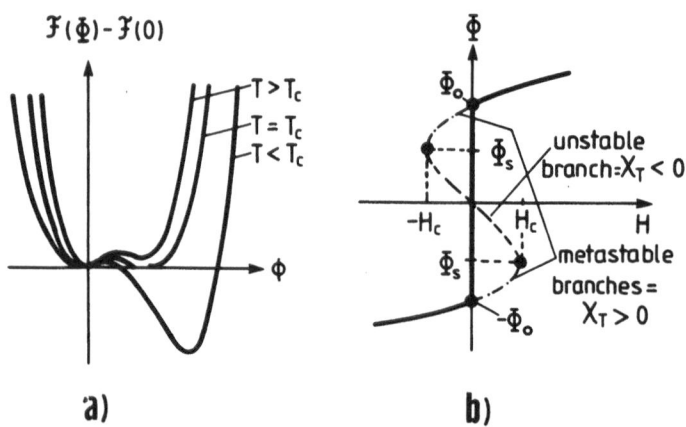

a) **b)**

<u>Fig. 3a</u> Variation of the free energy at a 1st-order transition due
to a cubic term.
<u>Fig. 3b</u> 1st-order transition due to variation of the field H at a
temperature less than the critical temperature of a 2nd-order transition ($\pm H_c$ are the corresponding limits of metastability).

severely modified by effects due to statistical fluctuations :
(i) metastable states and spinodals are no longer well defined;
(ii) terms $w\phi^3$ do not guarantee a 1st order transition; the character of the transition rather is given by the fixed point values
(u^*, v^*, w^*) to which the "coupling constants" (such as u, v, w)
tend upon renormalization group transformations (if $w \neq 0$ but $w^*=0$
the transition is of second order) ; (iii) parameters $w=0, u>0$ do
not guarantee a 2nd order transition , one may get $u^*<0$ ("fluctua-
tion-induced" 1st-order transition). In most cases, however, the
Landau rules determining the order of a transition do work.

The appropriate structure of the Landau expansion is found noting that F must be invariant against all operations of the symmetry
group $\underline{G}_o$ of the disordered phase. In the ordered phase, some symmetry elements of $\underline{G}_o$ fall away (spontaneously broken symmetry), the
remaining ones form a subgroup $\underline{G}$ of $\underline{G}_o$. The invariance of F must
hold for terms ϕ^k of any order k separately. Consider e.g. a cubic
ferroelectric, with the polarization vector $\vec{P}$ being the order parameter : F must be given by :

$$F=F_o + \int d\vec{x}\{\frac{1}{2}\chi_{el}^{-1}\vec{P}^2 \quad + \frac{1}{4}[U(P^2)^2 + U'(P_x^4+P_y^4+P_z^4)] + \frac{C}{2}(\nabla P_x)^2 +$$

$$(\nabla P_y)^2 + (\nabla P_z)^2]\} . \tag{10}$$

While the quadratic term in the general case involves the inverse
of the dielectric suceptibility tensor $(\chi_{el}^{-1})_{\alpha\beta} P_\alpha P_\beta$, this term
is completely isotropic in the cubic case. Inversion symmetry re-
quires invariance against $-\vec{P}\rightarrow\vec{P}$, and hence no cubic term is possible.
The quadratic term then contains two "cubic invariants". In gene-
ral all terms allowed by symmetry actually will occur, of course.

One may also consider transitions between two different struc-
tures with symmetry group $\underline{G}_1$, $\underline{G}_2$, which do not have a group-subgroup
relation between each other. Then it is not possible to describe
the transition by an order parameter which is zero in the one phase
and non-zero in the other, and the transition necessarily is of 1st
order. However, it is then often possible to incorporate this situation into Landau theory by considering the more general situation
involving a third structure $\underline{G}$, of which both $\underline{G}_1$ and $\underline{G}_2$ are subgroups.
Example : weakly anisotropic Heisenberg antiferromagnets in a field
have for weak fields a uniaxial antiferromagnetic structure (n=1)
(order parameter M_s^z) and for strong fields a spin-flop structure
(n=2 component, ordering of the perpendicular components M_s^x, M_s^y). In
Landau theory, all 3 components of the sublattice magnetization
must be taken as the order parameter :

$$((M_s^\perp)^2 = (M_s^x)^2 + (M_s^y)^2),$$

$$F=F_o+\int d\vec{x}\{\frac{1}{2} \; r_{\parallel}\;(M_s^z)^2+\frac{1}{2}r_{\perp}\;(M_s^{\perp})^2+\frac{1}{4}\;r_{\parallel}(M_s^z)^4+\frac{1}{4}U_{\perp}\;(M_s^{\perp})^4+\frac{1}{4}U(M_s^z)^2(M_s^{\perp})^2+$$

$$\text{gradient terms}\}\,.$$

$$(11)$$

The group $\underline{G}_o$ is the disordered paramagnetic phase to which both antiferromagnetic structures have 2nd-order transitions at $T_N^{\parallel}$(H), $T_N^{\perp}$(H). These lines join at a *"bicritical point"* which at the same time is the end point of the first-order transition line between the two ordered phases (Fig. 4a). For stronger anisotropy $T_N^{\parallel}$(H) rather has a *tricritical point* and $T_N^{\perp}$ (H) terminates at the 1st-order line in a *critical end-point* (Fig. 4b) [3].

While these phase diagrams refer to lattices which can be decomposed into two sublattices (e.g.sc, bcc), fcc antiferromagnets (with 4sc sublattices) may have a different phase diagram (Fig. 4c) [4]. One ordered phase has two sublattice magnetizations positive, two negative, the other one three sublattice magnetizations positive, one negative. All transitions are of first order and meet in a *triple point*.

In the space of *intensive* thermodynamic variables (such as H, T) both 1st and 2nd order transitions appear as lines in a phase diagram (Fig. 4a-c). If one uses an *extensive* variable, it will have a jump at a first-order phase transition : this "forbidden region" yields a *regime of two-phase-coexistence* in the phase diagram (Fig. 4d). For mixtures the composition is held fixed and hence it is natural to use it as independent variable, rather than the conjugate intensive quantity (chemical potential difference). The A_3B-structure on the fcc lattice nicely illustrates the above classification concepts : as order parameters, we may take the sublattice concentration differences, $\phi_i=C_B^{(i)}-C_B$, $i = 1,2,3,4$. Hence ϕ is a 4-component vector. Obviously, there is no symmetry $\phi_i \rightarrow -\phi_i$, and hence third-order invariants should exist, implying a 1st-order transition as observed (e.g. Cu_3Au, Ni_3Pt).

Another useful classification of structural transitions in solids considers the type of movements the atoms undergo there (Fig.5). Both for unmixing and sublattice ordering of mixtures, thermally activated diffusion processes are required, and the kinetics of these transitions is correspondingly slow. For other transitions, the order is described by periodic lattice distortions where atomic displacements are comparable to those of lattice vibrations. Short wavelength distortions may give rise to *"antiferrodistortive"* and *"antiferroelectric"* orderings, as exemplified by the Perovskites $SrTiO_3$, $PbZrO_3$. Long-wavelength distortions corresponding to optic phonons give rise to *"ferroelectric"* orderings, while those corresponding to acoustic phonons give rise to *"ferroelastic"* or *"martensitic"* ordering.

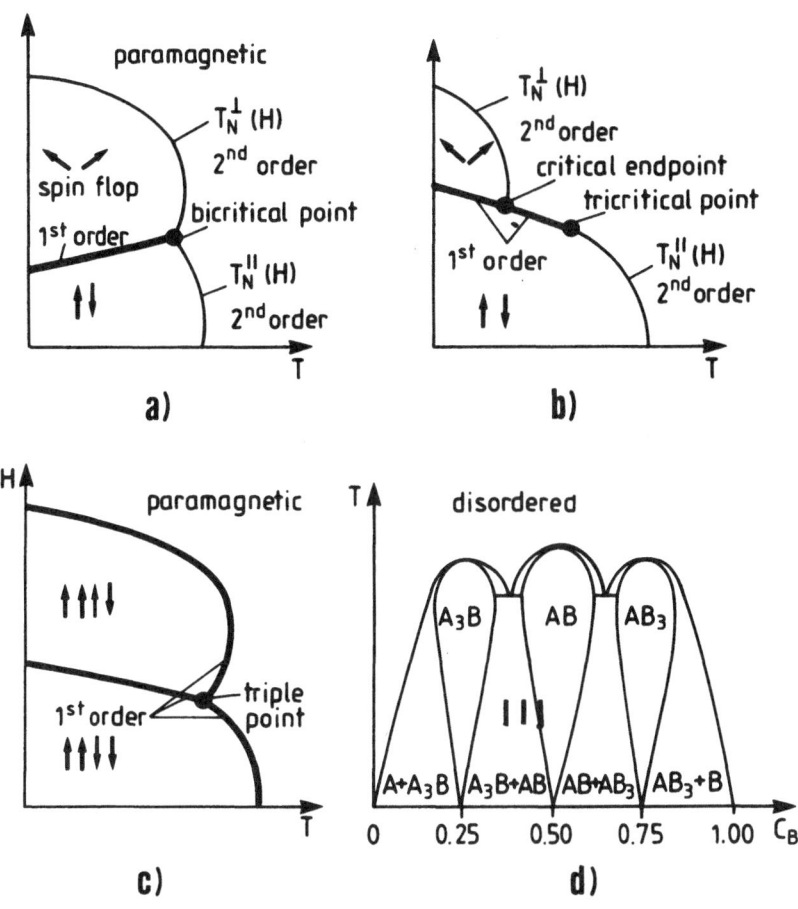

Fig. 4 Phase diagrams of antiferromagnets in a magnetic field, for the sc (or bcc) lattice with weak (a) or strong (b) anisotropy, the strongly anisotropic (Ising) fcc lattice (c) and the equivalent fcc binary mixture (d). Type of order is schematically indicated.

A further distinction concerns the effective single-particle potential seen by the atoms undergoing the distortion. If the ordered structure is doubly degenerate, the atoms can sit in the right or the left minimum of a double-minimum potential below T_c. If the potential above T_c is essentially the same, and only the distribution of the atoms over the minima is more or less random, the transition is called of "*order-disorder type*" (this happens e.g. for hydrogen-bonded ferroelectrics and is analogous to the sublattice

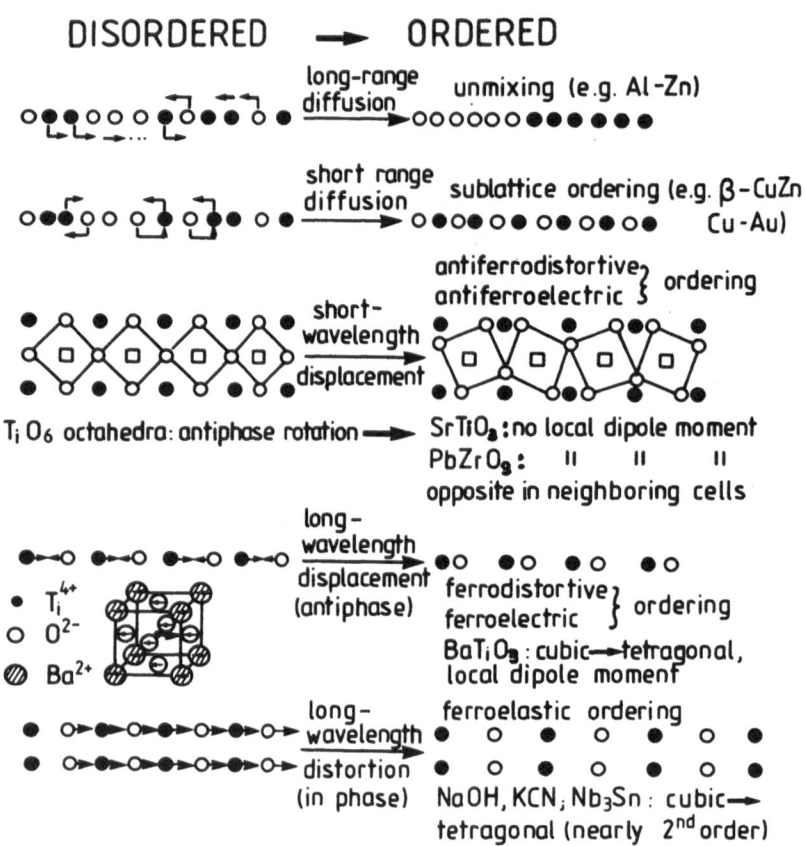

Fig. 5 Atomic movements in structural phase transitions.

ordering case.) If the potential changes above T_c to a single-potential form, the transition is called "*displacive*".[c] The average local displacement then may increase continuously at the transition (Fig. 6a). Since displacements $u_i(\vec{x})$ can be related to phonon normal coordinates $Q_{\vec{k},\lambda}$ (via $\vec{e}_\ell(k,\lambda)$ = phonon polarization vector):

$$\vec{u}_\ell(\vec{x}) = (NM_\ell)^{-1/2} \sum_{\vec{k}\lambda} \exp(i\vec{k}\cdot\vec{x}) \, \vec{e}_\ell(k,\lambda) Q_{\vec{k},\lambda} ,$$

M_ℓ = mass of the atom ℓ in the unit cell. $\qquad\qquad$ (12)

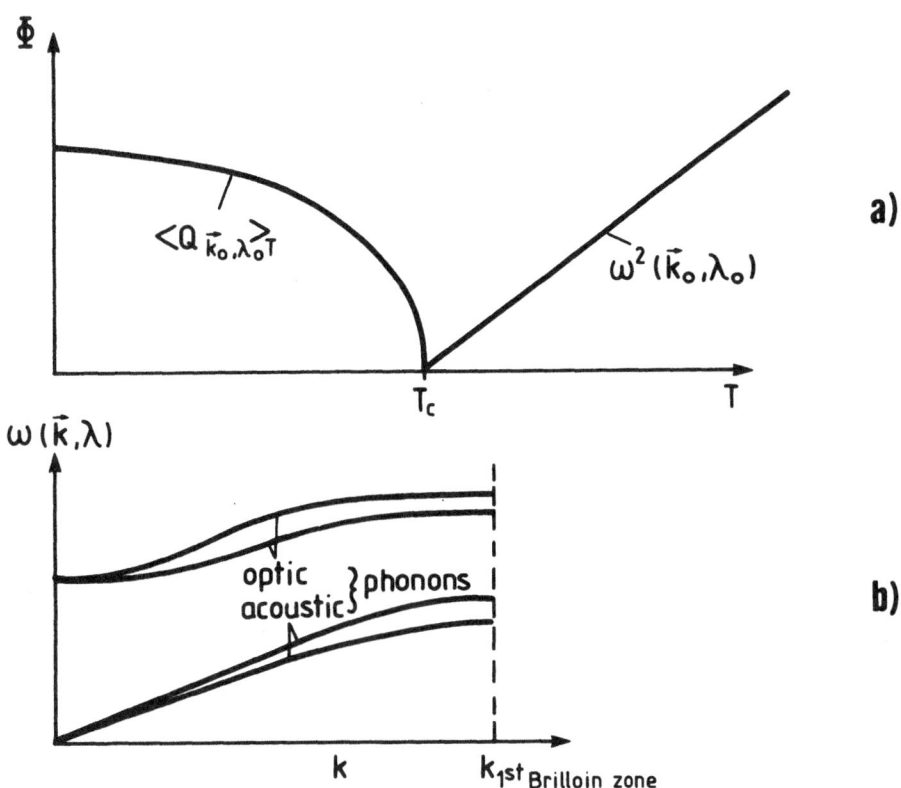

Fig. 6 a)Schematic temperature variation of order parameter and square of soft mode frequency at a displacive transition. b) Schematic phonon spectrum at $T>T_c$. For $SrTiO_3$, $\vec{k}_o=\Pi(\frac{1}{2},\frac{1}{2},\frac{1}{2})/a$, $T_c=106K$; for ferroelectric $Pb_5Ge_3O_{11}$, $\vec{k}_o=0$ [optic phonon], $T_c\cong45oK$; for martensitic $In-25at\%Tl$, $\vec{k}_o=0$ [acoustic phonon], nearly softening of $C_{11}-C_{12}$ at $T_c=195$ K; LaO_5O_{14}: $\vec{k}_o=0$ [acoustic phonon], softening of C_{55} at $T_c=400K$, where the structure changes from orthorombic to monoclinic [5].

$\langle Q_{\vec{k}_o}, \lambda_o \rangle_T$ plays the role of an order parameter (component) for a displacive transition. In mean-field theory, the associate eigenfrequency vanishes at the transition ("*soft phonon*"). If this happens for a phonon with wavevector at the Brillouin zone edge, we have an antiferroelectric order (if the phonon is polar, i.e. producing a local dipole moment) or antiferrodistortive order (for nonpolar phonons). Soft optic phonons at the Brillouin zone center give rise to ferroelectric (or ferrodistortive) orderings, and soft acoustic

ones to ferroelastic ordering (Fig. 6b). Softening at intermediate
points in the Brillouin zone may lead to incommensurate superstructures.

Note that the $Q_{\vec{k},\lambda}$ are defined such that the enthalpy U of the
crystal in quasiharmonic approximation is diagonalized, ($\vec{R}_i{}^\ell$ = coordinate of ℓth atom in the ith cell) :

$$U = U_o + 1/2 \sum_{\substack{\text{lattice} \\ \text{cells i,j} \\ \ell,\ell'}} \frac{\partial^2 U}{\partial R_i^{\ell} \partial R_j^{\ell'}} \vec{u}_\ell(\vec{R}_i)\vec{u}_{\ell'}(\vec{R}_j) =$$

$$U_o + 1/2 \sum_{\vec{k}\lambda} \omega^2(k,\lambda) \, Q_{-\vec{k},\lambda} \, Q_{\vec{k},\lambda} . \tag{13}$$

Thus, in the spirit of the Landau expansion, the coefficient
$\omega^2(k_o,\lambda_o)$ in front of the square of the order parameter corresponds
to the term $r = r'(T-T_c)$; the linear temperature variation of $\omega^2(k_o,\lambda_o)$
was already anticipated in Fig. 6. Again this meanfield picture
must be modified : obviously, anharmonic terms must be included near
and below T_c in Eq. 13. Then the modes are coupled to each other,
and the soft mode frequency does not really vanish at T_c, but rather
either becomes overdamped, or it stays finite and a central peak
develops which is still incompletely understood [6] . Also in the
regime where $\omega(\vec{k}_o,\lambda_o)$ is still well defined, its temperature dependence is more complicated. Due the coupling between $\omega(\vec{k}_o,\lambda_o)$ and
acoustic phonons, the ordering also induces an elastic distortion
of the crystal. Sometimes also a polarization may be indirectly
induced ("improper ferroelectrics").

The softening of $\omega(\vec{k}_o,\lambda_o)$ near T_c does *not* mean that the $\vec{u}_\ell(\vec{x})$
may become very large. In fact, the mean-square displacement even
is only weakly temperature-dependent [7] : $\langle \vec{U}_\ell^2 \rangle_T - \langle \vec{U}_\ell^2 \rangle_{T_c} \propto (T/T_c - 1)^{1-\alpha}$
(energy-like singularity). But the correlation function $\langle \vec{U}_\ell(o)\vec{U}_\ell(\vec{x})\rangle$
becomes long-ranged.

For phase transitions where acoustic modes soften the order
parameter is the strain tensor ε_{ik} [8]:

$$\varepsilon_{ik}(x) = 1/2 \left(\frac{\partial u^i}{\partial x_k} + \frac{\partial u^k}{\partial x_i} + \frac{\partial u^m}{\partial x_k}\right), \quad [x_i, \; x_k = x,y,z], \tag{14a}$$

$$F = F_o + \int d\vec{x} \left[\frac{1}{2} C_{ik\ell m} \, \varepsilon_{ik} \, \varepsilon_{\ell m} + \frac{1}{3} C^{(3)}_{ik\ell mrs} \, \varepsilon_{ik} \, \varepsilon_{\ell m} \, \varepsilon_{rs} \right.$$

$$\left. + \frac{1}{4} C^{(4)}_{ik\ell mrsuv} \, \varepsilon_{ik} \, \varepsilon_{\ell m} \, \varepsilon_{rs} \, \varepsilon_{uv} + \ldots + \text{gradient terms} \right]. \tag{14b}$$

Here $C_{ik\ell m}$ are the elastic constants, and $C^{(3)}_{ik\ell mrs}$, $C^{(4)}_{ik\ell mrsuv}$ analogous coefficients of anharmonic terms. For a cubic crystal, the quadratic term can be simply expressed in terms of the three independent elastic constants C_{11}, C_{12}, C_{44} as :

$$1/6 \ (C_{11}+C_{12}) \ (\varepsilon_{xx} + \varepsilon_{yy} + \varepsilon_{zz}) + 1/3 \ (C_{11}-C_{12}) \ \times$$

$$(\varepsilon^2_{xx} + \varepsilon^2_{yy} + \varepsilon^2_{zz} - \varepsilon_{xx}\varepsilon_{yy} - \varepsilon_{xx}\varepsilon_{zz} - \varepsilon_{yy}\varepsilon_{zz}) +$$

$$1/2 \ C_{44} \ (\varepsilon^2_{xy} + \varepsilon^2_{xz} + \varepsilon^2_{yz}).$$

Acoustic phonons near $\vec{k}=0$ are described by $\omega(\vec{k},\lambda)=v_\lambda k$, with v^2_λ being proportional to the appropriate elastic constant (or combination there of). For most elastic transitions, symmetry permits some nonzero $C^{(3)}_{ik\ell mrs}$, and hence leads to 1st-order transitions. Elastic distortions at these transitions may be very small (Nb_3Sn, V_3Si : $\varepsilon_{ik} \sim 10^{-4}$, nearly 2nd order); large distortions occur for transitions between phases for which no group-subgroup relation exists, e.g. martensitic transformations in Fe-C, Fe-Ni : $\varepsilon_{ik} \sim 10^{-1}$; these are instabilities of metastable phases, for stable phases unmixing would occur.

The softening of phonons at structural transitions is an expression of the "*critical slowing down*" of the dynamics at 2nd order phase transitions (in isotropic magnets magnon modes soften near T_C or T_N, etc.) . In mean-field theory a deviation of the order parameter $\Delta\phi(\vec{x},t)$ from equilibrium will relax to zero, if the order parameter is not conserved , (Γ_0 = rate factor) :

$$\frac{\partial}{\partial t} \Delta\phi(\vec{x},t) = - \Gamma_0 \frac{\delta \mathcal{F}}{\delta\phi(x)} = - \Gamma_0 \{ (r+3u\phi^2_0)\Delta\phi(\vec{x},t) - C\nabla^2[\Delta\phi(\vec{x},t)] \} \ .$$

$$(15)$$

Assuming that $\Delta\phi(\vec{x},t)$ was produced by a field $\Delta H_{\vec{k}}$ for $t<0$, which is switched off at $t=0$, one finds (cf. Eqs. 5 and 6) :

$$\Delta\phi_{\vec{k}}(t)/\Delta\phi_{\vec{k}}(0) = \exp [-\omega_{\vec{k}} t], \qquad \omega_{\vec{k}} = \Gamma_0/\chi_{\vec{k}} \ . \qquad (16)$$

Since $\chi_{\vec{k}=0}$ diverges at T_c (Eq. 7), the characteristic frequency vanishes there proportional to $\chi^{-1}_{\vec{k}=0}$. Again this "conventional slowing down" is a too simplified picture of critical dynamics : in general one has "dynamic scaling":

$$\omega_{\vec{k}} = k^z \ \tilde{\omega} \ (k\xi), \qquad (17)$$

184

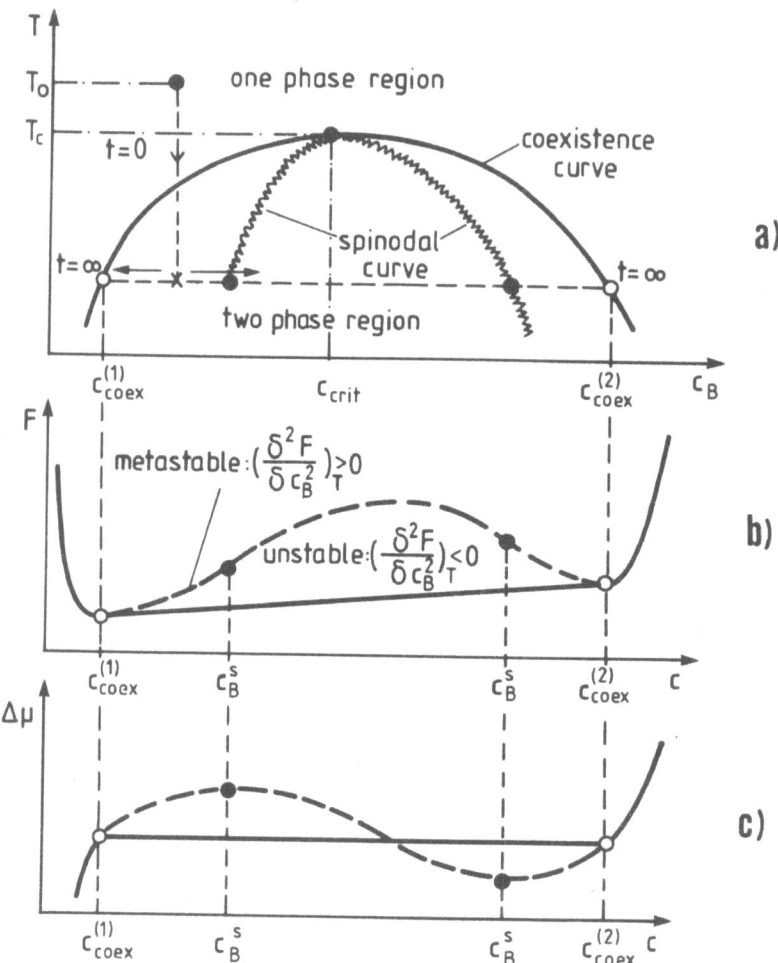

<u>Fig. 7</u>   Phase diagram of a binary mixture indicating a quench at t=o
to an unstable state at temperature T(a), free energy (b) and chemical potential difference (c) at that temperature.

where the "dynamic exponent" z is again universal within each "dynamic universality class" [9]; but usually z is not simply γ/ν as
Eq. 16 would imply.

    We now consider the *kinetics of first-order transitions*, starting
with the kinetics of unmixing (Fig. 7a) where the main concepts are
most easily explained. Suppose a system is suddenly quenched at t=0
from the one-phase region to a temperature underneath the coexistence
curve. The homogeneous state there is unstable, equilibrium consists
of a mixture of macroscopic regions with compositions $c_{coex}^{(1)}$, $c_{coex}^{(2)}$.

The amounts of these phases are inversely proportional to the distances from the coexistence curve and, hence, vary linearly with c_B. Therefore F also is linear (Fig. 7b), and the chemical potential differences $\Delta\mu$ of both phases (relative to the chemical potential μ_A of pure A) are equal (Fig. 7c; note the analogy to Fig. 3b, Φ_0 corresponds to $c_{coex} - c_{crit}$, H to $\Delta\mu$) :

$$\Delta\mu^{(1)} = (\partial F/\partial c_B)_T\Big|_{c^{(1)}_{coex}} = \Delta\mu^{(2)} = (\partial F/\partial c_B)_T\Big|_{c^{(2)}_{coex}}$$

(18)

The basic idea now is to introduce a free energy of one-phase states in the two-phase region (Fig. 7b , broken curve). This "analytic continuation" has two parts; a "metastable" and an "unstable" one. The curve separating them in the phase diagram is the "spinodal" (Fig. 7a). Two different mechanisms then are postulated for the phase transformation : in the unstable region, concentration fluctuations of *small amplitude* but *long wavelength* ("*homophase fluctuations*") will increase with time rather than decrease, and this "spinodal decomposition" leads to an inhomogeneous distribution of concentration (Fig. 8a). In the metastable region, a large amplitude ("heterophase") fluctuation is needed as a nucleus of the other phase in order that phase separation starts. In solids this *"homogeneous nucleation"* often is less important than "heterogeneous nucleation", i.e. microdomain formation at *surfaces, dislocations, grain boundaries* etc., where the *nucleation energy barriers* may be greatly *reduced* or even absent.

These submicroscopically small nuclei then grow by diffusion of atoms from the matrix to their surface, and thus the supersaturation of the matrix is reduced. A further *domain coarsening* then is obtained because the smallest domains re-dissolve then in the matrix again; thus the largest ones can grow further (*Lifshitz-Slyozov mechanism* [10]).

The continuum theory of spinodal decomposition [11] starts from the *conservation law* of the average concentration, $(1/V)\int c(\vec{x},t)d\vec{x}=c_B=$ constant:it is expressed as a *continuity equation* between *concentration density* $c(\vec{x},t)$ and *concentration current* $\vec{j}(\vec{x},t)$, $\partial c(\vec{x},t)/\partial t+\nabla\vec{j}(\vec{x},t)=0$. Assuming that the deviations from thermal equilibrium are small enough,$\vec{j}$ is proportional to the gradient of a local chemical potential difference $\mu(\vec{x},t)$, M=mobility :

$$\vec{j} = -M\nabla\mu(\vec{x},t); \ \partial c(\vec{x},t)/\partial t = -M\nabla^2\mu(\vec{x},t).$$ (19)

To find $\mu(\vec{x},t)$ the free energy density $f_{inh}(\vec{x})$ of an inhomogeneous system is introduced as :

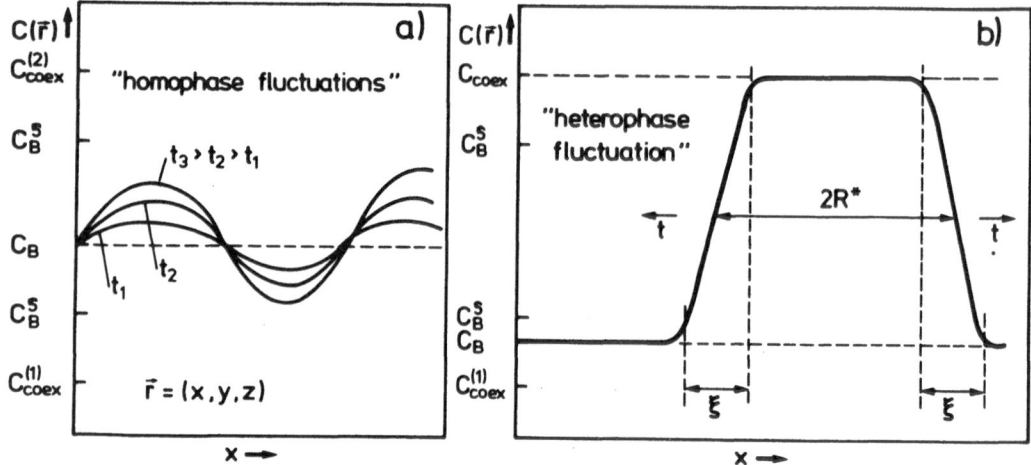

Fig. 8   Unstable fluctuations in the two-phase regime.

$$F = \int d\vec{x} f_{inh}(\vec{x}) = \int d\vec{x}\{f[c(\vec{x})] + C[\nabla c(\vec{x})]^2\}. \quad (20)$$

(C=coefficient of gradient energy, which accounts for the energy excess of domain walls etc.)

Here f is the density of the continuation of the free energy of one-phase states in the two-phase region. Now Eq.18 is generalized to the inhomogeneous case by the functional derivative $\mu(\vec{x}) = \delta F/\delta c(\vec{x}) = (\partial f/\partial c)_T - C\nabla^2 c$; hence Eq. 19 becomes:

$$\partial c(\vec{x}, t)/\partial t = M\nabla^2[(\partial f(c)/\partial c)_T - C\nabla^2 c]. \quad (21)$$

Of course, near T_c Eq. 20 is equivalent to the Landau theory, Eq. 3, with the order parameter $\phi = c_B - c_{crit}$ (cf. Fig. 1a, 7a). Then Eq. 21 becomes:

$$\frac{\partial}{\partial t}(c - c^{crit}) = M\nabla^2[r(c - c^{crit}) + u(c - c^{crit})^3 - C\nabla^2(c - c^{crit})].$$
$$(22a)$$

For describing the *kinetics of the formation of ordered structures* out of disordered ones by *long-wavelength unstable fluctuations,* the analogous equation is ($-M\nabla^2$ is replaced by the rate factor Γ_o for non-conserved order parameters):

$$\delta\phi(\vec{x}, t)/\partial t = -\Gamma_o[r\phi(\vec{x}, t) + u\phi^3(\vec{x}, t) - C\nabla^2\phi(\vec{x}, t)]. \quad (22b)$$

Near equilibrium ($\phi(\vec{x},t)=\phi$ $+\Delta\phi(\vec{x},t)$ $\Delta\phi(\vec{x},t) \rightarrow 0$);we could linearize Eq. 22b to find Eq. 15; for phenomena far from equilibrium as considered now it is essential to keep the full nonlinear term. Only during the initial stages one might try to linearize Eqs 21 and 22a around the initial state, $(\partial f/\partial c)_T = (\partial f/\partial c)_T|_{c_B} + (c-c_B) (\partial^2 f/\partial c^2)_T|_{c_B}$, to find:

$$\partial(c-c_B)/\partial t = M\nabla^2[(\partial^2 f/\partial c^2)_T|_{c_B} - C\nabla^2](c-c_B). \qquad (23)$$

By Fourier transformation, $\delta c(\vec{k},t) = \int d\vec{x} \exp(i\vec{k}\cdot\vec{x}) [c(\vec{x},t)-c_B]$, Eq. 23 is solved as :

$$\delta c(\vec{k},t) = \delta c(\vec{k},0) \exp\{-Mk^2 t[(\partial^2 f/\partial c^2)_T|_{c_B} + Ck^2]\}. \qquad (24)$$

The equal-time structure factor $S(\vec{k},t) \equiv \langle \delta c(-\vec{k},t) \delta c(\vec{k},t) \rangle_T$ then is:

$$S(\vec{k},t) \equiv \langle \delta c(-\vec{k},0) \ \delta c(\vec{k},0) \rangle_T \exp\{R(k),t\} ,$$
$$R(k) = 2Mk^2 [(\partial^2 f/\partial c^2)_T|_{c_B} + Ck^2] \qquad (25)$$

Thus, the fluctuations $\langle \delta c(-\vec{k}0) \delta c(\vec{k}0) \rangle_T \equiv \langle \delta c(-\vec{k}) \delta c(\vec{k}) \rangle_{T_0} = S_{T_0}(\vec{k})$ contained in the initial state at T_0 will decrease to zero, if the amplification factor $R(q)$ is negative. Amplification of fluctuations occurs only for c_B being in the unstable regime : $(\partial^2 f/\partial c^2)_T|_{c_B} < 0$ and for *long wavelengths* $\lambda = 2\Pi/q$ in the range :

$$0 < k < k_c , \qquad k_c = 2\Pi/\lambda_c = \sqrt{-(\partial^2 f/\partial c^2)_T|_{c_B}/C} . \qquad (26)$$

Maximum growth would occur for $k = k_{max} = k_c/\sqrt{2}$, while $S(\vec{k},t)$ at k_c would be time-independent.

While some experimental evidence for this behavior has been seen in quenched amorphous alloys such as Al- 22at%Zn [12] , even there the growth of $S(\vec{k}_{max},t)$ is much slower than exponential. In fact, the nonlinear terms will limit the exponential growth; the various modes $\delta c(\vec{k},t)$ are no longer independent as in Eq. 24, but rather strongly coupled. Due to this coupling, there is neither a time-independent k_c (where $dS/dt = 0$) nor a time-independent k_{max}; rather one has $k_{max}(t) \rightarrow 0$ as $t \rightarrow \infty$, as the formed inhomogeneous concentration distribution coarsens to form macroscopic domains. For unmixing, one expects $k_{max}(t) \propto t^{-1/3}$; domain linear dimensions $L(t)$ grow as $L(t) \propto t^{1/3}$ [10], while for nonconserved one-component orderings $L(t) \propto t^{1/2}$.

In Eq. 25 $S(k,t)$ for $k > k_c$ decays to zero, as fluctuations in the final state are neglected. To include those, a *random force* $\eta(x,t)$ is added on the right hand side of Eq. 22a, which describes gaussian noise but satisfies the conservation of concentration :

$$\langle \eta(x,t)\eta(x',t')\rangle_T = \langle \eta^2 \rangle_T \nabla^2 \delta(\vec{x}-\vec{x'})\delta(t-t'), \quad \langle \eta^2 \rangle_T = k_B T M . \quad (27)$$

Note that it is characteristic of fluctuations near thermal equilibrium that their strength $(\langle \eta^2 \rangle_T)$ is linked to a kinetic coefficient via an "Einstein relation". In the linearized case, one finds the equation of motion for the structure factor [13]:

$$dS(\vec{k},t)/dt = -2Mk^2 \{ [\partial^2 f/\partial c^2)_T|_{c_\beta} + Ck^2] S(\vec{k},t) - k_B T\}. \quad (28)$$

For $(\partial^2 f/\partial c^2)_T|_{c_\beta} > 0$ [stable or metastable region] Eq. 28 has a stationary solution $S(\vec{k},\infty) = S_T(\vec{k})$ of Ornstein-Zernike form, Eq. 6, $S_T(\vec{k}) = (k_B T/C) (\xi^{-2}+k^2)^{-1}$, with $\xi = [C/ (\partial^2 f/\partial c^2)_T|_{c_\beta}]^{1/2}$. At the spinodal curve, both ξ and λ_c (Eq. 26) diverge : in mean-field theory, this curve just is a line of critical points. Going beyond mean-field theory, one no longer has a well-defined spinodal c_s, however, but rather a concentration region δc_s over which a *gradual transition from nucleation to spinodal decomposition* occurs [14,15]. Similarly, in other first-order transitions there is also no sharply defined limit of metastability beyond which a system transforms by coherent growth of long-wavelength unstable modes. Rather the picture is always as follows : close to the first order transition (i.e. for small undercooling), only very rare fluctuations are unstable, namely heterophase nuclei of radius $R \gg \xi$ (Fig. 9a, upper part). These fluctuations need to overcome a free energy barrier $\Delta F^* \gg k_B T$, and hence occur as "nucleation events" which are statistically independent of each other, and hence give rise to the formation of an (incoherent) domain pattern. For larger undercooling of the first-order transition, ΔF^* decreases, R is no longer much larger than ξ, and the density of unstable fluctuations (which grow, in the average) is much larger. The transition from metastable to unstable states then means that ΔF^* is comparable to $k_B T$, R comparable to ξ, the size of typical fluctuations in the system. Now the growing fluctuations can no longer be well approximated as independent nucleation events : due to the depletion zone around each growing "cluster" of the new phase these clusters will have an effective excluded-volume interaction with each other, and hence even a quasi-periodic distribution of concentration may result (Fig. 9a, lower part). Clearly, this description is not essentially different from the description in terms of a "wavepacket" of unstable long-wavelength fluctuations. In fact, the structure factor $S(\vec{k},t)$ resulting from the picture of growing droplets with depletion zones around them

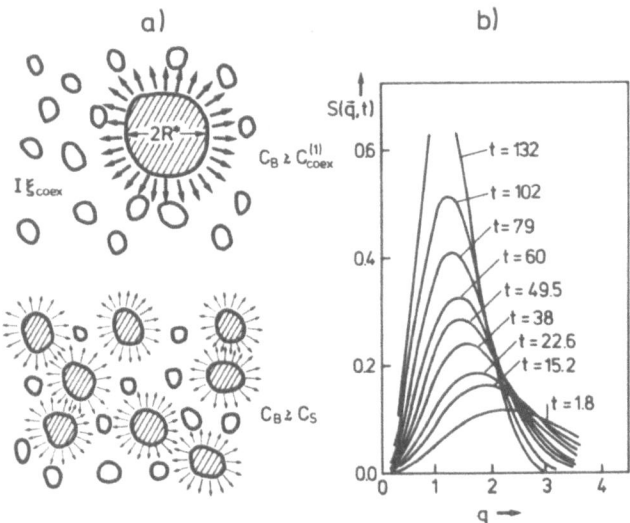

Fig. 9a Schematic "snapshot pictures" of fluctuations of an unmixing system in the metastable (upper part) and unstable regime
(lower part; unstable fluctuations are shaded).
Fig. 9b Structure factor $S(\vec{q},t)$ plotted vs. wavevector $\vec{q}$(normalized by a scale factor) for various times after the quench from infinite temperature to $T=0.6T_c$ for a three-dimensional Ising model
at $c_B=0.1$ [14] .

(Fig.9b) [14] is qualitatively indistinguishable from that resulting
from an approximate nonlinear theory for the early stages of spinodal decomposition [16]. In the latter theory, the nonlinear term
resulting from $(\partial^2 f/\partial c^2)_T$ is factorized in a clever way; this leads
to an equation similar to Eq. 28, but the constant $(\partial^2 f/\partial c^2)_T|_{c_B}$ now
is replaced by a timedependent function $A(t)$ which depends on $S(\vec{k},t)$
itself and must be determined self-consistently [16]. This theory
can be generalized to the kinetics of the formation of ordered superstructures as well [17], and, as expected for the theories valid near
T_c, can be cast in scaled universal form without adjustable parameters [16,17] extending dynamic scaling concepts (Eq. 17) to the regime far from equilibrium [18]. This theory shows, in agreement with
computer simulations [19], that the linear theory (Eqs. 23 and 28)
involving exponential growth (Eq. 25) always is a very poor approximation and nonlinear effects must already be included in the early
stage kinetics. On the other hand, this theory still contains a
sort of spinodal curve (which is not located at $(\partial^2 f/\partial c^2)_T|_{c_B}=0$ but
rather closer to the coexistence curve), where the theory breaks
down [14] : in the metastable region, it only describes the relaxation towards the metastable state, but cannot describe its decay by
nucleation. Thus this theory cannot describe the

above gradual transition between nucleation and spinodal decomposition and it also cannot describe the late stages of the phase transformation. In contrast, the phenomenological "cluster dynamics" theories [14,18,20-23], i.e. generalized nucleation theories, easily incorporate the late-stage Lifshitz-Slyozov growth laws [10], and predict again a scaling description [20]:

$$S(\vec{k},t) = [k_m(t)]^{-3} \tilde{S} (k/k_m(t)) . \qquad (29)$$

As in critical phenomena, where the justification for scaling and universality is the occurence of a large length ξ, on the scale of which details on the scale of interatomic distance are averaged over, here the characteristic length $k_m^{-1}(t)$ also is very large in the late stages. The predicted scaling function $\tilde{S}(x)$ [22] is in very good agreement with experiments both for solid and liquid mixtures. Also the anomalously large undercooling near T_c could be explained by this approach which considers the interplay of nucleation and growth of already nucleated clusters [21,23]. On the other hand, there is still some uncertainty concerning the theory of the nucleation rate J (= number of clusters larger than the critical size nucleated per second and cm^3, ν_A = attempt frequency) :

$$J = \nu_A \exp [-\Delta F^*/k_B T] . \qquad (30)$$

Neither the prefactor ν_A nor the energy barrier ΔF^* can in general be calculated reliably from first principles [23-25]. In the "capillarity approximation, ΔF^* is computed from the competition of bulk and surface terms as $(g(c) \equiv f(c) - \delta\mu c, \delta\mu = \Delta\mu - \Delta\mu coex)$ as :

$$\Delta F(R) = [g(c_{coex}^{(2)} - g(c_B)] \frac{4\Pi R^3}{3} + f_{int} 4\Pi R^2 = -\delta\mu [c_{coex}^{(2)} - c_{coex}^{(1)}] \frac{4\Pi R^3}{3}$$

$$+ f_{int} 4\Pi R^2. \qquad (31)$$

Here a spherical shape of the formed domains is assumed (which is true in fluids, but in solids only near T_c), and f_{int} is the interface tension of a flat planar interface between coexisting phases. The critical radius R^* and associate energy barrier ΔF^* are then found from $\partial (\Delta F)/\partial R|_{R*} = 0$, $\Delta F^* = \Delta F(R^*)$. One finds that $\Delta F^*/k_B T = x_0^2/x^2$, where x_0 is a constant (universal near T_c) and $x = (2/\beta)[c_B - c_{coex}^{(1)}]/[c_{coex}^{(2)} - c_{coex}^{(1)}]$. Fig. 10 shows that this expression for ΔF^* overestimates the free energy barrier in the regime of interest. Unfortunately, away from T_c no such general conclusions on ΔF^* can be drawn.

It is evident from this discussion that the understanding of the kinetics of first-order transitions is still rather incomplete. This is particularly true for the liquid-solid transition for which

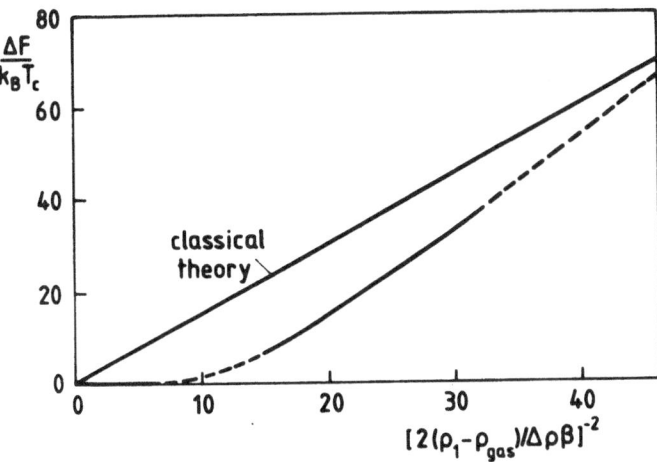

Fig. 10 Free energy barrier ΔF^* against nucleation in the critical region, according to the "classical" capillarity approximation and according to Monte-Carlo simulation for a simple cubic Ising model [25]. Note that this function should be universal near T_c.

concepts such as long-wavelength instabilities have been suggested only very recently [26].

We finally briefly discuss *transitions in systems driven far from equilibrium*. Consider, for example, a fluid held between parallel plates, where due to a temperature gradient $\Delta T/h$ a heat transport from the lower plate to the upper one is maintained (Fig. 11a) [27]. The characteristic quantity is the Rayleigh number R= $g\alpha h^3 \Delta T/k\nu$ (g= gravitational accelleration, α=isobaric thermal expansion coefficient, k=thermal diffusivity, ν= kinematic viscosity, h=distance between plates); if R exceeds a critical value R_c, convection sets in and a roll pattern is formed. Then the effective thermal conductivity $\lambda_{eff}=\dot{Q}h/\Delta T$, ($\dot{Q}$=power density of the heating), exceeds its normal value λ(which appears in the case of pure heat conduction), and the "Nusselt number" $N=\lambda_{eff}/\lambda$ exceeds unity. Related hydrodynamic instabilities occur in a fluid held in between two concentric cylinders, the outer of which is held at rest while the inner is rotating (Taylor-Couette instability) [27], or in nematic

192

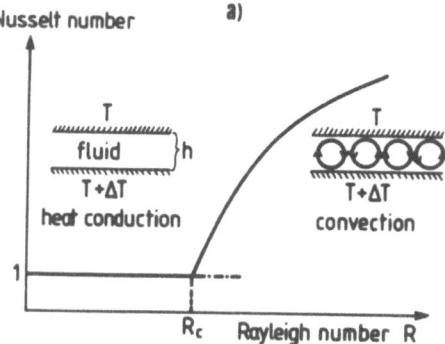

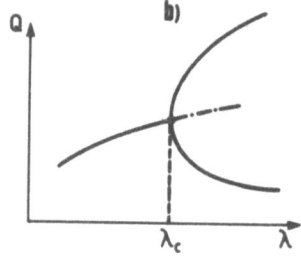

Fig. 11a Nusselt number N plotted vs. Rayleigh number R, indicating
the onset of the convective Rayleigh-Bénard instability (schematic).
Fig. 11b Bifurcation of a quantity Q upon variation of a parameter
λ.

liquid crystals.

Clearly, the behavior in Fig. 11a is reminiscent of a thermal second-order phase transition, the region $R<R_c$ corresponding to the disordered phase $(T>T_c)$ and N being proportional to the square of an order parameter. In fact, one may draw analogies to much wider classes of phenomena : "bifurcations" (Fig. 11b) where a quantity Q (some macroscopic variable) obtained as the solution of some non-linear equation turns unstable when a parameter λ is varied beyond a critical value λ_c and other stable branches appear [28] . Typically, λ is the strength of some external driving force, so that $\lambda=0$ corresponds to thermal equilibrium, while for $\lambda>0$ one has a *"nonequilibrium steady state"*; for $\lambda<\lambda_c$ this state is uniform while at λ_c a *"dissipative structure"* forms (the roll pattern in the example of Fig. 11a). Other examples are lasers near threshold, instabilities in systems undergoing chemical reactions, etc.. [28] . An example for a dissipative structure induced by laser irradiation is the inhomogeneous state in superconducting films, e.g. [29] . Also the laser-induced switch of amorphous $GeSe_2$ films to a darker state with higher absorption coefficient and the oscillatory transmittance observed there [30] can perhaps be interpreted along similar lines [31] .

We here briefly describe the treatment of the Bénard problem [27] . In the Boussinesq approximation the hydrodynamic equations are written with temperature-independent coefficients α, κ, ν in the dimensionless form (P=ν/κ=Prandtl number, p= pressure) :

$$\frac{\partial \vec{u}}{\partial t} + (\vec{u}.\nabla)\vec{u} = -\nabla p + \nabla^2\vec{u} + \sqrt{R}T\hat{z}, \quad \nabla.\vec{u}=0 , \qquad (32a)$$

$$P\left(\frac{\partial T}{\partial t} + \vec{u}.\nabla T\right) = \nabla^2 T + \sqrt{R}(\vec{u}.\hat{z}), \qquad (32b)$$

$\hat{z}$ being a unit vector in the z-axis, $\vec{U},T,p$ describe the deviations of the velocity-, temperature- and pressure profiles across the layer from the profiles in pure heat conduction. In Eq. 32, lengths are measured in units of the layer thickness h, times in units h^2/ν, temperatures in units $\Delta TP/\sqrt{R}$. One may also add random noise terms on the right hand side of Eq. 32 , but unlike Eq. 27 in the general case there is no Einstein relation (or fluctuation-dissipation theorem) linking the noise to the transport coefficient of the fluid. It is clear that due to the nonlinear terms Eq. 32 may have very complicated solutions , in fact chaotic turbulent motions do occur under suitable conditions for $R>R_c$.

Just as in the case of the nonlinear equations describing unmixing kinetics (Eq. 22a) or ordering (Eq. 22b), one starts by a *"linear stability analysis"* [32] , omitting the nonlinear terms and studying the decay rate of modes for various wavevectors $\vec{k}$. For Eq. 32, the analysis is simplest for an infinitely extended horizontal layer with plane free upper and lower surfaces, i.e. the boundary conditions :

$$u_z = T = 0, \quad \partial_z u_x = \partial_z u_y = 0, \quad \text{for } z = 0 \text{ and } z = 1. \tag{33}$$

Neglecting nonlinear terms Eqs. 32 are then rewritten; Eq. 32a couples T to u_z only :

$$\frac{\partial}{\partial t} \nabla^2 u_z = \nabla^4 u_z + \sqrt{R} \left(\frac{\partial^2 T}{\partial x^2} + \frac{\partial^2 T}{\partial y^2} \right); \quad P \frac{\partial T}{\partial t} = \nabla^2 T + \sqrt{R} u_z . \tag{34}$$

Introducing Fourier transforms [$\vec{k} = (k_x, k_y)$ is the horizontal wavevector, $\vec{x} = (x,y)$ the horizontal position vector^y] :

$$u_z(\vec{r},t) = \sum_{n=1}^{\infty} \int d\vec{\kappa} \, u_{zn}(\vec{\kappa},t) \, \sin(n\Pi z) \, \exp(i\vec{\kappa}.\vec{x}), \tag{35a}$$

$$T(\vec{r},t) = \sum_{n=1}^{\infty} \int d\vec{\kappa} \, T_n(\vec{\kappa},t) \, \sin(n\Pi z) \, \exp(i\vec{\kappa}.\vec{x}) . \tag{35b}$$

With $k^2 = \kappa^2 + (n\Pi)^2$ one obtains :

$$\frac{\partial}{\partial T} u_{zn}(\vec{\kappa},t) = -k \, u_{zn}(\vec{\kappa},t) + \sqrt{R} \kappa^2 \, k^{-2} T_n(\vec{\kappa},t), \tag{36a}$$

$$\frac{\partial}{\partial t} T_n(\vec{\kappa},t) = P^{-1} \sqrt{R} \, u_{zn}(\vec{\kappa},t) - P^{-1} k^{-2} T_n(\vec{\kappa},t). \tag{36b}$$

The instability sets in as soon as one of the eigenvalues of the two coupled linear equations goes to zero, i.e. as soon as the determinant of the coefficient matrix becomes equal to zero; this happens first for n=1 at the value of R given by :

$$R = R_c = k^6/\kappa^2 = (\kappa^2 + \Pi^2)^3/\kappa^2 = 27\Pi^4/4 \quad ; \quad (\kappa_c = / 2). \tag{37}$$

Hence one expects that a roll pattern characterized by a wavevector κ_c will form for R slightly above T_c. Of course, nonlinear terms again are needed for a more complete analysis.

From the above equations it is evident that instability occurs at zero frequency, i.e. the system wants to form a time-independent roll pattern. This is analogous to the "soft-mode instability" of second-order phase transitions. For $R > R_c$ and small P, it happens however that the time-independent roll pattern is unstable against *oscillating perturbations*, travelling along the roll axis; their frequency varies (for P→O) as $\omega = \sqrt{3} \; \Pi (R/R_c - 1)^{1/2} \kappa$. This is a "*hard mode*"-instability.

As a general rule, it can be said that for transitions in driven systems one must consider macroscopic degrees of freedom rather than the microscopic ones treated for thermal transitions : consequently, the effect of thermal fluctuations , leading to non-mean-field-like critical behavior at thermal transitions , is negligible in driven systems, their transitions are Landau-like. On the other hand, effects of boundary conditions are much more important. Note also that the driving force in Eq. 12 results from a thermodynamic potential (free energy functional) while for driven systems, where the final state is not in thermal equilibrium, a corresponding functional need not exist.

REFERENCES

1. Stanley, H.E : An Introduction to Phase Transitions and Critical Phenomena (Oxford, Oxford University Press, 1971).
2. Wilson, K.G. and Kogut, J., Phys. Rev. Rep. 12 (1974) 75.
3. Kerszberg, M., Mukamel, D., Rohrer, H. and Thomas, H., Jr., Appl. Phys. 50 (1979) 1836.
4. Binder, K., Z. Phys. B45 (1981) 61.
5. Gebhardt, W. and Krey, U. : Phasenübergänge (Braunschweig, Vieweg, 1980).
6. For a review, see K. Müller, in Dynamical Critical Phenomena (C.P. Enz, ed) (1979) 210.
7. Meissner, G. and Binder, K., Phys. Rev. B12 (1975) 3948.
8. Cowley, R.A., Phys. Rev. B13 (1976) 4877; Folk, R., Iro, H. and Schwabl, H., Z. Physik B25 (1976) 69; see also De Raedt, B., Binder, K. and Michel, K.H., J. Chem. Phys. 75 (1981) 2977.
9. Hohenberg, P.C. and Halperin, B.I., Rev. Mod. Phys. 49 (1977) 435.
10. Lifshitz, I.M. and Slyozov, V.V., J. Phys. Chem. Solids 19 (1961) 35.
11. Cahn, J.W. and Hilliard, J.E., J. Chem. Phys. 28 (1958) 258; 31 (1959) 688.
12. Agarwal and Herman, Scripta metall 7 (1973) 503.
13. Cook, H.W., Acta met. 18 (1970) 297.
14. Binder, K., Billotet, C. and Mirold, P., Z. Physik B30 (1978) 183.
15. Binder, K., Phys. Rev. B8 (1973) 3423; Langer, J.S., physica 73 (1974) 61.
16. Langer, J.S., Baron, M. and Miller, H.D., Phys. Rev. A11 (1975) 1417.
17. Billotet, C. and Binder, K., Z. Physik B32 (1979) 195.

18. Binder, K. and Stoll, E., Phys. Rev. Lett. 31 (1973) 47.
19. Binder, K., Kalos, M.H., Lebowitz, J.L. and Marro, J., Adv. Coll. Interface Sci. 10 (1979) 173.
20. Binder, K. and Stauffer, D., Phys. Rev. Lett. 33 (1974) 1006; Binder, K., Phys. Rev. B15 (1977) 4425.
21. Langer, J.S. and Schwartz, A.J., Phys. Rev. A21 (1980) 948.
22. Rikvold, P.A. and Gunton, J.D., to be published.
23. Binder, K. and Stauffer, D., Adv. Phys. 25 (1976) 343; Binder, K., J. de Phys 41, C4-51 (1980).
24. Langer, J.S., in Systems Far From Equilibrium, L. Garrido, ed., (Berlin, Springer, 1980).
25. Furukawa, H. and Binder, K., Phys. Rev. A (June 1982, in press)
26. Binder, K., J. de Phys. 41 (1980) C4-75; Brown, A., Unger, C. and Klein, W., to be published.
27. Whitehead, J.A., Jr., in Fluctuations, Instabilities, and Phase Transitions (I. Riste, ed.) (New York, Plenum, 1975) 153; Ahlers, G., ibid. 181; Graham, R., ibid. 215.
28. Glansdorff, G. and Prigogine, I., Thermodynamic Theory of Structure, Stability and Fluctuations (New York, Wiley, 1971); see also Lefever, R. and Graham, R., in 27 .
29. Scalapino, D.J. and Huberman, B.A., Phys. Rev. Lett. 39 (1977) 1365.
30. Hajto, J., J. Phys. 41 (1980) C4-63.
31. Fazekas, P., Phil. Mag. (1981).
32. Chandrasekhar, C., Hydrodynamics and Hydrodynamic Stability (Oxford, Clarendon, 1961).

THEORY OF CRYSTAL GROWTH

H. MÜLLER-KRUMBHAAR

Institut für Festkörperforschung
Kernforschungsanlage Jülich   5170 Jülich, W-Germany.

1. INTRODUCTION

The aim of this lecture is a concise summary of the basic concepts of modern theory of crystal growth [1-4]. Detailed quantitative calculations require quite a number of material parameters as an input, for example the lattice symmetry of a crystal, since obviously we cannot hope to explain macroscopic crystalline structure starting from microscopic quantum mechanical formulation (at least in most cases!).

Because of limited space, therefore, I will outline the various mechanisms of crystallization stressing the underlying principles, rather than discussing a special material in all its details. My hope though is that also the reader interested in a specific aspect of the whole complex will find this to be a useful guideline, helping him to find his way to the more specialized literature.

One of the main problems in explaining any specific phenomenon of crystal growth is to discriminate between important and less important parameters of the system under consideration. The structure of this article tries to take this into consideration by starting from interface structures at equilibrium in section 2, then incorporating dynamics by looking first at the immediate neighborhood the crystal surface in section 3, and finally extending the view into the bulk of the adjoining material by presenting the most dramatic effects of material and heat transport in the bulk-material in section 4, which are responsible for the macroscopic patterns of crystallization found in nature and experiment.

The special topic of this conference, the formation of a very thin non-crystalline layer on the surface of a crystal by laser-irradiation, is taken care of in section 2.4, 3.2, and section 4.3. There, the phenomena characteristic for double-interface systems and hydrodynamic phenomena important for this application are discussed. Both fields also are subject to intense activities in the theoretical physics community.

As a last point I would like to remark that all the present concepts should be used with great care for problems of crystallization from the melt, at least on atomic length scales. The point here is that we do not really have a generally accepted theory of freezing, comparable to anything that is known about phase transitions in lattice systems like magnets. The continuous symmetry of space makes it very hard to formulate tractable models. Recent progress in two-dimensional melting [5], however, demonstrating the importance of topological defects (dislocations, disclinations), gives already an idea on how the spontaneous breaking of the continous spatial symmetry of a liquid might occur to form a periodically ordered solid state. Despite this deficiency, present crystal growth theory has produced at least the working concepts to explain quite an impressive number of phenomena occuring as matter changes its state of aggregation.

2. STRUCTURE OF INTERFACES

2.1 Phenomenological models

The main virtue of phenomenological models is the possibility of combining a few simple subsystems to make predictions on the complicated behavior of the full system. The starting point for crystallization models usually is the experimentally observed phase diagram [16]. In Fig. 1 we have a schematic diagram for a simple one-component solid. The three phases solid, liquid, gas join each other along coexistence lines, where two phases separated by a flat interface may coexist for arbitrary times. T_c is the critical temperature, T_{tr} is the triple-point temperature and T_R is an (eventually existing) roughening temperature, to be discussed in the next subsection.

Phase diagrams for more complicated two-component systems are shown in figs. 2, 3. Fig. 2 gives the standard plot of temperature versus relative concentration with a cigar-shaped two-phase region: at a given temperature a liquid with a larger concentration C may coexist with a solid of a lower concentration of one component relative to the other. In Fig. 3 a more complicated diagram of a "peritectic" alloy is shown, where at sufficiently low temperatures a liquid may be in contact with either a disordered solid or a disordered solid may be in contact with an ordered solid with super-structure

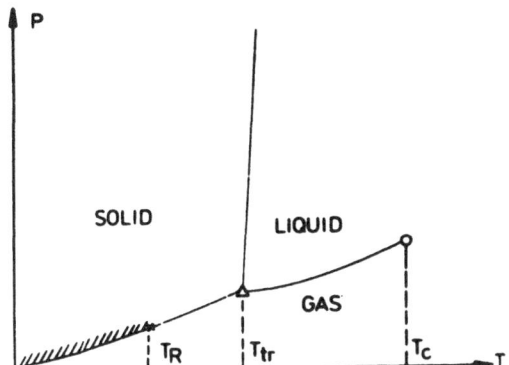

Fig. 1 Phase diagram pressure vs. temperature for a one-component material. The solid lines joining at the triple point T_{tr} are co-existence lines. For $T < T_R$ (roughening temperature) the hatched region indicates metastability of the crystal surface against growth.

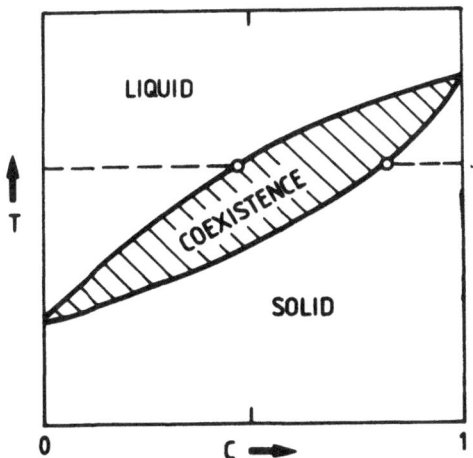

Fig. 2 Standard phase diagram temperature vs. relative concentration of a simple two-component system.

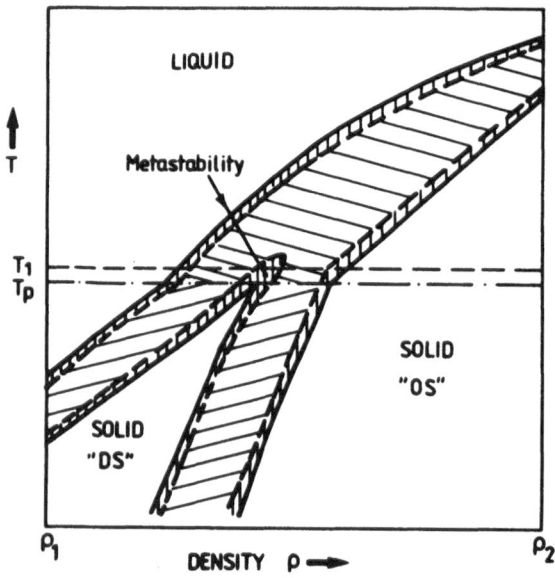

Fig. 3 Central part of the phase diagram of a "peritectic". T_P is the peritectic temperature, where the liquid can coexist with two different solid phases. At T_1 the solid DS-phase (see chapter 3.2) is metastable. The horizontal axis often represents a relative concentration rather than a density. The reason for this choice is given under [33].

such as a doubled unit cell. This diagram has actually resulted from a microscopic model hamiltonian [7] and we will return to it in section 3.

On the phenomenological level' we will model such phase diagrams by a suitably chosen functional for the free energy, where the temperature dependence of the parameters is adjusted to reproduce the phase diagram. Of course we also need a set of *order parameters* [8] ζ_i to discriminate between the different phases. For a two-component system a natural choice for a single order parameter ζ may be the concentration. The coexistence region then is characterized by a free energy which has two equivalent minima as a function of ζ, the one with smaller value of ζ being interpreted as solid, the other one as liquid. Such a functional, known as the Ginzburg-Landau free energy [8], usually is written as:

$$ F = F_0 \int d^3x \; \{ \tfrac{1}{2}(\vec{\nabla}\zeta)^2 - \tfrac{a}{2} \zeta^2 + \tfrac{g}{4} \zeta^4 + r \zeta \; \dots \} \tag{1} $$

where $a(T)$, $g(T) > 0$. r plays the role of a chemical potential. The integral goes over the whole system. In the homogeneous case, $(\vec{\nabla}\zeta)^2 = 0$ and, with $r = 0$, $F(\zeta)$ has two equivalent minima $\zeta_0 = \pm\sqrt{\tfrac{a}{g}}$, corresponding to two coexisting phases of different concentrations $c = c_0 + \zeta_0$. If the two phases are simultaneously present, the system is inhomogeneous ($\vec{\nabla}\zeta \neq 0$) and produces an interface. Performing the variational derivative :

$$ \frac{\delta F}{\delta \zeta} = 0 \tag{2} $$

to minimize the free energy, one obtains explicitly :

$$ \frac{d^2\zeta}{dz^2} + a\zeta(z) - g \zeta^3(z) = 0 \tag{3} $$

by just keeping the variation of ζ in z-direction, setting $r=0$. The solution is :

$$ \zeta = \zeta_0 \tanh (z/\xi); \qquad \zeta_0 = \sqrt{\tfrac{a}{g}} \tag{4} $$

where $\xi=\sqrt{2/a}$ is the width of the interface [9,10].

For a simple monoatomic system one needs a different quantity as an order parameter. One possibility is to relate ζ to the local density $\rho(\vec{x})$ of the system. In a perfectly ordered solid, the density $\rho(\vec{x})$ varies periodically from lattice plane to lattice plane around some average density, as one knows from computer-simulations, while in the homogeneous liquid it is constant $\rho(\vec{x}) = \rho_0$ (see Fig.4).

202

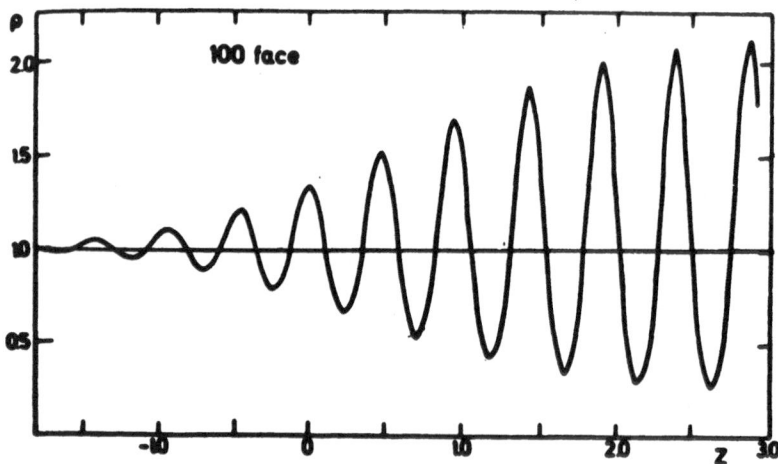

Fig. 4 Liquid-solid interface after [11a]. Plotted is the local density ρ across a (100) interface, where $z \to -\infty$ corresponds to liquid, $z \to +\infty$ to solid.

The envelope over the maximal values of $\zeta(z)$ (where averaging was performed in the plane parallel to the interface) again can be fitted by the Ginzburg-Landau from (1) and (3) to define an interface width.

This approach can be refined by extending (1) to a multiple-order-parameter form [11]. The presently available results for the interface structure gives the qualitative picture of Fig. 4. The quantitative agreement with computer-simulations [12], however, is not yet satisfactory. Related approaches [13], to include the presumably important rotational degrees of freedom of small liquid clusters into the order parameter concept, are very promising. In particular, for the understanding of the kinetics of interface displacement they are an essential ingredient for a satisfactory theoretical description.

2.2 Lattice models

In this section, I shortly introduce a few basic results for surface structures, obtained from the analysis of lattice models. These models are primarily good for the interface between a solid and a vapor or a dilute solution. The solid-liquid interface requires atomistic models as a starting point.

The simplest types of lattice models [14,15] for a solid-vapor interface are the solid-on-solid (SOS) and the discrete Gaussian (DG) models. The Hamiltonian is written as:

$$\widetilde{H} = \frac{1}{T} J \sum_{<ij>} [h_i - h_j]^m; \quad m = \{ \begin{matrix} 1, & SOS \\ 2, & DG \end{matrix}, \tag{5}$$

$$h_i = \{-\infty \ldots, -1, 0, 1, 2, \ldots \infty\},$$

(J is the positive energy constant). The SOS-model (m=1) with additive energies is the more intuitive one, the DG-model (m=2) is better to handle. The variables h_i denote the atomic height of the interface above some reference plane h=0, the summation runs over all nearest-neighbor pairs in a two-dimensional lattice. These models just describe the interface, ignoring bulk properties of the solid or vapor.

Just for completeness I would like to mention a slightly different type of lattice models [16] which also permit a discussion of bulk-properties. In the simplest case, the Hamiltonian is:

$$\hat{H} = \frac{1}{T} J \sum_{<ij>} n_i n_j - \Delta\mu \sum_i n_i, \tag{6}$$

$$n_i = \{0, 1\},$$

where the first sum runs over all neighboring pairs in a *three-dimensional* lattice, the second runs over all sites. The local variables n_i give the occupation of the site i; $\Delta\mu$ is the difference in chemical potential between a filled and an empty site. These models are identical to the well-known Ising models for magnetic systems. This model is the direct lattice-analogy to the continuum model (1). My main interest in this lecture, however, is on interface properties. Since the interface in this model (6) appears as a three-dimensional collective phenomenon, its properties are much harder to calculate. I therefore will dwell on the simpler models of type (5), where the interface is at most a two-dimensional object by definition.

One of the main successes of these models (5) is the detailed prediction of a roughening transition at a temperature T_R. (This is in contrast to a previous speculation based on models with inadequate internal symmetry). The DG-model can be mapped onto a two-dimensional model of a neutral Coulomb-gas [17,18] with logarithmic interaction on a lattice :

$$\widetilde{H}_{CG} = \frac{T}{J} \sum_{<ij>} q_i q_j \ln (r_{ij}) + \mu_o \sum_i q_i^2 , \tag{7}$$

where q_i has the same set of values as h_i, but here representing charged particles. μ_o acts as a chemical potential with a value fixed by the duality-transformation from (5) to (7). Note that the temperature here is reciprocal to the temperature in (5). This Coulomb-gas was known to have a phase-transition, which is easily understood in the limit of large μ_o. There, one has only a single pair of unit charges, $q_1 = 1$, $q_2 = -1$. The energy U of such a pair increases logarithmically with distance, $U \sim \ln r_{12}$, the entropy at a given energy is proportional to the logarithm of the distance, $S \sim \ln r_{12}$. The resulting free energy F of such a pair, therefore is:

$$F = U - TS \sim \ln r_{12} - T_{CG} \ln r_{12}, \tag{8}$$

where multiplicative constants have been suppressed. (As in (7) compared with (5), $T_{CG} \sim T^{-1}$).

Clearly, at small T_{CG} the two charges are tightly bound, while at $T_{CG} > 1$ the entropy term in (8) dominates and the charges become unbound in order to minimize F with respect to r_{12}. The first situation corresponds to an insulating charged system, the second to a metallic. The latter one is the low-temperature "plane" phase for the surface, the first the high-temperature "rough" phase. This concept also extends to μ_o-values required for the quantitative equivalence between (5) and (7). Thus, the existence of a roughening transition for the DG-model is established. Consequences will be discussed in the section on interface kinetics. In Figs. 5a, b computer generated interfaces with surface steps are shown : a) below, b) above the roughening temperature.

Another implication of this Coulomb-gas is very important. For the question on how a two dimensional atomistic system melts, the importance of dislocations in form of Burgers-vectors is well established. The re-parametrisation of such an atomistic system into an (irrelevant) harmonic elastic part and a (relevant) system of dislocation-vectors gives for the latter the Hamiltonian [19]:

$$H = J \sum_{<ij>} \{ \vec{b}_i \vec{b}_j \; \ln(r_{ij}) - \frac{(\vec{b}_i \vec{r}_{ij})(\vec{b}_j \vec{r}_{ij})}{r_{ij}^2} \} + \mu_o \sum_i b_i^2 \tag{9}$$

where the Burgers-vectors b_i have integer absolute values. A renormalization group calculation [19] just as for (7) gives completely analogous behavior, while computer simulations [20] on atomistic models in contrast indicated a discontinuous rather than a continuous transition. A way out of this puzzle was shown by a recent computer simulations [5] of Eq. 9, yielding a continuous transition for small μ_o but a discrete (first order) transition for large μ_o. This mechanism is overlooked by the renormalization group calculation. The still missing link here (in contrast to the roughening problem) is

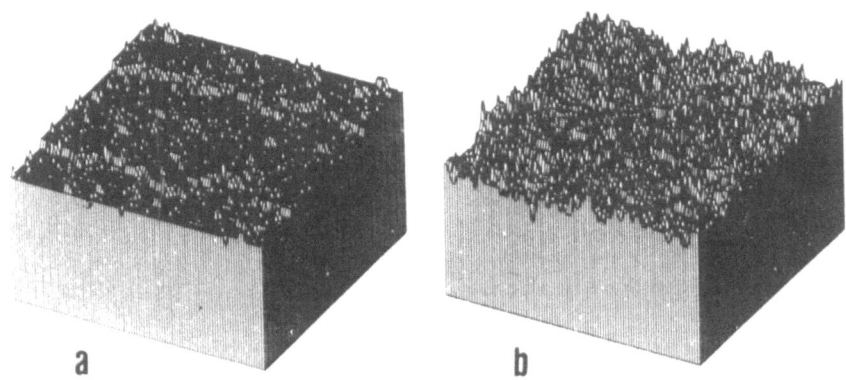

Fig. 5 Computersimulation of a solid-vapor interface at two temperatures a) below b) above the roughening temperature. Two initially
given surface steps are visible in a) but not in b), even though
still present.

the yet unknown absolute value for μ_o for most atomistic systems.
Computer simulations in the near future should be able to evaluate
this quantity in even small atomistic model systems, finally solving
the problem of melting in two dimensions. The mechanisms of the two
different predicted possibilities of melting transition are substantially different. The first one (like in the Coulomb-gas) predicts
melting via "dislocation-unbinding", i.e. a dissociation of pairs
of opposite Burgers-vectors, while the second one predicts [5,21]
just the opposite, namely the formation of interlinked dislocation
networks or small-angle grain boundaries. Such a dislocation vector
system is shown [5] for two temperatures below and above the melting
point in figs. 6a, b. If the latter mechanism is generally correct
(as indicated qualitatively by computer-simulations) this also is
presumably the mechanism for the generally believed first order melting in three dimensions. But then we are left with the very hard
problem to analyze the cooperation of topological defects (dislocations) in three dimensions ! Since this problem is completely open
we cannot say that there exists a sound theory of melting in three
dimensions. Hence our knowledge of the formation of crystals from
the melt is very vague.

2.3 Atomistic models

The difference between a solid and a liquid is the existence
of a nonzero or zero shear modulus. From the last remarks in the
preceding section it is clear that no reliable theory for freezing

206

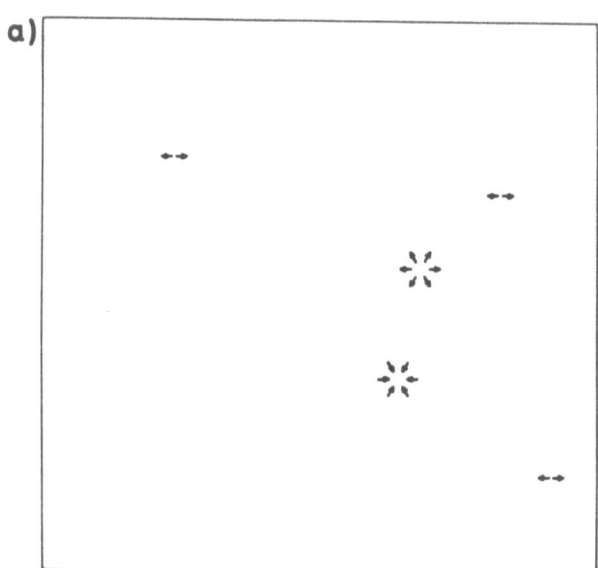

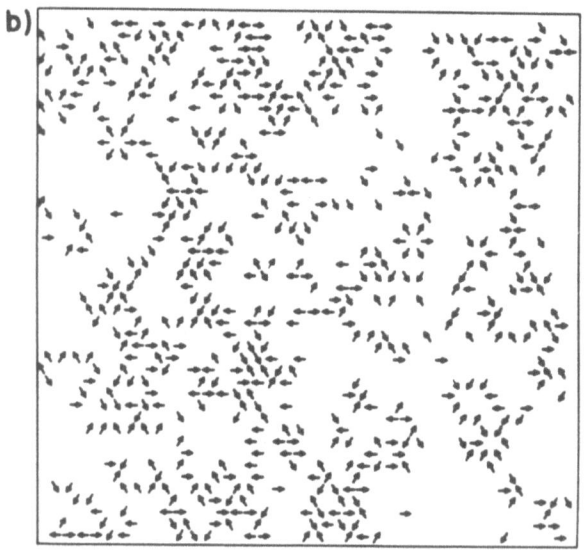

Fig. 6 Dislocation vectors in a two-dimensional system, a) below the melting point ("solid"), b) above the melting point ("liquid"). The formation of a dislocation-network in b) is clearly visible [5].

and crystallization on an atomistic basis exists so far. This statement applies on a global scale, i.e. for the existence and the temperature dependence of an order parameter. This order parameter
(shear modulus) relies on orientational order of neighboring pairs
of atoms relative to distant pairs. In other words, it requires
the evaluation of specific four-point correlation functions, theoretically and experimentally. Nothing is available so far. On a local
scale, however, one is in a slightly better situation for practical
purposes. There one may use the translational order or periodic
lattice as a representation for the solid which then has no counterpart in the liquid. Nevertheless one should keep in mind that the
so-called radial distribution function (pair correlation) shows no
qualitative difference in solid and liquid phase.

With these precautions in mind one may at least study certain
features of the solid-liquid interface. The best models for the
local atomistic density $\rho(\vec{r})$, treating solid and liquid within one
formalism [11] , today have the essential deficiencies that they
do not recover the "zero-density" right in between atomic layers of
the ordered solid and substantially overestimate correlation lengths
inside the liquid normal to the solid-liquid interface (Fig. 4).

The most reliable first-principles calculation so far treats
the solid-surface as a three-dimensionally structured hard wall,
without any feedback from the liquid in contact with this wall.
This simplification at least has the advantage that a recent theory
[22] for a Lennard-Jones liquid in contact with a three-dimensional
"stiff" Lennard-Jones solid gives excellent quantitative agreement
with computer-simulations for the liquid structure. As an example
of the state of the art we give in Fig. 7 a comparison between computer simulations and a recent theory [22] for a Lennard-Jones
liquid in contact with the (100) face of a solid. The agreement is
excellent, in particular considering the fact that the symmetry and
lattice constants in a (100) face are substantially different [23] from
a (111) face. The latter is very close to the short-range structure
of the liquid while the former generates non-trivial phases in the
liquid away from the interface. The calculations are rather lengthy,
therefore I am not giving any explicit formulas here. A selfconsis-
tent extension of this theory to include feedback onto the solid
seems possible. At present, however, we are restricted to computer
simulations which are excessively time - consuming. Some progress
has been made in the understanding of epitaxial layers [24] adsorbed on a solid surface or in the question of surface-reconstruction
[25] for the solid-vapor interface. Both subjects are topics on
their own and will not be followed here in any further detail.

2.4 Interacting interfaces

In many crystal-growth set-ups the interface of the growing
solid is only at first glance identifiable as a single object.

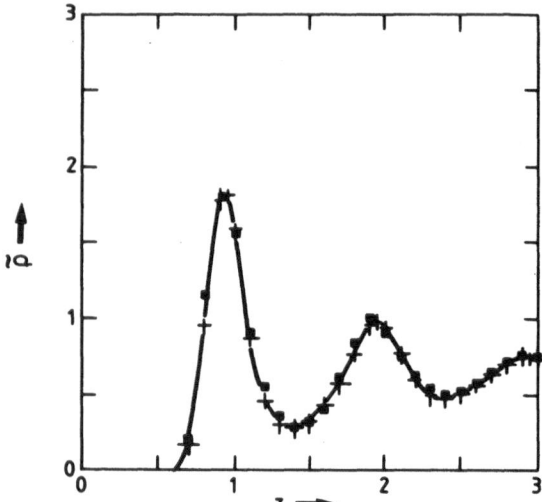

<u>Fig. 7</u> Structure of a liquid in contact with a (100) solid wall,
both subsystems having Lennard-Jones pair potentials. Plotted is
the local density of the liquid in conveniently normalized units
versus distance z from the first solid layer. The black dots are
results from computer simulations, the solid line through the crosses is a new theory [22] without adjustable parameters.

The free surface of a solid in contact with its vapor for example
often develops a "liquid" layer of a few lattice planes thickness.
The melt in contact with a facetted crystal may be more or less
wetting the facet. For a reliable theory - even on the phenomenological level - one therefore should treat the liquid close to the
solid differently than in the bulk, as a laminar layer is formed.
As a third example, a fluid condensing on a solid will behave differently, depending on the wetting properties of the condensed phase.

On a phenomenological level, all these effects may be summarized [26] in a model for two interacting interfaces, the one being
rigid ("surface"), the other one fluctuating ("interface") in structure and distance relative to the first one.

We will discuss the phenomena involved starting from a recent
phenomenological theory [27] for "surface melting". (Previous

theories based on atomistic model are more detailed but miss the universal aspects of the effects common to all the above examples mentioned). Assume a half-infinite system to be given (z⩾0) with a free surface at z=0. Define a local order parameter $\zeta(z)$ which discriminates between the solid ($\zeta \approx 1$) and the liquid phase ($\zeta \approx -1$). The system then can be described by a free energy F:

$$F = \int_0^\infty dz \{ \frac{1}{2}(\frac{\partial \zeta}{\partial z})^2 + A \ [-\zeta^2 + \frac{1}{2}\zeta^4 - \gamma (T) \zeta] + B(\zeta+1)^2 \ \delta(z)\}. (10)$$

The first term under the integral gives the contribution of the solid-liquid interface to the free energy. The term proportional to the constant A has a double-minimum structure for $\zeta = \pm 1$ (at $\gamma(T)=0$). $\gamma(T)$ is negative at low temperatures favoring $\zeta \approx +1$ or the solid phase and changes sign at the melting temperature T* ($\gamma(T*) = 0$), thereafter becoming positive or favoring the liquid phase $\zeta \approx -1$. This corresponds to the first-order melting transition in the bulk of the material. The last term, proportional to the constant B, finally acts only on the surface. It reduces the tendency for the system to order at the surface because of the missing bonds of the surface layer.

A typical profile [27] for the order parameter $\zeta(z)$ is sketched in Fig. 8. The interface width is denoted by ξ, its distance from the surface by ℓ. As $\gamma(T)$ increases towards zero with increasing temperature T, the bulk tends to undergo the first-order melting transition. But the behavior of the surface depends on the value of the ratio R=B/A. For small B/A the surface transition is also of first order, driven by the bulk transition. Above a critical value R*, however, the surface undergoes a critical pre-melting transition as $\gamma=0$ is reached. While the interface width ξ is always finite and determined by the bulk parameters in (10), the width of the liquid layer ℓ increases continuously [28] as :

$$\ell \sim [\gamma (T)^{-1}].\tag{11}$$

(In the mean-field approximation (10) the divergence is only logarithmical). Thus the interface becomes de-pinned and the order parameter on the surface shows melting as a continuous transition, even though the bulk transition is still first order ! This predicted effect, therefore, is the first indication for the existence of critical precursor effects for a first order transition.

In reality, however, one might also have to include the lattice structure of the solid. This was also done recently [29] on the basis of a field theory for an interface pinned to a surface, subject to a periodically varying bulk-potential. The energy functional describing the behavior of a two-dimensional fluctuating interface in a three dimensional system is :

210

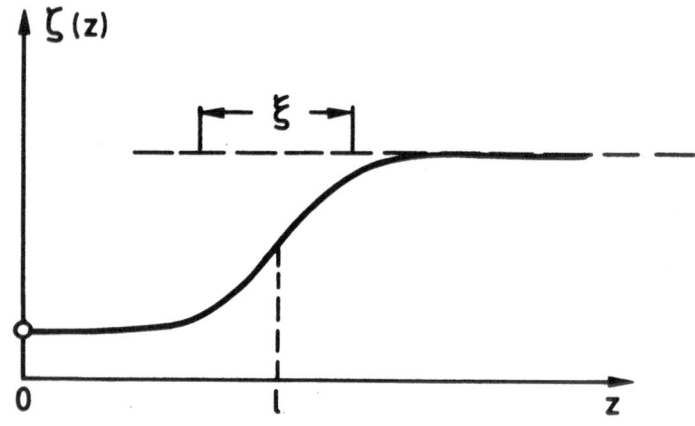

<u>Fig. 8</u> Order-parameter profile $\zeta(z)$ near a surface z=0. The fluc-
tuating interface of width ξ is located around z=ℓ. At the depinning
transition, $\zeta(0) \rightarrow 0$ and $\ell \rightarrow \infty$.

$$E = \iint_{-\infty}^{+\infty} d\vec{x} \; \{(\vec{\nabla}z(\vec{x}))^2 + u(z) + \mu z(\vec{x}) - g_o \cos{(z)}\} \,, \qquad (12)$$

where $z(\vec{x})$ denotes the local deviation of the interface from a
reference plane, u(z) is the potential attracting the interface to
the surface (corresponding to term B in(10)), the cos-term represents the periodic lattice structure. The chemical potential μ con-
trols the transition from solid (μ large) to liquid (μ small). The
functions u, μ and g_o are assumed to depend on temperature and hence
determine the phase diagram.

The behavior of (10) is recovered as a special case under the
following conditions. The function u is strongly positive for z<0
and has a small negative tail for z>0. The bulk transition takes
place at some $\mu=\mu_o$, therefore $\mu_o-\mu \sim \gamma$ and finally $g_o=0$. A renormalization group calculation [29] led to the result (11), identifying
$\ell = \langle z(\vec{x})\rangle$. (An order parameter profile $\zeta(z)$ as from (10) of course
cannot be obtained from (12), since the latter formulation concen-
trates on the interface-fluctuations only. Problems of epitaxial
superstructures of course also fall under this constraint.) It
should be clear at this point that this transition (11) is what

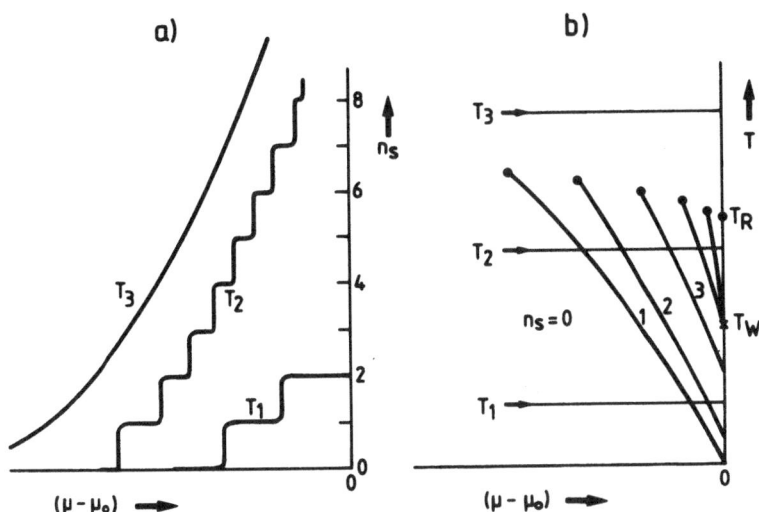

Fig. 9a Surface coverage n_s in numbers of adsorbed layers versus deviation of chemical potential from the (two-phase bulk-) coexistence value μ_0. At three different temperatures T_1, T_2, T_3 different behavior is observed : partial "wetting", layer-wise complete wetting, continuous complete wetting.
Fig. 9b The same diagram as a), but temperature plotted instead of n_s.

Cahn [30] called "critical wetting" of a surface by some condensate.

Another virtue of the formulation (12) is that it allows one to calculate global phase diagrams for this double-interface system, including critical wetting, pre-wetting, incomplete wetting, layering and interface roughening in a fully consistent way [29] . In Figs. 9a, b two–phase diagrams are sketched as an example for the complicated processes ocurring. As μ approaches the coexistence value μ_0 at a low temperature T_1 the interface separates from the surface step by step to a final distance $\ell(\ell\sim n_s{<}\infty)$ leaving a finite coverage n_s of molten phase near the surface (Fig. 9a). At an intermediate temperature T_2, above the wetting temperature T_W but below the roughening temperature T_R, these layering transitions continue to infinity, building an infinitely thick layer of molten material by a sequence of first order transitions. At high temperatures $T_3{>}T_R$, T_W the thickness of the molten layer increases continuously as one approaches the coexistence line μ_0. This again corresponds to (11).

The sequence of T_R and T_W shown here of course is not unique but depends also on the parameters entering (12). For this and other details the reader is referred to the original paper. Experimentally not much is known about these effects. At least the phenomenon of layering and partial wetting, temperature T_1 in Fig. 9, was experimentally found [31] in adsorption of ethylene on graphite. For the question of materials behavior under laser irradiation this whole complex presumably is of great importance.

3. INTERFACE KINETICS

3.1. Lattice models

The presently only microscopic approach to describing crystal-growth kinetics is based on lattice models [2,32-37]. A continuum approach, such as the Navier-Stokes equations [38] for fluid motion, is only possible for the large scale behavior but not for the important processes taking place on the atomic scale of a typical interface thickness. The results obtained from such a microscopic lattice approach of course will serve later as boundary conditions for macroscopic transport equations.

Lattice models as described in section 2.2 have to be generalized in order to account for dynamics. The way this is done in the following results from the observation, that crystal growth generally takes place on time-scales large compared to phonon frequencies (10^{-13}sec). The phonons in the crystal therefore may be assumed to serve as a heat bath, inducing stochastic processes for the dynamics of surface atoms, giving for example rise to surface diffusion or solid-state diffusion. In the neighboring vapor, on the other hand, a single atom covers large distances on the time scales of crystal growth. One therefore may safely assume that the vapor also thermalizes the atoms coming from and going into the solid surface very quickly, such that one may forget about the microscopic dynamics in the gas. With a little more care the same argument again holds for crystallization from a dilute solution. The dynamics of growth from the melt, conversely, is absolutely not understood on the atomistic level.

We will, therefore, concentrate here on the dynamics at a crystal-vapor interface. A convenient starting point to describing stochastic dynamics on a microscopic scale is the Master-Equation approach. Denoting the local height of a two dimensional interface above a reference level by h_i ("i" running over a two dimensional array of lattice-positions), we may characterize a specific configuration of the interface by the set $\{h_i\}$ of all local height variables. The energy of this configuration $H(\{h_i\})$ then is obtained by performing the sum (5). The probability $P(\{h_i\})$ of observing a specific configuration is determined by the Boltzmann-distribution.

At temperatures relevant for crystal growth we may usually ignore quantum effects.

We want to include surface diffusion as well as adsorption – desorption processes on the interface [35]. A single adsorption process then is characterized by the addition of one atom on column j : $h_j \rightarrow h_j + 1$; desorption correspondingly is $h_j \rightarrow h_j - 1$. An event of surface diffusion is the exchange of one atom between two columns at nearest neighbour distance d : $(h_j, h_{j+d}) \rightarrow (h_j-1, h_{j+d}+1)$ and vice versa. As crystallization goes on with time, the probabilities $P(\{h_i\})$ for the configurations become time-dependent : $P(\{h_i\}; t) = P(h_1, h_2 \ldots, h_n; t)$. The Master-Equation formulates now the rate of change with time of these probabilities :

$$
\begin{aligned}
\frac{d}{dt} P (\{h_i\}; t) = & \sum_j \{-P(h_1, h_2 \ldots, h_j, \ldots; t) \; W_1 (h_j \rightarrow h_j \pm 1) \\
& + P(h_1, h_2 \ldots h_j \pm 1 \ldots; t) \; W_1 (h_j \pm 1 \rightarrow h_j)\} \\
+ \sum_{j,d} \{ & -P(\ldots h_j, h_{j+d} \ldots, t) W_2 (h_j, h_{j+d} \rightarrow h_j \pm 1, h_{j+d} \mp 1) \\
& + P(\ldots h_j \pm 1, h_{j+d} \mp 1 \ldots, t) \; W_2 (h_j \pm 1, h_{j+d} + 1 \rightarrow h_j, h_{j+d})\}.
\end{aligned}
$$

$$(13)$$

The index d in the second sum goes over nearest neighbors; W_1 is the conditional probability for a single adsorption-desorption event for a given configuration $\{h_i\}$; W_2 is the conditional probability for a diffusion event, involving sites j and $j+d$.

This equation is intuitively clear. The probability for observing a specific interface configuration $\{h_i\}$ decreases proportional to the probability of $\{h_i\}$ being realized, multiplied by a jump-probability $(W_1(h_i \rightarrow \ldots)$ or $W_2(h_{j+d} \rightarrow \ldots))$ to leave this configuration. $P(\{h_i\}, t)$ conversely increases with time proportional to the probability-flow from neighboring configurations into the one under observation.

The virtue of this equation is that one immediately can formulate equations of motion for expectation values of arbitrary functions $f(\{h_k\})$ of the variables h_k :

$$\langle f(\{h_k\})\rangle \equiv \sum_{\{h_i\}} P(\{h_i\}, t) \; f(\{h_k\}), \qquad (14)$$

where the sum goes over all configurations of the system.

Taking for example $<f> = <h_k(t)>$ as the expectation value for the time development of the interface height at site m, eq. 13 yields with (14) :

$$\frac{d}{dt} <h_k(t)> = g_k(<h_\ell>,<h_\ell h_m>,<h_\ell h_m h_n>,\ldots) , \tag{15}$$

which means that evolution with time of $<h_k(t)>$ is a function g_k of expectation values of the higher moments $<h_\ell h_m>$, etc. But for those moments one also can formulate equations of motion in the same way :

$$\frac{d}{dt} <h_\ell h_m> = g_{\ell m}(<h_r>,<h_r h_s>,<h_r h_s h_q>,\ldots) \tag{16}$$

and so on. One therefore obtains an infinite hierarchy of equations of motion which require for practical calculations a truncation at some level. An alternative way of studying such a system is via Monte-Carlo simulation [39]. Hereby the full hierarchy is simulated directly on the computer by a stochastic process identical to eq. 13. Any closure approximation to truncate the hierarchy (15), (16) then can be tested via computer simulation and thus one may obtain very reliable methods to analyze any specific model defined by (13).

This is the formal machinery. The physics now enter via the transition probabilities W_1, W_2 which have to be constructed according to additional knowledge about the mechanism of the single particle processes. In other words : knowing the mechanism that makes a single particle to be adsorbed in a given (fixed) surface configuration or to jump to a neighboring position, the master-equation (13) automatically takes care of the complicated collective features of many particles being adsorbed or diffusing on the surface.

One condition of course has to be generally fulfilled by the transition probabilities. In thermal equilibrium we know that $P(\{h_i\})$ has to reproduce the Boltzmann distribution, in order to have a zero on the left-hand side of (13). This detailed balance condition implies for the adsorption-desorption probabilities :

$$\frac{W_1(h_j \rightarrow h_j+1)}{W_1(h_j+1 \rightarrow h_j)} = \exp\{-\frac{1}{T}[H(h_j+1) - H(h_j)]\} , \tag{17}$$

where $H(h_j)$ is the energy of the system with the particular value h_j at site j and $H(h_j+1)$ is the same, but with h_j increased by one unit. This fixes the transition probabilities apart from an arbitrary prefactor that may depend on the sites neighboring h_j, but not on h_j itself. For surface diffusion a similar relation holds.

I will now describe results [35] obtained for an application based on the Bethe-pair-approximation for the hierarchy (15), (16)

which allowed us to explicitly calculate rates of crystal growth
from the vapor, including both adsorption-desorption and surface
diffusion. The explicit equations are rather lengthy even though
they incorporate only expectation values $\langle h_k \rangle$ and $\langle h_k h_{k+d} \rangle$, where
h_{k+d} is a site neighboring h_k. Thus one has two coupled nonlinear
differential equations which can be solved with fairly little amount
of computing and even analytically for a number of important limiting cases.

For growth from the vapor (or dilute solution) we may safely
assume, that the impingement frequency ν_o of atoms at the surface
is simply controlled and known from the pressure in the vapor, independent of the surface structure :

$$W_1(h_j \to h_{j+1}) = \nu_o . \tag{18}$$

In thermal equilibrium, the detailed balance condition (17)
requires the desorption probability to be the same on the average.

In order to have the crystal growing we have to deviate from
thermal equilibrium e.g. by a pressure change, i.e. we have to introduce a difference $\Delta\mu$ in chemical potential across the interface.
As the simplest guess the "Wilson-Frenkel" theory assumes that the
surface structure is not changed on the average during growth, the
only change being a modification of the adsorption-rate by $\Delta\mu$:

$$W_1(h_j+1 \to h_j) = \gamma_o\, e^{-\Delta\mu/T}. \tag{19}$$

The resulting "Wilson-Frenkel" rate of growth V_{WF} then is the
difference between (19) and (18) :

$$V_{WF} = \nu_o(1-e^{-\Delta\mu/T}) . \tag{20}$$

For most interesting cases, however, this assumption of equilibrium structure of the interface is not correct. Even for sufficiently rough interfaces (high growth rates or high temperatures),
our Bethe-approximation [35] gives a correction :

$$V = V_{WF} \left\{ 1 - \frac{zJ}{2T} \int_o^{2\pi} \frac{d^2q}{(2\pi)^2} \frac{\zeta(\vec{q})}{1 + \ell_D^2 \zeta(\vec{q})} \right\}, \tag{21}$$

where :

$$\zeta(\vec{q}) = 1 - \frac{1}{z} \sum_d e^{i\vec{q}\vec{R}_d} .$$

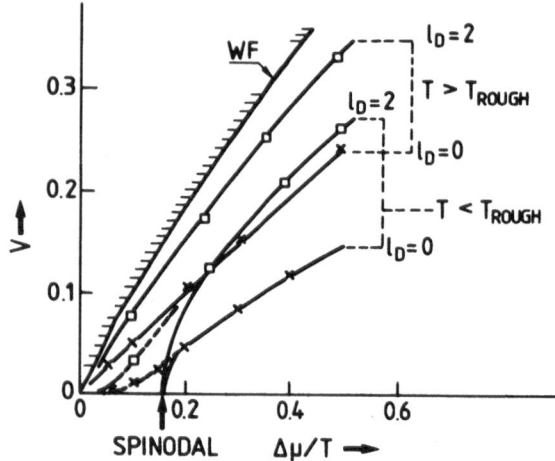

Fig. 10 Growth rate V (in units of impingement rate) versus chemical potential difference $\Delta\mu/T$. Full lines are theory [35], compared
with Monte-Carlo data. Parameters of the curves are temperature T
and diffusion length l_D. WF is the limiting Wilson-Frenkel approximation of crystal growth.

Here R_d is a nearest neighbor vector in the plane; z is the number
of neighboring sites in the plane (z=4 on a square lattice); ℓ_D is
the diffusion length (average range of atoms hopping over the surface), to be calculated also within this theory. J is the bond
energy per atom. We see that the Wilson-Frenkel rate is only the
upper limit for high temperatures $T\to\infty$ or high surface diffusion
$\ell_D\to\infty$.

A more detailed comparison of computer-simulation results and
our Bethe-approximation is given in Fig. 10. We see that for temperatures above the roughening temperature T_R the agreement is perfect. For lower temperatures we have excellent agreement outside
the "spinodal" regime. This deserves some further comment. At
low temperatures,in thermal equilibrium the interface is fairly
flat with only a few holes and a few ad atoms present. In order to
advance the interface by one lattice unit one has to form large
clusters of ad-atoms at an intermediate state. These clusters represent a gain in free energy ΔF proportional to the area $=\Pi R^2\Delta\mu$
(R is a typical cluster radius), but a loss in ΔF because of the
broken bonds on the circumference $= 2\pi R\gamma$, where γ is the edge free-

energy for this boundary. The formation of a cluster of given radius therefore costs an amount ΔF of free energy :

$$\Delta F = 2\Pi R\ \gamma - \Pi R^2\ \Delta\mu .\qquad(22)$$

This function first increases linearly up to a maximum at $R^* = \gamma/\Delta\mu$ and then decreases quadratically for $R \to \infty$. The advancement of the interface, therefore, has to encompass a "nucleation-barrier", i.e. a critical cluster of radius R^* has to be formed by thermal fluctuations, before the spreading of the cluster leads to a gain in free energy and hence forms a new layer on the surface. This is the essence of nucleation theory [40].

Any truncation of the hierarchy (15), (16) misses this subtle point but produces a growth rate $V = 0$ for small values of the driving force $\Delta\mu < \Delta\mu^*$. On the other hand, nucleation theory works well just inside this region and overlaps to a sufficient degree with the Bethe-approximation just outside $\Delta\mu^*$. Combining these two approaches we have, therefore, obtained a very satisfactory theory of crystal growth for arbitrary temperatures, provided the lattice-model representation is a meaningful basis.

3.2 Phenomenological models

The detailed lattice models with their advantage of allowing for very detailed calculations sometimes require too much effort if one is primarily interested in a macroscopic effect of a complicated system. There one better uses first the phenomenological approach, leaving detailed microscopic calculation of the system parameters to a later time.

A convenient starting point for phenomenological equations of motion is the time dependent Ginzburg-Landau equation [41]:

$$\frac{\partial \zeta(\vec{x},t)}{\partial t} = -\kappa \frac{\delta G(\zeta(\vec{x},t))}{\delta \zeta} ,\qquad(23)$$

where ζ is the order parameter and G is the Gibbs free energy functional (I write G rather than F to indicate non-conservation of the order parameter ζ in the subsystem under consideration). The intuitive meaning of this equation is that the system tries to gain free energy the fastest way possible by following the negative "gradient" of G in the function space. The kinetic coefficient κ sets the time scale. (For cases with conserved order parameter, like particle diffusion, κ has to be replaced by $\kappa\vec{\nabla}^2$...).

Simplifying the spatial structure $\zeta(\vec{x})$ to one dimension $\zeta(z)$, we immediately obtain from the free energy (1) (with deliberately fixed constants) :

$$G = \int dz \, \{ \frac{1}{2} \, (\frac{\partial \zeta}{\partial z})^2 - \frac{1}{2} \, \zeta^2 + \frac{1}{4} \, \zeta^4 - \Delta\mu \, \zeta \} \; ; \tag{24}$$

the equation of motion:

$$\kappa^{-1} \, \frac{\partial \zeta}{\partial t} = \frac{\partial^2 \zeta}{\partial z^2} + \zeta - \zeta^3 + \Delta\mu \, , \tag{25}$$

which has the stationary solution:

$$\zeta(z,t) = \delta - \tanh(\frac{z-Vt}{\xi}) \, , \tag{26}$$

$$V = \sqrt{\frac{9}{2}}\kappa\Delta\mu \, ; \; \xi = \sqrt{2} \, ; \; \delta = \frac{\Delta\mu}{2} \, ; \; |\Delta\mu| \ll 1 \, .$$

This corresponds to an interface between the phases $\zeta \approx +1$ and $\zeta \approx -1$ of width ξ, moving at constant velocity V.

This equation is thus generally valid for crystals growing normal to an interface at temperatures above the roughening transition, where a linear relation $v \sim \Delta\mu$ holds between the growth rate and the driving force $\Delta\mu$ across the interface.

I will now sketch the application of this formulation to the more complicated problem of the dynamics of two cooperating interfaces, whose statics were described in section 2.4. (The explicit calculation [33] in fact was performed using the master-equation formalism, but a fully analytical calculation on the basis of (23) is in progress and will be available soon). As an example we will discuss the phenomenon of *kinetic disorder* [42, 43, 33, 34] . This occurs in two-component systems where a solid grows from a liquid and the solid consists of two components A and B which tend to arrange in an ordered way with A and B atoms alternating in the lattice. Depending on the growth rate, the forming solid will turn out to be ordered or disordered.

A possible phase diagram is depicted in Fig. 3 which is called a "peritectic" mixture. ρ measures the density in arbitrary units. At high temperatures, the ordered solid (OS) can coexist with the liquid (L), at low temperatures the disordered solid (DS) can coexist either with the liquid or with the ordered solid, depending in which range the actual density (depending on pressure) is given. The particular growth forms under consideration will take place at a temperature T_1 shortly above the point where the OS-L two-phase region splits into the OS-DS and the DS-L regions. While the system is solidifying the OS-phase has the lowest minimum of the free energy, the liquid is next higher and the DS-phase is highest. We will, however, assume, that we work close enough to the peritectic splitting point, such that the DS may exist as an extremely long living metastable phase, once it could be prepared. (This is the usual case for amorphous quenched systems).

In the Ginzburg-Landau formulation (23) we have now two order parameters $\zeta(z,t)$ to describe the difference between the solid and the liquid phase $\eta(z,t)$ to describe the difference between the ordered and the disordered solid. Accordingly we have two equations of motion to solve :

$$\frac{\partial}{\partial t} \zeta = - \kappa_\zeta \frac{\delta G(\zeta,\eta)}{\delta \zeta},$$
(27a)

$$\frac{\partial}{\partial t} \eta = - \kappa_\eta \frac{\delta G(\zeta,\eta)}{\delta \eta}.$$
(27b)

The free energy G contains both for ζ and η a term like (24) plus a term that couples ζ with η. The philosophy for this coupling is that sublattice ordering requires the mother-phase to be solid. This suggest a structure like:

$$\lim_{s \to \infty} (1- \tanh [s(\frac{\eta}{\zeta} - 1)])^{-1},$$
(28)

which becomes infinite for $\eta > \zeta$ and is zero otherwise.

The detailed form of this term of course depends on the phase diagram and on the precise definition of the order parameters, in particular with respect to discriminating between solid and liquid. (Our previous analysis [33] has circumvented this point by directly constructing a lattice model of type (6) with three possible states for each site : liquid, A or B).

The results of the dynamic analysis of (27) are as follows. For growth rates V below a critical value v*, the system with two interfaces - between L and DS and between DS and OS - keeps the two interfaces tightly coupled. In this way an ordered solid grows out of the liquid, with only a thin intermediate layer of essentially disordered solid. Beyond the critical growth rate v* , however, the interface between OS and DS cannot keep up with the interface between DS and L. Therefore the DS-phase increases continuously in thickness and is the dominating phase at the end of the growth process. One may argue that the OS-DS interface, although slower, still keeps advancing. But in a typical experimental set-up cooling will take place usually through the solid phase and hence the OS-DS interface ultimately will come to rest.

I have emphasized this effect because of its obvious relevance for laser annealing techniques being discussed at this conference. There, one of the goals is to produce homogeneous solid mixtures out of systems that will not deliberately mix in the solid phase.

The sketched mechanism – as simple as it is – provides at least a sufficient condition to reach the mentioned goal. Sufficient and not necessary, since the change of kinetic coefficients κ_ζ, κ_η with temperature has not been discussed. Almost certainly, κ_ζ will be less sensitive to temperature changes than κ_η, which certainly varies exponentially with temperature, since solid-state reordering is a thermally activated process.

So far, I have essentially neglected the three dimensional aspects of interface profiles, concentrating only on the local profile in direction normal to the interface. In the remaining of this section I will shortly comment on generalizations, regarding macroscopic deviations of interfaces from a plane.

Being interested in the macroscopic three-dimensional shapes of such interfaces one may simplify further by neglecting the finite thickness ξ of the interfaces. The free energy of the system then is assumed to depend on the local position, the direction and the curvature of the interface. If the interface encloses a fixed volume of one phase (a crystal) it will in thermodynamic equilibrium tend to assume a shape that minimizes the interface free-energy. If the interface is forced to advance due to a difference in chemical potentials, the kinetics of interface motion can be described by a generalized time-dependent Ginzburg-Landau equation [41] (see Fig. 11):

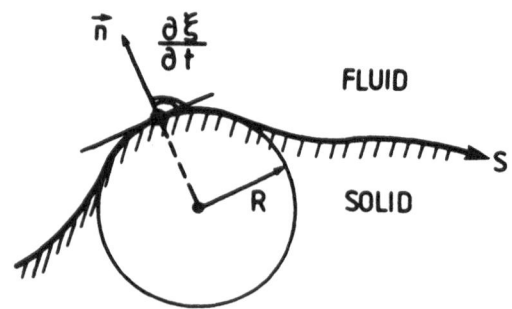

Fig. 11   Scheme for the definition of parameters for the equation of motion of a smooth interface, eqs. 29 and 30.

$$v_\perp \equiv \frac{\partial \, \xi \, (\vec{S})}{\partial t} = - \, \kappa (\vec{\Theta}) \, \frac{\delta G}{\delta \, \xi(\vec{S})} \tag{29}$$

where $v_\perp$ is the normal velocity of the interface, ξ is the normal displacement of the interface, $\vec{S}$ is a point on the interface, $\vec{\Theta}$ is the orientation of the interface relative to crystallographic axes and κ is the mobility of a plane interface moving in direction $\vec{\Theta}$.

The evaluation of the variation gives in three dimensions explicitly:

$$v_\perp = \kappa(\Theta_1, \Theta_2) \, [\Delta\mu - R_1^{-1} \, (\gamma + \frac{\partial^2 \gamma}{\partial \Theta_1^2}) - R_2^{-1} \, (\gamma + \frac{\partial^2 \gamma}{\partial \Theta_2^2})] \tag{30}$$

where $\Delta\mu$ is the difference in chemical potential (per unit volume) between the two phases, γ is the surface free energy, R_1 and R_2 are the two principal radii of curvature and Θ_1, Θ_2 the corresponding angles relative to crystallographic orientation.

The stationary solution $v_\perp = 0$ of course gives the shape of the critical nucleus for an anisotropic system. This kinetic equation has been compared with Monte-Carlo simulations of surface-spirals and was found to work already on length scales of only a few angstroms. An example for a surface spiral calculated from (30) is given in Fig. 12.

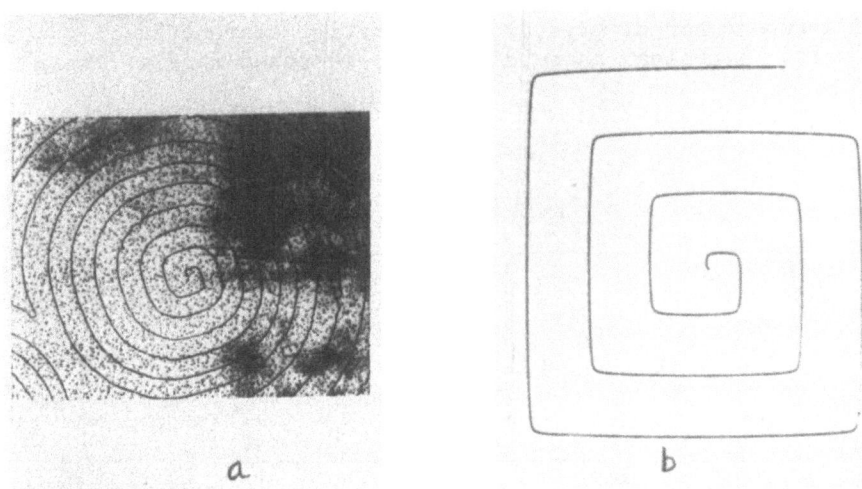

Fig. 12 Surface spirals formed on crystal surfaces around screw dislocations. a) experiments by B. Lampert et al. b) theory (cited in [44]).

The analysis of this nonlinear equation of motion (5) is analytically possible only in a few cases, among which the problem of shape conservation during growth has been solved [41]. A crystal or, in two dimensions, an adsorbed cluster or surface spiral preserves its shape only if the anisotropy of the kinetic coefficient κ is equal to the anisotropy of the surface tension γ. If κ is less anisotropic, the resulting structure rounds off at large diameters and sharpens in the opposite case. The disappearence of anisotropy of surface spirals [44] with increasing temperature allows us also to give excellent estimates for the value of the "roughening" temperature.

4. TRANSPORT PROCESSES

4.1 Diffusion

In the preceding sections I have discussed the structure and the kinetics of interfaces, ignoring the question on how matter and heat are transported [45] towards and away from the interface. This chapter on transport processes will discuss the effects of interface dynamics coupled to the transport mechanisms in the bulk of the material.

The first of the important effects, segregation and constitutional supercooling, have simply to do with the fact that the interface between solid and nutrient phase serving as a boundary condition changes its position with time. This is the so-called "Stefan"-problem of a moving boundary condition. As an example we study the growth of a two-component crystal with relative concentration C from the melt. A typical phase diagram is sketched in Fig. 13 a.

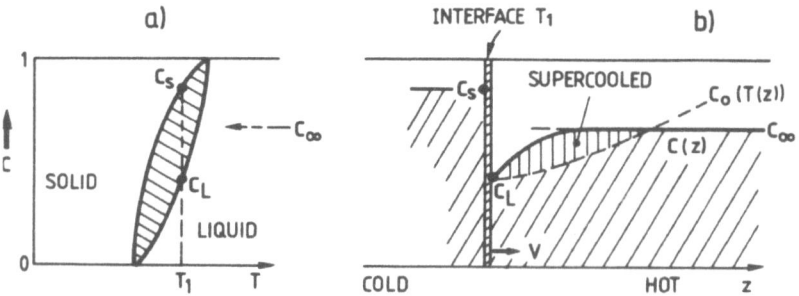

<u>Fig. 13</u> Scheme for the explanation of constitutional supercooling (see text).

In mathematical terms it starts with a diffusion equation for the solute within the solvent:

$$\frac{\partial C}{\partial t} = D \, \nabla^2 \, C \tag{31}$$

where C is the concentration of the solute and D the diffusion coefficient in the liquid.

It is convenient to re-phrase (31) in a coordinate system fixed relative to the position of the solid-liquid interface with a velocity v in z-direction :

$$\frac{\partial C}{\partial t} = D \, \nabla^2 \, C - v \, \frac{\partial C}{\partial z} \, . \tag{32}$$

The velocity v now in turn depends on the concentration at the interface, given by the conservation law of matter:

$$D \, \frac{\partial C}{\partial z} \, \Big|_{z=0} = V(C_{solid} - C_{liquid}) \, \Big|_{z=0} \, , \tag{33}$$

everything taken on the interface z = 0.

As solidification proceeds, the concentration profile in the liquid assumes a very slowly varying exponential form away from the interface. The concentration at the interface is close to the equilibrium value, far away it converges to the initial value before growth was started by lowering the temperature. In the usual set-up, the solid is kept cooler than the liquid (Fig. 13b) and the concentration $C(z,t)$ at the interface is always close to the equilibrium value $C_0(T)$. Ahead of the interface the theoretical equilibrium concentration increases with distance as the temperature increases (Fig. 13a, b). The actual concentration $C(z,t)$ however is completely determined by the diffusion process (32) (33) and thus may be higher than the corresponding equilibrium concentration. This is called "constitutional supercooling". The liquid, therefore is not in a stable equilibrium state but is unstable against deformation of the interface or even formation of microcrystals ahead of the interface. Both processes generally lead to a destruction of the crystalline perfection. The deformation of the interface into dendritic structures will be studied below.

One can, however, avoid undesirable constitutional supercooling by requiring the obvious condition for stability at the interface :

$$\frac{\partial c_o}{\partial z} \equiv \frac{\partial c_o}{\partial t} \frac{dT}{dz} \geqslant \frac{\partial c}{\partial z} \; ; \qquad z=0$$

or

$$\left(\frac{\partial c_o}{\partial T}\right) \nabla T \geqslant \frac{c_s - c_L}{D} \; v, \qquad\qquad\qquad (34)$$

i.e. by reducing the growth rate under otherwise fixed conditions.

The effect of constitutional supercooling on the other hand may be quite desirable for the purpose of homogenizing compound material by laser annealing. No precise data however seem to exist so far.

4.2 Dynamic instabilities : flat surface, dendrites, eutectics

4.2.1 Flat surface instability

As is clear from the diffusion equation (32) and the boundary condition (33), the moving boundary problem is essentially non-linear, thus allowing for considerably more complicated behavior than ordinary linear diffusion problems.

The basic condition for an instability of the solid-liquid boundary against wiggly deformations [46] was already indicated in the preceding section : a supercooled region of liquid ahead of the interface. Let us assume now for simplicity that the temperature field is practically constant, since heat transfer is generally faster than transfer of matter. As a convenient notation let us introduce the field:

$$u(z,t) = \frac{c_o - c(z,t)}{c_\infty - c_o} \qquad\qquad\qquad (35)$$

where c_o is the equilibrium concentration in the liquid at the interface, c_∞ the concentration at infinity, both of which are constants. The field $u(z,t)$ therefore is o at the flat interface, decaying to -1 at infinity. Let us finally assume, that the typical wiggles have linear dimensions $\lambda = 2\Pi/k$ small compared to the range of the diffusion field decaying exponentially with distance z from the surface.

This allows us to ignore the explicit time dependence in the diffusion equation (32), called the "quasi stationary" approximation. It says that the diffusion-field adjusts itself instantaneously to the varying interface structure.

The diffusion equation to be solved is therefore simply :

$$0 = \nabla^2 u(\vec{x},t) + \frac{2}{\ell} \frac{\partial u}{\partial z} \tag{36}$$

where $\ell = \frac{2D}{V}$ is the diffusion length encountered in the solution to (36):

$$u_0 = \exp(-\frac{2}{\ell} z) - 1. \tag{37}$$

But now we are interested in deformations of the interface. At this point the surface-tension enters as a restoring force just as in (30). We therefore have to modify the boundary-condition $u_\Theta = 0$ into :

$$u(z=0,t) = -d_0 R^{-1} \tag{38}$$

where R is the local radius of curvature of the interface, and $d_0 = \gamma(T_M c_p /L^2)$ is the "capillarity length" being proportional to the interface-tension γ. (T_M is the melting temperature, c_p the specific heat and L the latent heat of melting).

Since the position of the boundary is not known a priori we also have to use the continuity equation (33) which here has the form :

$$V_\perp = -\vec{n} D \vec{\nabla} u, \text{ (at the boundary)}, \tag{39}$$

where $V_\perp$ is the velocity in direction $\vec{n}$ normal to the boundary.

This equation also is the source of the variation of the inter-face-form with time, despite the quasi-stationary approximation (36). The mathematical problem is by (36), (38) and (39) now completely defined.

The perturbation-solution [46] of this non-linear system starts from a plane interface with small periodic deformations $\rho(\vec{x},t)$ in z-direction normal to the interface. I just state the results of the fairly simple calculation linear in ρ.

Fourier-decomposition of $\rho(\vec{x},t)$ yields :

$$\rho(x,t) = \rho_k e^{ikx+\Omega_k t}, \tag{40}$$

keeping only one coordinate x parallel to the plane interface. The spectrum Ω_k furthermore is :

$$\Omega_k \sim v \; k(1- \frac{1}{2} d_o \ell \; k^2),$$
(41)

increasing linearly with k up to a maximum, then decreasing below zero at $k_s = 2\Pi/\lambda_s$, where :

$$\lambda_s = 2\Pi\sqrt{\frac{1}{2} d_o \ell}$$
(42)

is called the stability length. Wiggles of wavelength larger than λ_s, therefore, tend to grow exponentially with time, while perturbations of smaller wavelength decay with time.

This dynamic instability is an important mechanism for the formation of macroscopic solidification patterns. In multicomponent materials they lead to more or less periodic variations of composition transverse to the growth direction. This phenomenon was also observed in laser-annealing experiments giving clear evidence of the existence of a constitutionally supercooled liquid layer.

4.2.2 Dendrites

If the instability of the planar growth front is strong enough such that the amplitude of the perturbation becomes substantially larger than the characteristic wavelength, the protuberances act as independent objects competing for the diffusion field. Small local differences in amplitude become amplified and an array of single spikes evolves.

These single spikes do not remain smooth but immediately develop tree-like structures. These are the dendrites [47] . Most industrially solidified alloys (even steel) grow via the dendrite mechanism, which determines the microcrystalline properties of the evolving solid.

Since the profile of a dendrite looks like a beared paraboloid (Fig. 14), the mathematical approach conveniently starts from a parabolic coordinate system which moves relative to the laboratory frame of reference. Otherwise the problem is quite similar to the one described under 4.2.1. We skip the complicated detailed analysis [48,46c] and sketch the approach in a greatly simplified version which shows at least the mechanism correctly. Assume that we have a needle-shaped crystal with tip-curvature R moving in tip direction. In quasi-stationary approximation the diffusion gradient (and thus the current of material at the spherical boundary of the tip) is :

$$|\vec{\nabla}u| \sim R^{-1}.$$
(43)

From the macroscopic equation (30) of an advancing interface, we know that there exists a critical radius R_c such that, if $R < R_c$,

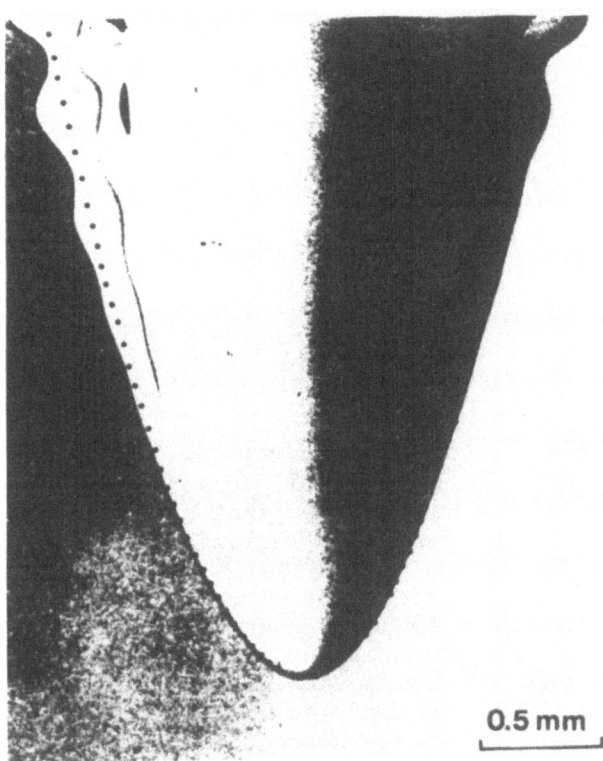

Fig. 14   Parabolic tip of a dendritic crystal [47].

surface tension hampers growth as also comes out of nucleation
theory (22). This gives the "Gibbs-Thomson"-factor $(1-R_c/R)$ which
restricts advancement of the tip to $R>R_c$. Combining this relation
with (43) we obtain the relation :

$$V \sim \frac{1}{R} (1- \frac{R_c}{R}) \qquad (44)$$

for the growth rate as a function of the radius of curvature R of
the tip. R_c of course is proportional to the surface tension of
the solid-liquid interface and inversely proportional to the supersaturation $c-c_o$ (see (35)) of the bulk liquid.

228

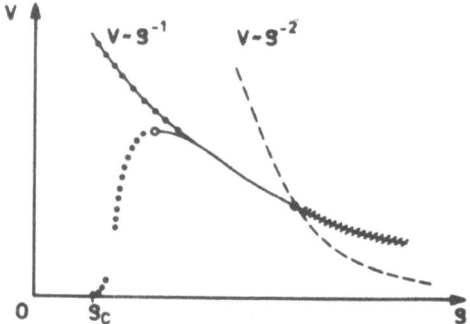

Fig. 15 Schematic plot of the velocity versus radius of curvature
for stationary dendritic growth (full curve). The broken curve gives
the scaling law for the point of marginal stability, the intersection with the full curve marks the region of operation of the side-
branching dendrite.

The exact form of $v = v(R)$ of course differs somewhat from (44)
but still has the same shape, shown in Fig. 15. At a first glance
one may be surprised that this does not predict directly the growth
rate or the size "R" of the dendrite, but only relates the two. For
allmost thirty years it was speculated, therefore, that the growth
rate would emanate from the competition principle that the fastest
dendrites ($v = v_{max}(R)$) would dominate the systems behavior. Experimentally [47] however, this is not observed but the growth rate is
typically an order of magnitude lower.

The principal answer [48,49] to the problem turned out to be
remarkably simple. The structure of the curve (Fig. 15) is basically
determined by two system parameters, the diffusion length $\ell=2D/V$ and
the critical radius R_c. But there is a third independent parameter
in the system, namely the capillarity length d_0 of surface fluctuations (38), which also can be re-phrased in terms of the stability
length λ_s (42).

Our detailed linear stability analysis then indeed revealed
that this stability length plays a crucial role in dendrite dynamics.

For $R \gtrsim \lambda_s$ the dendrite is unstable against splitting of the
tip into finer structures, while for $R \lesssim \lambda_s$ the dendrite is very
weakly stable. Directly at $R \approx \lambda_s$, side-branches are periodically
formed giving rise to the tree-like structure (Fig. 14) of the dendrite as it grows, the overall structure of the dendrite preserving
its shape as time goes on. Our previous conjecture, based on this
linear stability analysis that the operating point of the dendrite
is close to $R \approx \lambda_s$ was recently confirmed by an analysis of a

simplified nonlinear model. This investigation reproduces the hypothesis "marginal" linear stability $R \approx \lambda_s$ within a few percent precision, based on a fully nonlinear analysis.

Our general prediction [48] that the dendritic growth law should
show the scaling-behavior:

$$vR^2 = \text{constant} \tag{45}$$

under varying external conditions is well confirmed experimentally
[47].

Another, even more general, feature is crystallizing from this
nonlinear analysis [49] concerning the problem : what sort of pattern
is selected by nature among a continuous family of possibilities ?
We find that among the possible nonlinear patterns of sidebranches
stationary in the laboratory frame of reference, the one is selected
which spreads out at the fastest rate thus dominating the systems
behavior. In a sense, this is again the original competition principle of maximum velocity, but here acting between different side-
branching states of a single dendrite rather than between different
dendrites. We are hopeful to extending this idea to other problems
of pattern selection in various fields of physics, chemistry and
possibly even biologic systems.

4.2.3 Eutectics

Very shortly I will finally sketch some basic ingredients for
the problem of pattern formation in eutectic compounds [50]. An
eutectic is a binary material with a phase-diagram like in Fig. 3,
but high and low temperatures reversed. The "DS"-phase in Fig. 3
then becomes the "liquid" at high temperatures. The point, where
three phases coexist at the same temperature – two solid phases of
different composition and one liquid phase – is called eutectic
point. Starting at a concentration identical to the eutectic concentration, the solid will tend to split in two phases. Practically
this leads to periodic structures separated by boundaries normal to
the solid-liquid interface, as shown in Fig. 16.

The principal reason for the formation of these lamellar structures is easy to understand [51]. Assume that the rate of solidification is prescribed. In order to have a driving force for the
advancement of the interface, its temperature must be lower by an
amount ΔT than the eutectic temperature. Assume a lamellar spacing
λ to result from the theory. The supercooling ΔT must be larger
for large lamellar spacing λ, since the separation of the homogeneous
liquid into two solid phases of different concentration proceeds via
diffusion. A given velocity requires a constant diffusion flux proportional to the gradient. An increase of λ, therefore, has to be
compensated by an increase of ΔT. On the other hand, each of the

230

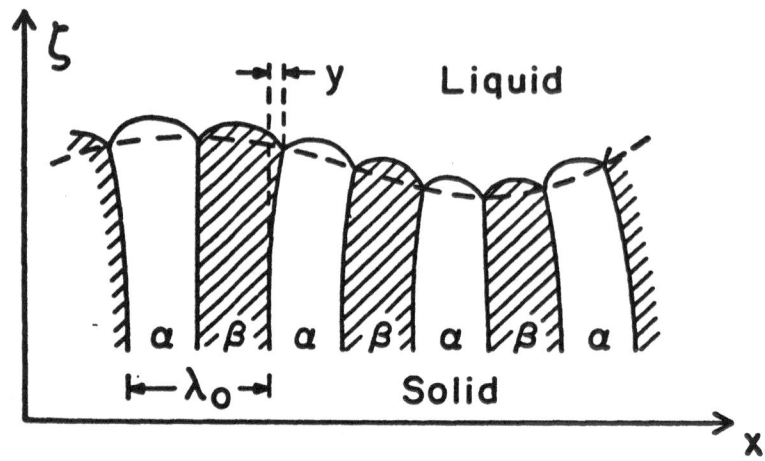

<u>Fig. 16</u> Schematic plot of a eutectic lamellar growth front [52].

lamellar regions has a small curved interface common with the liquid. The curvature, proportional to λ^{-1}, in turn produces a backdriving force on the solid.

Combining these two effects, one obtains for the supercooling a relation :

$$\Delta T \sim (\frac{\lambda}{\lambda_s} + \frac{\lambda_s}{\lambda}); \quad \lambda_s = \sqrt{\frac{Dd_o}{V}} \qquad (46)$$

where D is the diffusion coefficient, d_o the capillarity length of section 4.2 and V the growth rate; λ_s is again the stability length defined in (42), acting here as an obvious scale factor.

Just as in the dendrite problem, this consideration does not predict a specific lamellar spacing. In contrast to the dendrite problem, the special point $\partial\Delta T/\partial\lambda = 0$ here seems to be related to the operating point of the eutectic system, predicting a growth law $(\lambda \approx \lambda_s)$:

$$\lambda \sim V^{-1/2} \qquad (47)$$

for the lamellar spacing. The problem, however, is far from being understood and is still under investigation [52].

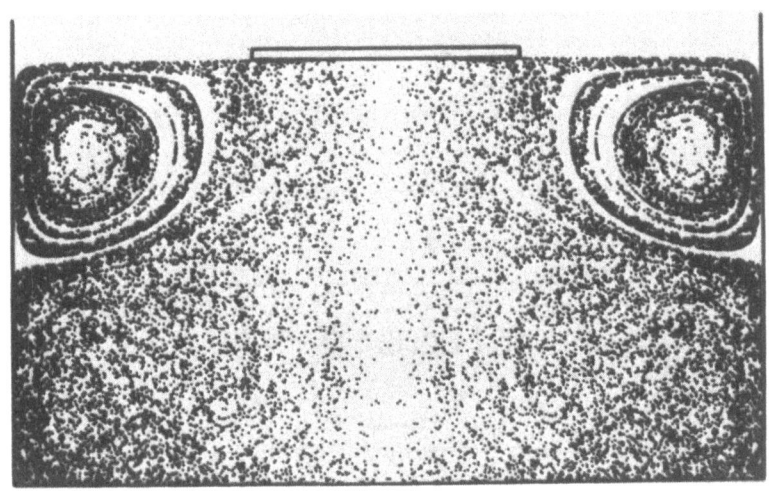

Fig. 17 Computer-simulation of hydrodynamic flow in a Czochralski-crystal-pulling apparatus. On the free surface the flow is directed outwards. The crystal is indicated by the flat rectangle. ([56], cylindrical symmetry assumed).

4.3 Hydrodynamic flow

Hydrodynamic flow [38] plays a very important role in industrial processes of crystal growth [53-55]. A standard instrumentation for example is the Czochralski-technique of pulling crystals from the melt, (Fig. 17). The plot here shows a computer-generated [54] flow-pattern. A crucible is filled with the hot melt, the crystal shown as a thin plate on top of the liquid is cooled and slowly raised. The cooling from the top via the crystal and the heating from the bottom immediately causes the so-called Rayleigh-Bénard instability to develop. The hot liquid generally is less dense than the cooler one and buoyancy drives the liquid upwards on the wall and downwards in the center. (The dots in Fig. 17 represent "painted" molecules in the liquid). This flow gives rise to undesirable periodic varia-tions of impurities in the crystal ("striations"). It is, therefore, highly desirable to gain a quantitative understanding of conserva-tion phenomena in crystal growth.

I will not go into any detail about hydrodynamic theory here, but merely make a few dimensional estimates for the influence of hydrodynamics on thin-layer growth, being the situation relevant for

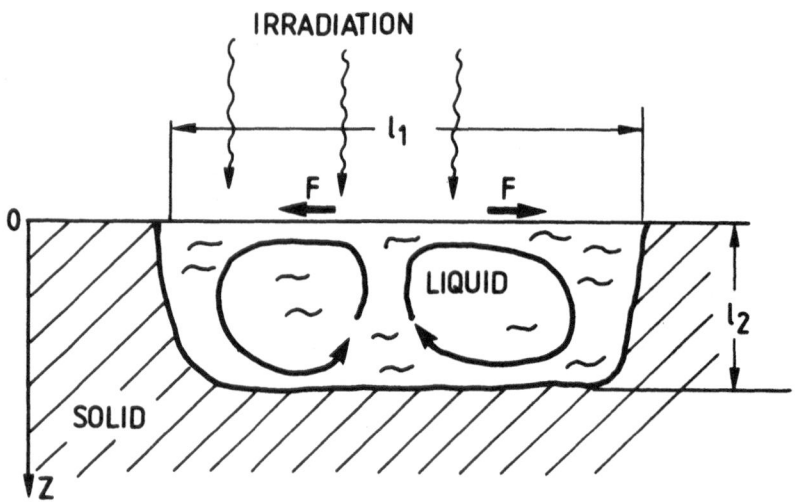

IRRADIATION

Fig. 18 Schematic plot of a thin liquid layer formed at the surface of a solid under laser irradiation.

laser-annealing experimental conditions. The experimental situation is sketched in Fig. 18, where I take the point-of-view that the material really melts, despite the possibly unusual states of matter to be encountered under this high intensity of electro-magnetic irradiation.

The first possibility of exciting hydrodynamic flow [55] in the thin layer is via the *density-difference* between the solid and the liquid phase. Let us ask whether there is a chance of turbulent, highly mixing flow in the liquid. This condition requires the dimensionless Reynolds number:

$$Re = \frac{\ell_2 v}{v} , \qquad (48)$$

to assume a value of at least about $Re \approx 1000$. v is here a typical velocity of the liquid relative to a rigid boundary and v is the kinematic viscosity. For silicon we have $v \approx 2 \times 10^{-3} \, cm^2/sec$. The velocity of the interface cannot possibly exceed the velocity of sound. The velocity of the liquid results from the density difference between liquid and solid of a few percent. Multiplication gives an effective velocity of about $v \approx 10^3$ cm/sec. Inserting these numbers into (48) we obtain $1000 \lesssim Re = \ell_2 \frac{10^6}{2}$ (cm-1) or

$$\ell_2 \gtrsim 10^{-3} (cm). \qquad (49)$$

This depth of the liquid layer seems to be already within the range of the most powerful laser pulses. Even if the turbulent regime (49) is not reached we already have excitation of flow at Reynolds numbers Re $\gtrsim$ 1. This condition seems to be fulfilled in most cases of laser annealing experiments.

As a second possibility of exciting hydrodynamic flow let us consider the thermocapillary or *Marangoni effect* [38]. It says that flow at an open surface will be induced if a gradient in the surface-tension is produced for example by inhomogeneous heating. This is of course the case under pointwise melting of a surface by laser irradiation, since in the center of the liquid surface the temperature is higher than on the sides. To estimate the effect in a simple fashion, let us start with the Navier-Stokes equation integrated over the depth ℓ_2 of the liquid:

$$\int_0^{\ell_2} dz\ \rho \dot{v} \simeq F + \int_0^{\ell_2} dz\ \eta\ \vec{\nabla}^2\ v \tag{50}$$

where v is the velocity of the flow, ζ is the density and η the viscosity. Here we have ignored the inertial term $(\vec{v}\vec{\nabla})\ \vec{v}$. F is the force per unit element of the surface area, produced by the lateral gradient $\Delta T/\ell_1$ along the surface, ΔT being of the order of 500 K. The change of surface tension per degree Kelvin is typically $d\gamma/dT \approx 10^{-1}$ g/sec^2.K . Replacing $\nabla^2 v$ by v/ℓ_2^2 and inserting in (50) we obtain for silicon approximately :

$$v \approx (\ell_2/\ell_1)\ .\ 5000\quad cm/sec\ . \tag{51}$$

Assuming a ratio of $\ell_2/\ell_1 > 0.1$ this velocity is sufficiently large to couple with the velocity of the advancing solid-liquid interface, thereby stirring up the melt.

There are of course other effects [53-55] of hydrodynamic instability possible, like the thermoconcentrational effect for inhomogeneously heated compound material. A more detailed investigation of the relevance of these mechanisms would require somewhat more experimental information. As a first impression, however, the two crude estimates given above already indicate the possible importance of hydrodynamic flow in these experiments.

ACKNOWLEDGEMENTS

I would like to thank Prof. F. Rosenberger for supplying me with materials-constants for the hydrodynamic estimates of section 4.3, on a sunny afternoon at Cargèse. I am also indebted to my institute colleagues Dr. D. Kroll and Dr. T. Meister for allowing me to use their unpublished data on interface structures.

REFERENCES

1. An up-to-date review of modern theoretical concepts of crystal growth will soon appear :
   Chernov, A.A., Müller-Krumbhaar, H., eds, "Modern theory of crystal growth" (Springer-Series : "Crystals; Growth, Properties and Applications", Heidelberg 1983. For published reviews see [2,3] and concerning applications see [4].
2. Müller-Krumbhaar, H., in "Current Topics in Materials Science", E. Kaldis ed., Vol. 1, p. 1, 1978 (Amsterdam, North Holland).
3. Rosenberger, F., in Proceedings of the NATO Advanced Study Institute, Erice, Aug. 1981; R. Kern and B. Mutafchiev eds. (D. Reidel publ. co., Dordrecht 1981).
4. Elwell, D., Scheel, H.J., "Crystal Growth from High Temperature Solutions" (Academic Press, London 1975).
5. Saito, Y., Phys. Rev. Letters 48 (1982) 1114 and Phys. Rev. B (to appear) 1982.
6. For an introduction into "phase diagrams" see F. Rosenberger, this volume.
7. Saito, Y., J. Chem. Phys. 74 (1981) 713.
8. The general theory of phase transitions is outlined by K. Binder, this volume.
9. Cahn, J.W., Hilliard, J.E., J. Chem. Phys. 28 (1958) 258.
10. Fisk, S., Widom, B., J. Chem. Phys. 50 (1969) 3219.
11a. Oxtoby, D., Haymet, A.D., J. Chem. Phys. 76 (1982) 6262.
11b. Ramakrishnan, T.V., Yussouff, M., Phys. Rev. B19 (1979) 2775; Ramakrishnan, T.V., Phys. Rev. Lett. 48 (1982) 541.
12a. Ladd, A.J.C., Woodcock, L.V., Mol. Phys. 36 (1978) 611.
12b. Toxvaerd, S., Praestgard, E., J. Chem. Phys. 67 (1977) 5291.
12c. Broughton, J.Q., Bonissent, A., Abraham, F.F., J. Chem. Phys. 74 (1981) 4029.
13a. Hess, S., Naturforsch, Z., 35a (1980) 69.
13b. Haymet, A.D., preprint.
14. Müller-Krumbhaar, H., in "Current Topics in Materials Science" Vol. 2, p. 115 (1977); Kaldis, E. and Scheel, H.J. eds (Amsterdam, North Holland).
15. Chui, S.T., Weeks, J.D., Phys. Rev. B14 (1976) 4978; Weeks, J.D. in "Ordering in Strongly Fluctuating Condensed Matter Systems", p. 293, K. Riste ed. (New York, Plenum Press, 1980).
16. For an introduction to Ising models see Stanley, H.E., "An Introduction to Phase Transitions and critical phenomena, (Oxford, Oxford University Press, 1971).
17. Kosterlitz, J.M., Thouless, D.J., J. Phys. C6 (1973) 1181.
18. Saito, Y., Müller-Krumbhaar, H., Phys. Rev. B23 (1981) 308.
19. Halperin, B.I., Nelson, D.R., Phys. Rev. Letters 41 (1978) 121, Phys. Rev. B19 (1979) 2457.
20. Abraham, F.F., Phys. Rev. Lett. 44 (1980) 463 and in Ordering in Two Dimensions, S. Simha ed. (Amsterdam, North Holland, 1980); Toxvaerd, S., Phys. Rev. Lett. 44 (1980) 1002;

Phys. Rev. A24 (1981) 2735;
van Swol, F., Woodcock, L.V., Cape, J.N., J. Chem. Phys. 23 (1980) 913;
Morf, R., Phys. Rev. Lett. 43 (1979) 931.
21. Chui, S.T., Phys. Rev. Lett. 48 (1982) 933.
22. Meister, T. (Inst. für Festkörperforschung, Jülich), unpublished.
23. Bonissent, A., in ref. 1 and 12c.
24. Reviews on phase transitions in adsorbed layers are given by Landau, D.P., chapter 9 of [39], and by Bak, P. in [1].
25. Tosatti, E., Solid State Comm. 25 (1978) 637;
Bak, P., Solid State Comm. 32 (1979) 581.
26. Wortis, M., Pandit, R., Schick, M. in "Melting, Localization and Chaos"; R. Kalia and P.D. Vashishta eds. (New York, North Holland, 1982).
27. Lipowsky, R., Phys. Rev. Letters, 1982 (to appear).
28. Kroll, D., Lipowsky, R., Phys. Rev. B 1982 (to appear).
29. Kroll, D., (Inst. f. Solid State Physics, Jülich) unpublished.
30. Cahn, J.W., J. Chem. Phys. 66 (1977) 3667.
31. Menaucourt, J., Thomy, A., Duval, X., J. Physique 38 (1977) 195.
32. Leamy, H.J., Gilmer, G.H., Jackson, K.A. in "Surface Physics of Materials"; J.M. Blakely ed., Vol. 1, p. 121, 1975 (New York, Academic Press).
33. Saito, Y., Müller-Krumbhaar, H., J. Chem. Phys. 74 (1981) 721 (Note that this theory holds for nonconserved order parameters only. If the order-disorder transition in the solid is accompanied by concentration change one has to include diffusional transport. A mere density change in contrast can propagate with sound velocity and thus is uncritical.)
34. Baikov, Yu.A., Zelenev, Yu.V., Haubenreisser, W., Pfeiffer, H., Phys. Stat. Sol. (a) 61 (1980) 435.
35. Saito, Y., Müller-Krumbhaar, H., J. Chem. Phys. 70 (1979) 1078.
36. Pfeiffer, H., Haubenreisser, W., Phys. Stat. Sol. (b) 96 (1979) 287.
37. Cherepanova, T.A., Kiselev, V.F.; Kristall und Technik 14 (1979) 545;
Cherepanova, T.A., Phys. Stat. Sol. (a) 58 (1980) 469.
38. Chandrasekhar, C., "Hydrodynamics and Hydrodynamic Stability", (Oxford, Clarendon, 1961).
39. For a review on computer-simulation in crystal growth see Müller-Krumbhaar, H., chapter 7 in "Monte-Carlo Methods in Statistical Physics", K. Binder ed. (Heidelberg, Springer-Verlag, 1979).
40. See Germain, P. in this volume . General reviews are given in A. Zettlemoyer ed, "Nucleation" (New York, Dekker, 1969); "Nucleation II" (New York, Dekker, 1976).
41. Müller-Krumbhaar, H., Burkhardt, T., Kroll, D., J. Crystal Growth 38 (1977) 13.
42. Chernov, A.A., Lewis, J.; J. Phys. Chem. Solids 28 (1967) 2185.
43. Cherepanova, T.A., van den Eerden, J.P., Bennema, P. , J. Crystal Growth 44 (1978) 537.

44. Müller-Krumbhaar, H., in Festkörperprobleme XIX, 1 (1979) J. Treusch ed. (Advances in Solid State Physics, Vieweg-Verlag, Braunschweig).

45. For an excellent review on diffusion in crystal growth see R.L. Parker in "Solid State Physics"; Ehrenreich, H., Seitz, F., Turnbull, D., eds. Vol. 25, p. 152 (1970).

46a. Mullins, W.W., Sekerka, R.F., J. Appl. Phys. 35 (1964) 444; 34 (1963) 323.

46b. A review is given by R.T. Delves in "Crystal Growth", B.R. Pamplin ed. (New York, Pergamon Press, 1975).

46c. A most recent review is given by J. Langer, Rep. Progr. Phys. 52 (1980) 1.

47. Huang, S.C., Glicksman, M.E., Acta Metall. 29 (1981) 701,717.

48. Langer, J.S., Müller-Krumbhaar, H., Acta Met. 26 (1978) 1681, 1689; Müller-Krumbhaar, H., Langer, J.S., Acta Met. 26 (1978) 1697, 29 (1981) 145.

49. Langer, J.S., Müller-Krumbhaar, H., Phys. Rev. A (1982) to appear.

50. Trivedi, R., J. Cryst. Growth 49 (1980) 219; Kurz, W., Fisher, D.J., Acta Met. 29 (1981) 11.

51. Jackson, K.A., Hunt, J.D., Trans. Met. Soc. of AIME 236 (1966) 1129.

52. Langer, J.S., Phys. Rev. Letters 44 (1980) 1023; Datye, V., Langer, J.S., Phys. Rev. B24 (1981) 4155.

53. An extensive review by Polezhaev will appear soon in [1]. See also the chapter by D.T.J. Hurle in [14].

54. Zierep, J., Oertel, H. Jr., Convective Transport and Instability Phenomena (Karlsruhe, G. Braun-Verlag, 1982).

55. Various hydrodynamic effects on the crystallization front are analyzed in Corriell, S.R., Turnbull, D., Acta Met. (1982) to appear; Corriell, S.R., Cordes, M.R., Boettinger, W.J., Sekerka, R.F., J. Crystal Growth 49 (1980) 13; Boettinger, W.J., Biancaniello, F.S., Corriell, S.R., Metall. Trans. A, 12A (1981) 321.

56. Mihelcic, M., Schroeck-Pauli, K., Wingerath, K., Wenzl, H., Uelhoff, W., van den Hart, A., J. Crystal Growth 57 (1982) 300.

DYNAMICAL PROCESSES DURING SOLIDIFICATION

J.H. Bilgram

Laboratorium für Festkörperphysik, Eidgenössische Technische Hochschule CH-8093 Zürich, Switzerland.

ABSTRACT

A model for the solid-liquid interface is presented. Its properties are compared with experiments. Pre-aggregation in the melt takes place in an interface layer during solidification. It leads to a slowing down of the thermal diffusion by about five orders of magnitude relative to the melt. The model and the experiments lead to a thickness of the interface layer of a few µm. Universal laws for free and directional solidification are proposed on the basis of the kinetics at the interface.

1. INTRODUCTION

The solid-liquid transition shows asymmetries which indicate that freezing and melting cannot be transformed into each other by a mere time reversal, e.g. :
- Crystals in contact with their pure melt may form facets during freezing, whereas facetting has never been observed during melting.
- The rates of dendritic solidification are by orders of magnitude smaller than the growth rates of liquid dendrites into superheated crystals.

There is evidence that, at least far from equilibrium conditions, the processes that are rate limiting are not the same for freezing and for melting [1,2] .

Two steps are necessary for the growth or for the melting of a crystal :

1. The transport of the latent heat away from the interface or to the interface.

2. A change in the arrangement of the molecules has to take place at the solid-liquid interface.

Theories which try to predict freezing rates simplify the problem assuming that only one of these two processes is rate limiting. As there are two possibilities, there exist two nearly independent sets of papers, both predicting growth rates of crystals on the basis of heat flow calculations or on the basis of atomic adsorption lkinetics. There are experimental techniques to compare these two approaches with measurements :

1.Free Solidification :

A crystal is immersed in a supercooled melt and the latent heat flows away from the solid-liquid interface into the melt. The negative thermal gradient in the melt destabilizes the shape of the interface. This leads to dendritic growth of substances with low melting entropy. The diffusion of heat away from the interface can be calculated in principle, but as the shape of the interface is not known, there are additional informations necessary.

2. Directional Solidification

A crystal is attached to a heat sink and the latent heat flows through the crystal. With this technique it is possible to obtain a stable planar interface and to measure the growth rate of a crystal in function of the supercooling at the interface.

In the second part of this paper growth rates of crystals and asymmetries of the freezing transition will be discussed, as they are predicted or observed in free solidification or in directional solidification. In the third part an experiment is presented where the kinetics of entropy fluctuations at the solid-liquid interface are studied. The experimental results can be described by a phenomenological model in paragraph Four. This model is able to predict growth rates of crystals for free and directional solidification and it explains the different rates which are observed for freezing and melting.

2. MACROSCOPIC GROWTH RATES

2.1 Free solidification

The rate of heat diffusion depends on the geometry of the interface, therefore the temperature field around a needle like crystal tip is different from the one in front of a planar interface. The diffusion laws for heat transport do not depend on the direction of the heat flow. From these two sentences we have to conclude : if heat diffusion is the only rate limiting process for the solid-liquid transition, then this phase transition is symmetrical with respect to freezing and melting. As heat transport depends on the shape of the interface, it is difficult to compare growth rates observed during planar growth with those measured during dendritic growth.

Theories on dendritic solidification are based on the assumption that the transition is governed entirely by heat flow (for a review see [3]). A simple approach to the calculation of rates of free solidification is the "spherical approximation" [4], where the shape of a dendrite is approximated by a rod with a spherical tip. Solving the equation of thermal diffusion in the melt surrounding the dendrite under the condition of heat balance at the surface of the growing crystal leads to following expression for the growth rate v of a tip with the radius ρ into a melt with a temperature T_∞ far away from the crystal, which has a melting temperature T_m:

$$v_{tip} \cong \frac{\alpha}{\rho} \left[\frac{(T_m - T_\infty)}{L/c_p} - \frac{2d_o}{\rho} \right]. \tag{1}$$

L designates the latent heat and c_p the heat capacity of the melt. α is the thermal diffusivity in the liquid and d_o the capillarity. $d_o = \gamma/(L/c_p)S$ where γ stands for the solid liquid interfacial energy and S is the melting entropy.
For large tip radii ($\rho > d_o$) the product ρv_{tip} is constant. For a tip radius comparable or smaller than the capilarity the growth rate decreases due to the Gibbs-Thomson effect. Thus eq. 1 leads, for a given supercooling, to a curve in the v-ρ field. More information is necessary to determine the growth rate of dendrites. If only thermal diffusion and heat balance were determining the growth rate, then the "criterion of maximum growth velocity" would be applicable. It implies that the dendrite shape leading to the maximum growth velocity is self stabilizing. The growth rates predicted with this criterion and eq. 1 are as much as seventy times larger than the growth velocities observed for dendritic growth in undercooled liquid metal systems [4]. In more realistic calculations the tip is assumed to have parabolic shape. This makes the value of v_{tip} to decrease some more than 50%, that means the dendrite grows more slowly than the simple cylindrical shape [5].
In addition to that, due to a changing curvature the surface of the dendrite is not isothermal. Numerical calculations of the growth rate of a nonisothermal parabolic dendrite tip have been done for the maximum growth velocity [6] leading to a power law:

$$v_{tip} \propto (T_m - T_\infty)^a, \quad \text{where} \quad a=2.65. \tag{2}$$

There exist very careful experiments on dendritic growth of succinonitrile [7,8]. These show that the criterion of maximum growth velocity is not applicable to freezing. In Fig. 1 a compilation of experimental data of dendritic growth of ice is shown [9]. Succinonitrile and ice have the same growth rate for a given supercooling of the melt. For both substances the experimental data can be fitted by a power law with a = 2.8. It has been observed that small traces of impurities in liquid succinonitrile lead to an increase in growth rate [8]. This observation cannot be understood

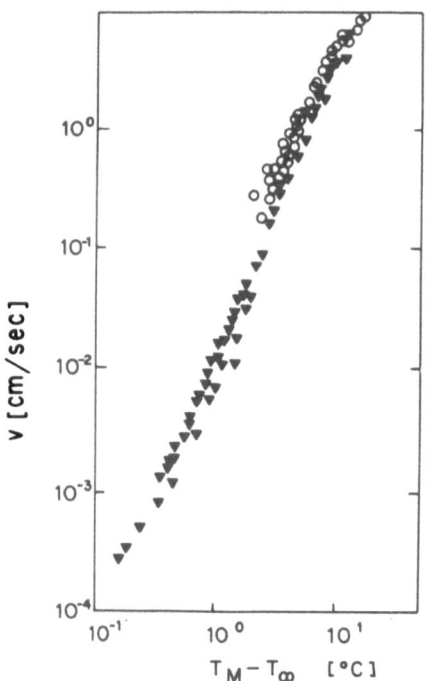

Fig. 1 Growth rates of ice dendrites into supercooled water. Data from several laboratories are compiled [9] .

on the basis of a solidification process which is rate limited by heat flow only.

Succinonitrile is a substance with low melting entropy. It does not form facets at all. Ice forms facets normal to the c-axis only. Dendrites grow along the a-axis. Salol has a high melting entropy. It forms many facets. Therefore for salol a growth behavior is expected which is different from the one of succinonitrile. In a free solidification experiment with salol, three different modes of stationary growth can be observed [10] :

1) at low supercooling $(T_m - T_\infty) < 3°$, an exponential growth behavior:

$$v_{tip} \propto e^{-b/T_m(T_m - T_\infty)};$$

2) for $3° < (T_m - T_\infty) < 10°$, a power law: $v_{tip} \propto (T_m - T_\infty)^{2.8}$,

where growth rate does not depend on crystal perfection (Fig. 2);

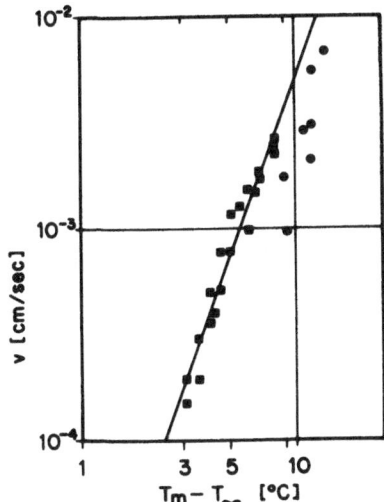

Fig. 2 Free growth of salol into supercooled melt. Squares represent the growth rate of a single crystal, circles represent multinuclear growth. Only the squares are used to fit the line. The
temperature dependence of the melt has been corrected [2,10] .

3) for high supercooling, multinuclear growth.
At $(T_m - T_\infty) = 12.5°C$, a maximum growth velocity of about 45 μm/s has
been measured. The growth rates observed are in agreement with earlier measurements [11] .

The behavior in mode 1 is expected for a facetting crystal.
Two dimensional nucleation is rate limiting. In the second mode a
similar growth law is observed as in dendritic growth. This indicates that the growth mechanism is no longer lateral. Therefore
the observation of the transition from growth mode 1 to mode 2 is a
quantitative proof for a transition to a normal growth mode.

Ice is the only substance where dendritic melting of a superheated crystal has been studied [12] . With the spherical approximation $(\rho > d_o)$:

$$v_{tip} = \frac{\alpha_{solid}}{\rho} \cdot \frac{(T_m - T_\infty)}{L/c_p}$$

and the criterion of maximum growth velocity:

$$\rho = \rho^* = \frac{2\gamma}{S(T_m - T_\infty)} \,, \tag{3}$$

one obtains, for $(T_\infty - T_m) = 0.3°$ C, a growth rate of 1.1 cm/s. This is by a factor of three higher than the rate of growth of liquid dendrites into superheated ice as measured by Käss and Magun ($v \cong 0.3$ cm/s) [12,1]. Considering that the spherical approximation leads to values which are more than a factor of two higher than the growth rates calculated for a parabolic tip, this simple estimation is in good aggreement with the experiment. From fig. 1 it can be seen that this growth rate of dendritic melting is by about three orders of magnitude larger than the growth rate of ice dendrites into supercooled water.

Combining eq. 1 ($\rho > d_o$) and eq. 3 one obtains a dimensionless rate $\tilde{v}_{tip}$ for dendritic melting :

$$\tilde{v}_{tip} = \frac{v_{tip}}{2\alpha/d_o} \propto \left[\frac{T_\infty - T_m}{2L/c_p} \right]^a \,. \tag{4}$$

This scaling law has been used by Langer et al. [13] to scale the rates of dendritic solidification of ice and succinonitrile. The solidification experiments lead to a = 2.8. The theories for dendritic solidification (for a review see [3]) are symmetrical for freezing and melting. These theories do not scale dendritic melting and dendritic freezing of ice. Hence one has to conclude that solidification is not a problem of heat transport and interface stability alone.

2.2 The rate of directional solidification

The kinetics of the arrangement of atoms at the planar surface of a growing crystal have been studied in several attempts where growth rates as a function of supercooling ΔT at the solid-liquid interface are calculated. Various approaches can be distinguished by their assumptions about the interface and the melt.

2.2.1 A sharp interface and a melt of non-interacting atoms

Jackson concluded from free energy considerations that substances with high melting entropy form flat interfaces (facets) whereas low melting entropy materials have surfaces which are rough on atomic scale. For a review see [14] . The growth on facets occurs layer by layer. On perfect crystals the growth rate is limited by two dimensional nucleation of a new layer leading to an exponential dependence of the growth rate on ΔT. Non-ideal crystals contain dislocations which provide growth sites and growth rate depends on the supercooling by a power law. On the rough surface of a crystal the growth proceeds normal to the interface. Phenomenological rate theories lead to a growth rate which is proportional to ΔT. The proportionality is called the kinetic constant of crystal growth. Depending on the derivation, this quantity contains the constants of thermaldiffusion and self-diffusion of the liquid, as well as sticking probabilities at the interface or jump probabilities across the interface, factors which cannot be measured in the experiment. There is a flexibility in choosing numbers which allows an estimation of the kinetic constant which reaches the order of magnitude of the measured values. These experimental values contain also some inaccuracy because it is difficult to measure the actual supercooling at the solid-liquid interface. This type of theory is developed for equilibrium conditions. It does not distinguish between freezing and melting.

2.2.2 A diffuse interface and a melt of non-interacting atoms

Cahn [15] proposed an interface which consists of several layers. This is in difference to Jackson's analysis which assumed only one layer which is partially filled. Cahn shows that for small supercooling the growth rate occurs by a lateral mechanism and growth rate increases about with the square of the supercooling. There is a critical driving force above which the growth mechanism changes to a normal one and the growth rate increases proportional to the supercooling [16]. The critical driving force is large for a perfectly smooth interface, decreasing as roughness increases. This model describes the transition from a lateral to a normal growth mechanism as it is observed during free solidification of salol.

2.2.3 A sharp interface and network melt

Crystal growth in covalent melts and glasses is generally controled by the rate of internal rearrangement rather than by heat transport. For a review see [17]. The crystallization rates are very sensitive to impurities which break bonds in the melt and lead to an increase in growth rate. If only rearrangement in the viscous melt is rate determining, no asymmetry between freezing and melting can be expected. In difference to that, for the solidification of a facetting crystal into a metallic melt with low viscosity, salol should be a transparent model substance. The fact that salol can be supercooled very easily indicates that liquid salol forms structures which have to be broken before freezing takes place.

3. ENTROPY FLUCTUATIONS AT THE SOLID-LIQUID INTERFACE

Free solidification as well as directional solidification experiments indicate that even for ice and succinonitrile the rate limiting process for freezing is not the heat transport alone. Freezing seems to be more time consuming than melting. It may be that there is a diffuse interface layer at the surface of a growing crystal as proposed by Cahn. There exist also molecular dynamic studies [18] and Ising type models [19] which predict interface layers. If there is a region close to a growing crystal surface where pre-ordering takes place, then entropy fluctuations may be expected there. Such fluctuations can be detected by Rayleigh scattering.

The isothermal compressibility and the thermal diffusivity are properties of the interface layer which can be studied by quasi-elastic light scattering. The total intensity of the light quasi-elastically scattered by a pure liquid is proportional to the isothermal compressibility χ_T, and the decay rate of entropy fluctuations is proportional to the thermal diffusivity. This decay rate corresponds to the linewidth Γ of the scattered light. If there is a pre-ordering of molecules in front of a growing crystal which influences compressibility and thermal diffusivity, then an increase of the scattered intensity and a decrease of the line width of the scattered light have to be expected. Light scattering experiments have been performed with H_2O [20,21], D_2O and Salol [22]. As will be shown in the next paragraph, the experimental results obtained with these substances can be scaled by a simple relation. In the following, only the experiments with H_2O will be described.

At the beginning of our early studies of entropy fluctuations at the solid-liquid interface, we did not expect a very strong effect and thus started with water. Due to the 4°C density maximum the Landau-Placzek ratio vanishes at a temperature close to the melting point. Hence pure water is the substance with the minimum scattering background. (Brillouin- and Raman scattering only). Dust free pure melt is prepared by zone refining, and the experiment is done in situ during zone refining, thus eliminating any possibility of contamination (Fig. 3). The crystal is lowered exactly with the growth velocity thus holding the solid-liquid interface at a stable position in the laboratory system. The surface of the growing crystal is illuminated by an argon laser. The scattered light is detected by a photomultiplier and the correlation function of the photoncounts is calculated by a digital correlator. A detailed description of the experimental setup is given in [20,21].
The experimental results for ice are [1,21,23]:
—No light scattering is observed at the interface at equilibrium conditions, or if the ice is melting;
—If growth rate exceeds a critical value of $v_{crit}=1.5$ μm/s, the onset of light scattering is observed. During a transient time the scattering power increases and the decay rate Γ decreases, until both quantities reach a stationary state after about half an hour (Fig.4). Probably the growth rate is not a good parameter to characterize the state of the interface, however it can be measured easily. The temperature close to the interface would be a more appropriate parameter, but it is difficult to measure without disturbing the system.

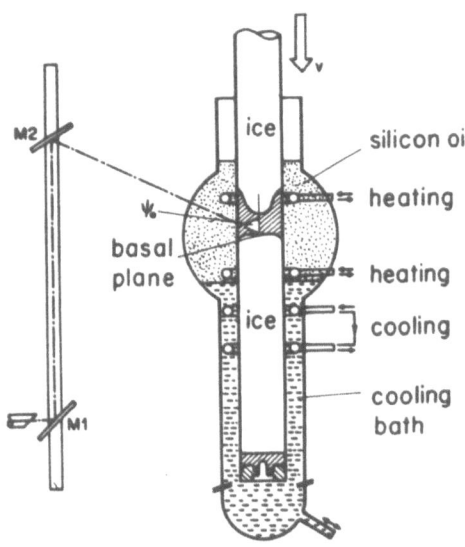

Fig. 3 Zone refining apparatus for the study of entropy fluctuations at the solid-liquid interface. The experiment is performed in a cold room at -18°C.

- The line width of the scattered light is proportional to the square of the scattering vector K :

$$\Gamma = D_i K^2. \tag{5}$$

That means : the decay process of the fluctuations can be described by a diffusion law. The corresponding diffusivity is D_i = 3.10^{-8} cm^2/s in the stationary state. The i stands for interface;
- The line width does not depend on the orientation of the scattering vector relative to the solid-liquid interface. The decay process is isotropic;
- In the stationary state, line width does not depend on growth rate;
- The fluctuation decay with a single decay rate. Only one relaxation mechanism is involved;
- The intensity of the scattered light is isotropic, indicating that the diameter of the scattering inhomogeneities is small compared to the wavelength of the scattered light;
- Measurements of the intensity of the scattered light as a function of the angle of incidence ψ_o are not compatible with model calculations assuming a corrugated interface or particles immersed in

246

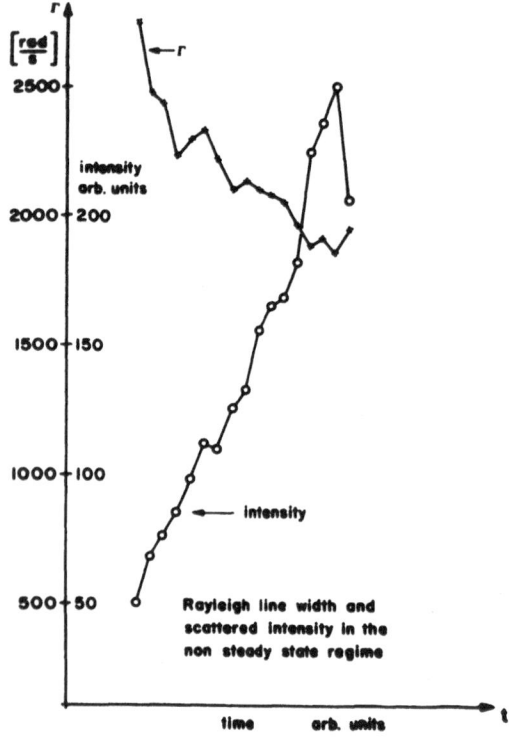

Fig.4 Transient phenomena during the onset of light scattering at the solid-liquid interface.

the melt
- Light intensities can be detected which are scattered by inhomogeneities with a Fourier component corresponding to a spatial wavelength of more than one μm perpendicular to the interface. The light is scattered in a layer of finite thickness which has been determined to be 1.7 μm < d < 6 μm.

4. THE MODEL

Not all the molecules (or atoms) in a liquid close to the freezing point are monomers, to a certain degree they are aggregated. It is assumed that the degree of aggregation is increased in the vicinity of the interface of a growing crystal. In analogy to the freezing process a change occurs in the volume and energy content during this aggregation. Hence an increase of the isothermal compressibility of the material in the interface layer is expected. Temperature fluctuations are no longer completely governed by Fick's law. They must be influenced by a term arising from the change in aggregation. With increasing temperature an increase of the number of broken bonds is expected. Therefore the heat flowing down the thermal gradient is reduced by an amount used up to break bonds. The form of the modified equation is well known for diffusion with a chemical reaction [24] :

$$\frac{\partial T}{\partial t} = \alpha \frac{\partial^2 T}{\partial x^2} - \frac{\partial B}{\partial t} . \tag{6}$$

We assume that breaking and forming of bonds is a reversible reaction and define a reaction constant R by putting :

$$B = R T . \tag{7}$$

Insertion of (7) into (6) leads to :

$$\frac{\partial T}{\partial t} = \frac{\alpha}{1+R} \frac{\partial^2 T}{\partial x^2} .$$

This equation can be interpreted as an equation for thermal diffusion with a diffusivity at the interface $D_i = \alpha/(R+1)$, which is reduced relative to the value α in the bulk melt far from the interface. If there occur several reactions with the reaction constants R_1, R_2, ... there will be observed only one relaxation process with a relaxation rate proportional to $D_i = \alpha/(\Sigma R_i + 1)$.
The reaction constant R contains two competing factors. The first one is due to the formation of bonds and is therefore proportional to the latent heat of melting L. The second one arises because the aggregates are hindered to form due to the energy γ of the interface between aggregated and non-aggregated molecules. This second factor also involves a correlation length d. For our model it is reasonable to assume that the characteristic length is the thickness of the interface layer. This suggest the following expression for the reaction constant : $R = L \frac{d}{\gamma}$. For $R \gg 1$ the thickness of the interface layer would then be :

$$d = \frac{\alpha}{D_i} \frac{\gamma}{L} . \tag{8}$$

Using eq. 8 and the measured value of D_i one can calculate the thickness of the interface layer. For ice one obtains $d \cong 4$ μm, which is indeed in the range determined experimentally. Since γ is roughly proportional to L [25] and since γ/L has the dimension of a length, γ/L can be used to scale the thickness of the interface layer. The dimensionless thickness of the interface layer is :

$$\tilde{d} \equiv \frac{d}{\gamma/L} = \frac{\alpha}{D_i} \ .$$

At the equilibrium conditions or during melting there is no pre-aggregation, hence $D_i = \alpha$ and $d = \gamma/L$.

It is a property of this model that pre-ordering of molecules in an interface layer slows down thermal diffusion. As heat transport is very important during solidification, the rate of solidification v is assumed to be proportional to D_i. The rate v is also proportional to the difference between the Gibbs free energies of the solid and the melt supercooled by ΔT. This leads to $v \propto L\Delta T/T_m$. A third factor influencing the rate of solidification arises from the solid-liquid interfacial energy γ. Since freezing cannot occur in infinitesimal steps it is always connected with fluctuations of the surface area of the crystal. A very high solid-liquid interface energy makes solidification impossible. Combining these three factors we propose for the rate of directional solidification :

$$v = D_i \ \frac{L}{\gamma T_m} \ \Delta T \ . \tag{9}$$

Intuitively, unity is used for a proportionality factor. In the following it will be shown that this leads to agreement with the experimental results. For the dynamical constant of crystal growth $\mu = \frac{v}{\Delta T}$ one obtains :

$$\mu = D_i \ \frac{1}{\gamma} \ \frac{L}{T_m} \ . \tag{10}$$

Combining eqs. 5 and 10 leads to $\Gamma = \mu(\gamma/S)K^2$. This is exactly the result which has been deduced earlier [26] under the assumption of a corrugated interface using Gibbs-Thomson equation. There are two possibilities to determine the diffusivity which is rate limiting for the solidification process : from the light scattering experiment $D_i = \Gamma/K^2$ and from directional solidification $D_i = \mu(\gamma/S)$. The two values are compared in Table 1.
Scaling eq. 9 with the thermal velocity $2\alpha/d_o$ one obtains the dimensionless rate of solidification :

$$\tilde{v} \equiv \frac{v}{2\alpha/d_o} = \frac{D_i}{2\alpha} \ \frac{\Delta T}{L/c_p} \ . \tag{11}$$

Free solidification is also influenced by the interface layer. For the spherical approximation we obtain from eqs. 1 and 3 ($\rho^* > d_o$):

$$v_{tip} = \frac{\alpha}{\gamma/L} \ \frac{1}{2} \ \frac{T_m - T_\infty}{T_m} \ \frac{T_m - T_\infty}{L/c_p} \ . \tag{12}$$

In the interface layer the effective thermal diffusivity is D_i

		ICE	SALOL
s	$J/cm^3 K$	1.1 [27]	0.324 [31]
L/c_p	K	80 [27]	60 [32]
α(liquid)	cm^2/s	$1.3 \cdot 10^{-3}$ [30]	$0.9 \cdot 10^{-3}$ [32]
α(solid)	cm^2/s	10^{-2} [30]	$1.7 \cdot 10^{-3}$ [32]
γ	J/cm^2	$2.9 \cdot 10^{-6}$ [28]	$2 \cdot 10^{-6}$ [22]
$\mu = v/T$	$cm/s \cdot K$	10^{-2} [29,30]	$2.3 \cdot 10^{-4}$ [33] 10^{-4} [22]
model $D_i = \frac{\mu \gamma}{s}$	cm^2/s	$2.6 \cdot 10^{-8}$	$1.5 \cdot 10^{-9}$ $6.5 \cdot 10^{-10}$
experiment $D_j = \Gamma/K^2$	cm^2/s	$3 \cdot 10^{-8}$ [21]	$\sim 10^{-9}$ [22]
$d = \frac{\alpha}{D_i} \frac{\gamma}{L}$	μm	4	88
d(measured)	μm	$1.7 < d < 6$ [23]	

Table 1 Physical parameters of the solid-liquid interface of H_2O and salol.

instead of α in the equilibrium. The thickness of the interface layer is $\frac{\alpha}{D_i} \frac{\gamma}{L}$ instead of $\frac{\gamma}{L}$ at equilibrium conditions. Replacing these quantities in eq. 12 we obtain :

$$v_{tip} \propto \frac{D_i}{\frac{\alpha}{D_i} \frac{\gamma}{L}} \frac{1}{2} \frac{T_m - T_\infty}{T_m} \frac{T_m - T_\infty}{L/c_p} \tag{13}$$

and the scaling law for dendritic solidification is :

$$\tilde{v}_{tip} = \frac{v_{tip}}{2\alpha/d_o} \propto \left[\frac{D_i}{2\alpha} \frac{(T_m - T_\infty)}{L/c_p} \right]^a , \tag{14}$$

with a = 2 for the spherical approximation. The experiment leads to a = 2.8. With this expression it is possible to scale the rates of free solidification of ice and salol (Fig. 5). During melting there is no clustering at the interface, hence $D_i = \alpha$ and eq. 14 changes to the expression which has been deduced for dendritic melting (eq. 4). During directional solidification heat flow is a linear process and one obtains, from eq. 14 with a = 1 and $\Delta T = (T_m - T_\infty)$, eq. 11 which has been deduced by simple dimensional analysis.

5. CONCLUSIONS

Directional solidification and dendritic growth cannot be understood on the basis of heat flow and interface stability considerations alone. Entropy fluctuations in an interface layer can be detected during solidification. The kinetics which determine the growth and the decay of these fluctuations are also rate limiting for the freezing process because the crystal is growing into this layer of polimerized material.

Table 1 compares the predictions of the model with the results as observed in the light scattering experiments and in crystal growth. The diffusivity D_i at the interface has to be compared with the thermal diffusivity of the melt which is in the order of magnitude of 10^{-3} cm^2/s. The self diffusivities of ice and water at the melting point are 10^{-11} and 10^{-5} cm^2/s respectively [27]. The value of $D_i \simeq 3 \times 10^{-8}$ cm^2/s is close to $\sqrt{D(ice) \times D(water)}$. This supports the model which assumes that aggregation of melt takes place at the interface. In contrast to the approaches discussed in section 2 of this paper, there are no fitting parameters involved in the model presented in section 4.

The essential number in solidification is α/D_i, the dimensionless thickness of the interface layer. It characterizes the reaction constant of the "pre-ordering" of molecules in front of the growing crystal and the slowing down of the thermal diffusion at the

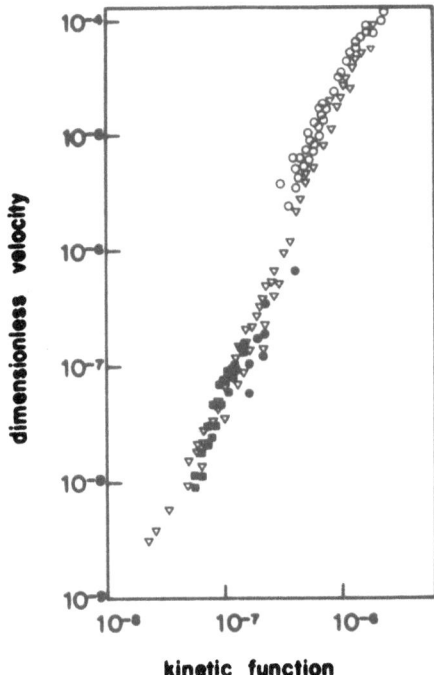

Fig. 5 The dimensionless growth rate $\tilde{v}_{tip} = \dfrac{V_{tip}}{2\alpha/d_o}$ is plotted versus the "kinetic function" $\dfrac{D_i}{2\alpha} \dfrac{(T_m - T_\infty)}{L/c_p}$ for ice dendrites (open symbols) and salol crystals (full symbols) according to eq. 14. The same experimental data are used as in Fig. 1 and Fig. 2.

interface. α/D_i can be determined by 4 different experimental methods which lead to consistent results :
1) measuring D_i in a light scattering experiment;
2) measuring the thickness of the interface layer;
3) measuring the dynamical constant of crystal growth;
4) measuring the rate of free solidification.

ACKNOWLEDGEMENTS

The author gratefully acknowledges the support of this work by Prof.
W. Känzig. He thanks his colleagues P. Böni and U. Dürig, who perform
the light scattering experiments, for helpful discussions during
the evolution of this paper. This work is supported by the Swiss
National Science Foundation.

REFERENCES

1. Bilgram, J.H., in "Nonlinear Phenomena at Phase Transitions and
    Instabilities" edited by T. Riste (New York, Plenum Press,
    1982), p. 343.
2. Bilgram, J.H., Ann. New York Acad. Sci. (in print).
3. Langer, J., Rev. Mod. Physics, 52 (1980) 1.
4. Chalmers, B., "Principles of Solidification" (John Wiley and
    Sons, New York, 1964).
5. Doherty, R., in "Crystal Growth" edited by Pamplin, B.R. (Oxford, Pergamon Press, 1975), p. 576.
6. Nash, G.E. and Glicksman, Acta Metallurgica, 22 (1974) 1283.
7. Huang, S.C. and Glicksman, M.E. Acta Metallurgica, 29(1981) 701.
8. Lappe, U., "Experimentelle Untersuchungen des Dendritischen
    Wachstums von Kristallen in unterkühlten Schmelzen" Berichte
    der Kernforschungsanlage Jülich 1671, Jülich 1980.
9. Sekerka, R.F., in "Proc. of the Darken Conference" (Monroeville
    United States Steel Res. Lab., Pa. 1976) p. 301.
10. Schneider, H., Diploma Thesis, ETH Zürich 1982.
11. Jackson, K.A., Uhlmann, D.R. and Hunt, J.D., J. Crystal Growth,
    1 (1967) 1.
12. Käss, M. and Magun, S., Z. Kristallogr, 116 (1961) 354.
13. Langer, J.S., Sekerka, R.F. and Fujioka, T., J. Crystal Growth,
    44 (1978) 414.
14. Jackson, K.A., in "Treatise on Solid State Chemistry, Vol. 5,
    Changes of State" edited by Hannay, H.B. (New York, Plenum,
    1975) p. 233.
15. Cahn, J.W., Acta Metallurgica, 8 (1960) 554.
    Hilliard, J.H. and Cahn, J.W., Acta Metallurgica, 6 (1958) 772.
16. Bolling, G.F. and Tiller, W.A., J. Appl. Phys., 32 (1961) 2587.
17. Turnbull, D. and Bagley, B.G., in : ref. 14, p. 513.
    Spaepen, F., Acta Metallurgica, 26 (1978) 1167.
18. U. Landmann, Cleveland, Ch.L. and Brown, Ch.S., Phys. Rev. Lett. 45 (1980) 2032.

19. Saito, Y. and Müller-Krumbhaar, H., J. Chem. Phys.,74 (1981) 721.
20. Güttinger, H., Bilgram, J.H. and Känzig, W., J. Phys. Chem. Solids, 40 (1979) 55.
21. Bilgram, J.H. and Böni, P., in "Light scattering in Liquids and Macromolecular Solutions" edited by Degiorgio, V., Corti, M. and Giglio, M. (New York, 1980, Plenum Press), p. 203.
22. Dürig, U. and Bilgram, J.H., in : ref. 1, p. 371.
23. Böni, P. and Bilgram, J.H., Helv. Phys. Acta 54 (1981) 266.
24. Crank, J., "The Mathematics of Diffusion"(London, Oxford University Press, 1975).
25. Hollomom, J.H. and Turnbull, D., Progr. in Metal Physics, Vol. 4, edited by Chalmers, B. (London, Pergamon Press, 1953), p. 333.
26. Bilgram, J.H., Güttinger, H. and Känzig, W. Phys. Rev. Letters, 40 (1978) 1394.
27. Hobbs, P.V., "Ice Physics"(Oxford, Clarendon Press, 1974).
28. Hardy, S.C., Phil. Mag. , 35 (1977) 471.
29. Hillig, W.B., in "Growth and Perfection of Crystals" edited by Doremus, R.H., Roberts, B.W. and Turnbull, D., (New York, Wiley, 1958) p. 350.
30. James, D.W. and Sekerka, R.F., J. Crystal Growth, 1 (1967) 67.
31. Pollatschek, H., Z. phys. Chem., A 142 (1929) 289.
32. Neumann, K. and Al-Yawir, D.M., J. Crystal Growth, 11 (1971) 323.
33. de Leeuw den Bouter, J.A. and Heertjes, P.M., J. Crystal Growth, 5 (1969) 19.

NUCLEATION IN CONDENSED MATTER

P. Germain[+] and K. Zellama[++]

+ Department of Physics, North Carolina State University
  Raleigh, NC 27650 USA
++ Groupe de Physique des Solides de l'E.N.S.,
  Université Paris VII Tour 23, 2 Place Jussieu,
  75221 Paris Cédex 05 France

1. DEFINITION OF THE NUCLEATION AND GROWTH PROCESS

Consider a first order phase transition or an amorphous to crystalline transformation $\alpha \to \beta$. Experience shows that such a transformation does not occur homogeneously. Small domains of the β phase appear at distinct points in α and then grow at the expense of α . This process is called nucleation and growth [1] . The kinetics of the phenomenon can be described by two constants : the rate of nucleation R^* (number of β domain appearing per unit time and per unit volume of the untransformed phase α) and the rate of growth after nucleation, $u(cm. s^{-1})$. The aim of nucleation theory is to calculate the nucleation rate R^*.

We shall present here a lecture on the theory of nucleation describing the nucleation phenomena involved during thermal annealing. Many theories of nucleation have been given in the literature; they are reviewed in references [1a, b, c] . In the present case where one expects to extend the concept and experiments to the study of nucleation induced by electron beam or laser annealing, we shall present the simplest theory. Particular emphasis will be given to nucleation at constant composition and the nucleation of precipitates will be only briefly reviewed.

2. HOMOGENEOUS NUCLEATION AT CONSTANT COMPOSITION

2.1 Thermodynamics barrier of the critical cluster

In the current theory of nucleation of crystals β in a condensed phase α the precursors of the β domains are assumed to be crystal clusters having the structure of the bulk phase β. One assumes that these β clusters are formed by the association of a few molecules and mix ideally with the molecules of the α phase. These clusters can be regarded as complexes characterized by standard thermodynamic coefficients.

Let ΔG_i be the *minimum standard* free energy of formation of a cluster of i molecules from the α phase. Although ΔG_i is positive when the clusters are very small, there is an entropy gain from mixing them with α phase molecules. If ideal mixing is assumed the net change in the free energy ΔG_m of a unit volume of α phase due to cluster formation is given by :

$$\Delta G_m = n_i' \Delta G_i + kT \left[n_i' \log \left(\frac{n_i'}{n_i' + n} \right) + n \log \left(\frac{n}{n_i' + n} \right) \right] \quad (1)$$

where n_i' = number of clusters/volume
n = number of molecules/volume in phase α.
The condition for equilibrium is that ΔG_m be a minimum. From this condition and as $n \gg n_i'$ the equilibrium number of clusters n_i is found to be :

$$n_i = n \exp \left[\Delta G_i / kT \right] . \quad (2)$$

To calculate ΔG_i we will assume that the free energy of formation G_i^β of a small β cluster containing i molecules is given by :

$$G_i^\beta = \mu_\infty^\beta i + \Sigma \sigma \boldsymbol{\lambda} , \quad \text{where}$$

μ_∞^β is the chemical potential of the phase β, σ = the surface tension of a cluster face of area $\boldsymbol{\lambda}$ and $\sigma \boldsymbol{\lambda}$ is summed over all faces. The ΔG_i is given by :

$$\Delta G_i = (\mu_\infty^\beta - \mu_\infty^\alpha) + \Sigma \sigma \boldsymbol{\lambda}, \quad \text{with} \quad (3)$$

μ_∞^α = chemical potential of the phase α.

The thermodynamic definition of the chemical potential allows one to write $\mu_\infty^\beta - \mu_\infty^\alpha$ as :

$$\mu_\infty^\beta - \mu_\infty^\alpha = v . \Delta G_v \quad (4)$$

where v is the volume of one molecule in the α phase and ΔG_v the variation of free energy per unit of volume; during the change from α to β. Then :

$$\Delta G_i = v.\Delta G_v i + \Sigma_\sigma \ell . \tag{5}$$

The first term (which is related to the volume) is proportional to i or to λ^3, where λ is a size parameter of the β cluster. The second term (which is related to the surface) must be proportional to λ^2 or to $i^{2/3}$.
We will write it as :

$$\Sigma_\sigma \ell = \Sigma \sigma a \; i^{2/3} \tag{6}$$

where a is a geometrical factor (in cm^2).

We will show later (section 5) that such a surface energy term proportional to the number of surface atoms $i^{2/3}$ is consistent with a nearest neighbor interaction model. Then :

$$\Delta G_i = v\Delta G_v i + \Sigma \sigma a \; i^{2/3} . \tag{7}$$

Stresses may appear in the cluster during nucleation, let E be the strain energy per atom of β.
Then ΔG_i takes the form :

$$\Delta G_i = v(\Delta G_v + E)i + \Sigma \sigma a \; i^{2/3} . \tag{7a}$$

E will be discussed in Section 5. These equations assume that the thermodynamic coefficients of the cluster are identical with those of the bulk crystal β and that the transformation strain energy is negligible.

The standard free energy of cluster formation is positive when i $\rightarrow$ 0, because in that case the surface free energy term $a\sigma i^{2/3}$ is predominant compared to the bulk term. Thus, for a crystallization, $(\Delta G_v < o)$, $\Delta G_i = f(i)$ goes through a maximum at constant cluster shape (fig. 1).

The coordinates of the maximum are :

$$\Delta G^* = 4(\Sigma a\sigma)^3 / 27(v\Delta G_v)^2 , \tag{8}$$

$$i^* = (2\Sigma a\sigma /3v \Delta G_v)3 . \tag{9}$$

In the particular case of a spherical cluster of radius r, if σ does not depend on the orientation, equation 7 can be written as a function of r.

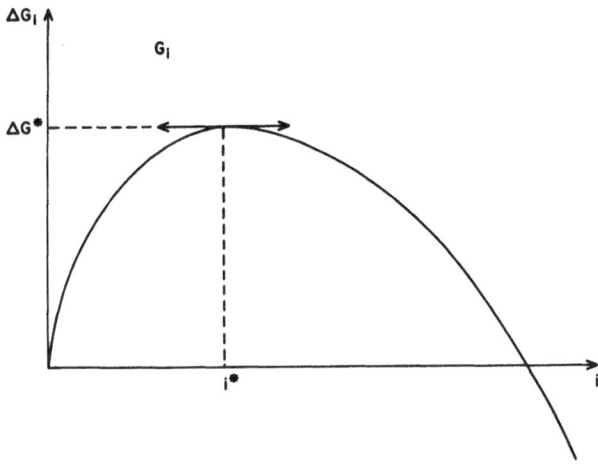

<u>Fig 1</u> Variations of ΔG_i vs. i

$$\Delta G_i \rightarrow \Delta G_{(r)} = \Delta G_v \cdot \frac{4}{3}\Pi r^3 + \sigma \cdot 4\Pi r^2 \ .$$

$\Delta G_{(r)}$ of course goes through a maximum for $r = r^*$:

$$r^* = - 2 \sigma/\Delta G_v \ .$$

A cluster containing i^* molecules will be called a "critical cluster" or a nucleus, and i^* and r^* will be respectively called the critical number of molecules and the critical radius. A cluster smaller than a nucleus will tend to dissociate since its free energy increases with the addition of new molecules $(\frac{d}{di} (\Delta G_i) > 0)$ (fig.1). On the contrary a cluster bigger than a nucleus will continue to grow since the addition of new molecules decreases the free energy $(\frac{d}{di}\Delta G_i < 0)$. Thus to become a nucleus a cluster must have a critical standard free energy ΔG^* greater than α phase (thermodynamic barrier). From eq. 2 the number of nuclei n_i^* at equilibrium is :

$$n_i^* = n \ exp[-\frac{\Delta G_i^*}{kT}] \ .$$

The coefficient σ has been theoretically identified as the macroscopic surface tension because, in condensed systems, experimental results provide values of $i^* \simeq 200$ to 300.

This identification is not always possible for it may be the case that the surface energy is minimized by configurations different from those existing in bulk phases.

2.2 Steady state rate of nucleation

In the particular case of bulk homogeneous nucleation previously described, formula 7 can be written :

$$\frac{\Delta G_i}{kT} = A \, i^{2/3} - Bi \tag{10}$$

with :

$$A = \frac{\Sigma a \sigma}{kT} , \tag{11}$$

$$B = \frac{v \Delta Gv}{kT} . \tag{12}$$

The function $\Delta G_i/kT$ versus i passes through a maximum $\Delta G_i/kT = 4 \, A^3/27 \, B^2$ at $i = 2A/3B$, and then decreases (fig. 1) without limit. Nucleation theory frequently leads to an expression of the form [10].

This expression, as well as the following theory is applicable to other cases such as heterogeneous nucleation. In the latter case one must only keep in mind that atoms available for nucleation are not in the bulk but on some surface. Turnbull [2] has set up a theory of steady state rate of nucleation, which we now present.

The steady state rate of nucleation for a given transformation corresponds to constant, equal net forward rates for the following set of reactions :

$$m\alpha_1 \underset{\leftarrow}{\overset{\rightarrow}{}} \beta_m$$

$$\beta_m + \alpha_1 \underset{\leftarrow}{\overset{\rightarrow}{}} \beta_m + 1$$

$$\beta_m + 1 + \alpha_1 \underset{\leftarrow}{\overset{\rightarrow}{}} \beta_m + 2$$

$$\beta_m + 2 + \alpha_1 \underset{\leftarrow}{\overset{\rightarrow}{}} \beta_m + 3 \tag{13}$$

where :

α_1 = an atom of phase α,

β_i = a cluster of phase β containing i atoms,

β_m = the smallest nucleus of phase β.

Turnbull points out that :

$$\ell in(\text{concentration of } \beta_i) = 0 \qquad (14)$$
$$i \to \infty,$$

since otherwise the percentage of untransformed α would be zero.

Consider the following reaction at steady state :

$$\beta_i + \alpha \underset{\leftarrow}{\rightarrow} \beta_{i+1} . \qquad (15)$$

Referring to phase α as the standard state, the free energies of $\beta_i + \alpha_1$ and β_{i+1} are respectively ΔG_i and ΔG_{i+1}. Assuming that the intermediate configuration between $\beta_i + \alpha_1$ and β_{i+1} corresponds to free energies greater than either ΔG_i or ΔG_{i+1}, each path by which β_i and α_1 can be combined to give β_{i+1} will have an intermediate configuration of maximum free energy. For one of these paths, there will be a particular intermediate configuration, called the activated complex, corresponding to the least maximum free energy. The difference between the free energy of the activated complex and the mean of ΔG_i and ΔG_{i+1} will be noted Δg^* (see fig. 2). Consider the forward reaction :

$$\beta_i + \alpha_1 \to \beta_{i+1} .$$

The rate of this reaction can be written [3]

$$R^+ = n_i a_1 i^{2/3} \left(\frac{kT}{h}\right) \exp\left(-\Delta g_1^*/kT\right). \qquad (16)$$

where :

n_i = steady state concentration of β_i nuclei,

$a_1 i^{2/3}$ = number of α atoms in contact with each β_i nucleus

so $n_i a_1 i^{2/3}$ is the number of α atoms ready for the reaction.

$\frac{kT}{h} \exp\left(-\frac{\Delta g_1^*}{kT}\right)$ is the reaction rate for the production of the activated complex with free energy increase Δg_1^* (see fig. 2).

In the same way Turnbull writes the rate of the reverse reaction $\beta_{i+1} \to \beta_i + \alpha_1$ in the form :

260

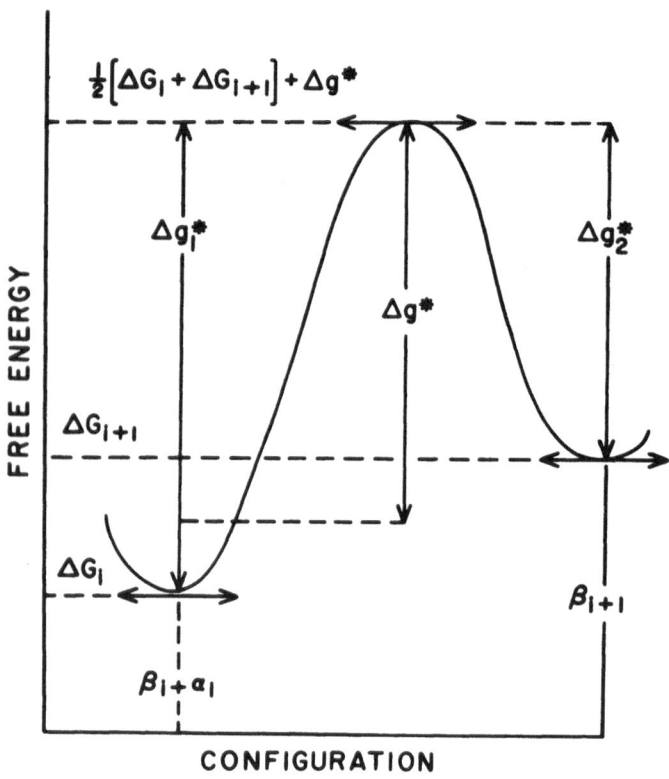

Fig. 2 Schematic representation of free energy of activation, from
ref. 2.

$$R^- = n_{i+1}a_2 i^{2/3}(kT/h) \exp(-\Delta g_2^*/kT), \qquad (17)$$

where $a_2 i^{2/3}$ is the number of β-atoms in contact with α per β_{i+1} nucleus. The net forward rate of the reaction is therefore :

$$R^* = R^+ - R^- = (kT/h)i^{2/3}[n_i a_1 \exp(-\Delta g_1^*/kT) - n_{i+1}a_2 \exp(-\Delta g_2^*/kT)] \qquad (18)$$

Turnbull obtains an approximate solution of equation 18 for n_i by making the following assumptions :
a) It is possible to admit that the number of α atoms in contact with a β_i nucleus is roughly equal to the number of β atoms in contact with α at the surface of a β_{i+1} nucleus; so he assumes that $a_1 = a_2 = a_0$. This is valid for all but the smallest nuclei.
b) n_i and ΔG_i are assumed smooth functions of i although they have only meaning for integral value of i. In addition $n_{i+1} - n_i$ is approximated by $dn_i/di \times i = dn_i/di$; similarly, he takes $\Delta G_{i+1} - \Delta G_i \approx \frac{d}{di}(\Delta G_i)$. The approximation is good for all but the smallest nuclei.
c) The quantities $\frac{1}{kT}\frac{d}{di}(\Delta G_i)$ and $\frac{1}{n_i} \times \frac{d n_i}{di}$ are both assumed to be small compared with unity. Again the approximation is valid except for the smallest nuclei.

Noting that :

$$\Delta g_i^* = \Delta g^* + \frac{1}{2}(\Delta G_i + \Delta G_{i+1}) - \Delta G_i$$

$$= \Delta g^* + \frac{1}{2}(\Delta G_{i+1} - \Delta G_i)$$

$$= \Delta g^* + \frac{1}{2}\frac{d}{di}(\Delta G_i)$$

$$\Delta g_2^* = \Delta g^* - \frac{1}{2}\frac{d}{di}(\Delta G_i)$$

$$R^* = (-a_0 \frac{kT}{h}) \exp\left(\frac{-\Delta g^*}{kT}\right) \times [n_i(\frac{2}{3}Ai^{-1/3}-B) + dn_i/di]\, i^{2/3} ; \qquad (19)$$

defining :

$$R = R^*/(\frac{a_0 kT}{h}) \exp(-\Delta g^*/kT), \qquad (20)$$

Eq. 19 reduces to the differential equation :

$$dn_i/di + (\frac{2}{3}Ai - \frac{1}{3} - B)n_i = -Ri^{-2/3}. \qquad (21)$$

The solution to Eq. 21 is :

$$n_i = \exp[-(Ai^{2/3} - Bi)] \times [-R\int_{i_0}^{i} \exp(Ai^{2/3} - Bi)i^{-2/3}di + n_{i_0}\exp(Ai_0^{2/3} - Bi_0)] \qquad (22)$$

where n_{i_0} is the steady state number of β_{i_0} nuclei.

The steady state concentration of subcritical nuclei containing say one-third the critical number of atoms does not differ appreciably from the equilibrium concentration assuming no nucleation. Taking $i_0 = i^*/3$

$$n_i = n\exp(-\Delta G_{i_0}/kT) = n\exp[-(Ai_0^{2/3} - Bi_0)]$$

where n is the number of atoms of untransformed α. The expression for the steady state number of β_i nuclei reduces to :

$$n_i = \exp(-\Delta G_i/kT) \times [-R\int_{i_0}^{i} \exp(\Delta G_i/kT)i^{-2/3}di + n] \quad .(23)$$

Since :

$$\lim_{i\to\infty} \exp(-\Delta G_i/kT) = \infty$$

and it is required that :

$$\lim_{i\to\infty} n_i = 0 \quad ,$$

it must follow that :

$$\lim_{i\to\infty} [-R\int_{i_0}^{i} \exp(\Delta G_i/kT)i^{-2/3}di + n] = 0 \tag{24}$$

and hence that :

$$R = n/[\int_{i_0}^{\infty} \exp(\Delta G_i/kT)i^{-2/3}di] . \tag{25}$$

We write (25) in the form :

$$R = \frac{n}{J} \quad .$$

Turnbull gives then an evaluation of the integral J. He remarks that for $0 < i_0 < i^*$ contributions of the integral come mostly from values of i near i^* , so he replaces $\Delta G_i/kT$ by a Taylor's expression about $i = i^*$; then he neglects terms after the third and obtains :

$$J = (\frac{9\pi}{A})^{1/2} i^{*-2/3} \exp(\frac{\Delta G^*}{kT}) . \tag{26}$$

From equations 20, 25 and 26, the steady state rate of nucleation is :

$$R^* = n^*(A/9\pi)^{1/2}(NkT/h) \times \exp[-(\Delta g^* + \Delta G^*)/kT] \tag{27}$$

R^* is in nuclei per second per mole of untransformed material, where n^* is the number of surface atoms in the critical size nucleus :

$$n^* = a_0 \times i^{*2/3}.$$

The quantity $n^*(A/9\Pi)^{1/2}$ proves to be within a factor of ten of unity for most nucleation problems of interest, giving :

$$R^* \simeq (NkT/h) \, \exp[\,-(\Delta g^* + \Delta G^*)/kT] \qquad (28)$$

nuclei per mole per second to an order of magnitude.

Equation (28) was derived for nucleation processes that do not require long-range diffusion. Δg^* is the free energy of activation for the short-range diffusion of atoms or molecules moving a fraction of an atomic distance across an interface to join a new lattice.

3. TRANSIENT NUCLEATION

We have previously assumed that the steady state concentration of β clusters exists at all times. This is unrealistic at the beginning of the transformation and there is a transient period during which the steady state concentration is established. There is no exact solution to this problem of transient nucleation. Kantrowitz [4] has proposed in the case of nucleation from vapor an expression of the transient rate of nucleation which probably gives the transient period within an order of magnitude.

We shall now transpose this calculation to condensed systems.

A net forward reaction rate can be written for each value of i (i.e. for each size of cluster). From equation 19 :

$$R_i^* = (-a_0 \frac{kT}{h}) \, \exp \, (-\frac{\Delta g^*}{kT}) \times [\, \frac{n_i}{kT} \frac{\partial \Delta G}{\partial i} + \frac{dn}{di}\,] \; i^{\,2/3} \qquad (29)$$

$$R_i^* = D(i) \times [\, \frac{n_i}{kT} \frac{\partial \Delta G_i}{\partial i} + \frac{dn_i}{di}\,] \qquad (30)$$

where $D(i)$ is a "diffusivity" term :

$$D(i) = (-a_0 \frac{kT}{h}) \exp \, [-\frac{\Delta g^*}{kT}] \times i^{\,2/3}. \qquad (31)$$

Equation 29 has previously been used under steady state conditions, but is still valid in the present case.

Now if we consider the cases where n_i is not steady in time, we readily get from (16) and (17) :

$$\frac{\partial n_i}{\partial t} = R_i^+ - R_{i+1}^- - (R_i^- + R_{i+1}^+) \quad \text{or} \quad \frac{\partial n_i}{\partial t} = R_i^* - R_{i+1}^* . \quad (32)$$

Assuming that i is not too small :

$$\frac{\partial n_i}{\partial t} = - \frac{\partial}{\partial i} R_i^* , \quad (33)$$

Substituting (30) into (33) :

$$\frac{\partial n_i}{\partial t} = - \frac{\partial}{\partial i} D(i) \left[\left(\frac{n_i}{kT} \frac{\partial \Delta G_i}{\partial i} \right) \right] - \frac{\partial}{\partial i} \left(D(i) \frac{\partial n_i}{\partial i} \right) \quad . \quad (34)$$

Following Kantrowitz, we shall now neglect the thermodynamic barrier term, the first term in equation 34, to obtain a simple solution which will be valid for the interval during which the nucleation rate is small compared to the steady state result. In addition we shall suppress the dependence of the diffusivity on the embryo size and adopt an average $-|D|$. Then equation 34 becomes :

$$\frac{\partial n_i}{\partial t} = |D| \frac{\partial^2 n_i}{\partial i^2} , \quad (35)$$

which is nothing but the diffusion equation. The order of magnitude of the (nucleation) time required for an appreciable nucleation to occur in a condensed phase α can be found by solving eq. 35 with the following boundary conditions :

1) A number of n_o embryos exist n(i = 0) for t = 0;
2) After embryos have grown to contain I molecules, they grow so rapidly that $n_I = 0$ for all t. To obtain the order of magnitude of the nucleation time it will be sufficient to take $I = 2i^*$;
3) n(i=0) is some constant n_o. Here we assume the existence of a constant number of molecules in the α phase, so we are using equation 35 only to very small values of i. The solution of eq. 35 with these boundary conditions is obtained as well as the nucleation rate (without the last term in eq. 34):

$$R_I = |D| \frac{\partial n_i}{\partial i} \Big)_{i=I} + \frac{2 |D| n_o}{I} \times \sum_{j=0}^{\infty} (-1)j \exp \left(- \frac{\Pi^2 j^2 Dt}{I^2} \right)$$

$$(36)$$

Eq. 36 has the same form as the result obtained by Kantrowitz [4] .

Using a Poisson summation formula this author replaces the series of eq. 36 with a form which converges much more rapidly.

$$RI = 2n_o \left(\frac{|D|}{\pi t}\right)^{1/2} \sum_{j=0}^{\infty} \exp\left[-\frac{I^2}{4Dt(2j+1)^2}\right] . \tag{37}$$

For small t one needs only retain the first term and then :

$$RI = 2n_o \left(\frac{D}{\pi t}\right)^{1/2} \exp\left[\frac{-I^2}{4Dt}\right] . \tag{38}$$

Large rate of nucleation only occurs for $t \sim \tau$ where :

$$\tau = \frac{I^2}{4|D|} \simeq \frac{i^{*2}}{|D|} = \frac{i^{*2}}{n^* \nu^1} , \tag{39}$$

$n^* =$ number of atoms at the surface $= a_o \, i^{2/3}$,

$\nu^1 = \frac{kT}{h} \exp\left[-\Delta g^* /kT\right] .$

In the case of nucleation from the gas phase, Kantrowitz neglects the thermodynamic barrier and compares his results with these of Probstein [5] who takes account of this term.

Characteristic times in these two theories agree within a factor of 50 %.

In the case of condensed systems, we have interpreted our experiments on amorphous germanium and silicon [6] and showed the coherence of the theory of steady state nucleation (eq. 27) with theory of transient nucleation (39).

4. HETEROGENEOUS NUCLEATION

There is much experimental evidence showing that, during a phase change, nucleation may be strongly promoted by a number of structural defects or impurities (see review in [1] , [7]).

4.1 Nucleation rate

Consider the formation (from α) of a small cluster β on the surface of an impurity particle s suspended in α , or on substrate s if α is deposited on it. Let i be the number of atoms of β.

The part of the substrate s where the nucleation will take place is AB (fig. 3A), its surface free energy is $i^{2/3} \Sigma_1$ as x $\sigma_{\alpha s}$ where Σ_1 is a summation on facets of the substrate. The surface free energy of the nucleus after nucleation is $i^{2/3}[\Sigma_1$ as $\sigma_{\beta s} + \Sigma_1 a_{\alpha\beta} \, \sigma_{\alpha\beta}]$, where Σ_1 is the summation on facets of the interface $\alpha\beta$.

266

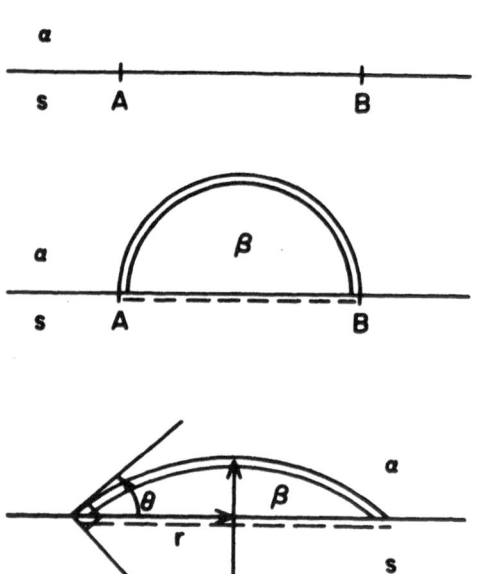

Fig. 3 a) 1. System before nucleation;
2. System after nucleation :
─────Boundary s/α free energy $\sigma_{\alpha s}$,
‐‐‐‐‐‐‐Boundary s/β free energy $\sigma_{s\beta}$,
══════Boundary α/β free energy $\sigma_{\alpha\beta}$:
b) Element of phase β forming (from α) with a contact angle
Θ on the surface of an impurity particle.

So the free energy of formation ΔG_s of the cluster β is, in general :

$$\Delta G_s = i^{2/3} \left[\sum_1 as(\sigma_{\beta s} - \sigma_{\alpha s}) + \sum_2 a_{\alpha \beta} \sigma_{\alpha \beta} \right] + i \, v \, (\Delta G_v) \quad . (40)$$

If :

$$W_s = \sigma_{\beta s} - \sigma_{\alpha s} \quad , \tag{41}$$

then :

$$\Delta G_s = i^{2/3} \left[\sum_1 a_s W_s + \sum_2 a_{\alpha \beta} \sigma_{\alpha \beta} \right] + i \, v \, \Delta G_v \quad . \tag{42}$$

Expression (40) shows that, whether or not the impurity surface (or substrate) catalyzes, β nucleation is determined by the magnitude of W_s. The impurity surface will be a preferred (relative to the bulk of α) β nucleation site if $W_s < \sigma_{\alpha \beta}$. The greater the intermolecular forces between β and s, the less is W_s. Eq. 41 shows that W_s can be negative. The radius of curvature r_s of the impurity surface may play an important role.

When $r_s \ll r$ (radius of the cluster), ΔG_s and W_s are function of r_s.

When $|r_s| \gg r$ and $\sigma_{\alpha \beta} > |W_s|$, eq. 42 shows that ΔG_s goes through a maximum ΔG_s^* for any cluster shape. In order to simplify the equation we shall assume that $\sigma_{\alpha \beta}$ and W_s do not depent on the boundary orientation. Then ΔG_s becomes:

$$\Delta G_s = A_{\beta s} W_s + A_{\alpha \beta} \sigma_{\alpha \beta} + vi \, \Delta G_v, \tag{43}$$

where :

$A_{\beta s}$ = area of the cluster in contact with s,
$A_{\alpha \beta}$ = area of the cluster in contact with α.

The cluster having a shape of a spherical cap with a characteristic angle Θ (fig. 3b) provides the lowest value of ΔG_s, for a fixed value of i. The equilibrium between s, α and β can never be rigorously realized if s has a perfect plane shape [8] but one can approximately relate Θ to W_s and $\sigma_{\alpha \beta}$ by :

$$\cos \Theta = - \frac{W_s}{\sigma_{\alpha \beta}} \quad . \tag{44}$$

From fig. 3b it follows that:

1) $A_{\beta s} = \Pi r^2$, $\tag{45}$

as

$$\frac{r}{R} = \sin \Theta, \tag{46}$$

$$A_{\beta s} = \Pi R^2 \sin^2 \Theta. \tag{47}$$

2) $A_{\alpha\beta} = R^2 2\Pi(1 - \cos \Theta),$ $\qquad$ (48)

3) $i \ v = \frac{4}{3} \Pi R^3 \ x \ \dfrac{2\Pi(1 - \cos\Theta)}{4 \ \Pi} - \dfrac{\Pi R^2 \sin^2\Theta \ \Theta x \ R \cos \Theta}{3},$

$$\tag{49}$$

substituting (47), (48), (49) into (43) :

$$\Delta G_s = [\ \Pi R^2 \ \sigma_{\alpha\beta} + \frac{\Pi R^3}{3} [\Delta G_v]] \ x \ [\ 2(1-\cos \Theta) - \sin^2 \Theta \cos \Theta]. \tag{50}$$

The derivation of (50) with respect to R provides ΔG_s^* :

$$\Delta G_s^* = \frac{16\Pi (\sigma_{\alpha\beta})^3 f(\Theta)}{3(\Delta G_s + E)^2} = \Delta G^* f(\Theta), \text{(taking account of strains)} \tag{51}$$

where :

$$f(\Theta) = (2 + \cos\Theta) (1 - \cos\Theta)^2 / 4 \tag{52}$$

and ΔG^* is the thermodynamic barrier to nucleation (at constant strain) in the body of α. Depending on W_s, the magnitude of $f(\Theta)$ ranges between unity and zero. The limiting values are $f(\Theta) = 1$ when $W_s = \sigma_{\alpha\beta}$, and $f(\Theta) = 0$ when $-W_s > \sigma_{\alpha\beta}$.

Now (50) can be written :

$$\frac{\Delta G_s}{kT} = 4 \ f(\Theta) \ \frac{\Pi R^2 \sigma_{\alpha\beta}}{kT} + \frac{\Pi R^3}{3} \ \frac{\Delta G_v \ x \ 4 \ f(\Theta)}{kT} \tag{53}$$

and identified with expression (10) :

$$\frac{\Delta G_s}{kT} = A_s i^{2/3} - B_s i. \tag{54}$$

Comparison of (53) and (54) immediately provides B_s and defines A_s in the case of heterogeneous nucleation. Then the formalism of 2.2 can be applied to obtain the steady state rate of heterogeneous nucleation under the form :

$$R_s^* = K_s \exp [-(\Delta G_s^* + \Delta g^*)/kT], \tag{55}$$

where :

$$K_s = n^* (\frac{\sigma_{\alpha\beta}}{kT})^{1/2} (\frac{2v}{9\Pi})^{1/3} [f(\Theta)]^{1/6} n_s v_1. \tag{56}$$

n^* is the number of molecules in the nucleus in contact with α and n_s is the number of molecules per unit area from α in contact with the impurity surface. A transient period exists in the heterogeneous nucleation as well as in the homogeneous case. An equation of the form (38) should stay valid with $\tau = i_s^{*2}/n_s^* v_1$. n_s^* is the number of atoms of the cluster which are at the external surface and which are available for the reaction.
i_s maximizes (54): $v_1 = kT/h$.

4.2 Nucleation on dislocations [9] .

Cahn calculates the thermodynamic barrier for nucleation on a dislocation, under the following assumptions :

The nucleus lies along the core of the dislocation; its cross section perpendicular to the dislocation is a circle. Cahn assumes incoherency between the initial phase α and the cluster β. The interface free energy is $\sigma_{\alpha\beta} = \gamma$. Phase α is assumed to be an isotropic elastic substance (valid for large nucleus).

The energy of the nucleus consists of three terms, a strain energy, a surface energy and a volume energy term. Nucleation is favored by the strain and volume energy term. The surface energy term tends to oppose nucleation. The strain and volume energy term are, respectively, dominant at low and high values of r. At intermediate values of r, the surface term under certain conditions can predominate over the others, resulting in a region where the free energy increases with r (curve A fig. 4a). The free energy presents a minimum which Cahn interprets as a subcritical metastable cylinder of phase β which surrounds initially the dislocation line. The radius of this cylinder r_o is obtained by minimizing the free energy per unit length :

$$G_D = -A \log r + 2 \Pi \gamma r - \Pi r^2 g + \text{const.}, \tag{57}$$

where :

$$A = \frac{Gb^2}{4\Pi(1-v)} \quad \text{for edge dislocation,}$$

$$A = \frac{Gb^2}{4\Pi} \quad \text{for screw dislocation,}$$

$$G = \text{elastic shear modulus,}$$

270

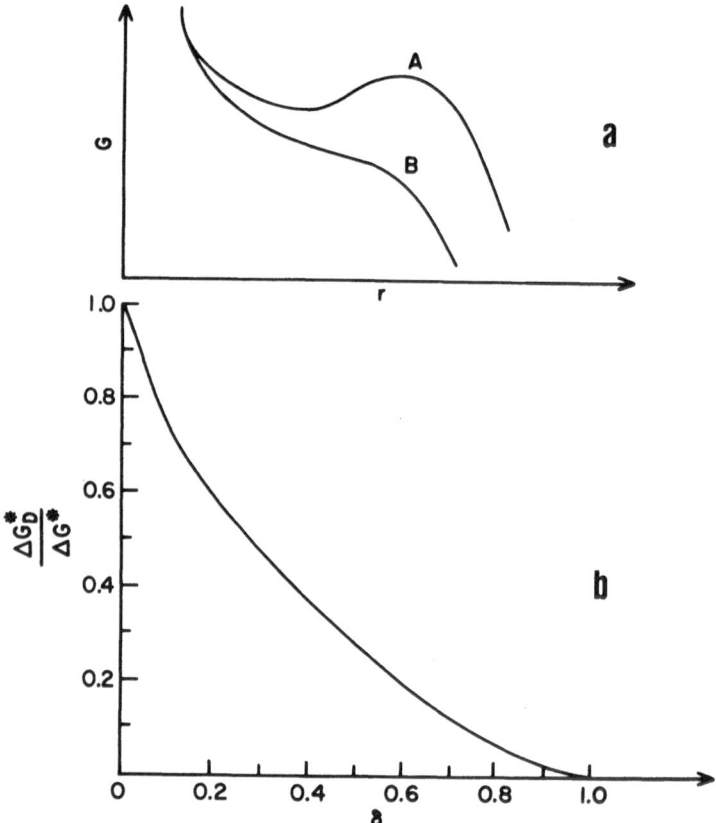

Fig. 4a Free energy per unit length of a cylinder surrounding a
dislocation from equation 57 for a system where the parameter
δ is less than 1 (Curve A) or greater than 1 (Curve B) [9].
Fig. 4b The ratio of the nucleation energy on a dislocation to the
homogeneous nucleation energy $\frac{\Delta G_D^*}{\Delta G^*}$ as a function of parameter δ
from Cahn[9].

b = Burgers vector,
v = Poisson ratio and
g = $-\Delta G_v$.

If the new phase is stable, then g > 0.

When the parameter $\delta = \frac{2Ag}{\Pi\gamma 2}$ is < 1, a minimum of the free energy $G_D(r)$ occurs for r = r_0. The value of r_0 is (curve A, fig. 4a) :

$$r_0 = \frac{\gamma}{2g} [1 - \frac{\sqrt{1-2Ag}}{\Pi\gamma 2}] . \tag{58}$$

When δ > 1, no minimum of G_D occurs (curve B fig. 4a). Cahn shows that r_0 is a weakly varying function of g. He considers further only the case where δ < 1 : there is a thermodynamic barrier to the nucleation of the phase β. The production of a β cluster requires energy, but beyond a certain size the cluster will keep growing. The maximum of energy which must be supplied depends on the size and shape of the cluster along the dislocation. Cahn defines the shape of the nucleus by the (unknown) function r(z) where z is the distance along the dislocation. The problem is to determine the shape of r(z) for which the thermodynamic barrier will be a minimum. The free energy of formation of the nucleus is given by :

$$\Delta G_D = \int_{-00}^{+00} [-A \log \frac{r}{r_0} + 2\Pi\gamma(r\sqrt{1+r'^2} - r_0) - \Pi g(r^2-r_0^2)] \, dz \tag{59}$$

where $r' = \frac{dr}{dz}$

The nucleus (defined by its shape and size) corresponds to a saddle point of ΔG_D : its shape is such that ΔG_D is minimum in energy with respect to shape; its size will be such that ΔG_D be a maximum as a function of size. Cahn determines the shape and size by applying the Euler-Lagrange equation to the integral with proper boundary conditions.

This provides ΔG_D^*, free energy of formation of the nucleus on the dislocation.

In fig. 4b, $\frac{\Delta G_D^*}{\Delta G^*}$ is plotted vs δ. (ΔG^* is the free energy of formation of a nucleus for homogeneous nucleation, previously determined (eq. 8)).

Cahn's model allows one to predict that the efficiency of dislocation to catalyze nucleation increases with increasing the parameter δ where :

$$\delta = \frac{2 \mid G_v \mid}{\Pi [\sigma_{\alpha\beta}]^2} \times \frac{Gb^2}{4\Pi(1-v)} \ .$$

Cahn estimates values of γ for which nucleation becomes copious. He finds for most systems $0.4 \leq \delta \leq 0.7$.

5. BOUNDARY ENERGY AND TRANSFORMATION STRAINS

There is no theory describing the free energy surface σ_{ac} for the case of a cristalline cluster c in an amorphous matrix a. Such a theory would apply to nucleation in amorphous germanium or silicon. It is possible, however, to give some general characteristics of σ valid in many cases. We give such properties at the beginning of this section. Then we review two examples of crystalline-crystalline interfaces which show the various phenomena involved in surface energies. In order to apply Becker's theory of nucleation, involving a thermodynamic barrier to nucleation, one needs to know interfacial free energies. Most of these energies have not been measured but it is possible to estimate them from the theory of solid interphase boundaries. There is a "chemical interfacial energy" σ_c^o and a "structural interfacial energy" σ_{st} we will discuss now.

In calculating σ_c^o, Becker [10] assumed structural continuity (coherence) across α-β interface. This expression is directly obtained from the definition of σ which is the difference in free energy per unit of surface with and without the boundary. The crystal structures of α and β are identical in symmetry and lattice parameter, σ_c^o can be estimated by nearest neighbor hypothesis. If α and β corresponds to particular atomic species and at T=0K :

$\varepsilon_{\alpha\alpha}$ = energy of an $\alpha\alpha$ bond,

$\varepsilon_{\beta\beta}$ = energy of a $\beta\beta$ bond ,

$\varepsilon_{\alpha\beta}$ = energy of an $\alpha\beta$ bond.

It follows that σ_c^o is :

$$\sigma_c^o = n_s z_s [\varepsilon_{\alpha\beta} - 1/2(\varepsilon_{\alpha\alpha} + \varepsilon_{\beta\beta})], \tag{60}$$

where n_s = number of atoms per unit area within one interfacial atomic plane,

z_s = number of $\alpha\beta$ bonds per atom α(or β) in the α-β interface.

We shall now obtain σ_{st} which is associated with any lack of coherence between α and β. We must remark that in the case of very

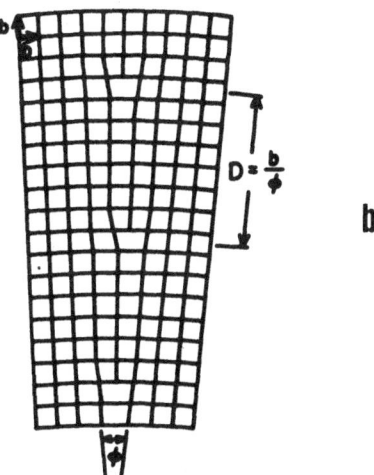

Fig. 5a Cube plane cut through two simple-cubic lattices of slightly different lattice parameters.
Fig. 5b Low angle grain boundary (after Burgers [13]).

small discrepancy the total surface free energy σ is :

$$\sigma = \sigma_c + \sigma_{st}. \tag{61}$$

We will now evaluate σ_{st} in two cases : first for the case where α and β are different, then for the case of grain boundary.

5.1 α-β Interphase

A large class of solid interphase boundaries can be described by dislocation models [Ref. 1a] .

Consider for instance Fig. 5a which represents an arrangement of molecules at the (100) interface between two cubic crystals α and β having different lattice parameters. One can imagine that the actual boundary would be composed of islands of good fit separated by dislocations. The disregistry d between the structures is $d = (a_\alpha^o - a_\beta^o)/a_\alpha^o$ where a_α^o and a_β^o are the nearest neighbor distances of the two structures at equilibrium. A good fit should be observed on island of linear dimension 1/d, so the linear density of dislocation at the boundary is d per unit length. If we call d' = d - e the disregistry between unstrained α and the β structure strained by an amount e, we can express σ_{st} as [11] :

$$\sigma_{st} = Ld' [M - \log d']. \tag{62}$$

L and M are function of the lattice constants of α and β. (The boundary entropy has been neglected.)

The structural free energy σ_{st} can be reduced if the cluster β is conveniently strained of an amount e for matching α more closely. Nabarro [12] has obtained the strain energy E per unit volume :

$$E = ce^2. \tag{63}$$

c is a function of the elastic modulus of α and β and depends on the shape of the cluster.

The condition of coherent nucleation is that e = d and the corresponding strain energy :

$$E = cd^2 \tag{64}$$

This value of E should be substituted in equations describing the nucleation rate (Eq. 7a).

5.2 Low Angle Grain Boundary

A model for low angle boundaries between adjoining crystallites has been proposed by Burgers [13] who described it by an array of dislocations. We show in Fig. 5b an example of this grain boundary model. A simple cubic lattice model is presented. The boundary occupies the (010) plane and divides two parts of the crystal that have a [001] common axis. The misorientation is described by a small rotation ϕ about this common axis of one part of the crystal with respect to the other. The array of edge dislocations describing this boundary is characterized by a spacing $D = b/\phi$ where b is the Burgers vector of the dislocations. Read and Shockley [14] give a general theory providing σ as a function of ϕ, in very good agreement with experimental results[15]. It is important to note that the region of elastic distorsion is confined to a slab of thickness equal to D and does not extend into the two cyrstals. Each dislocation is surrounded by its strain field and by the strain fields of the two neighboring dislocations. These last two strain dislocations are equal and have opposite sign and cancel each other. Thus the strain energy of dislocations results only from its own strain field. In this approximation for one dislocation, the strain energy per unit length is $\frac{Gb^2}{4\Pi(1-v)}$ Log $(\frac{\eta D}{b})$ where G is the elastic shear modulus, v is the Poisson ratio and $\eta \simeq 1$. As there are $1/D = \phi/b$ dislocations per unit length of the boundary, σ can be expressed as :

$$\sigma(\phi) = -(\frac{Gb}{4\Pi(1-v)}) \phi(\log \phi + \log \eta). \qquad (65)$$

We note that $\sigma(\phi=0) = 0$; for increasing values of ϕ, $\sigma(\phi)$ is an increasing function of ϕ, reaches a maximum σ_m for $\phi = \phi_m$ and then declines. The previous equation can be expressed as :

$$\sigma/\sigma_m = \frac{\phi}{\phi_m} [1 - \log \frac{\phi}{\phi_m}]. \qquad (66)$$

This equation has been compared with experimental results of relative grain boundary energy as a function of ϕ [15]. The agreement is fairly good for $0 < \phi < 30°$.

6. HOMOGENEOUS NUCLEATION OF PRECIPITATE

We only briefly review here the homogeneous nucleation of precipitates, describing crystallization observed during thermal annealing. In fact these phenomena are too complicated to allow a simple study of nucleation induced by electrons or laser annealing.

The homogeneous nucleation of a precipitate β in supersaturated solution α can be described by the formal theory presented in section 2.2 except that N in (27) should be replaced by the concentration in α of the precipitating component. In precipitation, the nucleus composition can be very different from that of the β phase at equilibrium, so Hobstetter [16] , and Hollomon and Turnbull [17] have extended the formalism of the theory. They have taken account of the variations of σ, ΔG_v and E with the cluster composition.

Assuming a binary system (containing atoms of species A_1 and A_2), let x be the mole fraction of component A_2 in α, and suppose the concentration of A_2 is greater in β than in α. Assuming negligible strain, one can express the free energy of formation of a cluster β of composition x (having an identical structure as α) under the form :

$$\Delta G_i = a\sigma(\Delta x) \, i^{2/3} + \left(\frac{i}{n}\right) \Delta G(\Delta x), \tag{67}$$

where:

$$\Delta x = x^1 - x,$$

x^1 = composition of the β cluster,

$\sigma(\Delta x)$ = surface free energy of the α-β boundary (equation 9.11 in [1a]),

$\Delta G(\Delta x)$ = the difference in the molar free energy of solutions of composition x^1 and x (eq. 6.5 in [1a]). Then the nucleation barrier is :

$$\Delta G^* = f(i^*, \Delta x^*). \tag{68}$$

i^* and Δx^* satisfy simultaneously :

$$\left[\frac{\partial}{\partial i}(\Delta G_i)\right]_{\Delta x} = 0, \tag{69}$$

$$\left[\frac{\partial}{\partial \Delta x}(\Delta G_i)\right]_i = 0. \tag{70}$$

ΔG^* obtained this way must be replaced in (27) and provides the nucleation rate.

7. CONCLUSION : COMPARISON WITH EXPERIMENTAL RESULTS ON GROUP IV SEMICONDUCTORS

We have presented here a classical theory of nucleation (giving a particular emphasis to nucleation at constant composition), valid for the case of thermal annealing. Germain and Zellama have realized experiments on crystallization providing the nucleation rate in the steady state regime on amorphous Germanium [18,19,20,21] and on amorphous silicon [22] ; these results have been successfully described by this nucleation theory.

We will show in other sessions that results on transient homogeneous nucleation [6a] as well as transient heterogeneous nucleation [6b] in group IV semiconductors can be interpreted by the same theory. Then we will try to give an interpretation to values of the surface free energy obtained from [18] and [22] for amorphous germanium and silicon.

The extension of this work to the case of electrons or laser induced nucleation is in its very infancy. As for experiments, Germain et al have obtained a partial result on a measurement of R^* under irradiation : the growth rate u has been measured under high energy electron irradiation and has proved to be ionization-enhanced. We have shown in this lecture that R^* is proportional to a jump frequency ν^j itself proportional to the growth rate.
Then this result on u is a partial result on R^*. More experimental results on R^* are needed.

REFERENCES

1a. Turnbull, D., Sol. Stat. Phys., 3 (1954) 226.
1b. "Nucleation", Edited by A.C. Zettlemoyer, (New York, Marcel Dekker, Inc. 1969).
1c. Binder, K. and Stauffer, D., Advances in Physics, 25 (1976) 343.
2. Turnbull, D. and Fisher, J.C., Journ. of Chem. Phys., 17 (1949) 71.
3. Glasstone, S., Laidler, K.J. and Eyring, H., "The Theory or Rate Processes" (New York, MacGraw Hill Book Company, 1941).
4. Kantrowitz, A., J. Chem. Phys., 19 (1951) 1097.
5. Probstein, R.F., J. Chem. Phys., 19 (1951) 619.
6a. Transient homogeneous nucleation in group IV semiconductors, Germain, P. and Zellama, K., to be published.
6b. Transient heterogeneous nucleation in group IV semiconductors, Germain, P., Zellama, K., to be published.
7. Gibbs, J.W., "Collected Works", 1 (New York, Longmans, 1931) p. 55.
8. Fisher, J.C. and Dunn, C.G. in "Imperfections in Nearly Perfect Crystals", Edited by W. Shockley, (New York, Wiley, 1952) 317-352.
9. Cahn, J.W., Acta Metallurgica, 5 (1956) 169.
10. Becker, R., Annales de Physique, 32 (1938) 128.

11. Brooks, H., "Metal Interfaces", American Society of Metals, Cleveland, Ohio (1952), p. 20-65.

12. Nabarro, F.R., Proc. Phys. Soc., London, 52 (1940) 90; Nabarro, F.R., Proc. Roy. Soc., 175 (1940) 519.

13. Burgers, J.M., Proc. Koninkl. Ned. Akad, Wetenschap, 42 (1939) 293; Proc. Phys. Soc. (London), 52 (1940) 23.

14. Read, W.T. and Shockley, W., Phys. Rev., 78 (1950) 275.

15. "Dislocation in crystals" by Read, W.T. (MacGraw Hill Book Company, Inc., 1953).

16. Hobstetter, J.H., Trans. Am. Inst. Mining Met. Engrs, 180 (1949) 121.

17. Hollomon, J.H. and Turnbull, D. in "Progress in Metal Physics" Edited by B. Chamlers and R. King, Vol. 4 (London, Pergamon, 1953), p. 333-388.

18. Germain, P., Zellama, K., Squelard, S., Bourgoin, J. and Gheorghiu, A., J. Appl. Phys., 50 (1979) 6986.

19. Germain, P., Squelard, S., Bourgoin, J. and Gheorghiu, A., Journ. of Ap. Phys., 48 (1977) 5.

20. Germain, P., Squelard, S., Bourgoin, J. and Gheorghiu, A., Journ. of Non-Cryst. Solids, 23 (1977) 93.

21. Gheorghiu, A., Squelard, S., Zellama, K., Germain, P. and Bourgoin, J., Revue de Physique Appliquée, 12 (1977) 721.

22. Zellama, K., Germain, P., Squelard, S., Bourgoin, J. and Thomas, P.A., J. Appl. Phys., 50 (1979) 6995.

23. Germain, P., Squelard, S. and Bourgoin, J., Journ. Non-Cryst. Sol., 23 (1977) 159.

24. Germain, P., Squelard, S. and Bourgoin, J., Radiation Effects in Semiconductors, n°31, Dubrovnik eds.N. Urli and J. Corbett.

TRANSIENT BULK INDUCED NUCLEATION IN AMORPHOUS GROUP IV
SEMICONDUCTORS

K. Zellama, Groupe de Physique des Solides de l'ENS
Université Paris 7, Tour 23, 2, Place Jussieu,
75251 Paris Cédex 05, France.
P. Germain, Department of Physics, North Carolina State
University, P.O. Box 5367, Raleigh, NC 27650, USA.
P.A. Thomas, Laboratoire de Physique des Solides,
Université Paris 6, Tour 13, 4, Place Jussieu,
75221 Paris, France.
A. Gheorghiu, Laboratoire d'Optique des Solides,
Université Paris 6, Tour 13, 4, Place Jussieu,
75230 Paris, France.

ABSTRACT

We present in this seminar an interpretation of our experimental
results on bulk induced transient nucleation and growth in amorphous
germanium and silicon with the theory of transient nucleation[1] .
The phenomena has been observed [2,3] in a temperature range of
660-678 K in germanium and of 833-873 K in silicon. Before the
steady state regime, a transient period of duration τ occurs. We
determine an expression for τ and show that it is thermally activated with an activation energy 1.8 eV for germanium and 3.66 eV
for silicon, very close to the respective growth rate activation
energies. From this, we obtain the number of atoms of the nucleus
of critical size for germanium and silicon. These values are in
agreement with those previously obtained from the study of the steady
state regime [2,3] .

1. INTRODUCTION

Crystallization in amorphous materials occurs through a nucleation and growth process [1]. We have studied crystallization in
amorphous germanium and silicon using conductivity measurements[2,3]
and have demonstrated that, after an induction period, a bulk induced
steady state regime of nucleation and growth occurs.

The interpretation of the steady state regime has provided coherent values of the growth rate V_g and the nucleation rate R^* for amorphous Ge and Si. The aim of this paper is to show that the observed induction period can be interpreted by the theory of transient nucleation [1].

We first recall elements of the theory of nucleation and growth. Then we present our experimental results on the induction period, and show that experimental results are correctly described by the theory of transient nucleation.

2. REVIEW OF THE THEORY OF TRANSIENT AND STEADY STATE BULK NUCLEATION

Crystal growth occurs through the jump of a molecule across an amorphous-crystalline phase boundary. The growth rate v_g (in cm s^{-1}) takes the form :

$$v_g = \delta \times \nu^1 ,\tag{1}$$

where δ is the jump distance of an atom at the interface and ν^1 is a frequency ; ν^1 is the product of a frequency for atomic vibrations and a thermally activated expression which describes the probability of the atomic jump at the amorphous-crystalline interface :
$\nu^1 = \frac{kT}{h}\exp [-\Delta g^*/kT])$.

The variation ΔG of the Gibbs energy involved with the transformation of i atoms from the amorphous phase into a crystallite of i atoms is written as the sum of bulk and surface terms; these terms being of opposite sign, ΔG presents a maximum ΔG^*. The steady state nucleation model presumes that crystallites are created when crystalline germs have grown to a critical size corresponding to the increase of free energy ΔG^*. Let the number of atoms in the crystallite of critical size be i* .

The steady state nucleation rate R^*, i.e. the number of stable crystallites which appear per unit time in an untransformed volume unity containing N atoms,is [1] :

$$R^* \simeq N \times \nu^1 \exp \left(\frac{-\Delta G^*}{KT}\right) .\tag{2}$$

In the steady state theory of nucleation summarized above and leading to equation 2, it is assumed that there exists at all times a steady state concentration of crystalline clusters (this concentration of clusters is a function of annealing temperature). However, at the beginning of the experiments, when the temperature of the material has just been raised to the annealing temperature, there must be a "transient" period before steady state conditions are reached. The theory of this transient phenomenon has been reviewed in a previous lecture [1]. A simplified expression of the nucleation rate R(t) vs time t which provides an order of magnitude of the transient period, τ, has been obtained. For condensed systems, the relation is :

$$R(t) = R^* \times \exp\left(-\frac{\tau}{t}\right), \tag{3}$$

where :

$$\tau \simeq \frac{i^{*2}}{n^*\nu} l . \tag{4}$$

n^* is the number of molecules at the surface of nucleus.

Note that $R(t) = R^*$ for $t \to \infty$. Comparison of (1) with (4) results in :

$$\tau \simeq \frac{i^{*2}\delta}{n^*} \times \frac{1}{v_g}, \tag{5}$$

or :

$$\tau = \frac{\lambda}{v_g}, \tag{6}$$

where :

$$\lambda = \frac{i^{*2}\delta}{n^*}. \tag{7}$$

The quantities i^* and n^* in equation 7 are not independent but correspond to the same crystallite of critical size. Therefore :

$$n^* \simeq i^{*2/3}. \tag{8}$$

Using (7) and (8), λ takes the form :

$$\lambda \simeq i^{*4/3} \times \delta. \tag{9}$$

Then :

$$i^* \simeq \left(\frac{\lambda}{\delta}\right)^{3/4}. \tag{10}$$

As the parameter i^* probably has weak dependence on T, [1], and δ is almost constant, then from the equation (9) λ depends very weakly on T. It follows from equation (6) that v_g and $\frac{1}{\tau}$ should have roughly the same activation energy.

3. EXPERIMENTAL RESULTS AND INTERPRETATION

Experimental conditions for the preparation of samples and the conductivity measurements are given in [2,3]. We only briefly recall them here. Amorphous germanium samples are made by evaporation (50 Å S^{-1}) in a 5 x 10^{-7} Torr vacuum on sintered alumina substrates. The substrate temperature is 300 K. The samples studied are not exposed to air atmosphere, so crystallization is studied in-situ.

Amorphous silicon samples are obtained by electron bombardment of a crystalline silicon target in ultrahigh vacuum (below 10^{-8} Torr). Deposition rate is 30 Å s^{-1}. The substrate temperature during deposition is $\simeq$ 670 K. The samples are exposed to air atmosphere between the preparation and the crystallization process [3].

Crystallization has been monitored by conductivity measurements made on amorphous germanium and silicon during isothermal runs. It has been shown that the volumic fraction crystallized x(t) at time t can be deduced from the variation of σ with t.

Letting x be the crystalline volume fraction, the variations of $y = \log [\text{Ln} (\frac{1}{1-x})]$ has been plotted vs log t, for amorphous germanium and silicon. Both figures 1a and 1b exhibit the same behavior.

During the lower times when x $\lll$ 1, $y = \log [\text{Ln} (\frac{1}{1-x})] \simeq \log x$ remains constant or varies very slightly. Then y rapidly increases as a t^3 function (previously interpreted as due to a bulk steady state nucleation and growth process presented in 2.2 of ref [1]). For each of these isothermal runs at temperature T, an induction time τ(T) can be deduced. In Tables 1 and 2, we give for amorphous germanium and silicon,respectively, the variations of τ^{Ge}_{BIC} (BIC = Bulk Induced Crystallization) and τ^{Si}_{BIC} as a function of T. We also recall $v_g^{Ge}(T)$ and $v_g^{Si}(T)$ previously obtained on the same kind of samples [2,3].

On figure 2, we have plotted the variations of $1/\tau^{Ge}_{BIC}$ and $v_g^{Ge}(T)$ vs. 1000/T, and on figure 3 the variation of v_g^{Si} $1/\tau^{Si}_{BIC}$ and $Vg^{Si}(T)$ vs. 1000/T. These quantities are thermally activated. The activation energy of $1/\tau^{Ge}_{BIC}$ and v_g^{Ge} are respectively 1.8 eV and 1.5 eV; the difference between these energies is small (17%). The activation energy of $1/\tau^{Si}_{BIC}$ and v_g^{Si} are respectively 3.66 eV and 3.2 eV; the difference between these two energies is also small (13%).This is in agreement with equation (6) (deduced from the theory of transient nucleation), which predicts that 1/τ and v_g

TABLE 1

VARIATION OF THE INDUCTION TIME τ_{BIC}^{Ge} FOR BULK INDUCED NUCLEATION

IN AMORPHOUS GERMANIUM

τ_{BIC}^{Ge} (hour)	3	2	1.3
T (K)	660	668	678
1000/T (K^{-1})	1.515	1.497	1.4749
Vg (Å x s^{-1})	1	1.3	2
τ_{BIC}^{Ge} Vg = λ(Å)	10800	9360	9360
1/τ_{BIC}^{Ge} s^{-1}	9.25 x 10^{-5}	1.38 x 10^{-4}	2.13 x 10^{-4}

TABLE 2

VARIATION OF THE INDUCTION TIME τ_{BIC}^{Si} FOR BULK INDUCED NUCLEATION

IN AMORPHOUS SILICON

τ_{BIC}^{Si} (hour)	10	4	2.5	2	1
T (K)	833	843	853	863	873
1000/T (K^{-1})	1.200	1.186	1.172	1.158	1.1454
Vg(Å x s^{-1})	0.38	0.6	0.86	1.3	2.1
τ_{BIC}^{Si} x Vg = λSi(Å)	13680	8640	7740	9360	7560
1/τ_{BIC}^{Si} (s^{-1})	2.77 x 10^{-5}	6.94 x 10^{-5}	1.1 x 10^{-4}	1.38 x 10^{-4}	2.77 x 10^{-4}

284

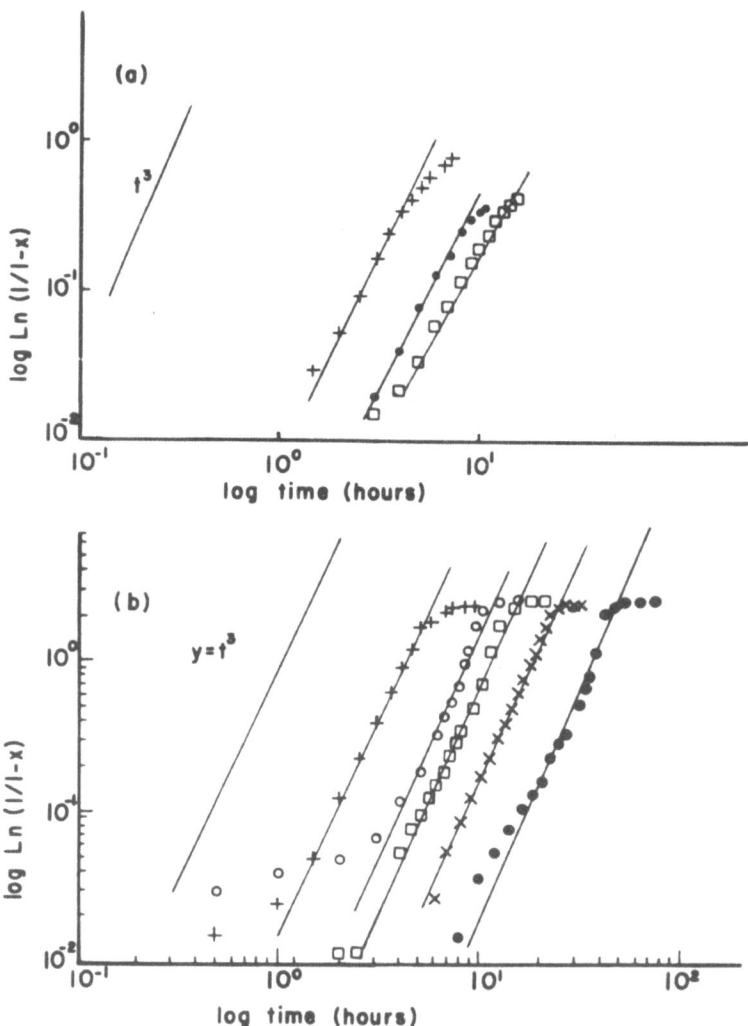

<u>Fig. 1 a</u> Variation of Ln(1/1-x) versus time at various temperatures
for amorphous germanium layers, where x = crystalline volume fraction.   Samples were crystallized in-situ [2]   : +678 K; ●668 K; □660K.
<u>Fig. 1 b</u> Variation of Ln(1/1-x) versus time at various temperatures
for amorphous silicon layers (series 2 [3]) : ● 833 K; x 843 K ;
□ 853 K; o 863 K; + 873 K.

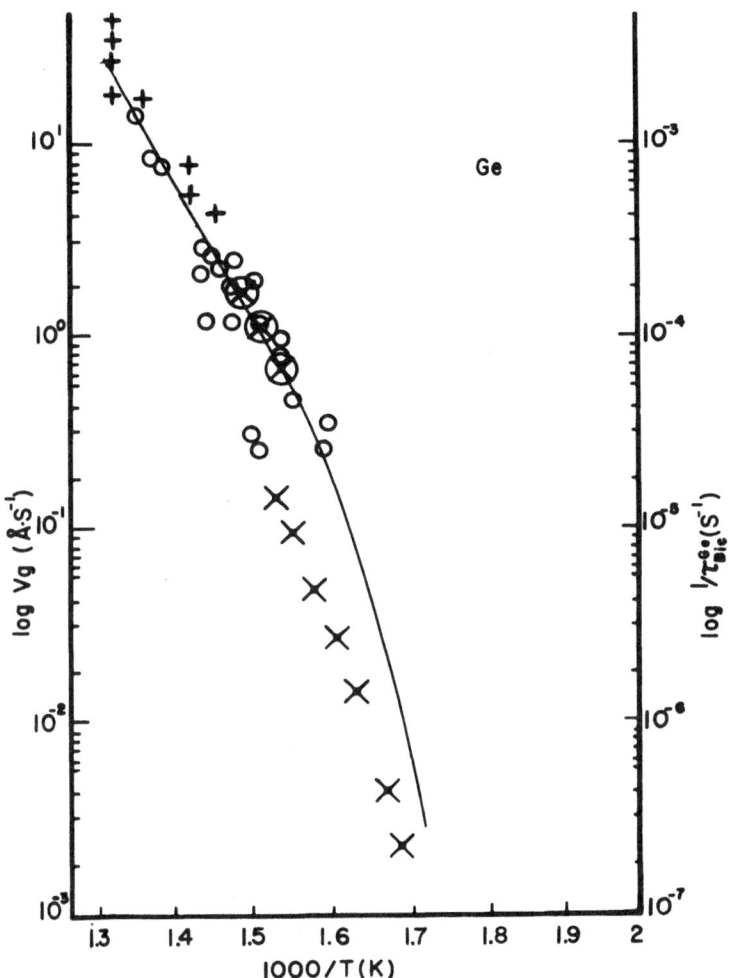

Fig. 2 Arrhenius plot of the growth rate v_g of crystallization in amorphous germanium layers : x [2] ; + data of Barna et al [4] ; o data of Germain et al [5]. Arrhenius plot of $1/\tau_{BIC}^{Ge}$: ⊗ present study.

286

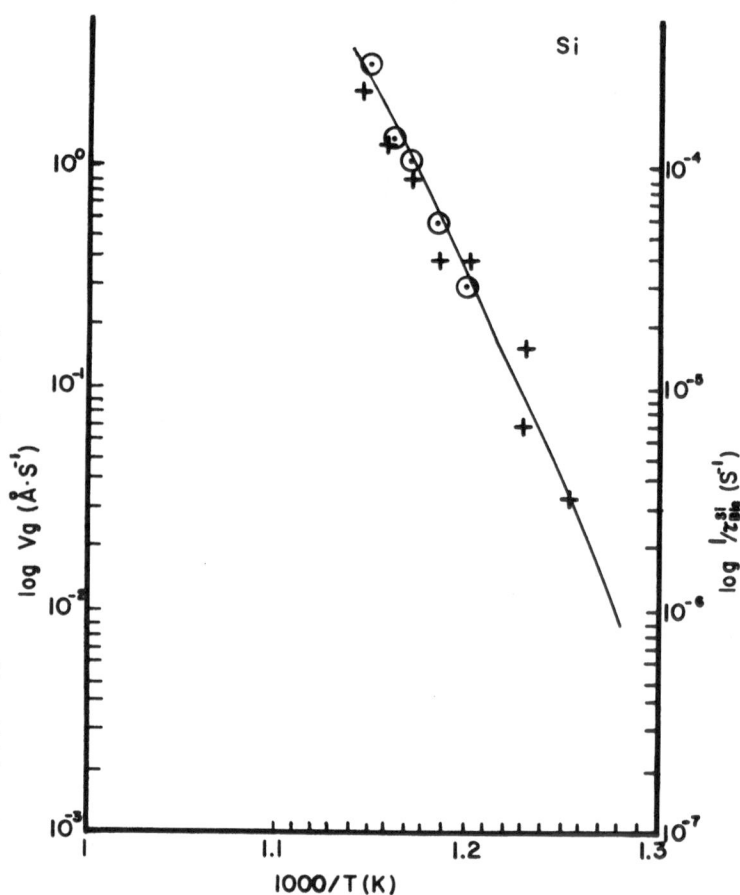

Fig. 3 Arrhenius plot of the growth rate v_g of crystallization in amorphous silicon layers : + data of Zellama et al [3]. Arrhenius plot of $1/\tau_{BIC}^{Si}$: ⊙ present study.

should have the same activation energy.

In the last line of Tables 1 and 2, we have tabulated, according to equation (6), the products $\lambda_{Ge} = \tau_{BIC}^{Ge} \ v_g^{Ge}$ and $\lambda_{Si} = \tau_{BIC}^{Si} \cdot v_g^{Si}$. λ_{Ge} appears to be defined within a factor of 10 percent in a temperature range of 10 K, its mean value $\bar{\lambda}_{Ge}$ is about :

$$\bar{\lambda}_{Ge} \simeq 9900 \text{ Å} . \tag{11}$$

$\bar{\lambda}_{Si}$ is defined within a factor of 2 in a temperature range of 40 K. Its mean value $\bar{\lambda}_{Si}$ is :

$$\bar{\lambda}_{Si} \simeq 9300 \text{ Å} . \tag{12}$$

Using equation (10) respectively with (11) and (12), we obtain the number of atoms in the nucleus of critical size in germanium i^{*Ge} and in silicon i^{*Si} :

$$i^{*Ge} = 430 \simeq 400 , \tag{13}$$

$$i^{*Si} = 415 \simeq 400 . \tag{14}$$

We have assumed here $\delta \simeq 3 \text{ Å}$.

Previous studies [2,3] of the steady state regime of this BIC allow one to obtain $v_g(T)$ and $R^*(T)$ and to deduce ν^1 and ΔG^*, as well as the order of magnitude of i^*. We only recall here the values of i^* :

$$i^{*Ge} \simeq 300 , \tag{15}$$

$$i^{*Si} \simeq 100 . \tag{16}$$

There is an agreement in order of magnitude of (13) and (14), obtained from the transient regime, respectively with (15) and (16), obtained from the study of the steady state regime. This shows the coherence of all these experimental results in amorphous group IV semiconductors with the two parts of the theory of bulk induced nucleation : transient bulk-induced nucleation and steady state bulk-induced nucleation.

The interest of this work is double : first, on a fundamental point of view, it is important to show the total coherence of our results (transient and steady state crystallization) with the theory. Second, for applications, it shows to what extent the growth of polycrystallites can be improved.

REFERENCES

1. Germain, P. and Zellama, K., Nucleation in condensed Matter, (Lecture presented in these proceedings).
2. Germain, P., Zellama, K., Squelard, S., Bourgoin, J. and Gheorghiu, A., J. Appl. Phys., 50 (1979) 6986.
3. Zellama, K., Germain, P., Squelard, S., Bourgoin, J. and Thomas, P., J. Appl. Phys., 50 (1979) 6995.
4. Barna, A., Barna, P.B. and Pocza, J.F., J. Non-Crystalline Solids, 8 (1972) 36.
5. Germain, P., Squelard, S., Bourgoin, J. and Gheorghiu, A., J. Appl. Phys. 48 (1977) 5.

CRYSTALLINE, AMORPHOUS AND LIQUID SILICON

M. Combescot and J. Bok

Groupe de Physique des Solides de l'Ecole Normale Supérieure, 24 rue Lhomond, 75231 Paris Cédex 05, France.

1. INTRODUCTION

The large amount of work done recently on laser annealing of
ion-implanted silicon has raised some fundamental questions about
transitions between various phases of silicon : ion-implanted layers
(which are amorphous), crystal and liquid. Most authors believe
that during laser annealing the surface layer melts and recrystallizes in times of order of hundreds of nanoseconds, i.e. short enough
to avoid diffusion of dopants.

In this lecture, we describe a new theory for the crystal-
liquid transition in silicon and more generally in covalent semiconductors with the diamond structure. Study of the electron-phonon
interaction has shown that, if a certain fraction α of the total
number of valence band electrons are excited into the conduction
band, the frequency of the TA phonon goes to zero and the crystal
should become fluid. This softening of the TA phonon mode has been
used to explain the variation of the energy gap with temperature.
We show that the softening of this TA phonon mode produces an instability leading to a first-order phase transition which, we suggest,
is the origin of melting.

In the second part, we discuss the crystallization of an amorphous layer during rapid heating. The amorphous state is metastable
and we compute the variation of crystallization time $\tau_c(T)$ with
temperature T. The various crystallization processes are : (1)
epitaxial regrowth which is controlled by the crystal growth velocity v given by an Arrhenius law with an activation energy
E_v ($v = v_0 e^{-E_v/kT}$) ; (2) volume nucleation which is controlled by the
volume nucleation rate $J = J_0 e^{-E_j/kT}$. In the first case, the

nucleation time is given by $\tau_g(T) = \frac{e}{v_o} e^{E_v/kT}$, in the second case
by $\tau_n = \frac{1}{v} (\frac{v}{J})^{1/4}$ where e is the thickness of the amorphous layer.
We show that volume nucleation is faster than epitaxial regrowth
if the layer thickness e is larger than $e_n \sim (\frac{v}{J})^{1/4}$. For T=1700 K,
e_n = 800 Å while for T = 1400 K, e_n is of order of 2000 Å. We conclude that with heating time scale of the order of hundreds of microseconds, amorphous layers crystallize before melting while,with
nanoseconds pulses, the heating is fast enough to melt the layer
directly from the amorphous state.

2. INSTABILITY OF THE ELECTRON-HOLE PLASMA IN COVALENT SEMICONDUC-
TORS

The interest of laser annealing in the manufacturing of semiconductor devices of submicronic size has produced a large amount
of work on that subject. Irradiation with laser pulses appears as
an interesting method for annealing of ion-implanted semiconductors
[1-8] . After some controversy about the actual mechanism of annealing most authors now agree that the sample melts and recrystallizes,
the diffusion of impurities being very rapid in the liquid phase
[6-8] . One of the problems raised by laser annealing is the role
played by the plasma of electron-hole pairs generated by the incident photons. The purpose of this paper is to show that,at thermal
equilibrium (i.e. even without laser), the e-h plasma coupled to
the phonons is unstable above a critical temperature T*. We suggest
that this instability might be the origin of melting of the tetrahedrally bonded covalent semiconductors like germanium and silicon.
One of the most striking consequence of our model is the value of
the intrinsic carrier density n that we find near melting. For silicon for instance our theory gives $n \sim 10^{21} cm^{-3}$ near melting,
while the classical formula gives $n = 2 \times 10^{19} cm^{-3}$.

Finally we show that the extra amount of e-h pairs created by
laser irradiation reduces the instability temperature (i.e. the
melting temperature). With the laser powers used in silicon annealing, we find that the reduction in melting temperature is very
small. A first account of this model has been published recently
[9] . L.R. Godefroy and P. Aigrain [10] were the first to propose
a theory relating the melting temperature to the softening of the
phonon frequency by the creation of electron-hole pairs (breaking
of covalent bonds). This effect has been computed by Heine and
Van Vechten [11] for silicon. In diamond-type crystalline structures, the transverse acoustic (T.A.) modes depend critically on
the covalent bonding without which the whole diamond structure goes
unstable. The shift in T.A. phonon frequencies due to the excitation of n e-h pairs per unit volume is calculated as :

$$\omega_i(n)' = \omega_{oi}(1- \alpha\frac{n}{n_o}) \qquad (1)$$

where n_o is the density of semiconductor atoms. For Si, $n_o = 5 \times 10^{22}$ cm^{-3} and α is given by :

$$\alpha = f_{cv} \frac{\varepsilon_o \varepsilon_o^*}{\varepsilon_o^* - \varepsilon_o}$$

where f_{cv} is a bond charge reduction factor for the fundamental gap ($f_{cv} = 0.85$ for silicon), ε_o^* is an optical dielectric constant for the β-tin structure ($\varepsilon_o^* = 24$) and ε_o is the optical dielectric constant ($\varepsilon_o = 13,3$ for Si at high temperature). With this model Heine and Van Vechten find $\alpha = 6.5$ for Si.

To find an instability of the plasma-phonon system we compute the free energy of a crystal containing N_a atoms in a volume V, at temperature T, in presence of N e-h pairs ($n_o = N_a/V$, $n = N/V$). The phonon contribution to the free energy has the form :

$$F_{ph} = \sum_i kT \, Ln \, (1 - e^{-\frac{\hbar\omega i}{kT}}) \sim kT \sum_i Ln \frac{\hbar\omega i}{kT} , \qquad (2)$$

at high temperature where $kT \gg \hbar\omega_i$. Using Eqs. 1 and 2, we find that the variation of the phonon free energy due to the creation of n e-h pairs is :

$$\Delta F_{e,ph} = N_a \, kT \, Ln \, (1 - \frac{\alpha n}{n_o}) \qquad (3)$$

which will contribute to the e-h pairs chemical potential, giving an additional term :

$$\frac{\partial (\Delta F_{e-ph})}{\partial N} = - \frac{\alpha kT}{1 - \alpha \frac{n}{n_o}} . \qquad (4)$$

If one neglects the Coulomb interaction, negligible compared to the kinetic part at density higher than 10^{19}cm^{-3}, the e-h pairs chemical potential $\mu = \frac{\partial F}{\partial N}$ is :

$$\mu(n,T) = E_{GO} - \frac{\alpha kT}{1 - \alpha \frac{n}{n_o}} + \mu_e + \mu_h$$

where E_{GO} is the band gap (1.17 eV for Si) and μ_e and μ_h are the chemical potentials of a free electron- and-hole gas of density n at temperature T. They are related to n by:

$$\frac{n}{n_G} = g_e \, m_e^{*\,3/2} \, (\frac{kT}{E_{GO}})^{3/2} \, F \, (\frac{\mu_e}{kT})$$

$$= g_h \, m_h^{*\,3/2} \, (\frac{kT}{E_{GO}})^{3/2} \, (\frac{\mu_h}{kT}) \quad , \tag{6}$$

with

$$F(y) = \int_0^\infty \frac{\sqrt{x} \, dx}{e^{x-y}+1} \tag{7}$$

and

$$n_G = 2\Pi \, (\frac{2mE_{GO}}{h^2})^{3/2} = 4.5 \times 10^{21} \, cm^{-3} \quad , \tag{8}$$

m_e^* and m_h^* being the electron and hole effective masses, respectively, and $g_{e,h}$ their degeneracies.

At thermal equilibrium, the e-h pair density adjusts itself such that the free energy is minimum, i.e. $\frac{\partial F}{\partial N} = 0 = \mu$. μ is negative for low n as usual but also due to the softening of the T.A. phonon modes for $n \lesssim n_0/\alpha$; so that, at low T, $\mu = 0$ for two densities as shown on Figure 1. The lowest one, associated to a minimum of F, corresponds to the well known intrinsic carrier density :

$$n(T) = n_G \sqrt{g_e g_h m_h^{*\,3/2} \, m_e^{*\,3/2}} \, e^{-E_G(T)/2kT} \quad , \tag{9}$$

with

$$E_G(T) \approx E_{G_O} - \alpha kT \quad . \tag{10}$$

The second zero of μ corresponds to a maximum of F and leads to an unstable state.

When T increases, this oscillation of F is smoothed and finally disappears above a certain temperature T^*, the equilibrium carrier density n(T) increasing up to n^* such that :

$$\mu(n^*, T^*) = 0 = \frac{\partial \mu}{\partial n}\Big|_{n^*, T^*} \quad . \tag{11}$$

At higher T, there is no longer a stable equilibrium density and the system undergoes a first order phase transition at T^*. This effect is due to the decrease of the T.A. phonon frequency and n^* has to be a sizeable fraction of n_0/α . α can be related to the variation of the gap via Eq. 10. Measurements of the temperature variation of $E_G(T)$ have been made up to 400°K [11] which is

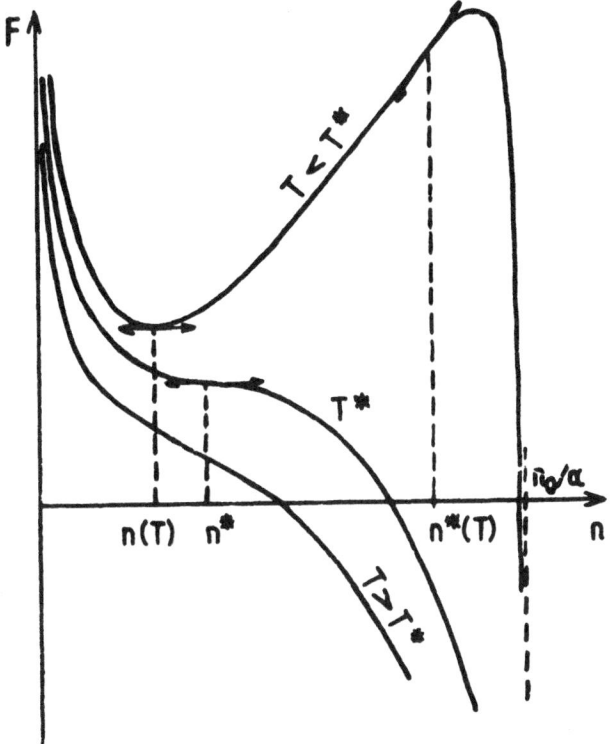

Fig. 1   Variation of the free energy F as a function of the e-h
pair density above and below the instability temperature T^* .

low compared to the Debye temperature of Si at 645°K. The fit of $E_G(T)$, made at low T, would give $\alpha \approx 6.5$ for the linear term, but the linear regime is not yet reached at 400°K and, in the range 1500-1700°K, α could be modified by the expansion of the crystal. The electron and hole effective masses are also expected to vary with T, but they have not been measured at high temperature. A solution for Eq. 11 is obtained for $T^* = 1700°K$ taking $\alpha = 10.8$ and the masses as for T = 0 or taking $\alpha = 7$, the masses having twice their values at T = 0; the equilibrium carrier density is found in all cases between 1 and $2 \times 10^{21} cm^{-3}$, i.e. outside the classical limit where eq. 10 is valid (one has $\mu/kT \approx 1$). For germanium, using the same procedure we find $T^* = 1210°K$ for $\alpha = 13.8$. This gives $n \approx 9 \times 10^{20} cm^{-3}$ at melting.

The effect of laser irradiation has been estimated in [9].

In summary, we have shown that, due to electron-phonon coupling, the tetrahedrally coordinated semiconductors undergo a plasma instability at high temperature which could be the cause of melting. For the case of silicon, we find that the electron-hole pair density near melting is much higher than the one predicted by the classical formula :

$$n = \sqrt{N_c N_v} \; \exp \left(\frac{-E_G(T)}{2kT} \right) .$$

The creation of a high density electron-hole plasma by an external source decreases the instability temperature so that the crystal will melt at lower T.

3. KINETICS OF CRYSTALLIZATION OF AMORPHOUS SILICON

Due to the considerable technological interest in producing low-cost silicon layers, a large amount of work has been devoted to the crystallization of amorphous silicon. The oldest way is to use a furnace in which amorphous silicon crystallizes in times of order an hour at a well defined temperature. A more recent technique is to use nanosecond pulse laser; most authors believe that, in that case, a surface layer melts and recrystallizes in times of order hundreds of nanosecond, i.e. short enough to avoid the diffusion of dopants. An interesting idea has been proposed by Van Vechten [6], who thought that the electron-hole plasma created by the laser absorption can produce a new "fluid-like" phase [7,9]. Unfortunately, the lasers used actually are not powerfull enough to modify the melting of the crystal [8] . In fact, lasers are just a way to heat the sample faster and more locally than in a furnace; but with them, the temperature at which silicon changes from amorphous to crystal is not directly defined (what can be measured for example is the time delay at which the reflectivity

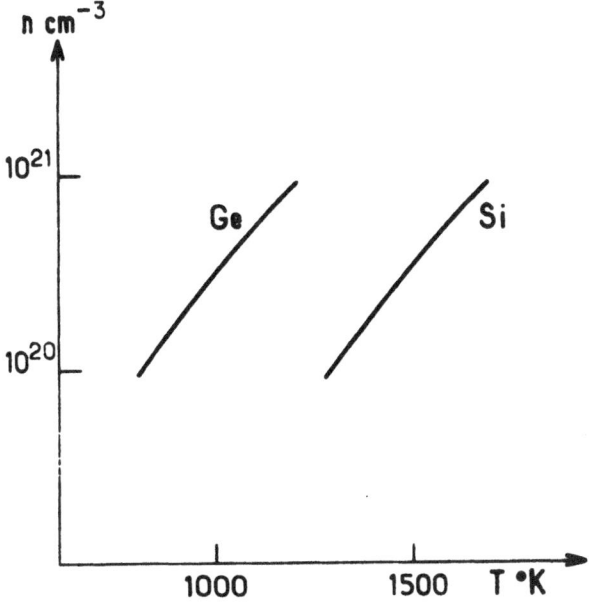

Fig. 2 e-h pair density versus temperature for germanium and silicon. The calculation has been made keeping the T = 0 values of
the effective masses and fitting α to obtain the observed melting
temperature. This gives α = 10.8 for Si and 13.8 for Ge.

changes which then has to be related to the real temperature).

We want to discuss the crystallization (or eventually melting)
of an amorphous silicon layer during rapid heating by a laser or
electron beam. As the amorphous state is metastable, its free
energy is higher than the one of the crystal. It has been speculated by some authors [12] that this leads to a lowering of the
melting temperature of the amorphous state T_m^* compared to that of
the crystal T_m. Several experiments have tried to measure this

difference in melting temperature but the conclusions are not clear
[13] . The amorphous state being unstable, it crystallizes during
rapid heating and the purpose of this paper is to estimate, using a
simple model, the crystallization time $\tau_c(T)$ at a temperature T.
If the heating time scale τ is smaller than $\tau_c(T_m^*)$, one can heat
the amorphous sample up to T_m^* without crystallization and see indeed its melting at T_m^*. But if τ is larger than $\tau_c(T_m^*)$, the amorphous layer crystallizes during the heating and the melting is seen
at T_m, as for a pure crystalline state.

In order to determine $\tau_c(T)$, one should consider the various
crystallization processes. For ion-implanted amorphous silicon
layers, one can think first of epitaxial regrowth. The crystallization time for such a process is :

$$\tau_g = \frac{e}{v(T)} = \frac{e}{v_o} e^{E_v/kT} \qquad (12)$$

where e is the layer thickness and v(T) is the crystal growth rate.
Low temperature measurements [14,15,16] of v have given $v \approx 10^{-7}$
cm s^{-1} for $T \approx 900°K$ and an activation energy E_v in the range 2.6
to 3 eV. So,for a 1000 Å thick layer, it would take, at 1700°K,
between 10^{-6} and 10^{-5} sec to crystallize. This would mean that with
nanosecond pulses, the silicon layer would surely melt in the amorphous state while on a 100 μsec time scale experiment, T_m^* could be
measured only if it is lower than 1400 to 1550°K (depending on the
value taken for E_v).

But epitaxial regrowth is not the only way to crystallize, and
if there is a faster one, it will be the one observed. Moreover,
it cannot play a role for amorphous layer deposited on a substrate.
Usually, in that case, the nucleation takes place at the interface
but the surface nucleation rate [17,18] turns to be so large that
the growth time τ_g is much longer that the time necessary to form
a polycrystalline film at the interface; and the observed crystallization time is in that case again τ_g.

Finally, one can think of volume nucleation. This process
would not need for the crystal to grow all over the thickness e and
can be faster than τ_g if the volume nucleation rate J is fast enough.
One can have an order of magnitude of the corresponding crystallization time τ_n saying that during τ_n, an amount $J\tau_n$ of critical
embryos has been created, at an average distance d from each other
such that $J\tau_n d^3 \approx 1$. On the other hand, the crystallization is
finished after τ_n if $d \approx v\tau_n$. This gives τ_n of order :

$$\tau_n \sim \frac{1}{v} \left(\frac{v}{J}\right)^{1/4} . \qquad (13)$$

As the nucleation rate J is of the form $J = J_0 \exp(-E_J/kT)$, Eq. 13 implies an activation energy for τ_n : $E_n = (E_J + 3E_v)/4$. This can be checked by comparing the activation energy $E_n \approx 3.3$ eV measured by Blum and Feldmann [19] with the value of the nucleation current ($E_J \approx 4.9$ eV and $J \approx 3 \times 10^7$ cm^{-3} s^{-1} for T = 600°C) measured by Koster [14] or Zellama et al [15] .

Eq. 13 tells that volume nucleation is faster than epitaxial regrowth if the layer thickness e is larger than $e_n \sim (v/J)^{1/4}$. For T = 1700°K, e_n is of order 800 Å while for T $\approx$ 1400°K, e_n is of order 2000 Å. As these values of e_n fall in the range of thicknesses of ion-implanted layers, one concludes that τ_n will not be orders of magnitude smaller than τ_g so that the melting temperature T_m^* of amorphous silicon should be reached, without precrystallization, with nanosecond pulses. We can also predict that with thick layers, one might see the change from epitaxial regrowth to volume nucleation if T_m^* is high enough. We want to emphasize at that stage that the exact temperature at which one switches from epitaxial to volume nucleation depends on the precise calculation of the prefactor term of τ_n which is not present in Eq. 13 but depends also, and much more crucially, on the value of the nucleation current. A small variation [20] on the measure of E_J has high implication on τ_n and as usual this nucleation current will be very sensitive to nucleation centers, i.e., the kind of amorphous material and the way it has been prepared.

Let us now end by looking at the latest experiment on the melting temperature of amorphous silicon by K.O.R.H. [16] . They follow the reflectivity of an ion-implanted silicon sample irradiated with a CW argon laser and, by using a reference time-dependent reflectivity curve in order to determine the temperature, they estimate that the amorphous sample can reach temperature as high as 1654°K without melting. We want to propose a theoretical way to determine the temperature which will lead to a somewhat lower value.

The temperature increase of a sample under a heat source $S(\vec{r},t)$ obeys the heat equation :

$$C_s \frac{\partial T}{\partial t} = \nabla(K_s \nabla T) + S(\vec{r},t) \tag{14}$$

where $C_s \approx 2.0$ J cm^{-3} K^{-1} for silicon and K_s decreases from 0.40 to 0.22 J cm^{-1} s^{-1} K^{-1} when T increases from 800 to 1700°K. As the heat diffusion length $\sqrt{4\chi t}$ (with $\chi = \frac{K_s}{C_s} \approx 0.1$ cm^2 s^{-1}) is of order 80μ for experiment with a 100μ sec time-scale, the laser penetration depth, which varies form 0.1 to 1μ for amorphous or crystalline silicon, can be neglected but one has to take into account the finite size of the laser spot, which for KORH's experiment has a radius $R_0 = 80\mu$. As the laser risetime is much shorter than 100μ sec, the heat source can be taken as:

$$S(\vec{r},t) = \frac{\mathcal{P}(1-\mathcal{R})}{\Pi R_0^{\,2}} \; \exp\left(-\frac{\rho^2}{R_0^{\,2}}\right) \; \delta(z) \; \Theta(t) \tag{15}$$

where $\mathcal{P}$ is the laser power, $\mathcal{R}$ the reflectivity, z the distance from the surface and ρ the distance from the center of the laser spot. With such a source, one can show [13] that the solution of the heat equation, for constant K_s, is:

$$T(\vec{r},t) = \frac{2\mathcal{P}(1-\mathcal{R})}{c_s \Pi^{3/2}} \int_0^t dt \; \frac{e^{-z^2/4\chi t}}{\sqrt{4\chi t}} \; \frac{e^{-\rho^2/R_0^2+4\chi t}}{R_0^2 + 4\chi t} \; . \tag{16}$$

So that the surface temperature, at the center of the laser spot varies with time as:

$$T(0,T) = \frac{\mathcal{P}(1-\mathcal{R})}{K_s R_0 \Pi^{3/2}} \; \text{Arc tg} \; \frac{\sqrt{4\chi t}}{R_0} \; . \tag{17}$$

There is no analytic solution for the non-linear heat equation obtained when the temperature dependence $K_s(T)$ is included. Beside exact numerical solutions, one can do perturbative type calculation and show [21] that the discrepancy with Eq. 17 is minimized if K_s is taken between $K_s(T_i)$ and $K_s(T_f)$ where T_i and T_f are the initial and final temperatures.

For $\mathcal{P} = 22.6$ W, $\mathcal{R} = 0.45$ and $K_s = 0.3$, one finds that the melting temperature is reached at the surface after 240 μsec in perfect agreement with KORH's figure 3. The same solution [17] gives at their time $t_g = 150$ μsec a surface temperature $T \approx 1580°K$, i.e. 30° smaller than their estimated value. The real surface temperature is probably even smaller because the regrowth time for the last 600Å, calculated at 1580°K, are in the range 3 to 30 μsec (for $E_v = 3.0$ to 2.6 eV), i.e.,somewhat smaller than the time necessary for the reflectivity to decrease.

It can also be noted that the interferences observed in the reflectivity implie that for a 1600 Å layer at 1580°K, the epitaxial regrowth is faster than volume nucleation.

In conclusion, we have shown that with heating time scale of the order of hundreds of microsecond, amorphous layers crystallize before melting, while with nanosecond pulses, the heating is fast enough to melt the layer in its amorphous phase.

REFERENCES

1. Auston, D.M., Surko, D.M., Venkatesan, T.N.C., Slusher, R.E., Golovchenko, J.A., Appl. Phys. Lett. 35 (1978) 437.
2. Tseng, W.F., Mayer, J.W., Campisano, S.U., Foti, G., Rimini, E., Appl. Phys. Lett. 32 (1978) 824.
3. Liu, P.L., Yen, R., Bloembergen, N., Hodgson, R.T., Appl. Phys. Lett. 34 (1979) 864.
4. Lo, H.W. and Compaan, A., Phys. Rev. Lett. 44 (1980) 6604; Aydinli, A., Lo, H.W., Lee, M.C., Compaan, A., Phys. Rev. Lett. 46 (1981) 1640.
5. Bentini, G.G., Cohen, C., Desalvo, A., Drigo, A.V., Phys. Rev. Lett. 46 (1981) 156.
6. Van Vechten, J.A., Tsu, R., Saris, F.W. and Hoohout, D., Phys. Lett. 74A (1979) 417 and 422.
7. Bok, J., Phys. Lett. 84A (1981) 308.
8. Combescot, M., Phys. Lett. 85A (1981) 308.
9. Combescot, M., Bok, J., Phys. Rev. Lett. 48 (1982) 1413.
10. Godefroy, L.R., Aigrain, P., Proc. Int. Conf. on the Physics of Semiconductors, Exter, 1962, Institute of Physics, London, 1962.
11. Heine, V., Van Vechten, J.A., Phys. Rev. B13 (1976) 1622.
12. Bagly, B.G., Chen, H.S., Laser Solid Interaction and Laser Processing (1978), p. 73.
13. Baeri, P., Foti, G., Poate, J.M., Cullis, A.G., Phys. Rev. Lett. 45 (1980) 2039; Kokorowki, S.A., Olson, G.L., Roth, J.A., Hess, L.D., Phys. Rev. Lett. 48 (1982) 498.
14. Koster, U., Phys. Stat. Sol. 48 (1978) 313.
15. Zellama, K., Germain, P., Squelard, S., Bourgoin, J.C., Thomas, P.A., J. Appl. Physics 50 (1979) 6995.
16. Kokorowski, S.A., Olson, G.L., Hess, L.D., J. Appl. Physics 53 (1982) 921.
17. For a nice review of crystallization processes in Germanium, see Barna, A., Barna, P., Bede, Z., Poeza, J.S., Pezsgai, I., Radnoczi, J., Thin Solid Film 3 (1974) 49 and Journal of non Crystalline Solid 8-10 (1974) 36.
18. Magarino, J., Kaplan, D. (private communication).
19. Blum, N.A., Feldmann, C., J. non cryst. Solids 11 (1972) 242.
20. All these estimates have been made using the steady-state nucleation current which is reached after a timelag τ_L; τ_L has been measured in silicon by Koster [7] between 550° and 650°. Using an Arrhenius law, we find $\tau_L \sim 10^{-5}$ s at T_m, which is of order the nucleation times calculated using the steady-state nucleation current. This may produce an increase, by a factor of the order of 2, of this estimated nucleation time.
21. More details on the resolution of the heat equation in cases relevant with laser heating of silicon will be published elsewhere.

OPTICAL PROPERTIES OF SEMICONDUCTORS

R.G. ULBRICH

Institut für Physik der Universität Dortmund
46 Dortmund 50, Germany.

1. INTRODUCTION

In this lecture we discuss optical properties of semiconductors in the framework of dielectric polarizability and the formulation of this quantity in terms of corresponding elementary excitations, like vibronic or electronic polarization waves. The lecture is organized as follows : a paragraph on the basic light-matter interaction reviews the concept of the macroscopic dielectric function, and its relation to transverse and longitudinal excitations in the crystal. The main emphasis is laid on electronic interband transitions and the band edge spectra in the semiconductors Si, Ge and GaAs, which are of common interest in the context of laser annealing. Excitonic effects are discussed in some detail. The paragraph on linear coupling of light waves with transverse polarization waves gives a description of resonance phenomena on a higher level of sophistication, in terms of so-called polaritons. Part 4 on probing techniques gives comments with respect to the usefulness of various kinds of experimental configurations and points out certain problems connected with some techniques. The last paragraph is more specialized and focusses on nonlinear effects in the optical response of semiconductors under conditions of high excitation. Some specific cases, like gain spectra of optically pumped electron-hole plasmas in direct gap semiconductors, the disappearance of sharp exciton line spectra at the Mott transition, plasmon light scattering, and lattice heating effects are discussed in greater detail. An attempt is made to relate these contemporary and very active areas of research with the current topic of this summer school. Some illustrations of what can be done are presented. The treatment is qualitative and not at all exhaustive and is intended to serve as a guide to key references and as an introduction to some of the concepts in the field of linear and nonlinear optical spectroscopy of semiconductors.

2. BASIC LIGHT-MATTER INTERACTION

A simple and qualitatively correct picture of the polarizability and the dielectric function connected with interband electronic transitions in a semiconductor which is driven by a weak (external) light field can be developed in steps : one first considers the interaction of one free atom with the electromagnetic field[1,2] The eigen-states of the unperturbed system are given by :

$$H_o \psi_i = E_i \psi_i . \tag{1}$$

For times $t \geq 0$ the interaction energy term is added :

$$H' = q_e \times (\vec{E}_{loc} \cdot \vec{r}), \tag{2}$$

i.e. the dipole interaction, where $\vec{E}_{loc}$ is the local electric field at the position $\vec{r}$ of the electron. After the application of first order perturbation theory recipes and the use of Fermi's Golden Rule, one finds the transition probability per time from the ground state $|0>$ with energy E_o to the excited state $|i>$ with energy E_i [2] :

$$W_{o \rightarrow i} = \frac{2\pi}{\hbar} |<0|H'|i>|^2 \delta(E_i - E_o - \hbar\omega) . \tag{3}$$

In a homogeneous ensemble of atoms and for sinusoidally varying field components $\tilde{E}(\omega)$ we may define a complex dielectric susceptibility $\tilde{\chi}_{eff}(\omega)$ in such a way that it connects modulus and phase of the *induced* macroscopic (i.e. averaged over the cell) polarization $\vec{P}_{ind}(\omega)$ with the *external* driving field :

$$\vec{P}_{ind}(\omega) = \tilde{\chi}_{eff}(\omega) \cdot \vec{E}_{ext}(\omega) . \tag{4a}$$

The induced polarization, i.e. the redistribution of the charges in the medium, gives rise to an induced electric field. The sum of both fields, external and induced, is the total electric field $\vec{E}_{tot}$ inside the medium, which enters Maxwell's equations for dielectrics [1]. The relation of this total electric field with the (external) driving field defines the dielectric function $\tilde{\varepsilon}(\omega)$: $\vec{E}_{ext}(\omega) = \tilde{\varepsilon}(\omega) \vec{E}_{tot}(\omega)$. In this case we have by definition $\varepsilon = (1-4\pi\chi)^{-1}$. If one defines the susceptibility such that :

$$\vec{P}(\omega) = \tilde{\chi}(\omega) \cdot \vec{E}_{tot}(\omega) , \tag{4b}$$

one gets the different relation $\varepsilon = 1+4\pi\chi$. The choice of the definitions has to do with the approach to the subtle problem of a correct quantum mechanical description and is explicitely discussed in [3]. We adopt here the latter procedure and calculate the induced polarization via eq. 3 and obtain [2] :

$$\widetilde{\varepsilon}(\omega) = 1 + 4\pi\widetilde{\chi}(\omega) = 1 + \frac{4\pi N\ q_e^2}{m_e} \sum_i \frac{f_{oi}}{\omega_{oi}^2 - \omega^2 - i\omega\Gamma_i} \ . \tag{5}$$

Here f_{oi} is the oscillator strength of the transition $o \rightarrow i$ (the electric field chosen parallel to x) :

$$f_{oi} = \frac{2m_e\ \hbar\ \omega_{oi}}{\hbar^2} <0 \mid x \mid i> \tag{6}$$

and $\hbar\omega_{oi}$ denotes the energy separation between ground state level $|0>$ and state $|i>$, Γ_i is the energy uncertainty ("width") of the level $|i>$, and N is the concentration of (non-interacting) atoms present.

If Γ_i is small compared with the transition frequency considered, the effect of interband transitions is almost purely *dispersive* for $\hbar\omega \ll \hbar\omega_{oi}$. If $\hbar\omega$ approaches "resonance" with one of the levels, $\hbar\omega \simeq \hbar\omega_{oi}$, the transitions become strongly absorptive, and the propagation of a light wave with frequency ω is appreciably damped in the medium. It is instructive to notice the formal correspondence of $\widetilde{\varepsilon}(\omega)$ with the classical expression for the response of an ensemble of oscillators $|1>$, $|2>$, ... $|i>$ with eigen-frequencies ω_{oi} and weight factors f_{oi} describing the coupling strength to an external perturbation with frequency ω (Fig.1).

In the case of a dense ensemble of atoms, when the index of refraction is appreciably different from unity (i.e. $\vec{E}_{ext} \neq \vec{E}_{tot}$), the dipole moment operator in eq. 2 has to take into account the effect of the local field due to the polarizations of surrounding atoms and becomes (in cubic crystals, assuming point dipoles as the sources of $\vec{P}$) :$H' = q_e x(\vec{E}.\vec{r}) x(n^2+2)/3$ [1]. In a solid the atoms are packed so close that the electron wave functions are spread out over more than one lattice site. Now the choice of the proper local field in eq. 2 (i.e. the formulation of the interaction Hamiltonian) represents a considerable problem [3]: the macroscopic polarization has to be taken as a weighted average of the microscopically varying polarization over the unit cell and one has to take into account the delocalization of the electrons.

The definition of the dielectric function eq. 5 can be generalized for the description of the response $\vec{P}$ of any homogeneous

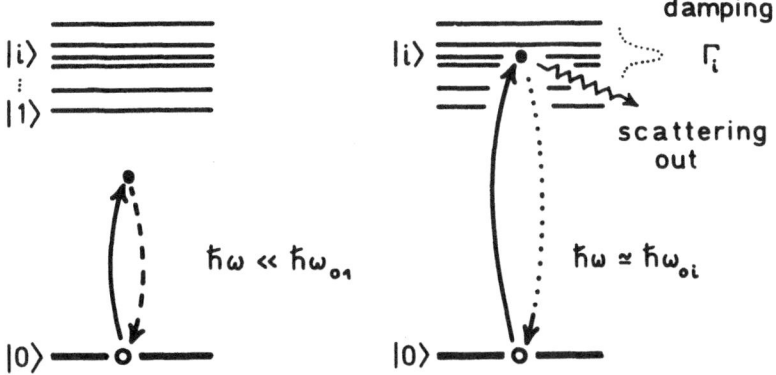

<u>Fig. 1</u> Optical transitions between ground state and excited states of an ensemble of atoms. The response of the system to a driving light field with frequency component ω is described by the complex dielectric function $\tilde{\varepsilon}(\omega)$; dispersive (left) and absorptive (right) behavior are controlled by resonance denominators and the damping coefficients Γ_i in eq. 5.

system of *interacting* atoms (the "dielectric solid") with respect to an electromagnetic field $\vec{E}$ of arbitrary time dependence, irrespective of the origin of $\vec{E}$, partially external or not [3,4]. If the fields and polarizations in the solid are decomposed into longitudinal and transverse components, one finds from Maxwell's field equations together with the constitutive matter equation: $\varepsilon(\omega)\vec{E}_{tot}=0$ for the bulk situation (infinite solid) that the *zeros* of the dielectric function give the eigen-frequencies ω_L of *longitudinal* excitations for vanishing wave-vector $\vec{k} \to 0$, (like LO phonons, or plasmons) and that the solutions of $\tilde{\varepsilon}(\omega) = c^2k^2/\omega^2$ represent the dispersion $\omega(\vec{k})$ of *transverse* excitations (like TO phonons, transverse excitons).

Real and imaginary part of $\tilde{\varepsilon}$ describe the dispersive and absorptive effects on the propagation of the coupled polarization and electromagnetic waves inside the medium; the spectral structure of both functions is contained in the sum of resonance denominators of eq. 5 and the "damping" of the waves in the dielectric is described phenomenologically by the (hitherto unspecified) "scattering out" process of Fig. 1, which appears as Γ_i in eq. 5.

If the external driving field is strong, i.e. comparable with the total electric field acting on the electron inside the solid in its ground state, the induced response P (see eq. 4) will no longer be linearly connected with that field because of two basic effects : i)multi-quantum processes can occur and the dielectric

304

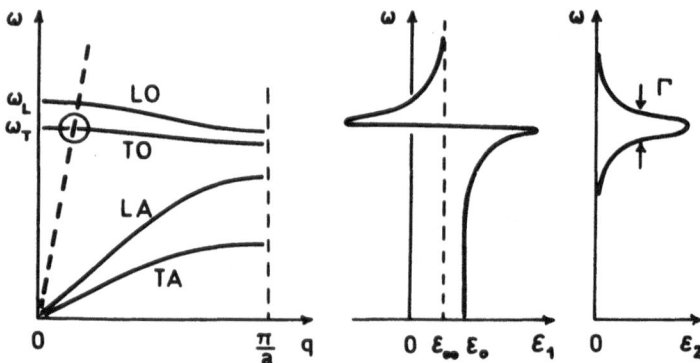

<u>Fig. 2</u> In polar semiconductors the lowest (and most prominent) re-
sonance of $\tilde{\epsilon}(\omega)$ is caused by transverse optical lattice vibrations
in the infrared spectral region ($\hbar\omega \sim 20 \dots 60$ meV). Phonon and
photon dispersion curves $\omega(\vec{q})$ and the resulting response functions
for dispersion and absorption are shown schematically (see [5]).

response eq. 5 may contain terms of higher order in the perturbation
H' ; ii) the population of elementary excitations in the crystal can
be significantly altered by the external stimulus and this effect
changes the susceptibility (see below, part 5).

The resonances in the dielectric function are connected with
transitions between the ground (or equilibrium) state $|0\rangle$ of the so-
lid and the excited states $|i\rangle$, separated by an energy $\Delta E = \hbar\omega_{oi}$.
In general the most prominent lowest resonance, at least in the
case of polar semiconductors (like zincblende or wurtzite materials),
is caused by the vibronic excitation of transverse optical phonons
which couple directly, via their dipole moment, to the light field
(analogous to eq. 2). The corresponding response functions can be
found classically [5] from spring-and-mass models and they give a
pronounced spectral structure of $\tilde{\epsilon}(\omega)$ in the infrared region (typ.
$\hbar\omega = 20 \dots 60$ meV), Fig. 2.

Of more interest in the present context are the electronic excitations and resonances in the visible part of the spectrum. Consider an intrinsic, i.e., undoped semiconductor at T = 0. It turns
out that the energetically lowest possible electronic excitations
above the ground state is a correlated pair of an electron and a
hole, a so-called "exciton" [4,6]. These excitons represent excited
states of the N-electron system. The single constituents alone,

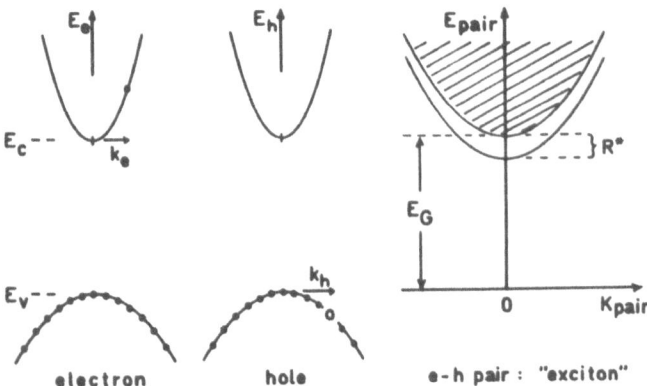

<u>Fig. 3</u> The lowest possible electronic excitation in an intrinsic semiconductor is an electron-hole pair, or "exciton". The total energy of the pair is lowered relative to the sum of single parti-cle energies (left and middle) because of the attractive e-h Cou-lomb interaction. R^* is typically 5 ... 50 meV in common semicon-ductors. E_G is the energy to create a pair at rest and infinite relative separation.

electron and hole, stand for the (N+1) electron problem ("conduc-tion band electron") and the (N-1) electron problem ("valence band hole"), Fig. 3. The correlation of the motion of e and h in the exciton state is, though in principle a complicated interplay bet-ween the direct Coulomb interaction and the exchange interaction among all the N electrons in the crystal [7], in a certain limiting case easy to describe : it resembles the binding of the two parti-cles through a simple Coulomb interaction potential of the form: $V_0 = - q_e^2/4\pi\varepsilon_{eff}|\vec{r}_e - \vec{r}_h|$, where ε_{eff} is an effective screening die-lectric constant close to the value of the low frequency limit: $\varepsilon(\omega=0)$: $\varepsilon_{eff} = \varepsilon(\omega=R^*/\hbar)$. In this case the spectrum of possible exciton energy eigenvalues is hydrogen-like with an effective Ryd-berg R^*:

$$E_n = E_G - \frac{R^*}{n^2} + \frac{\hbar^2 K^2}{2 M^*} , \tag{7}$$

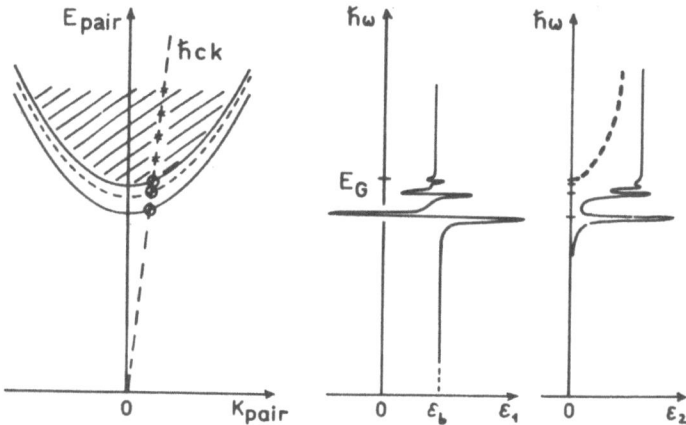

<u>Fig. 4</u> The near-band-edge optical spectra of all direct gap semi-
conductors with dipole-allowed transitions are characterized by a
series of dispersive resonances in ε_1 (middle) and absorptive peaks
in ε_2 which are due to bound e-h pair states ("exciton series").
Correlated e-h pair continuum states form the smooth spectra above
E_G. An independent particle description would predict a square-
root dependence of ε_2 (dashed curve) which is never observed in
experiment.

where E_G is defined through the two conditions $n \to \infty$ (i.e. ioniza-
tion limit, $|\vec{r}_e - \vec{r}_h| \to \infty$) and $\vec{K} = 0$, total wavevector of the pair
zero. In other words, E_G is the energy to create an electron-hole
pair at rest and infinite relative distance. The second term is
the *relative* e-h interaction with $R^* = (R/\varepsilon_{eff}^2)(\mu/m_e)$, where μ is
the reduced exciton mass $\mu^{-1} = m_e^{*-1} + m_h^{*-1}$ and $R = 13.6$ eV. The
third term denotes the kinetic energy contained in the translational
motion of the pair. This hydrogen-like behavior of relative e-h
motion with an exciton Bohr radius a_B is found in materials like Si,
Ge, and GaAs with large ε_0, relatively small bandgap and, as a consequence, delocalized wavefunctions for electron and hole relative
motion ($|\vec{r}_e - \vec{r}_h| \gg a$, where a is the lattice constant). In this
limit the exchange interaction is small compared with the direct
e-h Coulomb interaction [7].

2.1 Optical Spectra in the Near Band Gap Region

Starting with the approach along eqs. 1-6 with proper wave-functions for the N-electron system one finds the dielectric response [4] for the excitation of excitons (Fig. 4). In the cases considered here the gap energies E_G are large compared with the optical phonon energies, so that the "excitonic series" of absorptive peaks in ε_2 and the dispersive resonances in ε_1 (see Fig. 4) are dominating the band edge spectrum of all semiconductors. (For a discussion of valence and conduction band symmetries and their connection with the free atom orbitals of the constituents see [8]). The oscillator strength of the n-th exciton line is in direct gap materials with dipole-allowed transitions [8] :

$$f_n = \frac{p^2}{E_G} \frac{\Omega}{\pi a_B^3} \frac{1}{n^3} , \tag{8}$$

where $p^2 = |<s \mid \hat{p}_x \mid p_x>|^2 x2/m_e$ is the reduced interband matrix element of the momentum operator $\hat{p}_x$ between p-like valence band states and s-like conduction band states, and Ω is the volume of the elementary cell. This expression is the analog of eq. 6. A typical value of p^2 is ~ 20 eV, and one obtains relatively small oscillator strengths $f_1 = 10^{-5}$ to 10^{-2}. The continuum absorption constant is given by :

$$\alpha_{cont} \sim \gamma \frac{\exp\gamma}{\sinh\gamma} \sqrt{\hbar\omega - E_G} , \tag{9}$$

with $\gamma = \pi \sqrt{R^*/\hbar\omega - E_G}$. α_{cont} is practically independent of $\hbar\omega$ if the condition $\hbar\omega - E_G \lesssim 5R^*$ is fulfilled [4]. The agreement between theoretical near-band-edge excitonic spectra $\varepsilon(\omega)$ calculated along these lines and the measured absorption and refraction spectra is excellent for the above mentioned semiconductors.

In semiconductors with "indirect" band structures (like Ge, Si, GaP), the lowest possible pair excitations occur with wavevector $\vec{K}_o \neq 0$ (K_o is typically 10^8 cm^{-1} at the edge of the Brillouin zone), so that photons with their relatively long wavelength ($K_{photon} = 3.10^5$ to 10^6 cm^{-1} inside the crystal) cannot couple efficiently to these "indirect" excitons (Fig. 5). Absorption and emission processes involving such excitons are mediated by phonons with appropriate wavevector and corresponding energies. In Ge, for instance, LA and TO phonons (and, for $\vec{K}_{pair} \neq \vec{K}_o$, also TA and LO) do participate in the optical transitions in the vicinity of E_G. The transition probabilities contain matrix elements of second order (they involve a photon and a phonon) and are accordingly much smaller than in the direct gap case [4]. Typical absorption coefficients close to the indirect gap are $\alpha \sim 1$ to 30 cm^{-1} and the energy dependence for an exciton band (Fig. 5) is : $\alpha(\hbar\omega) \sim (\hbar\omega - E_G)^{1/2}$,

308

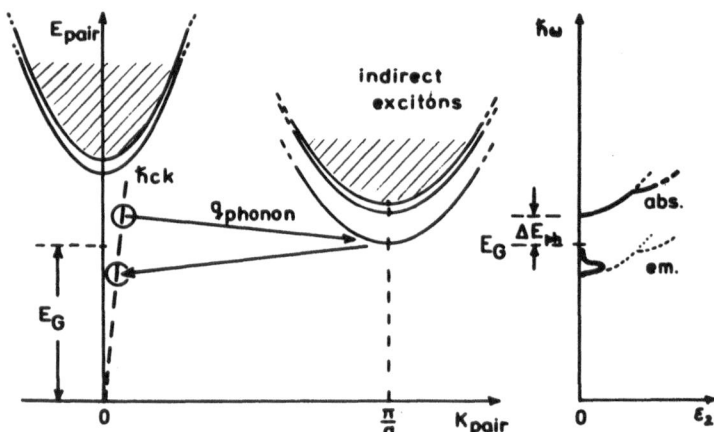

<u>Fig. 5</u> Absorption and emission processes involving indirect exci-
tons are accompanied by phonons with appropriate wavevector and
energy. Instead of discrete exciton lines one observes edge contours in the absorption spectrum ε_2 (schematic).

for the pair continuum $\sim(\hbar\omega-E_G)^{3/2}$. The sharp exciton line structures are blended and only edge contours remain. Anisotropy of
$E(\vec{K}-\vec{K}_0)$ around the band minimum complicates the actual shape of the
absorption lines. In the well-characterized indirect gap materials,
like Ge, Si and GaP the details of $\tilde{\varepsilon}(\omega)$ close to E_G are now well
understood [9].

All band-edge optical spectra of semiconductors exhibit exciton effects. The correlated e-h pair motion leads to a drastically
enhanced transition probability at the gap energy, causes discrete
line series in direct allowed transitions below E_G, and shows up
in *edge* series for indirect transitions. In all cases the underlying density of pair states is fundamentally different from the
single particle description, where $D(E)\sim\sqrt{E-E_G}$ [6]. This is the
consequence of the long range Coulomb interaction between electrons
and holes.

The structures in band-edge absorption spectra at elevated
temperatures are broadened due to exciton-phonon coupling ("phonon
collisions"). In addition, there is a rigid shift of the spectra
towards lower energy because of the gap shrinkage induced by the

phonons [10,11]. At sufficiently low temperatures the residual linewidths are determined by extrinsic effects, like doping and crystal imperfections. Semiconductors with high structural perfection and low doping levels, such that $a_B N_{imp} \lesssim 10^{-3}$ (a condition which can be readily achieved in Ge, Si, CdTe, CdS, GaAs etc.) will exhibit linewidths of exciton absorption $\Gamma = 10^{-5}$ to 10^{-4} eV at helium temperature ! For such crystals the linewidths are usually constant up to $T \sim 20$ K and proportional to T for higher temperatures with absolute values of Γ typically much below kT (Fig. 6).

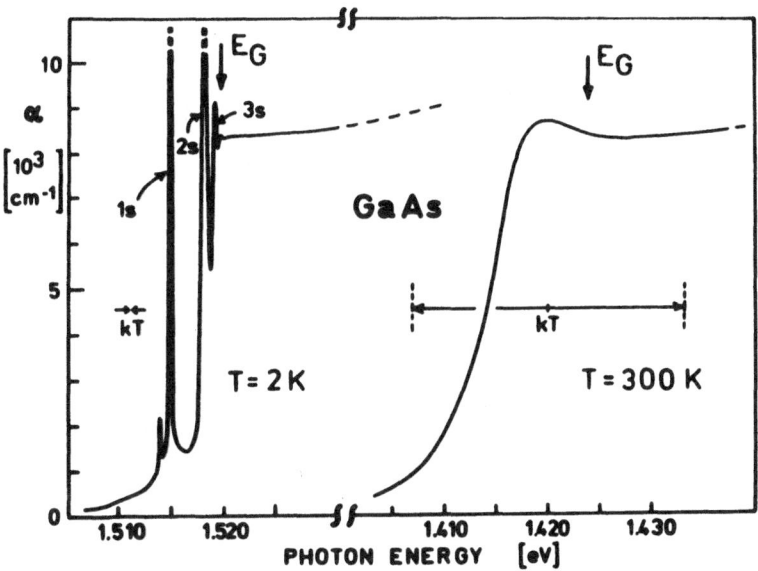

Fig. 6 Near-band-edge absorption spectra of high-purity GaAs (thickness 4 µm). The sharp exciton line spectrum with 1s, 2s and 3s excitons resolved at low temperature (left) is broadened by phonon collisions at room temperature (right). The corresponding values for both thermal energies kT are indicated. Note the change in energy scale : the band gap energy E_G varies with temperature.

2.2 Optical Spectra above the Band Gap

For incident light $\hbar\omega > E_G$ the dielectric response function eq. 5 contains resonant terms which are closely spaced in energy and give rise to continuum absorption and dispersion. Typical spectra for ε_1 and ε_2 reveal structure due to van Hove-singularities in the e-h combined density-of-states which are reminiscent of the single-particle band structures $E_c(k)$ and $E_v(k)$ [12-14]. The effect of correlation between electrons and holes has an appreciable influence on the energetic position of these structures [7] as compared with simple independent particle descriptions. Because of the high damping ($\varepsilon_2 \gg 1$), most experimental spectra were not obtained from light transmission but rather from reflectivity data and subsequent evaluation of $\tilde{\varepsilon}(\omega)$ from Kramers-Kronig transformations [12-14]. The overall behavior of the above-band-gap spectra is similar to an oscillator with effective transition frequency centered at the energy difference between the average conduction band and valence band (one-electron) energies. The dispersive behavior of $\tilde{\varepsilon}(\omega)$ of crystalline Si, where this effective "interband oscillator" has an energy of ~4 eV, is shown as an example in Fig.7.

2.3 Collective Electronic Excitations in Doped or Optically Excited Semiconductors

Semiconductors may contain free (mobile) charge carriers :
i) doping with impurities can produce a plasma of mobile charges of one type (n- or p-type material) on a background of fixed (stationary) charges of the opposite type; ii) in pure crystals the optical or thermal excitation of free e-h pairs above the band gap generates a "bipolar" plasma, where *both* constituents, electrons and holes, are mobile. In full analogy to metals, these plasmas of free carriers can support collective excitations, so-called plasma oscillations. Their response to a total longitudinal electric field is characterized by [3] :

$$\varepsilon(\omega) = 1 + \frac{\omega_p^2}{\omega(\omega + i\Gamma/\hbar)} \tag{10}$$

and $\omega_p = 4\pi q_e^2 N/(\varepsilon_o m^*)$ is the classical plasma frequency at zero wavevector. The quanta of these oscillations, plasmons, do not couple directly (at least for normal incidence in cubic materials) to a transverse electromagnetic field, i.e. do not contribute directly to $\tilde{\varepsilon}(\omega)$ in optical spectra. The presence of a rough surface or defects in the bulk can give rise to structure in the optical spectra at $\hbar\omega_p$ [12]. Here the main interest in plasmon excitation lies in its potential for probing purposes via inelastic light scattering and the extraction of plasma parameters n_e, n_h and T (see below).

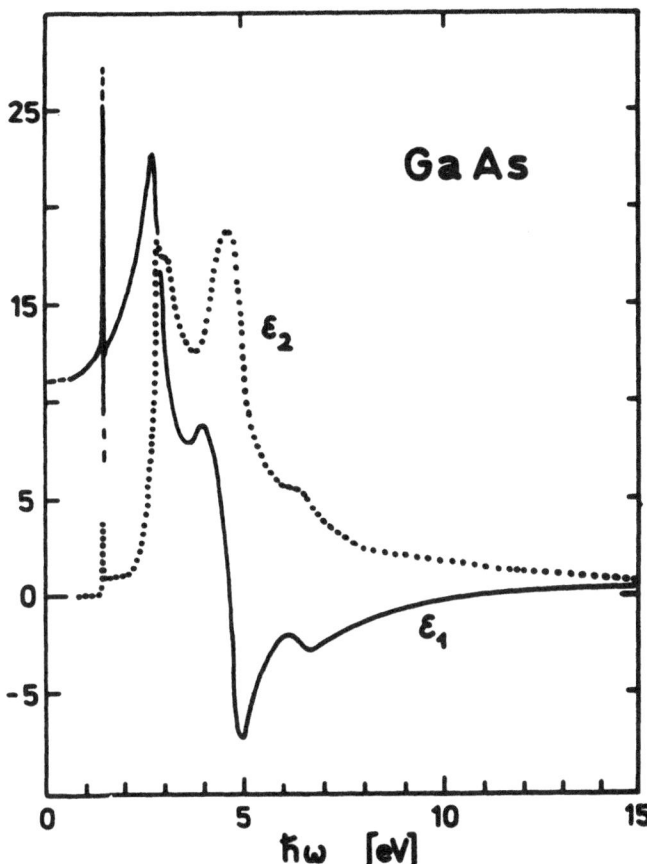

Fig. 7 The dielectric function $\tilde{\varepsilon} = \varepsilon_1 + i\varepsilon_2$ of GaAs (after [14]). The sharp exciton resonance at $\hbar\omega = E_G = 1.5$ eV has been included. The overall spectral dependence is approximately like that of an effective "interband oscillator" with energy ~ 4 eV (see text).

3. POLARIZATION AND LIGHT WAVES COUPLED : THE POLARITON APPROACH

In a more refined description of the optical properties of solids the starting point is not the "unperturbed" system eq. 1, but rather the system of electrons and ions with their mutual interactions *plus* their coupling to the omnipresent electromagnetic field. This approach in terms of mechanical polarization waves (of excitonic or vibronic origin) *coupled* with light waves leads to qualitatively new dispersion curves $\omega(\vec{K})$ which characterize the bulk dielectric medium. The corresponding quantum of elementary excitation has been named phonon- or exciton-"polariton".

The proper description of such coupled-mode propagation in a classical framework has been done first for optical phonons [17]. The quantum mechanical treatment gives results in exact correspondence [18,19]. The dispersion $\omega(\vec{K})$ for transverse polaritons is implicitely given by the usual condition $\varepsilon(\omega,\vec{K}) = c^2K^2/\omega^2$. The retardation due to finite light velocity c leads to two "photon-like" branches well below and above the resonance of $\tilde{\varepsilon}$, and a region of coupled-mode behavior around the resonance. The most relevant parameter in this approach is the coupling strength of the light field to the transverse excitation, which can be conveniently expressed in terms of the energy difference E_{LT} between the zeros of $\varepsilon(\omega, \vec{K}\rightarrow 0)$ at $\omega=\omega_L$ and the transverse resonance frequency ω_T in the explicit wave-vector dependent dielectric function:

$$\varepsilon(\omega,\vec{K}) = \varepsilon_{background} + \frac{2 \hbar\omega_T \ E_{LT}}{\hbar\omega_T^2(\vec{K}) - \hbar\omega^2 - i\omega\Gamma} \quad . \qquad (11)$$

The dispersion relations which are obtained from eq. 11 by setting it equal to c^2K^2/ω^2 (transverse waves, see above) or to zero (longitudinal waves) are given schematically in Fig. 8. They describe all possible wave excitations (eigen-modes) of the infinite bulk medium.

In the polariton approach the process of light absorption in the resonance is not only characterized by the coupling parameter of the mechanical excitation (phonon or exciton) to light, i.e. f_{oi} in eq. 5 or E_{LT} in eq. 11. Now the scattering-out rate, resp. the damping constant Γ in eq. 11, determines exclusively the damping of the polariton propagation in the crystal. It follows that the range of useful applications of the schemes is bound to the conditions (i) $E_{LT} \ll \Gamma$, or (ii) $E_{LT} \gtrsim \Gamma$. The former "overdamped" case can be handled with the simple oscillator scheme and phenomenological Γ (eqs. 1-5), the latter case of "small damping" *has* to be treated in the polariton framework to obtain agreement with experimental findings [20]. This is simply a consequence of the hierarchy of coupling strengths, and on which level the (irreversible) damping process enters the description.

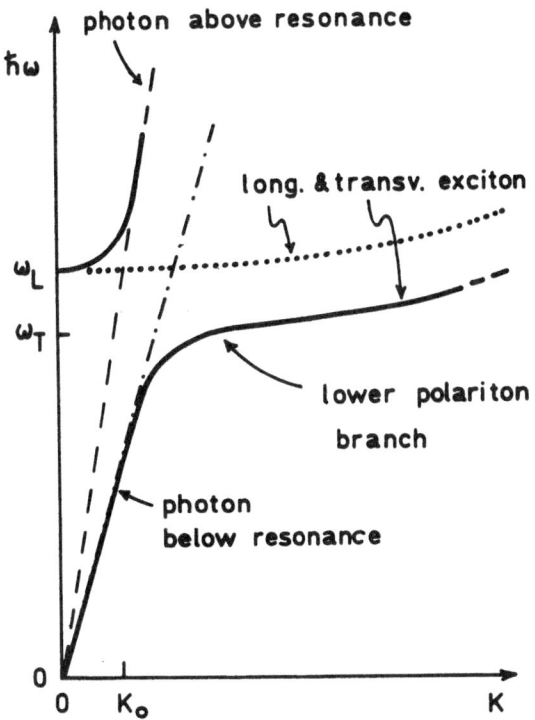

<u>Fig. 8</u> Dispersion curves $\omega(\vec{K})$ for exciton-polaritons (full lines, no damping). Close to the resonance frequency ω_T and the corresponding light wavevector K_0 the photon and exciton dispersions are strongly modified. Longitudinal excitons ($\vec{P} \parallel \vec{K}$, dotted curve) do not couple to light.

If the dielectric function eq. 11 of the infinite medium depends explicitly on the wavevector $\vec{K}$ of the coupled mode, the response $\vec{P}(\vec{r},t)$ on a field $\vec{E}(\vec{r}',t')$ is in general non-local, i.e. $\vec{P}$ depends on the stimulus at all other space points $\vec{r}'$ and earlier times t' :

$$P(\vec{r},t) = 1/4 \; \pi \iint [\, \widetilde{\varepsilon}(\vec{r}-\vec{r}',t-t') - 1]\; \vec{E}(\vec{r}-\vec{r}',t-t') \; dt' \; d\vec{r}'.$$

$$(12)$$

This poses a subtle problem in the treatment of ordinary reflection, transmission and absorption processes which, of course, require a semi-infinite (or plane-parallel slab) geometry : the usual procedure of finding the boundary conditions from Maxwell's equations, namely $\vec{E}_{tan}$ and $\vec{H}_{tan}$, $\vec{D}_{normal}$ and $\vec{B}_{normal}$ continuous at the boundaries, is no longer sufficient to find the connection between the external and interval fields for the case of non-local response, eq. 12. Literature refers to this fundamental problem as "additional boundary conditions" (ABC). Its solution, since it requires in principle the explicit and microscopic knowledge of $\tilde{\varepsilon}(\vec{r},\vec{r}',t,t')$ in the vicinity of the solid surface, is still not complete neither for optics on dielectric nor on metal surfaces or interfaces [21,22].

The coupled-mode approach can be generalized to other cases of interacting excitation, e.g. longitudinal optical phonons and longitudinal plasmons in semiconductors [23].

4. PROBING TECHNIQUES

Experimental configurations for measuring the linear optical response $\tilde{\varepsilon}(\omega)$ of a given sample are usually based on the following four basic methods : (i) direct access to bulk *dispersion* ε_1 by measurements of phase velocities $v_{ph} = \omega/K = c/n = c/\sqrt{\varepsilon_1}$ (e.g. via beam deflection after transmission through a prism-shaped sample; or counting interference fringes in Fabry-Perot etalons filled with the dielectric); closely related are group velocities $v_{gr} = \partial\omega/\partial\vec{K}$, measured in pulse propagation through plane-parallel slabs ; (ii) direct measurement of bulk absorption (i.e. ε_2) on sufficiently thin samples in transmission configuration. The absorption coefficient is $\alpha = (K/n) \varepsilon_2$. For reasons of stray light suppression the sample thickness t should fulfil $\alpha t \lesssim 6$; (iii) measurements of specular reflectivity on plane surfaces ; (iv) measurements of elastic ($\omega\to\omega$) or inelastic ($\omega\to\omega'$) light scattering, eventually with well-defined wavevector transfer ($k\to k$) in small-angle resp. forward scattering; $\Delta k \leq 2\,k_{light}$ for backward scattering .

Methods (i) and (iii) are conceptually simple. (iii) has been universally applied to all materials, whereas (i) can only be used in spectral regions of small damping. Absorption measurements (ii) are more difficult and require considerable effort to prepare thin samples which ensure the dynamical range of the experiment with respect to stray light and also possible inelastic contributions to the transmitted signal due to fluorescence. The three methods are restricted to probe transverse excitations at small and fixed wavevectors, which are not independently varied nor measured directly in the sample. Methods (i) to (iii) are in fact spectroscopies working only in ω-space and leave it to the proper choice of the model to find the true constitutive relation $\tilde{\varepsilon}(\omega,\vec{K})$.

By far the most versatile and powerful experimental method which involves light in the optical or near infrared part of the spectrum is (iv). One can determine dispersion relations and occupation numbers of elementary excitations (like excitons, phonons, plasmons, etc.) in semiconductors by inelastic light scattering in suitable geometries [23,24]. As an example, carrier distribution functions in n-GaAs have been studied in detail as a function of doping, temperature and applied static electric fields with this technique [25]. The domain of single-particle excitations, ranging from non-degenerate Maxwell-Boltzmann to degenerate distributions, and also collective excitations (plasmons and coupled LO-phonon-plasmon modes) could be covered with essentially one experimental set-up : a 1.06 μm cw-Nd:YAG laser light source and a double grating spectrometer with photon-counting detector [25].

More recent applications have used pulsed lasers and time-resolved detection of inelastic light scattering spectra : these techniques offer potential to investigate in a complementary way the full dynamic aspect of the excitations involved, i.e. to measure directly the relaxation times of specific decay channels in the time domain down to resolutions of the order of 10^{-12} sec [26].

5. NONLINEAR EFFECTS

The optical spectra of semiconductors are, in general, dependent on the degree of excitation. One may distinguish extrinsic or intrinsic situations : an intense enough external driving field, which feeds (selectively) energy into the electronic system, will cause a highly non-equilibrium situation; if excitations are already present in the crystal in thermal equilibrium at finite temperature T, then all the states ψ_i of eq. 1 will be modified. A simple example for the latter intrinsic case is the temperature dependence of E_G in all semiconductors, which exhibits a behavior $E_G(T) \simeq E_G(0) \times (1-T^2/T_0^2)$ due to electron-phonon interactions in second order [27]. The former case constitutes the rapidly growing field of nonlinear optics and leads to an enormous variety of effects [28-30]. When the induced dielectric polarization is large and eq. 2 can no longer be treated as small perturbation to H_o, the dielectric function depends on $\vec{E}$. The relevant parameter is obviously $|\vec{E}|/|\vec{E}_{atom}|$, where $|\vec{E}_{atom}|$ is the average field seen by the electrons in the unperturbed crystal. The problem can be formulated phenomenologically as a Taylor series in terms of the nth-order dielectric susceptibilities χ_n [28] :

$$\vec{P}_{tot} = \chi_1 \vec{E}_1 + \chi_2 \vec{E}_1 : \vec{E}_2 + \dots \tag{13}$$

where the χ_n are tensors of rank (n+1) and couple the driving fields $\vec{E}_i$ together to produce the induced polarization. The χ_n are related to multi-quantum processes of corresponding order. As an example, sum or difference frequency generation via coupling of two light waves $\vec{E}_1(\omega_1)$ and $\vec{E}_2(\omega_2)$ are mediated through terms

$$\chi_2(\omega_1,\omega_2;\omega_3=\omega_1\pm\omega_2).$$

In a microscopic description of the resonance behavior of χ_n one considers in addition to eqs. 1 and 2 also the *changes* in occupation of the various excitations (excitons, free e-h pairs, plasmons, phonons) under the influence of the driving light wave. Of central importance is the question of coherence between the driving field and the induced excitations and their momentum and energy relaxation times. Small damping of intermediate states involved in the multi-quantum transitions can lead to enormous resonance enhancement factors in certain terms of the series eq. 13 [31]. The number of experimental and theoretical investigations of non-linear optical properties of semiconductors is currently growing at a rapid pace; the field is far from maturity. Some special cases of interest, especially in the context of stimulated emission from optically excited semiconductors, have been worked out thoroughly [32,33].

5.1 Gain spectra

When the density of e-h pairs in a semiconductor is increased to such a high value that discrete bound states of e-h pairs ("excitons") are no longer stable because of the screening of the mutual e-h Coulomb attraction, these mobile carriers form a qualitatively new state : the electron-hole plasma (EHP). In the indirect semi-conductors Ge and Si a first-order phase transition between the low-density excitonic "gas" phase and a "liquid" EHP phase of high density n_0, the so-called electron-hole liquid (EHL), has been firmly established in quasi-equilibrium experiments (steady state optical pumping) below a critical temperature T_c [34]. In Ge one finds: $n_0(T=0) = 2.4\times10^{17}cm^{-3}$, $T_c = 6.5$ K; in Si : n_0 (T=0) = 3.3×10^{18} cm^{-3}, and $T_c = 25$ K [34].

In direct gap materials, like GaAs or CdS, the situation is less clear because of the short total lifetimes of e-h pairs which are typically 10^{-9} sec and even less under conditions of gain and subsequent stimulated emission from the pumped sample : presumably macroscopic phase boundaries (which show up through droplet formation in Ge and Si) are not formed on such a time scale [33]. Nevertheless, a wealth of spectroscopic data in the near-band-edge region exists where ε_2 can become negative when the pump intensity is strong enough and the electron populations close to the band edges are "inverted" in the direct gap case [32,33]. The detailed analysis of luminescence- and gain-spectra has led to the construction of phase diagrams which describe thermodynamical (quasi-) equilibrium properties of the system of e-h pairs at given mean pair density n_{pair}, fixed volume and temperature : possible phases are the gas of excitonic molecules, excitons, ionized excitons (i.e. continuum e-h pairs), and the EHL (Fig. 9) [33].

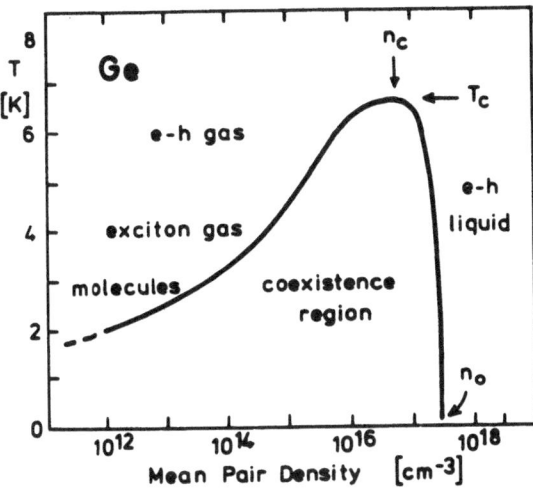

<u>Fig. 9</u> Phase diagram of the system of photoexcited e-h pairs in quasi-equilibrium in Ge (after [34]).

At T = 0 a homogeneous bipolar plasma of e-h pairs of stationary density n_{pair} will occupy conduction and valence band states up to Fermi energies E_F (e) and E_F (h). The gap energy E_G (n) depends on n_{pair} and is reduced ("gap shrinkage") because of exchange and correlation energy effects in the e-h system. Effective masses m_e^* and m_h^* of carriers are only weakly affected, at least for moderate Fermi energies in the meV range [34]. As a consequence of this population inversion one observes in the direct gap materials negative ε_2, i.e. gain, for $E_G(n) < \hbar\omega < E_G(n) E_F(e) + E_F(h)$ (Fig.10). The experimental gain spectra give specific information on plasma parameters and are reasonably well described by theory [35].

The occurrence of net gain with concurrent stimulated emission via superradiance or even laser action with feedback in the excited semiconductor volume represents a very efficient channel for energy relaxation of the electron-hole system : the total e-h recombination lifetime can be drastically shortened by this effect and estimates of stationary or transient carrier densities from generation/recombination rate equations have to be done with care.

Apart from the gain mechanism due to a degenerate bipolar plasma there exists a variety of other exciton-related recombination processes, which may also lead to gain at characteristic energies below the gap : exciton-exciton scattering, exciton-LO phonon scattering, exciton-electron scattering. These processes were observed

318

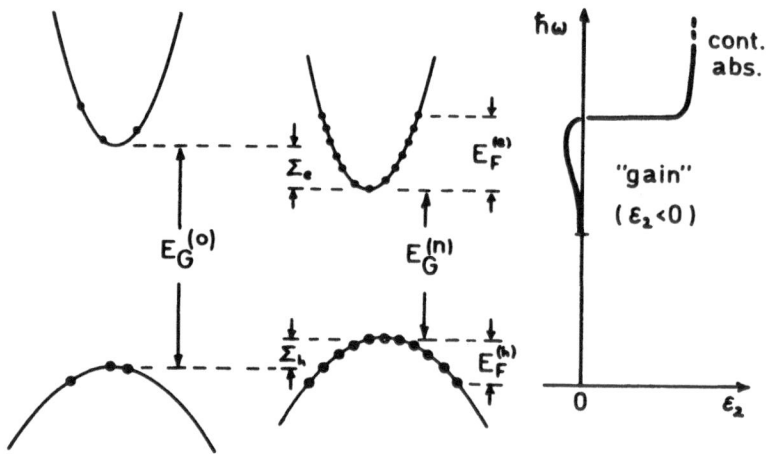

<u>Fig. 10</u> At sufficiently high pair densities the electron and hole
distribution functions become degenerate and one observes in direct
gap semiconductors "gain" (i.e. negative ε_2). Due to exchange and
correlation effects of e and h, the gap energy shrinks with increasing density (terms Σ_e and Σ_h).

and have been analyzed in detail in some of the large band gap semiconductors like CdS, ZnO, CuCl, ZnTe, ZnSe [32].

5.2 <u>Exciton Spectra at High Excitation Densities</u>

The absolute energetic position of the lowest discrete exciton
state is not affected by screening due to the presence of an e-h
plasma of sufficiently low density, because the *reduction* in binding
energy relative to the band edge $E_G(n)$ and the gap *shrinkage*
$E_G(0) - E_G(n)$ cancel each other quite perfectly [36]. At higher
e-h pair densities the discrete bound states become unstable, the
condition for this transition is : $E_G(n) = E_{1s}$, Fig. 11. In the
case of static Debye-Hückel screening the corresponding density is
estimated (from Mott's criterion that the exciton Bohr radius a_B
equals the screening length at the transition density n_{Mott}) [34] :

$$n_{Mott} = \frac{\varepsilon_o \, k \, T_{plasma}}{4\pi q_e^2 \, a_B^2} .$$

(14)

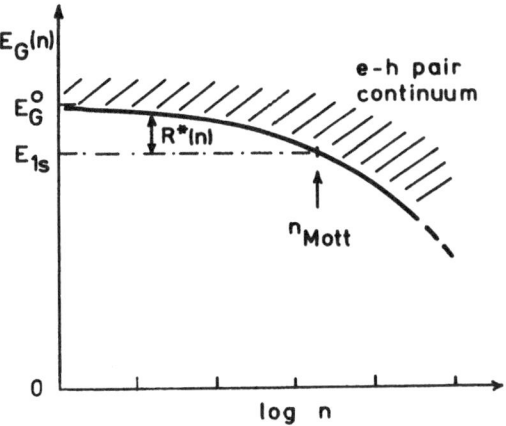

<u>Fig. 11</u> The absolute energetic position of the lowest exciton re-
sonance E_{1s} does not change appreciably with increasing pair density:
the band gap reduction cancels almost completely the decrease of
relative e-h binding energy R^*. The bound exciton states become
unstable in equilibrium above the Mott density, where $E_{1s} \geq E_G(n)$;
(see text).

For the degenerate Thomas-Fermi screening case ($T_{plasma}=0$) the thermal energy kT_{plasma} should be replaced by $2 E_F/3$ (where E_F is the
Fermi energy of the lighter particles, typically the electrons).

In direct gap materials with higher gap energies (like CdS,
CuCl) and relatively small a_B the exciton resonance is persistent
up to much higher pair densities and gives rise to a whole class of
new phenomena in the optical spectra *before* the formation of the
EHP sets in. The main effect is the occurence of a relatively stable biexciton state, composed of two excitons with a relative binding energy E_B of the order of several meV (in analogy with the hydrogen molecule). This biexciton resonance at an energy $2(E_G-R)-E_B$
mediates spectral features in absorption, dispersion and reflectivity spectra as a function of excitation light intensity [32].
These effects can be described in terms of eq. 13 as four-wave mixing
via the χ_3 term which can become multiple resonant in ω_1, ω_2 *and* ω_3
(with respect to ingoing and outgoing photon energies and the inter-
mediate exciton and biexciton states) and reach considerable values
[31,32,41].

If pair densities are expressed through the mean interparticle
distance r_s (in units of a_B), then $r_s \sim 4$ at the Mott transition at
low T for the typical semiconductors discussed here. The theoretical

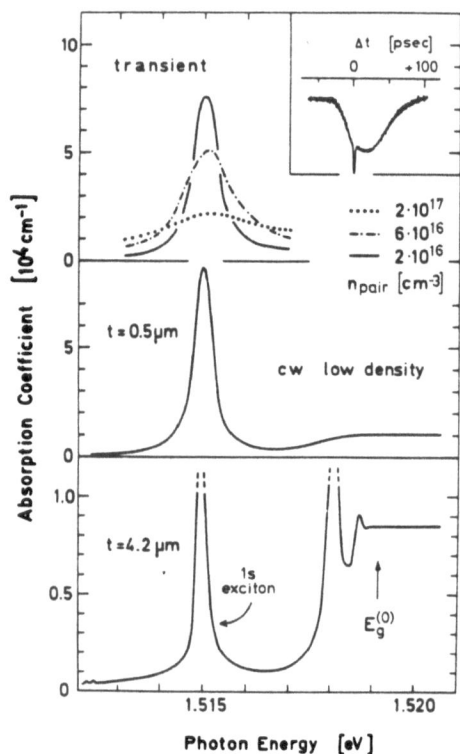

Fig. 12 Transient absorption spectra of a thin GaAs platelet at low temperature in the near-band-gap region after deposition of a density n_{pair} of 1s excitons with an 8 psec light pulse. The broadening and final bleaching of the 1s exciton absorption peak occurs above $n_{pair} = 6 \times 10^{16}$ cm^{-3}. The insert shows the time evolution of the peak absorption as a function of time delay between excite and probe pulses. (After [38]).

treatment of the ground state properties of the bipolar plasma is especially difficult for $r_s \cong 1$ [34] . Not much is known, therefore, on the transition region from strongly correlated e-h pair ("exciton") behavior to plasma (or metal-like) spectra at higher densities.

Picosecond excite-and-probe experiments have been performed to find out the screening influence of an optically injected hot e-h plasma of high density on the near-band-edge absorption spectrum in

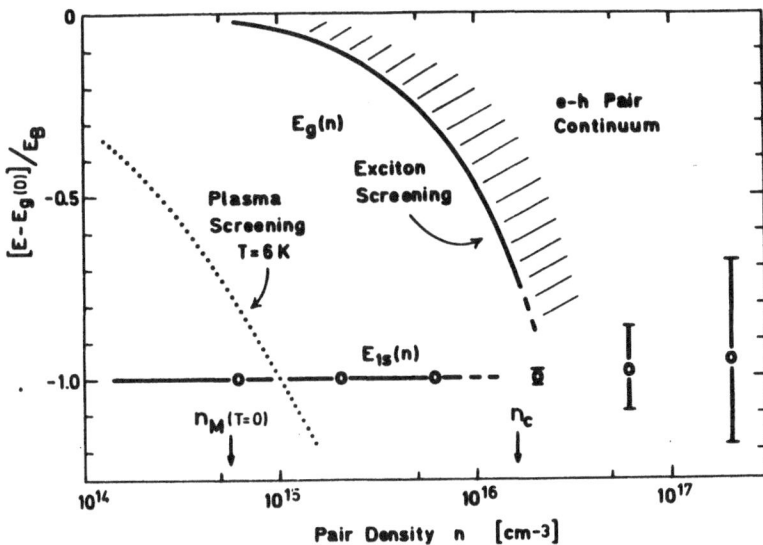

Fig. 13   Measured 1s exciton absorption peak position (circles) and excess halfwidths (vertical bars) as a function of deposited pair density. Mott density, the theoretical dependence of the band gap $E_G(n)$ for the plasma screening mechanism ([36]) and the less effective exciton screening are indicated. (After [38]).

GaAs [37]. Recent studies of transient absorption spectra under conditions of resonant optical excitation of discrete 1s excitons have indicated the existence of a well-defined exciton phase up to pair densities two orders of magnitude higher than the Mott criterion, eq. 14 [38] (Figs. 12, 13). Under steady state conditions, the optical spectra change from exciton-dominated line (resp. edge) spectra to smoother and temperature-broadened single particle behavior at much lower pair densities which in turn were found to be in qualitative agreement with eq. 14 [39, 40].

Further experimental and theoretical work is clearly needed to understand the optical properties of semiconductors for pair excitation densities $r_s \approx 1$. This is the lower end of the carrier density regime in which pulsed laser annealing processes do occur and where optical techniques are among the most promising tools for the investigation of the non-thermal melting process : temporal resolution down to fractions of a psec and lateral spatial resolution of several µm (and below µm in penetration depth) are feasible in practice.

322

5.3 Plasma light scattering

A direct method to determine experimentally the density and temperature of a carrier plasma in the wide density range 10^{16}cm^{-3} to 10^{19} cm^{-3} is the above-mentioned technique of inelastic light scattering [17,19]. This method has been applied recently for the in-situ diagnostic of optically excited e-h bipolar plasmas in semiconductors at low temperatures [42, 43]. Intraband scattering of free carriers as well as plasmon scattering were observed and allowed the assignment of plasma density and temperature (Fig. 14).

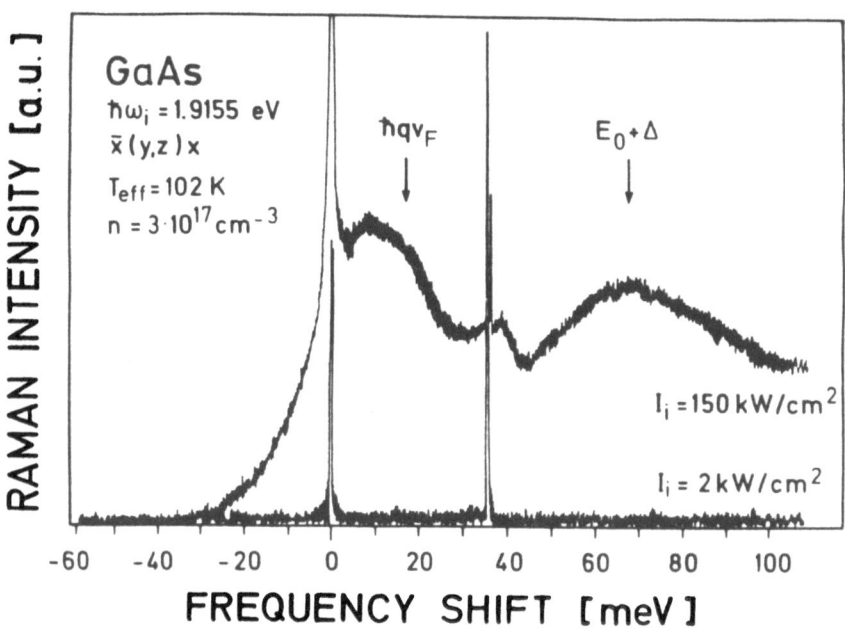

Fig. 14 Light scattering spectra from an optically excited e-h pair plasma of density 3×10^{17} cm^{-3} (upper trace; lower trace at small density for comparison). In addition to the sharp LO phonon Stokes peak, structure due to single particle excitations up to energy $\hbar v_{Fermi} \times 2 k_{light}$ is observed together with the plasmon/LO-phonon coupled mode at ~39 meV. The broad band around $E_0+\Delta$ is due to luminescence involving split-off valence band holes. (After [42]).

5.4 Lattice heating

The transfer of energy from the optically excited e-h pairs to the surrounding thermal reservoir is in most cases established by phonons (an exception is the stimulated emission of light, see above). The steady state carrier-to-phonon energy relaxation rates are quite large and range from 10^6 eV/sec at low carrier kinetic energies E_{kin} up to more than 10^{13} eV/sec for $E_{kin} > \hbar\omega_{LO}$ in a typical polar semiconductor like GaAs [44]. The details of carrier-phonon coupling and its dependence on carrier density (because of screening) play the crucial role in the balance between the energy stored in the electron system and the lattice [45,46].

There exists no direct probing technique for measuring phonon distribution functions over a wide range of wavevectors and frequencies; for this reason the investigation of the coupled transport problem (i.e. e-h pairs and phonons together) is formidable. In light scattering experiments the ratio of Stokes/anti-Stokes intensities has been measured to obtain occupation numbers for near-zone-center ($\vec{k}_{phonon} \sim 2\,\vec{k}_{light}$) optical phonons in Si and the TO and LO phonons in zincblende compounds [47,48]. It is questionable, however, to infer from such a selective measurement at essentially one wavevector the validity of the concept of a lattice temperature. It is highly probable that the phonon distributions are *not* in equilibrium under conditions of high excitation, and cannot be described with one temperature parameter [49]. Very long lifetimes of near-zone-edge TA phonons have been found at low temperatures in several zincblende compounds [50]. The TA phonon branch, corresponding to a shear motion of adjacent atoms, plays the crucial role in the melting process.

Non-equilibrium phonon distributions have been modeled in great detail in the context of high-field charge transport in semiconductors [44]. Many of the results obtained there may be applicable to the present problem.

6. SUMMARY

Linear optical spectra of semiconductors provide key information on elementary excitations ("phonon", "exciton", etc.) which stand for the complete dynamics of nuclei and electrons in the solid. Nonlinear optical spectroscopy probes the changes of the optical properties induced by the driving light field itself. The common basis for description is the frequency- and wavevector-dependent dielectric function. An inherent limitation is obvious : only transverse excitations with small wavevectors couple directly to light. The near-band-edge optical spectra of semiconductors are dominated at low temperature by correlated electron-hole pair excitations; plasma effects and electronic gap shrinkage come into play at higher carrier densities; phonon sidebands and phonon-induced gap reduction occur at higher lattice temperatures. Light scattering techniques are the most versatile tool to investigate in detail the internal structure of an excited semiconductor.

REFERENCES

1. Jackson, J.D., "Classical Electrodynamics" (New York, Wiley, 2nd edition) Ch. 7.5.
2. Wooten,F., "Optical Properties of Solids" (New York, Academic Press, 1972) Ch. 3.
3. Nakajima, S., Toyozawa, Y., Abe, R., "The Physics of Elementary Excitations" (Berlin, Springer, 1980) Ch. 2.
4. Semiconductors and Semimetals, ed. R.K. Willardson and A.C. Beer (New York, Academic Press, 1967) Vol. 3, Ch. 6,7 and references therein.
5. Ibid., Vol. 3, Ch. 1.
6. Dow, J.D., in : New Developments in the Optical Properties of Solids, ed. B.O. Seraphin (Amsterdam, North Holland, 1976) Ch.2.
7. Hanke, W., in : Advances in Solid State Physics, Vol. XIX, ed. J. Treusch (Vieweg, Braunschweig, 1979) p. 43.
8. Bassani,F. and Pastori-Parravicini, G., "Electronic States and Optical Transitions in Solids" (Oxford, Pergamon Press, 1975) Ch. 4.
9. Frova, A., Thomas, G.A., Miller, R.E. and Kane, E.O., Phys. Rev. Lett. 34 (1975) 1572;
   Thomas, G.A. et al., Phys. Rev. B4 (1976) 1692.
10. Varshni, Y.P., Physica (Utrecht) 39 (1967) 149.
11. Cohen, M.L., Chadi, D.J., in : Optical Properties of Solids ed. M. Balkanski (Amsterdam, North Holland, 1980) Vol. 2, Ch. 4B.
12. Petroff, Y., ibid., Ch. 1.
13. Greenaway, D.L., Harbeke, G., "Optical Properties and Band Structure of Semiconductors" (Pergamon Press, Oxford, 1968) Ch. 4 and 5.
14. Ehrenreich, H., in : Proc. Int. School of Physics E. Fermi, Course XXXIV, ed. J. Tauc (New York, Academic Press, 1966) p.106; Phillips, J.C., ibid. p. 155.
15. For a discussion of modelling $\tilde{\epsilon}(\omega, K)$ for dielectrics in the nearly-free-electron scheme, see : W. Jones, N.H. March, "Theoretical Solid State Physics" (Wiley, New York, 1973) Vol. 1, Chap. A2.6.
16. For a recent review on Inelastic Electron Scattering Spectroscopy, see : S.E. Schnatterly, in : Solid State Physics, 34 (1979) 275.
17. Born, M., Huang, K., "Dynamical Theory of Crystal Lattices" (Oxford, Clarendon Press, 1954) Ch. II.8.
18. Hopfield, J.J., in : Proc. Int. School of Physics E. Fermi, Course XLII, ed. R.J. Glauber (New york, Academic Press, 1969),

p. 340 and references therein.

19. Polaritons, ed. E. Burstein and F. de Martini (New York, Pergamon Press, 1974).
20. Ulbrich, R.G. and Weisbuch, C., in : Advances in Solid State Physics, Vol. XVIII, ed. J. Treusch (Braunschweig, Vieweg,1978) p. 217.
21. Zeyher, R., in : Recent Developments in Condensed Matter Physics, ed. J.T. Devreese (New York, Plenum Press, 1981) p. 807.
22. Stahl, A., Uihlein, C., in : Advances in Solid State Physics, Vol. XIX, ed. J. Treusch (Braunschweig, Vieweg, 1979), p. 159; Forstmann, F. and Gerhardts, R.R., ibid. Vol. XXII (1982) p.291.
23. Platzman, P.M., Wolff, P.A., "Waves and Interactions in Solid State Plasmas", in : Solid State Physics, Suppl. 13 (New York, Academic Press, 1973) Ch. 34.
24. Light Scattering in Solids, Vols. I, II, III, ed. M. Cardona, G. Güntherodt (Berlin, Springer, 1982).
25. Mooradian, A., in : Laser Handbook, Vol. 2, ed. F.T. Arecchi and E.O. Schulz-Dubois (Amsterdam, North Holland, 1972), Ch.E8.
26. Kuhl, J. and Bron, W.E., in : Proc. 16th Int. Conf. Physics of Semiconductors, 1982, Montpellier (to be published).
27. Heine, V. and Van Vechten, J.A., Phys. Rev. B13 (1976) 1622.
28. Bloembergen, N, "Nonlinear Optics" (New York, Benjamin, 1965); see also Baldwin, G.C., "An Introduction to Nonlinear Optics", (New York, Plenum Press, 1969).
29. Fröhlich, D., in : Advances in Solid State Physics, Vol. XXI, ed. J. Treusch (Braunschweig, Vieweg, 1981), p. 363.
30. Levenson, M.D., "Introduction to Nonlinear Laser Spectroscopy", (New York, Academic Press, 1982).
31. Maruani, A. and Chemla, D.S., J. Phys. Soc. Japan 49 (1980) 585; Abram, I. and Maruani, A., Phys. Rev. B26 (1982) 4759.
32. Klingshirn, C. and Haug, H., Physics Reports 70 (1981) 315.
33. Göbel, E.O., Mahler, G., in : Advances in Solid State Physics, Vol. XIX, ed. J. Treusch (Braunschweig, Vieweg, 1979), p. 105.
34. Rice, T.M., in : Solid State Physics Vol. 32, ed. H. Ehrenreich, F. Seitz, D. Turnbull (New York, Academic Press, 1977), p. 1; Hensel, J.C., Phillips, T.G., and Thomas, G.A., ibid, p. 88.
35. Haug, H., in : Advances in Solid State Physics, Vol. XXII, ed. P. Grosse (Braunschweig, Vieweg, 1982), p. 149.
36. Zimmermann, R. et al., phys. stat. Sol. (b) 90 (1978) 175.
37. Shank, C.V., Fork, R.L., Leheny, R.F. and Jagdeep Shah, Phys. Rev. Lett 42 (1979) 112.
38. Fehrenbach, G.W., Schäfer, W., Treusch, J. and Ulbrich, R.G., Phys. Rev. Lett. 49 (1982) 1281.
39. Asnin, V.M., Fiz. Tverd. Tela 15 (1973) 3298.
40. Jagdeep Shah, Leheny, R.F. and Wiegmann, W., Phys. Rev. B16 (1977) 1577.
41. Grun, J.B., J. Phys. Soc. Japan A49 (1980) 563.
42. Romanek, K.M., Göbel, E.O., Conzelmann, H. and Nather, H., ibid. p. 523; Romanek, K.M., Nather, H. and Göbel, E.O., Solid State Commun.

39 (1981) 23.

43. Pinczuk, A., Jagdeep Shah, Gossard, A.C., and Wiegmann, W., Phys. Rev. Lett. 46 (1981) 1341.

44. Conwell, E.M., "High Field Transport in Semiconductors", Solid State Physics Suppl. 9 (New York, Academic Press, 1967).

45. Collet, J., Cornet, A., Pugnet, M. and Amand, T., Solid State Commun. 42 (1982) 883.

46. Yoffa, E.J., Phys. Rev. B21 (1980) 2415.

47. Jagdeep Shah and Mattos, J.C.V., Proc. 3rd Int. Conf. on Light Scattering in Solids, ed. M. Balkanski, R.C.C. Leite, S.P.S. Porto (Paris, Flammarion, 1975), p. 145.

48. Compaan, A., Lo, H.W., Lee, M.C. and Aydinli, A., Phys. Rev. B26 (1982) 1079.

49. For a short review of phonon transport and decay in the context of carrier transport, see : Ulbrich, R.G., J. de Physique C7 (1981) 423.

50. Ulbrich, R.G., Narayanamurti, V. and Chin, M., Phys. Rev. Lett. 45 (1980) 1432.

RECOMBINATION MECHANISMS IN SEMICONDUCTORS

Michel VOOS

Groupe de Physique des Solides* de l'Ecole Normale Supérieure, 24, rue Lhomond, 75005 Paris, France.

ABSTRACT

We give here a brief survey of the main radiative recombination processes in semiconductors. The Auger non-radiative recombination mechanism is also considered.

1. INTRODUCTION

Electron-hole recombination in semiconductor crystals has been
the subject of an enormous number of experimental and theoretical
studies which have been very fruitful for our understanding of semiconductor physics and have led to important applications such as
semiconductor lasers. Clearly, so many articles have been published
on this subject up to now that it is not possible to review here all
the interesting results which have been reported. For this reason,
we have chosen, more or less arbitrarily, to cover only some aspects
of electron-hole recombination and to exclude some important topics
such as processes involving defects, dislocations or deep centers,
for example.

We consider, at first, radiative recombination processes of
free electrons and holes excited into an initial state corresponding
to any accessible region of the Brillouin zone. Through subsequent
interaction with the crystal lattice these carriers relax to states
at or near the bottom of the conduction band and the top of the valence band for electrons and holes, respectively. Relaxation to

* Laboratoire associé au C.N.R.S.

328

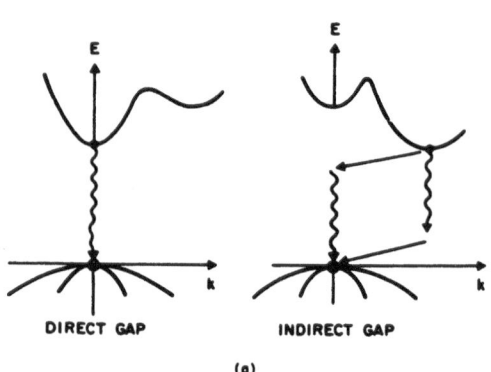

MOMENTUM SPACE

DIRECT GAP   INDIRECT GAP

(a)

Fig. 1a Schematized band to band radiative recombination for direct
(left) and indirect (right) band gap semiconductors : an electron
in the conduction band recombines with a hole in the valence band,
and a photon is emitted. For indirect gap material, band extrema
do not lie at the same point in k space, and the difference in crystal momentum is taken up by emission of a phonon, so that the energy
of the emitted photon is $h\nu = E_g - h\omega$, where E_g is the band gap
energy, and $\hbar\omega$ the phonon energy. Two possibilities for phonon-
assisted recombination processes are shown here.

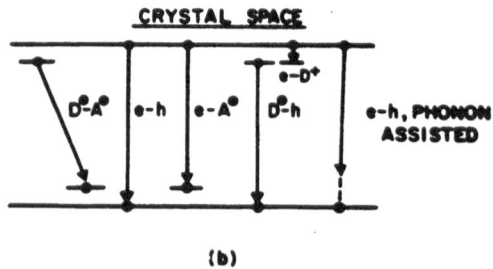

CRYSTAL SPACE

D^0-A^0 e-h e-A^0 D^0-h e-h, PHONON
ASSISTED

(b)

Fig. 1b Description in real space of various possible channels for
radiative recombination, including band-to-band mechanisms and processes involving impurities. Impurity states are located in the
band gap and can trap carriers which can then recombine radiatively
with free carriers (e-A^0, D^0-h) or with carriers bound to other
impurities (D^0-A^0).

the crystal ground state can then result from the radiative recombination of these excited carriers. Several possible mechanisms for radiative recombination are illustrated in Fig. 1 in the case of direct and indirect band gap semiconductors. In Fig. 1a a photon is emitted through band-to-band recombination in both cases. Considering that total momentum should be conserved and that a photon carries away negligible momentum, radiative recombination in an indirect band gap semiconductor implies that the usually large crystal momentum associated with a carrier, for instance at the band edge, is transferred to the lattice, typically through phonon emission. Fig. 1b illustrates some additional radiative recombination mechanisms involving shallow impurities (donors and/or acceptors) so that the energy of the emitted photon is, of course, smaller than the band gap E_g. Furthermore, the Coulomb interaction between a free electron and a free hole can lead to the formation of a free exciton, namely of a two-particle complex which can be viewed as a positronium -like bound state. The motions of the electron and hole paired into a free exciton is completely correlated and the energy of this state is reduced relative to that of the uncorrelated free carrier states by an amount which is the exciton binding energy R^*, thus equal to E_g-E_O if E_O is the free exciton energy. In addition to R^*, another important parameter for free excitons is their Bohr radius a_O. Excitons can also be bound to a shallow impurity in the case of slightly or moderately doped materials. More complex pairings of carriers have also been observed and studied. Indeed, with increasing density, free excitons interact and new bound states can occur. The first one, which has been proposed by Lampert [1] in 1958, is the excitonic molecule which results from the binding of two free excitons. The second one, which has been introduced by Keldysh [1] in 1968, is the electron-hole liquid corresponding to the condensation of free excitons, considered as a gaseous phase, into the so-called electron-hole drops. These two additional bound states are characterized, in particular, by their binding energy which is necessarily larger than that of free excitons. All these complexes can emit photons, as a result of the radiative recombination of the involved electrons and holes. It is frequent that, in a given experimental situation, the excited carriers relax radiatively through more than one of the processes briefly described above. The radiative recombination spectrum can thus be rich in details and gives a great deal of informations about the fundamental processes and interactions influencing carriers, being, as a consequence, a good method to characterize semiconductors.

Other important recombination processes, though less studied and less understood than the radiative ones, are the non-radiative recombination mechanisms. Among them, perhaps most relevant to laser annealing, is the Auger effect where excited electrons and holes recombine non-radiatively, the released energy being given to a third particle.

We will review here some fundamental aspects of radiative recombination in semiconductors, which, we think, are of particular
interest. In the case of non-radiative recombination, we will focus
on the Auger effect because of the essential importance of this mechanism at high excitation.

2. RADIATIVE RECOMBINATION IN SEMICONDUCTORS

From an experimental point of view, fundamental radiative recombination investigations require a source of excitation which is
very often a CW laser, a spectrometer, a detector followed by an
appropriate electronic equipment to process and to record the data.
Many studies have to be done as a function of temperature, from liquid helium temperatures up to room temperature, implying to use an
optical cryostat. It is also worth noticing that luminescence time
decay measurements can give interesting informations and, in this
case, pulsed lasers are very convenient to excite the samples under
study.

2.1 Band-to-band recombination

When a photon of energy greater than the band gap E_g is absorbed, free electrons and holes are created and the simplest radiative recombination mechanism in intrinsic semiconductors is the band-
to-band one, where a free electron recombines with a free hole
yielding a photon of energy $h\nu$, as schematized in Fig. 1a. The
corresponding luminescence spectrum depends on the carrier distribution in the conduction and valence bands and on the selection
rules for momentum conservation. In direct gap semiconductors, the
electron and hole wave vectors should be the same, the photon wave
vector being small enough to be neglected. If the conduction and
valence band extrema are at k=0 and assuming spherical surfaces of
constant energy, the luminescence spectrum $I_0(h\nu)$ is given by the
following relation [2], if the electron and hole effective masses
are constant :

$$I_0(h\nu) = A \nu^2 (h\nu - E_g)^{1/2} f_e f_h \qquad (1)$$

where A is a constant and $f_e (f_h)$ is the electron (hole) distribution
function. For Maxwell-Boltzmann distributions :

$$I_0(h\nu) = A \nu^2 (h\nu - E_g)^{1/2} \exp(-(h\nu - E_g)/k_B T) \qquad (2)$$

where T is the temperature.

Fig. 2 shows [2] a luminescence line corresponding to band-to-
band recombination in pure InSb at 4.2 K. The theoretical line shape is obtained from Eq. 1 and the agreement with theory is quite

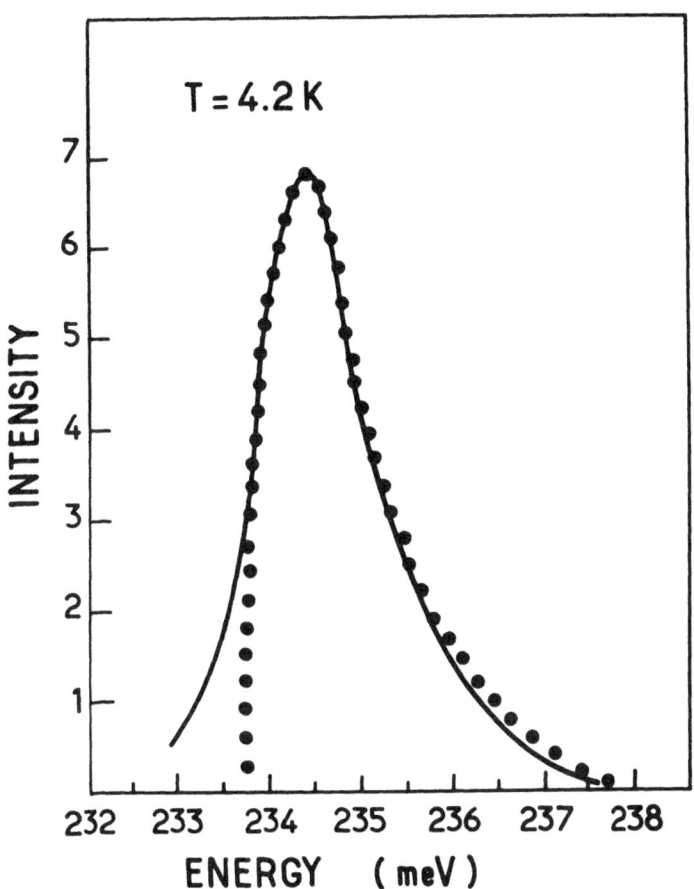

Fig. 2 Experimental (solid line) and theoretical (dotted line) band-to-band recombination line in InSb at 4.2 K (from [2]).

good. The low-energy cut-off gives of course the band gap E_g.

In indirect gap materials, the situation is more complicated because phonon emission is required to conserve the momentum, as illustrated in Fig. 1a. In doped materials, the conservation of momentum can also be insured through impurity scattering. The phonon being able to take on a range of momenta without appreciable change in energy, any electron can practically recombine with any hole. The luminescence spectrum is thus calculated by integrating over the electron and hole distributions [3] :

$$I_o(h\nu) = B\int_o^\infty \int_o^\infty n(E_e)n(E_h)f_e(E_e)f_h(E_h)\delta(h\nu-E_g-E_e-E_h+\hbar\omega)dE_e dE_h$$

(3)

where B is a constant ; $n(E_e)$, $(n(E_h))$ the density-of-states in the conduction (valence) band, $E_e(E_h)$ the electron (hole) kinetic energy, and $\hbar\omega$ the phonon energy.

At low temperatures and low or moderate excitation intensities, free electrons and holes become bound to impurities or bind into free excitons and the corresponding recombination dominates the luminescence spectrum. Band-to-band recombination becomes usually significant at temperatures corresponding to $k_B T$ larger than an energy E_b characteristic of the binding, depending of course on the material. Indeed, in this case, due to thermal dissociation, the population of electrons and holes in the conduction and valence bands is significant. Band-to-band luminescence has been detected in a number of semiconductors such as InSb [2], CdS [4], GaAs [5] for direct gap semiconductors; Ge [6] , Si [7] for indirect ones. The temperature range over which it can be observed is obviously different for each semiconductor but, to our knowledge, few experiments have been done at temperatures larger than 300 K. For instance, in Ge data have been obtained [6] up to 313 K.

The radiative annihilation of electron-hole drops in Ge and Si is an interesting example of band-to-band radiative recombination [8] . This condensed phase which occurs at low temperatures looks like a liquid metal and is constituted by a degenerate neutral two-component plasma of free electrons and holes in the conduction and valence bands, respectively. The energy of the liquid phase is smaller than that of free excitons and the occurence of electron-hole drops can be viewed, at least in first approximation, as a phase transition between the free exciton gas and liquid drops analogous to ordinary gas-liquid transformations. This is the only case where the above condition, $k_B T > E_b$, is not necessary to observe band-to-band recombination, because the liquid phase corresponds to the lowest bound state in usual intrinsic semiconductors. We are not going to discuss here this phase transition in detail but it provides a nice illustration of band-to-band recombination

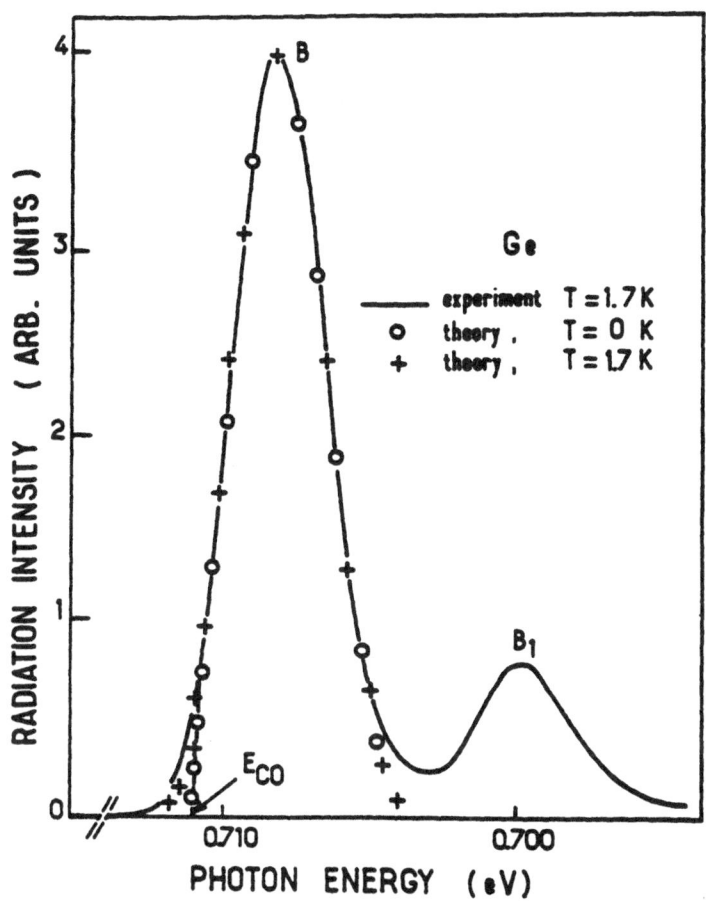

Fig. 3 Experimental shapes of the LA and TO phonon assisted radiative recombination lines of electron-hole drops in Ge. Also shown
are theoretical fits of the LA phonon assisted line (from [3]).

in indirect gap semiconductors. Fig. 3 shows [8] a typical luminescence spectrum at 1.7 K characteristic of electron-hole drops.
Two emission lines, B and B_1, can be seen and correspond to the radiative annihilation of electrons in the conduction band and holes
in the valence band with emission of an LA and a TO phonon, respectively. Indeed, Ge is an indirect gap material with [9] the valence band at k = 0 and four conduction bands at the boundaries of
the Brillouin zone in the (111) (L point) and equivalent directions,
so that phonons are emitted to conserve momentum. Also shown in
this figure are theoretical fits [8] of the B line shape at two temperatures using Eq. 3 with n_o, the electron-hole pair density,
equal to 2.4×10^{17} cm^{-3}, which fixes obviously the electron and hole
Fermi energies in the conduction and valence bands.

To conclude, we would like to emphasize the following general
features : i) At high temperatures (T>300K) one can expect to detect
essentially band-to-band recombination in usual semiconductors;
ii) Even at rather low temperature, band-to-band recombination is
likely to be observable, provided the excitation is sufficiently
high.

2.2 Free exciton recombination

In semiconductors, free excitons are described in the Wannier-
Mott approximation [10] , free electrons and holes being treated as
nearly independent particles with opposite charges and submitted
to Coulomb interaction. This results in a lowering of the energy
of the electron-hole pair bound in a free exciton with respect to
E_g. The exciton binding energy is [11]:

$$R^* = E_g - E_o = \frac{\mu e^4}{2\hbar^2 \varepsilon^2} \tag{4}$$

where E_o is the ground state energy of a free exciton, ε is the dielectric constant and $1/\mu = 1/m_e^* + 1/m_h^*$, m_e^* (m_h^*) being the appropriate electron (hole) effective mass. Excitons have also excited
states similar to those of hydrogen-like atomic systems up to the
continuum of states in the single-particle band of the semiconductor. Furthermore, free excitons can move in the crystal with an
associated mass M equal to ($m_e^* + m_h^*$). The total exciton energy is
thus $E(k) = E_o + \hbar^2 k^2 / 2M$, where k is here the exciton wave vector.

In direct gap semiconductors, free excitons can interact with
photons to form a mixed state, the polariton [12-14] . The exciton
state is split into longitudinal and transverse excitons due to the
exchange interaction. Only the transverse exciton couples to photons, resulting in two coupled modes generally known as the upper
and lower polariton branches (UPB and LPB), as schematized in Fig.4.
The upper polariton branch energy at k=0 is the longitudinal exciton one, E_L, the transverse exciton lying at a lower energy E_T.

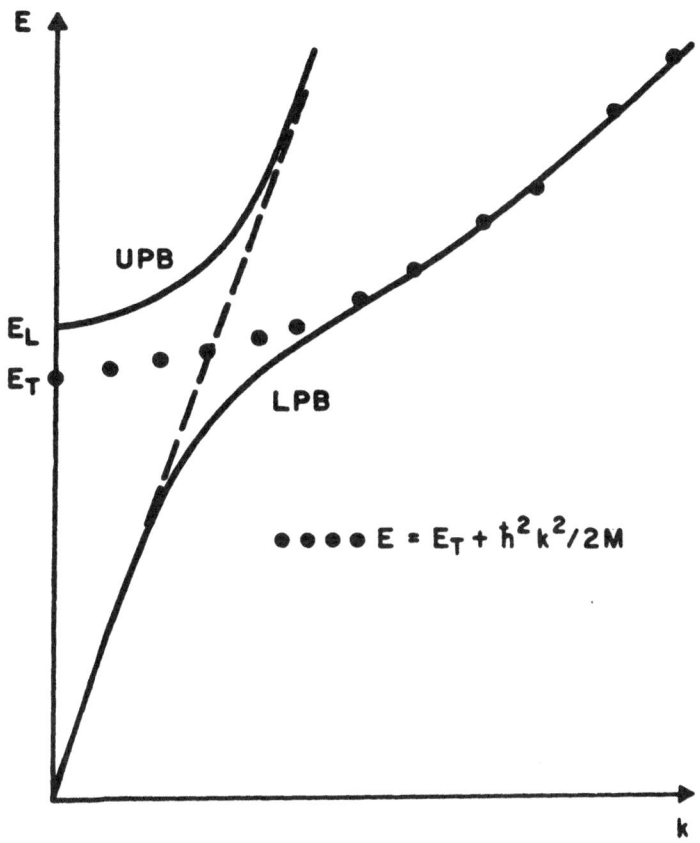

Schematic polariton curve showing the coupling between free excitons (dotted curve) and photons (dashed line) whose dispersion relations are respectively $E = E_T + \hbar^2 k^2/2M$ and $\hbar c k / \varepsilon^{1/2}$.

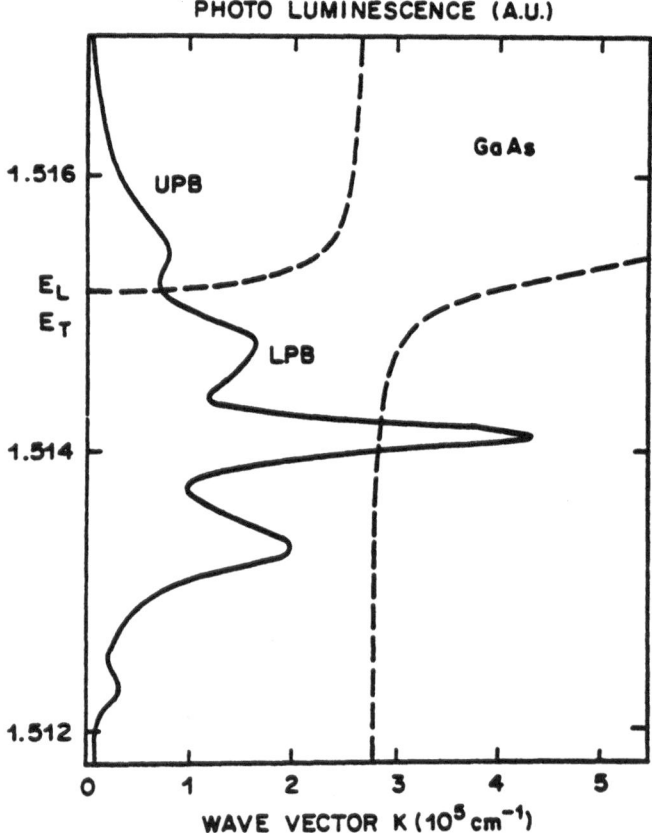

Fig. 5 Polariton luminescence spectrum is GaAs at 2 K (solid curve) corresponding to the UPB and LPB. The dashed lines represent the polariton dispersion curve. The low energy part of the spectrum corresponds to luminescence involving impurities (from [19]) .

Excitation of free excitons in a crystal corresponds to the transformation of a photon at the surface in a polariton-mode propagating in the semiconductor. Radiative recombination corresponds to
the inverse mechanism : a polariton which propagates toward the
semiconductor surface transforms into a photon. Polariton luminescence has been observed in many materials at low temperatures :
CdS [15] , GaAs [16] , CuCℓ[17] , ZnSe [18] , for example. A typical spectrum [19] is given in Fig. 5 in the case of GaAs. Note,
finally, that the calculation of the polariton lineshape is complicated and this problem is in fact not yet completely solved.

In indirect gap semiconductors, the situation is simpler because the large value of the momentum due to the band structure pre-
vents mixed states to occur, and the polariton formalism cannot
apply in this case. The electron and the hole involved in a free
exciton can recombine radiatively with emission of a phonon to conserve momentum, as in band-to-band recombination. In pure Ge, for
instance, the luminescence spectrum [20,21] is dominated at low
temperatures by emission lines corresponding to the radiative annihilation of free excitons with emission of LA, TO and TA phonons.
Each line occurs at an energy $E_0 = E_g - R^* - \hbar\omega$, where $\hbar\omega$ is the
energy of the phonon involved in the observed spectrum ($R^* = 4.15$
meV in Ge). The LA phonon-assisted emission line of free excitons
in Ge is shown [22] in Fig. 6. The line shape is easy to calculate
and is given by [22] :

$$I_0(E) = C(E-E_0)^{1/2} \exp (-(E-E_0)/k_B T) \qquad (4)$$

where C is a constant, and $E = h\nu$ the energy of the emitted photon.
Some discrepancy between experiment and theory can be seen in Fig.6.
The agreement can be improved by introducing a Gaussian broadening,
$\delta = 0.31$ meV in the theoretical line shape which is then given
by I(E) in Fig. 6 and by taking into account the effect of the
electron mass anisotropy on the degeneracy of the hole involved in
the exciton. This effect results in a splitting [23] of the exciton ground state into two levels distant [24] of ~1 meV. The luminescence spectrum corresponding to this split level lying at higher
energy can be calculated according to Eq. 4 and is given in Fig. 6
by I'(E). In Fig. 6 is also shown [22] I(E) + I' (E), and it can
be seen that the agreement between experiment and theory is satisfying, yielding E_0 = 713.2 meV. In Si [20] , the situation is comparable, R^* being equal to 14.7 meV [25] .

Finally, it is worth mentioning that the radiative recombination
of free excitons has been observed in a number of other indirect
gap semiconductors such as GaP [26] and AgBr [27] for instance.
In direct as well as in indirect gap materials, free excitons are
usually studied at low temperatures but they can be expected to
be observable up to temperatures corresponding roughly to $k_B T \sim R^*$.

338

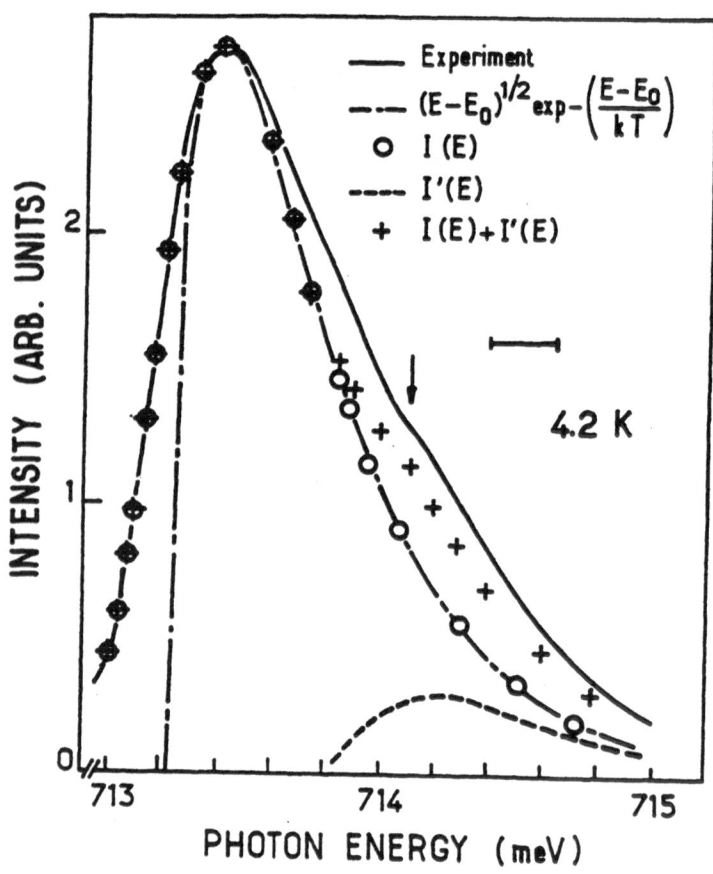

Fig. 6 Experimental and theoretical shapes of the LA-phonon-assisted recombination line of free excitons in Ge (from [22]).

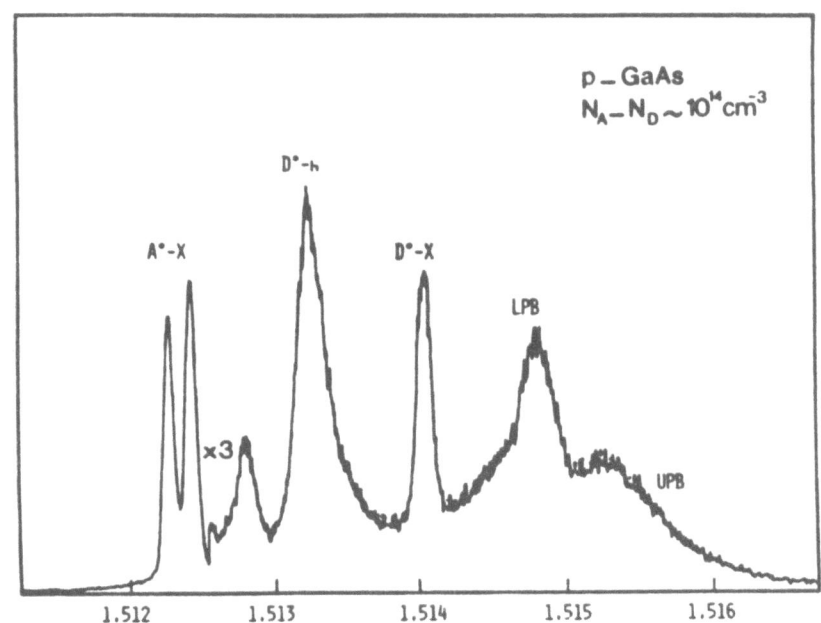

PHOTOLUMINESCENCE (A.U)

p ─ GaAs
$N_A - N_D \sim 10^{14} cm^{-3}$

D°-h

A°-X

D°-X

LPB

×3

UPB

1.512   1.513   1.514   1.515   1.516

Fig. 7   Typical luminescence spectrum of GaAs at 2K   (From [19]).

2.3   Band—to—impurity transitions

In Sections 2.1 and 2.2 are described intrinsic recombination
processes but, at low temperature and excitation intensity, they
are seldom predominant in luminescence spectra. Indeed, impurities
are generally unavoidable in most semiconductors and, even at low
impurity densities, they have often an important influence on the
emission spectra. For example, Fig. 7 shows [19] a typical radiative recombination spectrum obtained at 2K in p-type GaAs with
$N_A - N_D \sim 10^{14} cm^{-3}$, where $N_A (N_D)$ is the acceptor (donor) concentration. Except for the high energy emission lines LPB and UPB, which
are respectively the lower and upper polariton branches, all the
spectrum corresponds to recombination transitions involving impurities. The understanding of these extrinsic recombination processes
is important to obtain information on impurities in semiconductors.

We consider here only shallow impurities, donors and acceptors
which can be generally described in the effective mass approximation [28] . When free carriers are excited, one can expect to observe, at low temperatures, luminescence corresponding to the radiative recombination of a free electron with the hole of a neutral

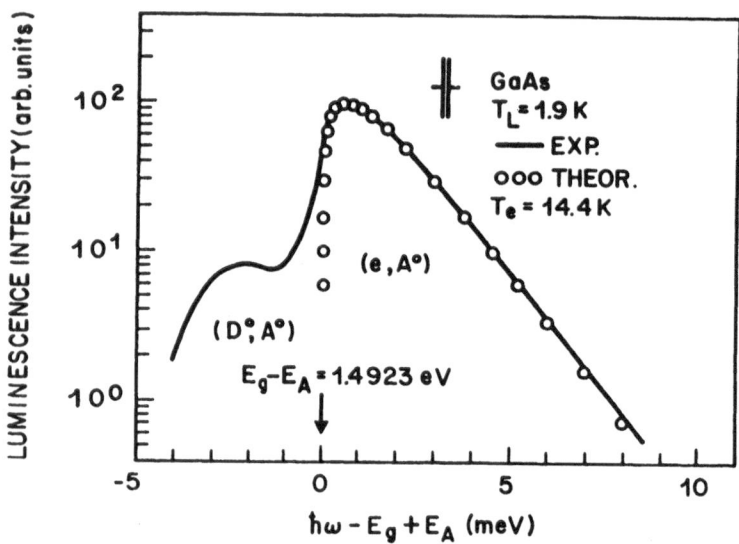

<u>Fig. 8</u> Experimental and theoretical line shapes for the (e,A°)
mechanism in GaAs (from [31]).

acceptor, while the free hole created in the valence band is then
captured by the acceptor. Similar radiative transitions between
neutral donors and free holes can be also detected. These transitions are usually called (e-A°) and (D°-h), respectively. For example, the (D°-h) transition can be seen in Fig. 7.

The line shape corresponding to these recombination mechanisms
has been calculated by several authors [29,30] and is, in fact,
quite analogous to that of band-to-band recombination, except that
it involves only one free carrier distribution function instead of
two. Fig. 8 presents the experimental and theoretical line shapes
[31] in the case of the (e-A°) transition as obtained in pure GaAs
at 1.9 K. The low-energy cut-off of the emission line gives (E_g-E_A)
where E_A is the acceptor binding energy and can be used to determine E_A. From the high-energy part of the line, one can deduce
the free carrier temperature which can be different from that of
the lattice and is here T_e = 14.4 K for the electrons.

2.4 Donor-acceptor pair transitions

When both donors and acceptors are present in a semiconductor, one can expect to observe donor-acceptor pair radiative transitions [32] where an electron bound to a donor recombines with a hole bound to an acceptor, yielding thus an emission line at lower energy than the previous (e, A°) and (D°, h) recombination mechanisms. The corresponding recombination probability depends of course on the overlap of the electron and hole wavefunctions and is rather small, because donors and acceptors can be somewhat remote. However, at temperatures such that $k_B T < E_A$ and E_D, where E_D is the donor binding energy, the carriers cannot escape from the impurities, and donor-acceptor transitions, often called (D°, A°) transitions, become an important recombination process.

In the initial state, impurities are neutral while they are ionized in the final state. Due to Coulomb interaction between the ionized impurities, the energy of the final state is lowered, leading to an energy of the emitted photon increased by the same amount. The photon energy is thus given by :

$$h\nu = E_g - E_A - E_D + \frac{e^2}{\varepsilon r} \qquad (5)$$

where ε is the dielectric constant, e the electron charge, and r the distance between donors and acceptors. The photon energy depends therefore on r and, for substitutional impurities, r changes discretely, so that one can observe a number of discrete sharp lines. For example about 300 lines have been detected [33,34] in GaP. For large values of r, one ends up with a quasi-continuum and one can only observe a broad band which, for instance, has also been put into evidence in GaP. Fig. 9 gives typical spectra obtained [35] in GaP for different impurities, showing simultaneously sharp lines due to close donor-acceptor pairs and a broad band due to distant pairs.

The radiative recombination probability depends, as already mentioned, on the overlap of the wavefunctions of the electron and the hole at a donor and an acceptor, respectively. Thus, it depends on r so that it decreases with increasing r, corresponding to decreasing photon energy. This results in two important features which are, in fact, clear signatures of this recombination mechanism :
i) When the excitation level is increased, the luminescence spectrum shifts to high energy as the low energy states, corresponding to large r, are saturated;
ii) In time-resolved luminescence experiments under pulsed excitation, the spectral weight shifts to low energy, corresponding to large r, when the time delay between the exciting pulse and the measurement time is increased.

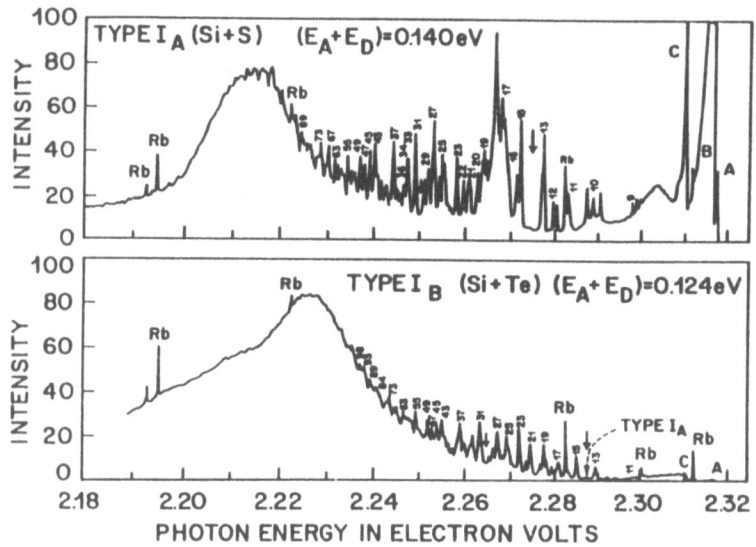

<u>Fig. 9</u> Donor-acceptor pair recombination spectra in GaP at 1.6 K. The lines labeled Rb correspond to calibration markers (from [35]).

Donor-acceptor pair recombination has been first identified [33] in GaP, and is now recognized as being a very important mechanism at low temperature in a number of semiconductors. It has, for example, been put into evidence unambiguously in GaAs [36] (see Fig. 8), InP [37], CdS [38], ZnS [39]. In Si, the broad band at low energy has been observed [40] in 1969 but the sharp line series has only been reported quite recently [41].

2.5 <u>Bound exciton recombination</u>

Another important extrinsic radiative recombination mechanism is the radiative annihilation of bound excitons. These complexes can be formed by the capture by an impurity of an electron or a hole followed by the capture of the oppositely charged particle, namely a hole or an electron, respectively. They can also be formed by direct capture of a free exciton. Bound excitons which can occur on neutral or ionized donors and acceptors have been first observed in Si [42].

At low temperatures, many semiconductors exhibit rather sharp lines due to the radiative recombination of an electron and hole involved in a bound exciton. For example, Fig.7 shows [19] emission lines, D°-X and A°-X, corresponding to the recombination of excitons bound to neutral donors and acceptors in GaAs, respectively. Other semiconductors present analogous features due to the luminescence of bound excitons : CdS [43] , GaSb [44] , GaP [45] ... Bound exciton luminescence lines are usually very sharp because no free particle is involved and there is no corresponding dispersion relation. The best experimental technique used to identify bound excitons has been, up to now, magneto-luminescence [43] where one takes advantage of the Zeeman effect. Bound exciton luminescence is generally observed only at low temperatures because the corresponding binding energy is usually small.

An interesting aspect of bound excitons is that Auger non-radiative recombination is likely to occur since three particles are involved. An electron and a hole recombine and the released energy is given to the third carrier. This has been put into evidence in the case [46] of sulfur donors in GaP for instance.

Bound excitons have been the object of a very large number of optical studies, and present many features that cannot be developed here. In particular one should mention excitons bound to isoelectronic impurities [47] which have been extensively investigated. Finally, it is certainly worth noting that the luminescence associated to the extrinsic radiative recombination mechanisms briefly described in this article tend to disappear when T is raised above the relevant binding energies and tend to saturate when the excitation is increased because the number of impurities is limited.

3. NON-RADIATIVE AUGER RECOMBINATION

The Auger effect is not the only non-radiative recombination mechanism taking place [48] in semiconductors, but it can be a very important one at high excitation and high temperature, i.e. in situations where most of the excited carriers lie in the conduction and valence band. There are several possibilities [49] of achieving Auger recombination and we consider here a rather simple one implying three particles : a free electron recombines with a free hole, the corresponding energy being given to a second electron or hole, as schematized in Fig. 10. Energy and momentum should be conserved, and there should be in the semiconductor band structure some appropriate state available for the third particle. Theoretical calculations [48] relative to this Auger recombination process will not be presented here because they are complicated and would require long developments.

344

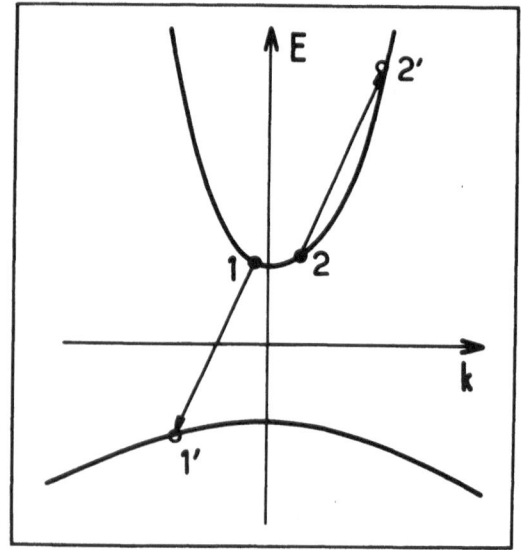

Fig. 10 Schematized Auger effect. A free electron (1) recombines non-radiatively with a free hole (1'), and the released energy is given to a second electron (2) which is sent to another available state (2') in the band structure.

If a density n of free electrons and holes is created by pulsed excitation, the decay rate of this population can be written as follows in the simplest situation :

$$\frac{dn}{dt} = -\frac{n}{\tau} \tag{6}$$

where τ is the total carrier lifetime given by $1/\tau = 1/\tau_R + 1/\tau_{NR}$, τ_R and τ_{NR} being the radiative and nonradiative lifetimes. For the Auger effect under consideration [49] here : $\tau_{NR} = 1/Cn^2$ so that at high excitation, the non-radiative term is usually dominant and $\tau \simeq \tau_{NR}$, leading to $dn/dt \simeq -Cn^3$, where C is the Auger recombination coefficient. Thus, in a luminescence decay time experiment under pulsed excitation corresponding to this situation, the decay rate, obtained from the observation of the time dependence of the luminescence signal emitted after the exciting pulse, is essentially governed by the Auger non-radiative recombination. Note that this implies that this kind of experiment is generally difficult because the luminescence is weak since the radiative recombination is only a small fraction of the total recombination.

Interesting results have been obtained [50] recently in undoped Si at 300 K from very careful luminescence decay rate measurements under high excitation provided by a pulsed laser. Clearly, in this case, the radiative recombination mechanism is essentially a band-to-band one and this work shows that the non-radiative recombination process involved in the experimental situation under consideration corresponds certainly to an Auger effect. Analysis of the data yields $C = 3.4 \times 10^{-31}$ cm^6 xs^{-1} consistent with previous values [51]. One can also compare this result with that obtained in the case of electron-hole drops in Si. As already mentioned drops occur only at low temperatures and their radiative recombination corresponds to a band-to-band recombination process. It is generally admitted that their non-radiative recombination which is dominant involves a three-particle Auger effect and, from the experimental data available [8], $C \sim 5 \times 10^{-31}$ cm^6 xs^{-1} at about 2 K, in satisfying agreement with the previous value. However, we think that one cannot consider that all the aspects of the Auger recombination are fully understood.

4. CONCLUSIONS

We wish to emphasize again that this brief review of recombination mechanisms in semiconductor crystals is uncomplete. Even for the topics discussed here, many available results have not been discussed and references are missing. This was unavoidable because this field of semiconductor physics has been very extensively studied for about the last 25 years.

Finally, we would like to stress that, in crystalline semiconductors at high temperatures and excitation levels, if luminescence can be detected, it is likely to be due to band-to-band recombination and also that one can expect the Auger non-radiative recombination mechanism to be important in such experimental conditions.

ACKNOWLEDGMENTS

The author would like to thank B. Etienne, Y. Guldner and P. Voisin for very fruitful discussions.

REFERENCES

1. Lampert, M.A., Phys. Rev. Lett. 1 (1958) 450;
    Keldysh, L.V., Proc. 9th Int. Conf. on Semicond. Phys.,
    Moscou (Nauka, Leningrad) 1968, p. 1303.
2. Mooradian, A. and Fan, H.Y., Phys. Rev. 148 (1966) 873.
3. See, for example, Benoit à la Guillaume, C. and Voos, M.,
    Phys. Rev. B7 (1973) 1723.
4. Reynolds, D.C., Phys. Rev. 118 (1960) 478.
5. Shah, J. and Leite, R.C.C., Phys. Rev. Lett. 22 (1969) 1304.
6. Haynes, J.R. and Nilson, N.G., in Radiative Recombination in
    Semiconductors (Dunod, Paris) 1964, p.21.
7. See, for example, Nilsson, N.G. and Svantesson, K.G., Sol. St.
    Commun., 11 (1972) 155.
8. For a review see, for example, Voos, M. and Benoit à la Guillau-
    me, C., in Optical Properties of Solids - New Developments,
    ed. B.O. Seraphin (Amsterdam, North-Holland, 1976), p. 143.
9. Herman, F., Phys. Rev. 95 (1954) 847.
10. See, for example, Knox, R.S., in Solid State Physics, Suppl.
    5, eds Seitz and Turnbull (New York, Academic Press, 1963).
11. For a discussion of the theory of excitons see, for instance,
    Knox, R.S. in Ref. 10; Elliot, R.J., in Polarons and Excitons,
    eds. Kuper and Whitfield (New York, Plenum Press, 1963), p.269;
    Dimmock, J.O., in Semiconductors and semimetals, Vol. 3, eds.
    Willardson and Beer (New York, Academic Press, 1967), p. 270;
    Barry, H., Bebb and Williams, E.W., ibid., Vol. 8, 1972, p.181.
12. Pekar, S.I., Sov. Phys. JETP 6 (1958) 785.
13. Hopfield, J.J., Phys. Rev. 112 (1958) 1555.
14. Hopfield, J.J. and Thomas, D.G., Phys. Rev. 132 (1963) 563.
15. Thomas, D.G. and Hopfield, J.J., Phys. Rev. 116 (1959) 573.
16. Weisbuch, C. and Ulbrich, R., Phys. Rev. Lett. 39 (1977) 654.
17. Suga, S., Cho, K., Heisinger, P. and Koda, T., J. Lumin. 12/13
    (1976) 109, and references therein.
18. Sermage, B. and Voos, M., Phys. Rev. B15 (1977) 3935.
19. Weisbuch, C., Thesis, Paris, 1977, unpublished.
20. Haynes, J.R., Lax, M. and Flood, W.F., J. Phys. Chem. Solids
    8 (1959) 392.
21. Benoit à la Guillaume, C. and Parodi, O., J. Electron. Control
    6 (1959) 356.
22. See, for example, Benoit à la Guillaume, C. and Voos, M., Sol.
    St. Commun. 12 (1973) 1257.
23. Zerdling, S.B., Lax, B., Roth, L.M. and Button, K.J., Phys. Rev.

114 (1959) 80.

24. Gross, E.F., Safarov, V.I., Titkov, A.N. and Shlimack, I.S., JETP Lett. 13 (1971) 235.
25. Shaklee , K.L., Leheny, R.F. and Nahory, R.E., Phys. Rev. Lett. 26 (1971) 888.
26. Wight, D.R., J. Phys. C1 (1968) 1759.
27. Von der Osten, W. and Weber, J., Sol. St. Commun. 14 (1974) 1133.
28. Kohn, W., in Solid State Physics, Vol. 5, eds Seitz and Turnbull (New York, Academic, 1957) p. 257.
29. Eagles, D.M., J. Phys. Chem. Solids 16 (1960) 76.
30. Dumke, W.P., Phys. Rev. 132 (1963) 1998.
31. Ulbrich, R., Phys. Rev. B8 (1973) 5719.
32. For a review, see Dean, P.J., in Progress in Solid State Chemistry, Vol. 8, eds. McCaldin and Somorjai (New York, Pergamon Press, 1973).
33. Hopfield, J.J., Thomas, D.G. and Gershenzon, M., Phys. Rev. Lett. 10 (1963) 162.
34. Trumbore, F.A. and Thomas, D.G., Phys. Rev. 137 (1965) A1030.
35. Thomas, D.G., Gershenzon, M. and Trumbore, F.A., Phys. Rev. 133 (1964) A269.
36. Leite, R.C.C. and Digiovanni, A.E., Phys. Rev. 153 (1967) 841.
37. Leite, R.C.C., Phys. Rev. 157 (1967) 672.
38. Colbow, K., Phys. Rev. B 139 (1966) 274.
39. Kukimoto, H., Shionoya, S., Koda, T. and Hioki, R., J. Phys. Chem. Solids 29 (1968) 935.
40. Kaminskii, A.S. and Pokrovskii, Y.E., Sov. Phys. Semicond. 3 (1969) 1496.
41. Ziemelis, U.O., Parsons, R.R. and Voos, M., Sol. St. Commun. 32 (1979) 445.
42. Haynes, J.R., Phys. Rev. Lett. 4 (1960) 361.
43. Thomas, D.G. and Hopfield, J., Phys. Rev. 128 (1962) 2135.
44. Johnson, E.J. and Fan, H.Y., Phys. Rev. 139 (1965) A 1991.
45. Thomas, D.G., Gershenzon, M. and Hopfield, J.J., Phys. Rev. 131 (1963) 2397.
46. Nelson, D.F., Cuthbert, J.D., Dean, P.J. and Thomas, D.G., Phys. Rev. Lett. 17 (1966) 1262.
47. For a review, see, for example, Dean, P.J., J. Lumin. 1/2 (1970) 398.
48. For more informations, see the Proceedings of the International Conference on Recombination in Semiconductors, Southampton, 1978, Sol. St. Electronics 21 (1978).
49. Haug, A., Sol. St. Electron. 21 (1978) 1281; Landsberg, P.T. and Robbins, D.J., ibid. 21 (1978) 1289.
50. Svantesson, K.G. and Nilsson, N.G., Sol. St. Electron. 21 (1978) 1603.
51. See, for example, Woerdman, P., Amsterdam, Thesis, 1971, unpublished; Nilsson, N.G. and Svantesson, K.G., Sol. St. Commun. 11 (1972) 155; Beck, J.D. and Conradt, R., ibid. 13 (1973) 93; Dziewior, J. and Schmid, W., Appl. Phys. Lett. 31 (1977) 346.

GENERATION, DIFFUSION AND RELAXATION OF DENSE PLASMAS IN SEMICONDUCTORS

Arthur L. Smirl

Center for Applied Quantum Electronics
Department of Physics, North Texas State University
Denton, Texas 76203 U.S.A.

ABSTRACT

In this article, we review the recent (since 1980) use of two-pulse-excitation-and-probe and three-pulse transient grating techniques to study the nonequilibrium carrier dynamics and active non-linearities in crystalline germanium at excitation levels approaching damage. An investigation of the dynamics of the Moss-Burstein shift of the absorption edge allows the identification of the onset of intervalence-band absorption, indicates a slow carrier cooling and suggests that nonlinear diffusion mechanisms are important at these high carrier densities. A simple parametric approach to the analysis is described that circumvents knowing the coupling constants at these carrier densities and that provides agreement with the tendencies in the data.

1. INTRODUCTION

In this lecture, we shall attempt to provide an introduction to the dynamics of high density laser-induced electron-hole plasmas in semiconductors. By high density, we shall often mean carrier densities in excess of 10^{20} cm^{-3}. In these cases, the optical excitation levels are within a factor of two or three of sample damage or melting threshold. We also remark at the outset that we shall study the response of crystalline (not amorphous) material.

Introductions to the fundamentals of optical excitation, diffusion and recombination have been provided by Ulbrich [1] , Dewel [2] and Voos [3] respectively, in earlier chapters of this same volume. Consequently, we shall assume that the reader is familiar

with those discussions and, instead, attempt to provide an introduction to picosecond spectroscopic techniques. Specifically, we
shall review a series of simple saturation and transient grating
measurements in Ge to illustrate, in a tutorial way, the difficulties in interpreting and extracting information from such experiments.

In general, in the experiments to be reviewed here, the direct
absorption of intense picosecond pulses with a wavelength of 1.06 µm
in Ge (where the photon energy $\hbar\omega_o$ is greater than the direct band
gap energy E_g) will induce large numbers of electrons to make transitions from the valence band to states high in the conduction band,
leaving behind holes in the valence band. Following such an absorption process, the photo-excited electrons are left with an excess
energy ΔE_e. These energetic electrons then will quickly relax by
various collisional processes (that have been reviewed in detail
by Ulbrich [1] and Voos [3] in earlier chapters) to the bottom of
the conduction band, where eventually they will recombine with holes
that have similarly relaxed to the top of the valence band. Also,
because $\hbar\omega_o > E_g$, the absorption depth for 1.06 µm light in Ge is
on the order of one micron. Consequently, the optically-generated
carriers are initially created near the front surface of the sample,
and they diffuse into the sample bulk while loosing their excess
energy to the lattice.

From the beginning, the motivation for these studies then has
been to determine :
(a) the number of excess electron-hole pairs created by the laser
    energy absorbed,
(b) the fraction of laser energy transferred directly to the carriers,
(c) the fraction eventually transferred to the lattice,
(d) the time scale on which this energy transfer occurs,
(e) and, since the carriers are initially deposited near the front
    surface of the sample, how far the carriers diffuse into the
    bulk while depositing their excess energy to the lattice.

After all, it is this balance between cooling rate and diffusion
rate that determines the energy per unit volume deposited in the
sample - and, thus, ultimately the damage or melting threshold.
We emphasize that we wish to answer these questions when the excitation levels are near (or above) the sample melting or damage
threshold. Since several other authors in this volume will discuss
results above the threshold for a phase transition, we shall limit
our discussions to carrier dynamics below this threshold. At these
excitation levels, nevertheless, we can produce carrier concentrations in excess of 10^{20} cm^{-3}. At these carrier densities and on
picosecond time scales, the complicated dynamics of the photogenerated carrier distribution are determined by the simultaneous interaction of a large number of processes. Moreover, on short time
scales, the carrier temperature will be elevated and as the carriers

give their excess energy to the lattice, the lattice temperature
will rise. All scattering rates are expected to be modified by the
effects of screening at these carrier densities. In addition, the
band structure will be altered by the large carrier densities and
the elevated lattice and electron temperatures (e.g. the bandgap
will narrow). Similarly, the diffusion rate will be modified by
the effects of screening, bandgap narrowing, and elevated lattice
and carrier temperatures (as we shall later discuss). In short,
this is a very difficult problem and, to this point, we have achieved
our objectives only in a very limited and qualitative sense.

Prior to 1980, several attempts had been made to separate and
identify the contributions of the various processes. Indeed, past
studies have demonstrated the importance of parametric coupling
[4-6] , Auger Recombination [7,8] , phonon-assisted carrier cooling,
[9,10] , intervalence band absorption, [8,11,12] , diffusion, [13]
and the dynamic Moss-Burstein shift of the absorption edge [5,8,9]
in determining the optical response of germanium. (For a survey of
this and other early work and for a summary of our understanding of
these interactions as of 1980, the reader may find the reviews by
Smirl [14,15] of interest).

In spite of the considerable effort expended during this period
(1974-1980), serious questions remained regarding the picosecond
dynamics of the photo-excited excess carriers and the precise mechanisms by which they relax. This was partially because this problem
was by its very nature a complicated many-body problem involving
the simultaneous interaction of a large number of processes but primarily because there was very little reliable data on which to base
a theoretical model. Before 1980 (to the author's knowledge), the
active medium in the laser systems used in these studies was neody-
mium-doped silicate glass. In short, these systems had a low repetition rate (typically one laser firing per minute for our systems),
and they always produced a multimode transverse profile when properly
mode-locked. These characteristics made data acquisition a tedious
and exasperating procedure. Moreover, because of the multimode nature
of the beam profile, the pulse fluence could not be accurately determined, regardless of the calibration procedure adopted. All fluences
reported in these early studies (including our own) should be viewed
with some skepticism. Because of the remaining questions regarding
the carrier dynamics and primarily because of the development of
improved laser glasses that allow construction of higher repetition
rate lasers at 1.06 μm with much improved beam characteristics, we
chose in 1980 to re-examine the excitation-and-probe response of
germanium. This has proved a fortunate choice. We have time-resolved many features in the germanium response not previously observed
or reported. For the purposes of this review, then, we shall res-
trict our discussions primarily to our own more recent (since 1980)
measurements and analysis, where all data was obtained using pulses
with a well-calibrated Gaussian spatial profile.

The organization of this review is as follows. Following a brief introduction to the generic experimental technique in Sec. 2, we then divide our review of the experiments and analysis into two parts. In the first of these (Sec. 3), we describe our use of the two-pulse excitation-and-probe technique in an extensive study of thin crystalline Ge wafers as a function of temperature, excitation level, sample thickness, and time delay. We also introduce a simple parametric analysis that produces agreement with all tendencies in the data but does not assume apriori knowledge of the coupling constants. Finally, (Sec. 4), we describe the use of a three-pulse transient grating technique to isolate the effects of nonlinear diffusion.

2. EXPERIMENTAL TECHNIQUES

In all of the studies to be reviewed here, we have used a variation of the so-called excitation-and-probe technique. Here, the germanium sample is first irradiated with an intense optical excitation pulse that causes a change in the transmission or reflection properties of the sample. This initial pulse is then followed, at some later time, by a much weaker probe pulse that monitors the change in transmission or reflection as the Ge returns to its equilibrium condition. Throughout the studies discussed here, we have used a number of modifications to this simple scheme. The details of each configuration will be provided as they are required.

The actual experimental apparatus that is common to both the two-pulse excitation-and-probe studies to be described in Sec. 3 and the three-pulse transient-grating experiments to be described in Sec. 4 is shown schematically in Fig. 1. The excitation source for the two-pulse experiments was mode-locked Nd : phosphate glass laser that produced pulses at 1.054 μm and for the transient grating measurements a mode-locked Nd:YAG laser at 1.064 μm. Consequently, for practical purposes, the excitation wavelengths of the two sources were equal. By careful alignment and aperturing, both lasers consistently produced pulses with a single transverse Gaussian spatial mode. An electro-optic shutter (EOS) selected a single pulse from the mode-locked train of pulses. Following amplification, this pulse was directed to a beamsplitter where it was divided into two parts, one part delayed with respect to the other, and the two parts (excitation and probe) recombined after focusing on the surface of a thin Ge wafer that was mounted in a closed-cycle refrigerator. All beam profiles (both before and after focusing) were determined by scanning the beam with a small pinhole and were checked by observing the profile on a vidicon detector. For the transient grating studies to be described in Sec. 4, a portion of the initial pulse was focused into a cell containing benzene to provide a third, time-delayed probe pulse at 1.55 μm by stimulated Raman scattering. This path was blocked in the initial two-pulse experiments described in

352

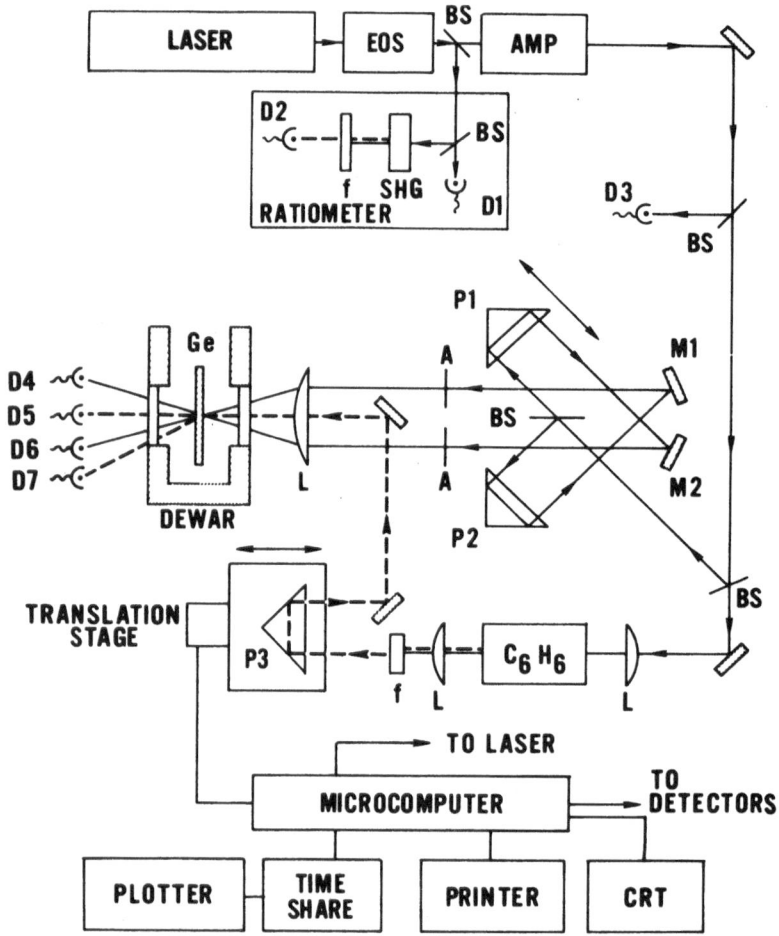

<u>Fig. 1</u> Schematic of experimental apparatus.

Sec. 3. Notice that a ratiometer [16] was provided to measure the pulsewidth on a shot-to-shot basis and that the entire apparatus was computer controlled.

The samples were in all cases produced from a 2 mm thick by 5 cm diameter slice of Czochralski-grown single-crystal germanium (ρ_{min} = 40 Ω-cm), cut with the $\langle 111 \rangle$ plane as the face. The crystal was mechanically polished on one side, etched and bonded to a fused-silica substrate with a cement that is highly transparent at 1.06 μm and that has approximately the same thermal expansion coefficient as germanium. The second surface was then polished until the sample thickness was approximately 10 μm and then etched with Syton HT-30 until the thickness was 5.7 μm, as determined by interferometric techniques. This was the "standard" thickness used in most studies reviewed here. When thinner samples were required, we subsequently ion-milled the sample described above to the desired thickness and once again etched the milled surface.

3. INTERBAND SATURATION

In this section, we review recent (since 1980) experiments in which we have used the simplest of the excitation and probe techniques in an extensive parametric study of thin wafers of single-crystal germanium as a function of sample temperature, sample thickness, sample surface preparation, excitation pulse energy, and time delay between the excitation and probe pulses using a Nd-glass laser with much improved beam characteristics. For these two-pulse experiments, the measured excitation beam diameter was 200 μm (FWHM) at the crystal surface; the probe beam was half this size to ensure that the center of the excited area was monitored. The maximum excitation fluence was 100 mJ/cm^2, corresponding to a peak irradiance of approximately 13 GW/cm^2. Although the widths of the optical pulses were regularly measured, the ratiometer shown in Fig. 1 was not employed in most of the studies described in this section.

Let us begin by considering the simplest possible saturation experiment. That is, initially, the probe pulse was blocked (as shown in the inset of Fig. 2), and the transmitted excitation-pulse energy was measured as a function of increasing incident pulse energy to determine the irradiation level required to induce a change in the transmission of the Ge sample. The sample transmission is seen to begin at its constant linear (Beer's law) value for low excitation energies then to increase by a factor of more than 30 with increasing excitation level. Notice that two abscissa scales are provided. The lower scale is in units of quanta (energy) in the excitation pulse – the quantity directly recorded by the detector. The measured spot size (200 μm, FWHM) was then used to convert the energy to peak incident fluence in units of mJ/cm^2, as shown on the upper scale. We emphasize that this conversion is meaningful only

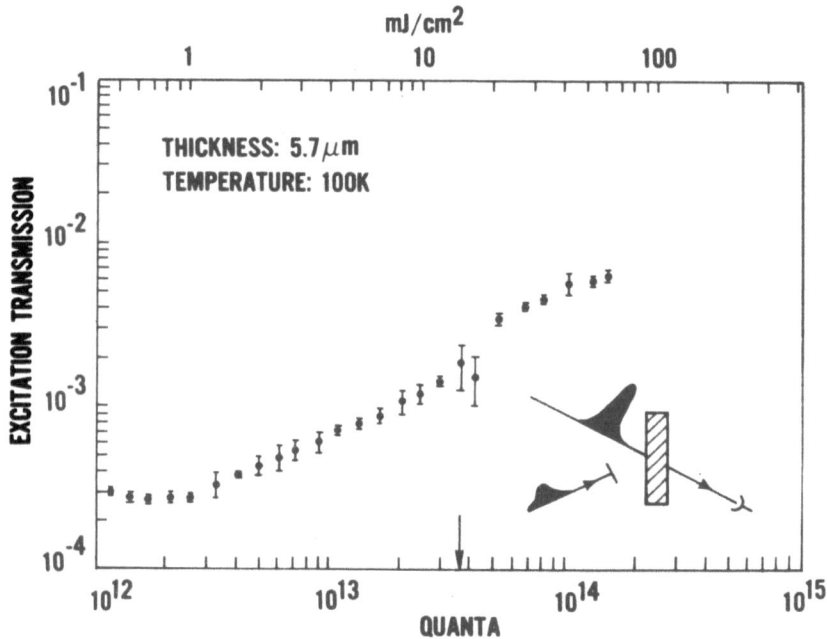

<u>Fig. 2</u> The energy transmission of the excitation pulse as a func-
tion of energy in that pulse.

for well-characterized spatial beam profiles.

Having demonstrated that we indeed could bleach the Ge transmission, the experimental procedure was then to measure the probe
transmission as a function of time delay between the two pulses
with exciting and probing energies fixed (inset Fig. 3). The excitation level (15 mJ/cm^2) was chosen to ensure an initial bleaching
of the Ge transmission and is indicated by an arrow in Fig. 2.
Typical results are shown in Fig. 3, where the probe transmission
is plotted as a function of time delay between the two pulses.

Inspection of this figure reveals two distinct features. The
most prominant of these is a rapid rise and fall in probe transmission (~2 psec, FWHM) located near zero delay. This structure is
caused by a self-diffraction of the excitation pulse into the probe

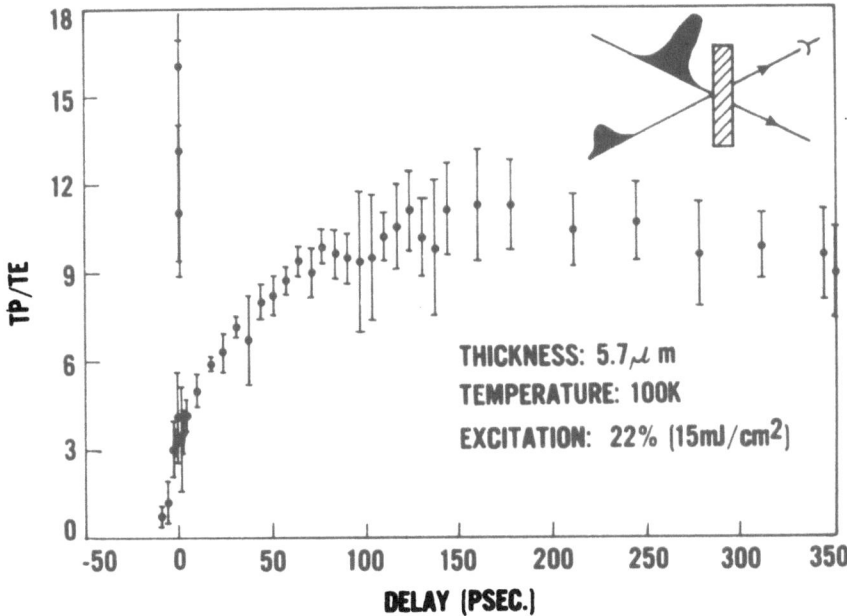

THICKNESS: 5.7 μ m
TEMPERATURE: 100K
EXCITATION: 22% (15mJ/cm²)

Fig. 3 Probe pulse transmission vs delay between the excitation pulse at 1.06 μm and the probe pulse at 1.06 μm for a sample temperature of 100 K and an excitation energy of 15 mJ/cm². The data are plotted as the ratio of probe pulse transmission to excitation pulse transmission, T_p/T_e. The error bars represent twice the statistical standard deviation.

direction when the two pulse overlap near zero delay. That is, when the pulses are spatially and temporally coincident, the two pulses interfere to produce a periodic modulation of the total electric field. Thus, the direct absorption of the pulses produces a spatially modulated excited-state carrier population, i.e. a free-carrier grating is produced. This grating self-diffracts a part of the excitation pulse into the direction of the probe. Such a self-diffracted signal is produced on the probe detector only so long as the two pulses overlap and interfere. Although this signal contains a wealth of useful information [17-20] we shall ignore it here and consider it an artifact of the measurement technique. This feature often will not be resolved in our figures.

The sharp spike near zero delay is followed by a gradual rise in probe transmission lasting more than one hundred picoseconds. For pedagogical purposes, this is the feature that we wish to

356

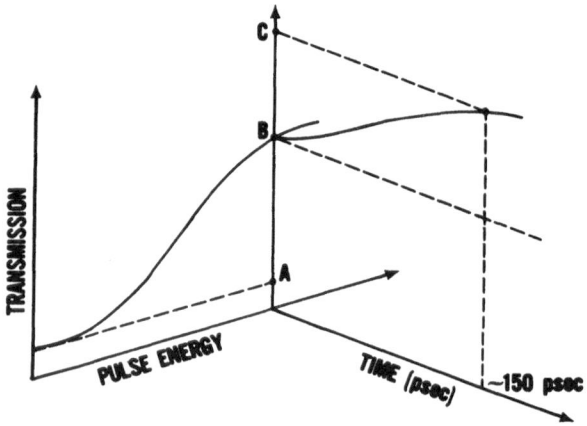

Fig. 4   Schematic of the saturation behavior of Ge.

emphasize in this section.

     Notice that we have actually plotted the ratio of probe trans-
mission T_p to excitation transmission T_e along the ordinate. We
have plotted this ratio to emphasize that the probe transmission at
longer delays is enhanced (in this case, by more than an order of
magnitude) over that of the excitation pulse. Since the excitation
pulse transmission was itself enhanced roughly by a factor of 5
(see Fig. 2), the overall bleaching of the Ge transmission is seen
to exceed a factor 50 for the experimental parameters used here.

     We emphasize this behavior by schematically sketching the
last two figures together in Fig. 4. First, the Ge transmission
is plotted as a function of excitation energy. As was shown in
Fig. 2, the single-pulse (or excitation) transmission was bleached
by a factor of 30 over Beer's law at the highest excitation levels.
In the two-pulse experiments (e.g., Fig. 3), we then chose the excitation level in the saturated-absorption regime, say at point B,
and tracked the transmission of the sample in time utilizing the
weak probe pulse. The sample transmission was observed to increase
for more than 150 psec following excitation. That is, the Ge

absorption was more nearly saturated 150 psec after excitation than immediately following it. This is the feature that we would like to understand here.

In this regard, since we are attempting to model a feature in the time-resolved saturable absorption of Ge, we should almost certainly begin by listing the possible absorption processes and their associated coefficients. The relevant absorption processes at this wavelength and at these excitation levels are :
(1) direct interband absorption (between the valence and conduction bands),
(2) phonon-assisted free-carrier absorption, and
(3) direct intervalence absorption (between the light-hole valence band and split-off valence band and between the heavy-hole valence band and splitt-off band).
Detailed discussions of the expressions for the various absorption coefficients are given elsewhere [10,21,22] and those discussions will not be repeated here. We merely wish to point out that, given the Ge energy band structure [23-31], these coefficients are well-known functions of *carrier number, carrier temperature,* and *lattice temperature.* If one knew the carrier distribution function at every position in the sample and for every instant of time, one could obviously track the Ge transmission in detail.

Quantitatively, one would like to take a "first principles" or fundamental approach to this problem. By such, we mean that we would like to begin with the Boltzmann equation, obtain rate equations for the parameters characterizing the distribution function (i.e., carrier number, temperature and Fermi energies), then use Fermi's Golden Rule to calculate the various scattering rates. In principle, of course, this is possible. For example, we can formally write a rate equation for the carrier density (see [22] for such an expression) that contains a carrier generation term due to the excitation pulse, various electron-hole recombination terms and a term describing the diffusion of carriers from near the sample surface into the bulk. One could also write a more complicated rate equation for the carrier temperature. Such an expression would contain the electron-phonon coupling constants, the Auger recombination rate, the diffusion coefficient and various absorption coefficients as principal parameters. These rate equations obviously would be coupled. In principle, all fundamental coupling coefficients needed for such a solution have been measured and are available in the literature , at least at low carrier densities.

In fact, Elci, Scully, Smirl and Matten [10] have taken such an approach to obtain satisfactory fits to the then available data. In their model, the dominant mechanism responsible for the rise in probe transmission with delay was a cooling of a hot electron-hole plasma following excitation. The key parameter, then, was the electron-optical phonon coupling constant, which we denote by Q_o.

Later, Leung and Scully [12] included diffusion in the same model
while de-emphasizing cooling and, again, obtained agreement with
existing data. Subsequently, Van Driel [32] modified the same initial model to include hot phonon effects (neglected diffusion) and
also obtained agreement with available data. The point is that
three studies that used basically the same approach produced satisfactory agreement with the data by emphasizing three different mechanisms (carrier cooling, diffusion and a hot-phonon bottleneck).
The calculations just mentioned (although they are tedious) are
fundamentally sound. Such a detailed first-principles approach,
however, is clearly impractical. The reason is that the various
coupling parameters required for such a solution are in question and
have not been measured satisfactorily at these excitation levels or
at these carrier densities. The electron-optical phonon coupling
constant (carrier cooling rate) and the diffusion coefficient are
of specific importance to our deliberations on laser annealing. We
illustrate this problem by briefly discussing these two parameters
below.

The laser-generated electrons are deposited into the conduction
band with an excess energy ΔE_e, as we have discussed previously.
These electrons can relax by intravalley optical phonon emission.
Similar comments, of course, apply to holes in the valence band.
Unfortunately, values for the electron-optical phonon coupling constant Q_0 for Ge are uncertain by a factor of 3, even at low carrier
densities [33-43]. Since the carrier energy relaxation rate is
proportional to Q_0^2, the time required for the carrier distribution
temperature to reach that of the lattice is uncertain by an order
of magnitude.

This uncertainty in the cooling rate is severely exasperated
by the presence of the dense electron-hole plasma. At the large
carrier densities reported here, the electron-phonon interaction
might be effectively screened [44], resulting in a decrease in
the carrier cooling rate. Moreover, as on the order of 10^{20} cm^{-3}
optically-created carriers cool, the carriers give their excess
energy to the optical phonon reservoir in a characteristic time τ_e;
the optical phonon reservoir, in turn, gives its excess energy to
the lattice (by decaying into long wavevector acoustic phonons) with
a time constant τ_p. Van Driel [32] has noted that the optical phonon lifetime for germanium is approximately 10 psec. This is relatively long compared to τ_e when the low density electron-optical
phonon coupling constant is employed. Van Driel has suggested that
this mismatch could result in a relaxation bottleneck for the hot
carriers caused by a buildup of the optical phonon population.
Although neither the hot-phonon bottleneck nor the screening effect
has been directly observed or quantified in Ge, it is clear that
Q_0 is not well enough known to allow a fundamental, precise calculation of the energy relaxation rate.

As we have discussed previously, the absorption depth for Ge at 1.06 μm is approximately 1 μm. Consequently, the direct absorption of intense pulsed laser radiation at this wavelength is accompanied by :
(1) large carrier densities and density gradients,
(2) large carrier temperatures and carrier temperature gradients for short times and
(3) large lattice temperatures and lattice temperature gradients for longer times.
Although highly interrelated, we attempt to briefly discuss the effect of each of these on the carrier diffusion rate separately.

The carrier densities generated here are sufficiently high that the single carrier picture of the Ge band structure may not be sufficient to describe the observed phenomena, and many-body effects may need to be included. Recent theoretical studies suggest that many-body effects could influence diffusion in two counteracting ways. Yoffa [44] has suggested that the electron-phonon interaction might be screened substantially (as we have already mentioned). This should result in an increase in the ambipolar diffusion coefficient. By contrast, exchange and correlation effects at high electron and hole densities can cause a reduction of the energy band gap in semiconductors. Since the band gap tends to narrow most where the carrier concentration is largest, the carriers induce a potential gradient that opposes the normal diffusion and that tends to produce self-confinement [45-46] .

For the period that the carrier temperature is elevated ($T_c \gg T_L$), there are again two counteracting mechanisms that modify the diffusion rate. First, one expects that the increased kinetic energy for the carriers will lead to an enhancement of the diffusion coefficient. In addition, Van Vechten and Wautelet [47] have predicted an anomalously large band gap narrowing that leads once again to an inhibition of the diffusion process.

If, on the other hand, the lattice is heated by the excitation pulse, one might expect two other effects to be simultaneously operative. The diffusion coefficient is known to decrease with increasing lattice temperature. Since, in this case, the lattice temperature will be highest where the carrier density is highest, the increased lattice temperature should have a tendency to slow the diffusion. Likewise, the energy band gap in Ge is known to decrease with increasing lattice temperature, again leading a potential well in the irradiated region that tends to confine the plasma [47,48] .

Hopefully, we have convinced the reader that diffusion under these conditions is complicated. We shall return to these discussions in Sec. 4.

The purpose of our rather lengthy discussions so far in this section has been two-fold. First, we intended to convince the reader that the carrier dynamics on picosecond time scales and at these excitation levels are complicated. Second, we hoped to convince the reader that a first-principles approach to this problem is not only tedious and complicated but is not feasible at this time because of the uncertainties in the fundamental constants that would be required for such a calculation. In the remainder of this sections, we describe a parametric experimental study and analysis that partially circumvents these problems to at least allow the identification of the mechanism dominating a particular experimental feature. Once the mechanism has been identified, in principle, quantitative rates eventually can be extracted at these carrier densities.

We review these parametric studies in some detail since no extended discussion of these results are presently available in the literature, although abbreviated accounts are contained in conference proceedings [49] or letter form [50] .

3.1 Parametric measurements

Experiments were performed in two separate, but related ways :
(1) for specific fixed excitation energies, the probe transmission was measured as a function of time delay after the excitation pulse and,
(2) for specific fixed delays, the probe transmission was measured as a function of varying incident excitation fluence.

In experiments of the first type, we found that the time required for the probe transmission to attain its maximum value depends strongly on the excitation pulse fluence (or irradiance). The transmission of the probe pulse was measured as a function of time delay for fixed excitation levels of 68 mJ/cm^2 ($\sim$9 GW/cm^2), 21 mJ/cm^2 ($\sim$3 GW/cm^2), 15 mJ/cm^2 ($\sim$2 GW/cm^2), 6 mJ/cm^2 ($\sim$0.8 GW/cm^2) and 3 mJ/cm^2 ($\sim$0.4 GW/cm^2). The results for three of these excitation levels are shown in Figure 5. Each data point shown is the average of ten laser shots. The error bars represent plus or minus one statistical standard deviation. At very high excitation levels of roughly 68 mJ/cm^2 (Fig. 5a), the probe transmission rises rapidly and is indistinguishable from a simple accumulation of electrons (holes) in the conduction (valence) band as a result of the finite width of the excitation pulse (i.e., a simple band-filling). Specifically, in the case of a band-filling by a cool carrier distribution, the rise in the probe-pulse transmission should be proportional to the integral of the excitation and probe pulse correlation function [51] . If the excitation pulse is attenuated, say by placing a 9% transmission filter in its path, the probe transmission rises much more slowly (Fig. 5b), taking approximately 200 psec to attain its peak value. At excitation levels below 6 mJ/cm^2, the rise again becomes increasingly rapid until at 3 mJ/cm^2 (Fig. 5c)

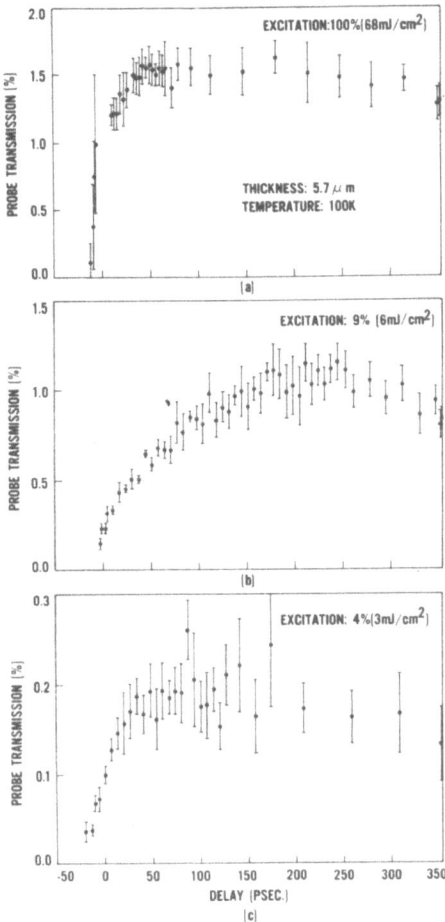

Fig. 5 The measured rise in the probe transmission with time delay for various fixed excitation levels : (a) 68 mJ/cm^2, (b) 6 mJ/cm^2, (c) 3 mJ/cm^2.

362

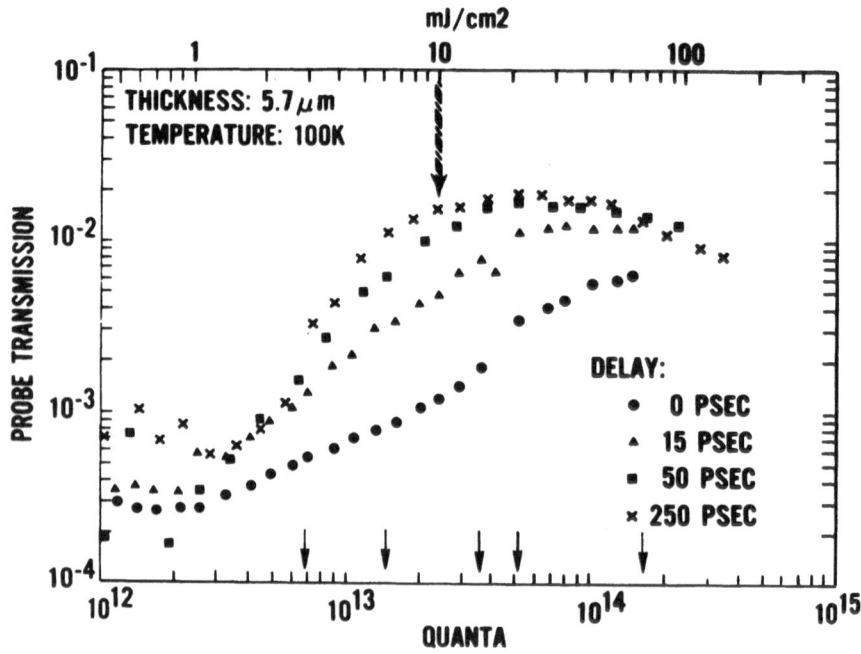

<u>Fig. 6</u> Measured probe transmission versus excitation fluence for various fixed time delays.

the rise has the form of an integration effect once again. All excitation levels, including those not shown, smoothly and reproducibly fit this trend. (No attempt was made to resolve the sharp spike in these studies).

In experiments of the second type, the probe transmission was measured as the excitation fluence was continuously varied for fixed delays of 0, 15, 50, 150 and 250 psec. The results for four of these delays are shown in Fig. 6 . For a fixed delay of 0 psec, the transmission is enhanced at high excitation fluences by more than a factor of 20 over the Beer's law value. If the delay between the two pulses is increased to 15 psec, the probe transmission is more readily enhanced at the lower excitation levels and exhibits a distinct leveling or saturation at 1% for the highest excitation fluences. This trend continues for time delays between 50 and 250 psec.

Phys. Rev. A24 (1981) 2735;
van Swol, F., Woodcock, L.V., Cape, J.N., J. Chem. Phys. 23 (1980) 913;
Morf, R., Phys. Rev. Lett. 43 (1979) 931.

21. Chui, S.T., Phys. Rev. Lett. 48 (1982) 933.

22. Meister, T. (Inst. für Festkörperforschung, Jülich), unpublished.

23. Bonissent, A., in ref. 1 and 12c.

24. Reviews on phase transitions in adsorbed layers are given by Landau, D.P., chapter 9 of [39], and by Bak, P. in [1].

25. Tosatti, E., Solid State Comm. 25 (1978) 637;
Bak, P., Solid State Comm. 32 (1979) 581.

26. Wortis, M., Pandit, R., Schick, M. in "Melting, Localization and Chaos"; R. Kalia and P.D. Vashishta eds. (New York, North Holland, 1982).

27. Lipowsky, R., Phys. Rev. Letters, 1982 (to appear).

28. Kroll, D., Lipowsky, R., Phys. Rev. B 1982 (to appear).

29. Kroll, D., (Inst. f. Solid State Physics, Jülich) unpublished.

30. Cahn, J.W., J. Chem. Phys. 66 (1977) 3667.

31. Menaucourt, J., Thomy, A., Duval, X., J. Physique 38 (1977) 195.

32. Leamy, H.J., Gilmer, G.H., Jackson, K.A. in "Surface Physics of Materials"; J.M. Blakely ed., Vol. 1, p. 121, 1975 (New York, Academic Press).

33. Saito, Y., Müller-Krumbhaar, H., J. Chem. Phys. 74 (1981) 721 (Note that this theory holds for nonconserved order parameters only. If the order-disorder transition in the solid is accompanied by concentration change one has to include diffusional transport. A mere density change in contrast can propagate with sound velocity and thus is uncritical.)

34. Baikov, Yu.A., Zelenev, Yu.V., Haubenreisser, W., Pfeiffer, H., Phys. Stat. Sol. (a) 61 (1980) 435.

35. Saito, Y., Müller-Krumbhaar, H., J. Chem. Phys. 70 (1979) 1078.

36. Pfeiffer, H., Haubenreisser, W., Phys. Stat. Sol. (b) 96 (1979) 287.

37. Cherepanova, T.A., Kiselev, V.F.; Kristall und Technik 14 (1979) 545;
Cherepanova, T.A., Phys. Stat. Sol. (a) 58 (1980) 469.

38. Chandrasekhar, C., "Hydrodynamics and Hydrodynamic Stability", (Oxford, Clarendon, 1961).

39. For a review on computer-simulation in crystal growth see Müller-Krumbhaar, H., chapter 7 in "Monte-Carlo Methods in Statistical Physics", K. Binder ed. (Heidelberg, Springer-Verlag, 1979).

40. See Germain, P. in this volume. General reviews are given in A. Zettlemoyer ed, "Nucleation" (New York, Dekker, 1969); "Nucleation II" (New York, Dekker, 1976).

41. Müller-Krumbhaar, H., Burkhardt, T., Kroll, D., J. Crystal Growth 38 (1977) 13.

42. Chernov, A.A., Lewis, J.; J. Phys. Chem. Solids 28 (1967) 2185.

43. Cherepanova, T.A., van den Eerden, J.P., Bennema, P., J. Crystal Growth 44 (1978) 537.

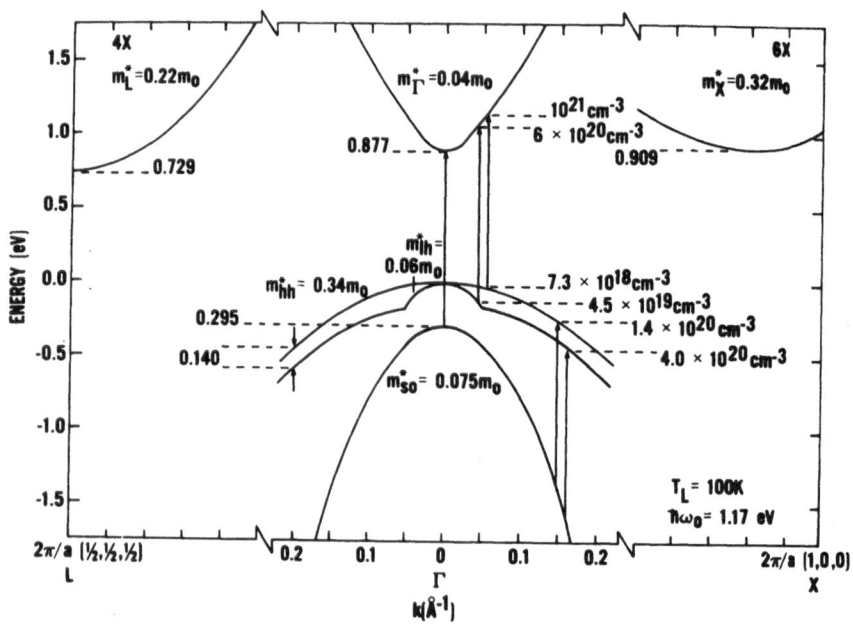

Fig. 7 Relevant features of Ge band structure at 100 K.

intervalence band transitions between the split-off band and the heavy and light hole bands.

Qualitatively then, the initial enhancement of the germanium transmission with increasing excitation level is caused by a depletion of the valence electrons in the heavy and light-hole valence bands. The turnover and eventual decrease in transmission is caused by the onset of direct intervalence band transitions at the very highest excitation fluences. The exact magnitude of the absorption change, of course, will depend on carrier temperature as well as carrier number.

A re examination of Fig. 6 shows that the measured probe transmission is in good qualitative agreement with these concepts. The transmission at 0 delay rises by a factor of approximately twenty at high intensities, presumably by band-filling. When the excitation fluence is in excess of 30 mJ/cm^2 (4 GW/cm^2) a slight leveling in the transmission can be observed at 1%. Complete saturation of the direct valence-to-conduction band transition alone by band filling would limit the transmission to approximately 40% because of

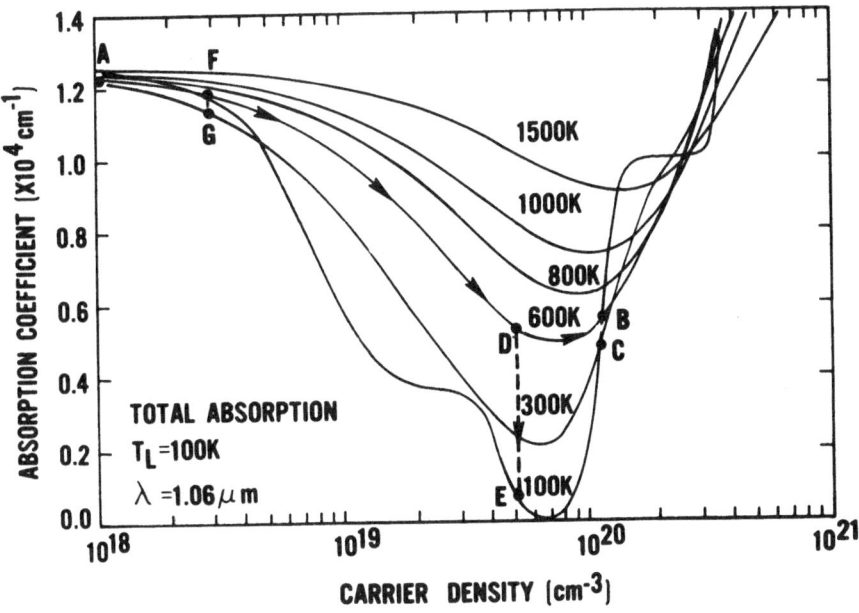

Fig. 8 Illustration of the parametric tendencies of the Ge transmission as a result of carrier cooling using the calculated dependence of the total absorption coefficient on carrier density and
carrier temperature.

reflection losses; however, the pulse transmission is limited by
the onset of another absorption process and does not begin to
approach this value. These effects can be seen more clearly by
examining the probe transmission at longer delays as shown in Fig.6.

We can gain a qualitative appreciation for the tendencies that
the germanium transmission should exhibit by considering the set of
parametric curves shown in Fig. 8. The total absorption coefficient,
including direct interband and intervalence band, is plotted versus
the optically-created carrier number for various assumed carrier
temperatures. The lattice temperature is held constant at 100 K.
As an example of how these curves can be used, consider the expected tendencies for a model that is "cooling dominated". That is,
suppose that 10^{20} cm^{-3} carriers are created by the excitation pulse
and that the carrier temperature is elevated during the excitation
process (the exact temperature is unimportant but the analysis detailed below shows that 600-700 K fits the data well). That is,

the evolution of the distribution is described by first walking along the path AB to a carrier density of 10^{20} cm^{-3}. Since we are considering a cooling model, we neglect recombination. As the carriers cool, we walk along path BC. Thus, we would expect little change in absorption or transmission as the carriers cool at this excitation level. For a lower excitation level, sufficient to create say 5×10^{19} cm^{-3} carriers (path AD), we would expect a large change with time as the carriers cool (path DE). For still lower excitation levels (path AF), again there is little change. These tendencies are in exact agreement with those reported in Fig. 5 and Fig. 6. Similar arguments can be made for a "recombination dominated" model. That is, we can neglect all cooling but allow the carriers to recombine following creation. Retracing the above arguments, we see that the tendencies are opposite to those observed.

To make these qualitative speculations more quantitative, we now describe a procedure for performing simple parametric calculations. What we wish to do is to quantitatively investigate the tendencies in the germanium transmission following excitation when only a single process is active. For example, we would like to set the diffusion and recombination rates to zero and to determine the evolution of the germanium transmission with time as the carriers cool and then to set the cooling and recombination rates to zero and study the effects of diffusion, and so on. To carry out the study that we have in mind, we must first determine the carrier density as a function of position x immediately following the excitation process. One way to estimate the initial carrier density is to assume that all of the optically-generated carriers are exponentially distributed away from the front sample surface according to Beer's Law. That is, the generation rate is given by:

$$G(x,y,t) = (1-R) \ \alpha_o I_e(x,t) \ \exp(-\alpha_o y)/\hbar\omega_o \qquad (1)$$

where α_o is the Beer's Law value for the direct absorption coefficient, I_e the irradiance of the excitation pulse,
R the reflection coefficient of Ge, and where x is taken along the sample face and y is in the direction of propagation. In this case, the carriers are localized to within roughly a micron of the sample surface. This certainly provides a good estimate for low excitation levels where there is little saturation of the direct absorption coefficient. Another estimate can be obtained by assuming that the carriers are uniformly created from the front to the back of the sample. Now, the carrier generation rate is given by:

$$G(x,y,t) = (1-R) \ I_e(x,t)/\hbar\omega_o \ell \qquad (2)$$

where ℓ is the sample thickness. This generation rate, when integrated over the temporal profile of the excitation pulse, provides a good estimate of the carrier density following excitation only

for the highest excitation levels, where the direct absorption
throughout the sample is expected to be bleached. Neither Eq. 1
nor Eq. 2 will allow an accurate estimate of the carrier distribution for moderate excitation levels. Moreover, an exact calculation of the carrier density immediately following the generation
process is futile, as we have explained. Such an approach requires
the knowledge of coupling constants that are as yet undetermined at
these carriers. Nevertheless, a better estimate of the initial
carrier distribution than allowed by either Eq. 1 or Eq. 2 can be
obtained by the numerical procedure outlined below.

To determine the approximate carrier density following the excitation pulse, the transmission of the excitation pulse is numerically calculated by solving the coupled differential equations for
the irradiance as a function of position and time and for the optically-created density as a function of position and time :

$$\frac{dI_e(\underline{r},y,t)}{dy} = - \alpha_T(\underline{r},y,t) \; I_e(\underline{r},y,t) \tag{3}$$

and

$$\frac{dn(\underline{r},y,t)}{dt} = \alpha_{DA}(\underline{r},y,t) \; I_e(\underline{r},y,t)/\hbar\omega_o \tag{4}$$

where the total absorption coefficient α_T includes direct and intervalence band absorption, i.e.:

$$\alpha_T(\underline{r},y,t) = \alpha_{DA}(\underline{r},y,t) + \alpha_{IVB}(\underline{r},y,t) . \tag{5}$$

We emphasize once again (see [22]) that these absorption coefficients are well-known functions of carrier number and carrier temperature. The numerical procedure for simultaneously solving Eq. 3
and 4 is illustrated schematically in Fig. 9. First, the Ge sample
is divided into a large number (N) of very thin wafers and the
optical pulse is likewise divided into M parts. The transverse beam
profile is taken into account by dividing both pulse and sample
into L concentric regions. Then Eqs. 3 - 5 are solved as each piece
of the excitation pulse is allowed to traverse the sample. In this
way, the jth part of the excitation pulse experiences the saturation
(carrier density) induced in the ith slice of the Ge by all of the
parts that preceeded it. The only free parameter in such a numerical calculation is the carrier temperature. *If* the carrier temperature *during* the excitation process can be taken to be a single constant value, then, the output from such a procedure will be:
(a) the excitation-pulse transmission and
(b) the carrier density versus position immediately following the

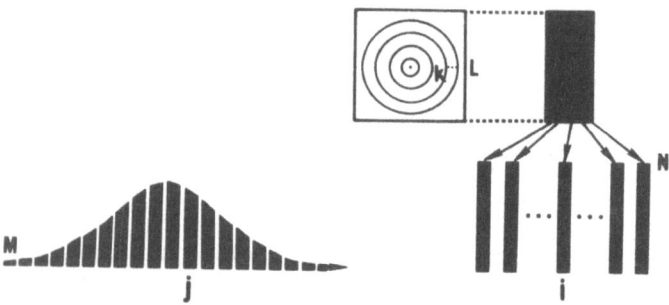

<u>Fig. 9</u> Schematic for the numerical calculation of excitation trans-
mission.

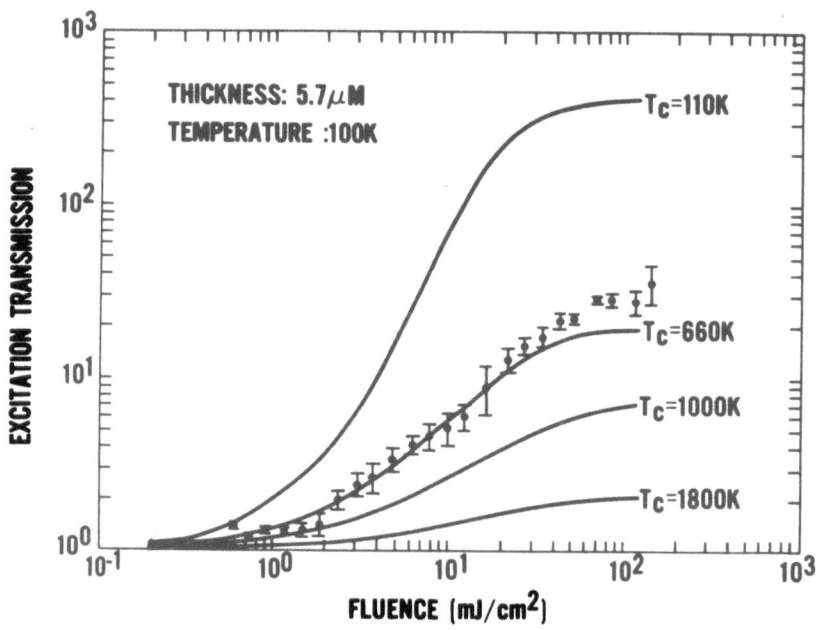

<u>Fig. 10</u> Calculated energy transmission of the excitation pulse as
a function of energy in that pulse for various assumed values of
the "characteristic temperature" during excitation. The dots are
the experimental data.

generating pulse for each excitation fluence and for *each* carrier
temperature assumed. The carrier temperature is adjusted to force
agreement between the calculated and measured excitation transmission (i.e., the 0 delay data of Fig. 6) as a function of excitation
level, as shown below. In other words, we use one piece of the data
shown in Fig. 5 and Fig. 6 to choose a "characteristic" temperature
during the excitation process.

The calculated excitation transmission as a function of excitation fluence is shown in Fig. 10 for several values of the characteristic carrier temperature. The experimental data is plotted for
comparison. The experimental transmission is too low for the
carriers to be at lattice temperature (100 K) and too high for the
carriers to have maintained all of their excess energy following
excitation (1800 K). A satisfactory fit is obtained for a characteristic temperature between 600 K and 700 K.

The choice of a characteristic temperature (e.g., 660 K) uniquely determines the carrier distribution for each excitation level
as a function of distribution through the sample as shown in Fig.11.
Notice that for the maximum fluences the carrier concentration is
roughly uniform from the front to the back of the sample; for the
lowest, the distribution is approximately exponential as expected
from Beer's law. The concentrations shown are those generated by
the center of the Gaussian spatial profile of the excitation pulse.
Also notice that the maximum carrier concentrations achieved in Ge
at this wavelength are on the order of 10^{20} cm^{-3}, because the direct
absorption coefficient saturates near this density (the exact value
depends on the carrier and lattice temperatures). Intervalence band
absorption is the dominant process for higher densities but would
not result in the creation of additional electron-hole pairs. For
shorter excitation wavelengths (e.g., 0.53 μm), larger densities
can be expected prior to damage.

We pause at this point to emphasize that we do not claim that
the carrier distribution can actually be characterized by a single
constant temperature throughout the generation process nor do we
claim that Fig. 11 represents the exact spatial distribution of
carriers. What we do claim is that this procedure provides a
"better" estimate of the initial spatial profile of the carrier concentration than assuming the concentration to be uniform or assuming
it to be exponential as determined by Beer's law. Furthermore, we
feel that the results shown in Fig. 10 do suggest that the carrier
temperature is elevated during the generation process. If this
temperature is taken seriously, it implies that approximately 70%
of the excess electronic energy has been given to the lattice by
the end of the excitation pulse (~10 psec). This has important
implications for our deliberations at this Advanced Study Institute.

Using these predicted carrier concentrations, one can now

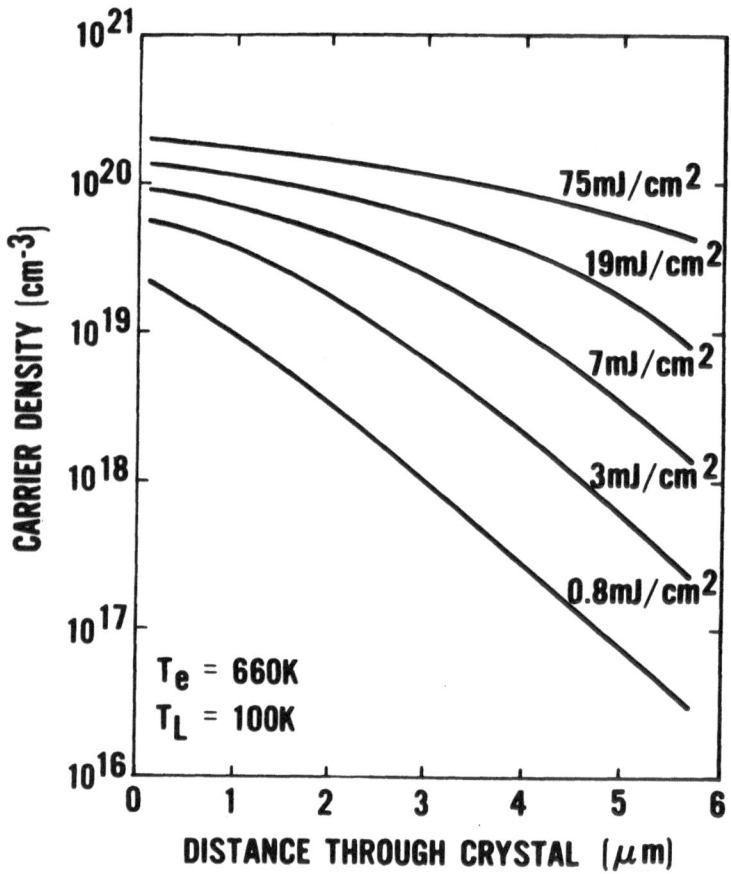

<u>Fig. 11</u> The calculated carrier densities as a function of position in the crystal, immediately following excitation, for various peak excitation fluences.

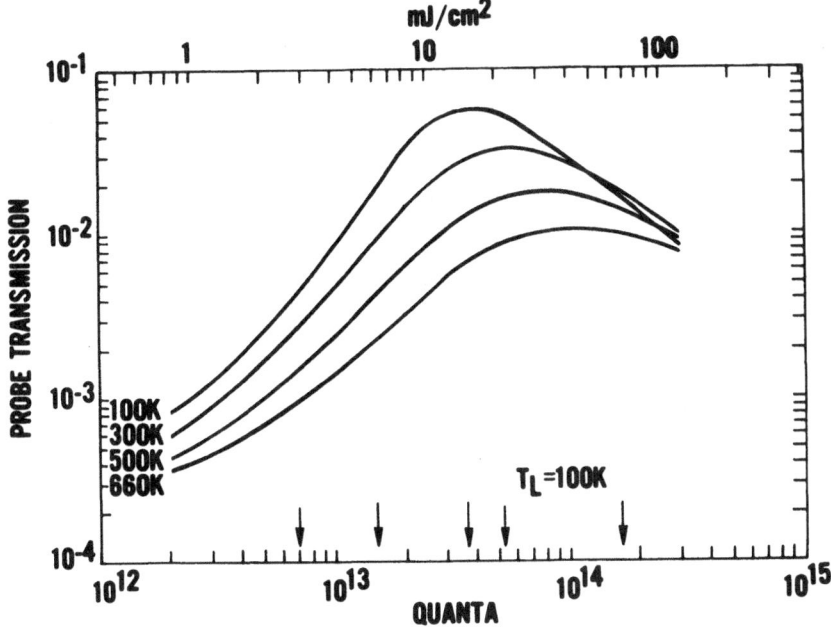

Fig. 12   The calculated probe transmission versus excitation fluence for various assumed carrier temperatures, using the initial distributions of Fig. 11.

calculate the expected tendencies in the probe transmission as single mechanisms, such as carrier cooling, recombination and diffusion, are allowed to individually determine the evolution of the distribution.

Using the predicted carrier concentrations determined by the passage of the excitation pulse, we then calculated the expected probe transmission assuming various values for the carrier temperature. Diffusion and recombination were neglected during this portion of the study. The results are shown in Fig. 12. Notice that the parametric dependence of the numerically-calculated probe transmission on temperature is identical to the measured dependence of the probe transmission on time (Fig. 6). If this correspondence is correct, it implies a slow cooling of the hot-carrier distribution that can exceed 200 psec. This is much slower than would be expected from a much more complicated, detailed calculation of the cooling rate using the accepted electron-optical phonon coupling

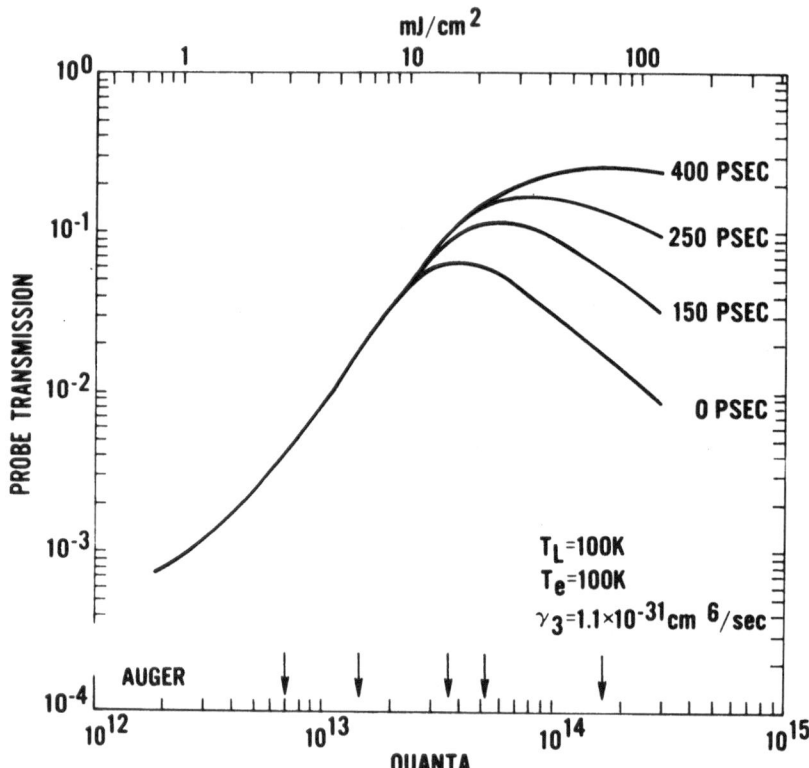

<u>Fig. 13</u> The parametric dependence of the probe transmission versus excitation level on carrier recombination.

constant [10,52] . However, possible slower cooling mechanisms have been suggested recently involving a hot optical phonon bottleneck [32] or a screening of the electron-optical phonon interaction [44], as we have discussed before. We also emphasize that it is possible that the carriers have cooled to within one optical-phonon energy of the lattice temperature and that the structure that is shown in Fig. 5 and Fig. 6 is the result of the final stages of cooling as a result of acoustic phonon emission. One would expect the latter to be slower. The presently available data does not allow us to distinguish among these three possibilities.

Notice that our simple parametric analysis avoided the complication of knowing the exact value of the coupling constants

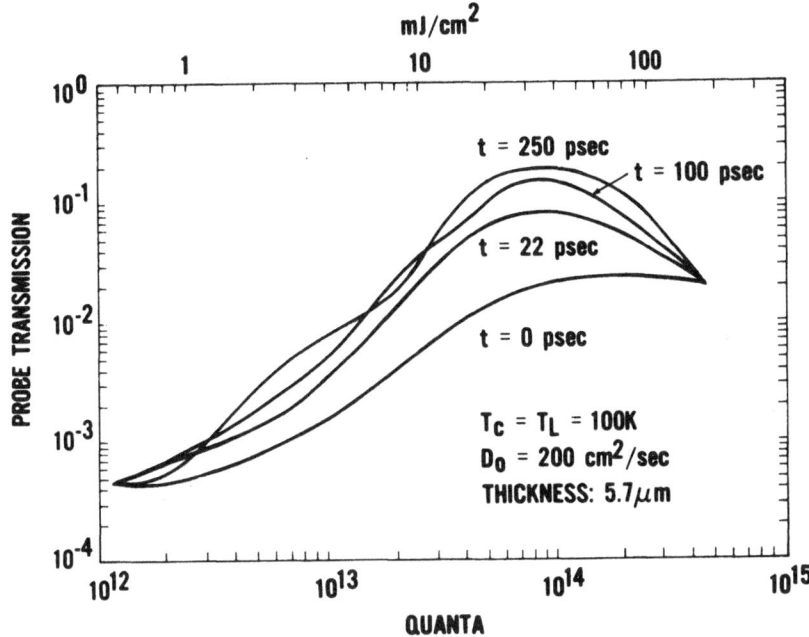

<u>Fig. 14</u> The parametric dependence of probe transmission versus excitation level on carrier diffusion.

or calculating the cooling rate. The tendencies in the Ge transmission were calculated as the carriers were allowed to cool following excitation; no mention was made of the time scale on which this cooling occured. This analysis only required a calculation of the absorption coefficients as functions of carrier density and carrier temperature.

A similar parametric study involving the effects of Auger recombination, in which the carriers were assumed to instantly cool to lattice temperature following the excitation pulse, produced a dependence of the probe transmission on time and excitation level in direct disagreement with experimental results (Fig. 13). Again we emphasize that the strength of the recombination is irrelevant in this approach; we simply are interested in parametric tendencies as the carriers recombine.

Many other mechanisms (including lattice heating and surface recombination) have been considered, some qualitatively and some

quantitatively. A model in which the carriers are assumed to have
an elevated temperature during the generation processes, are assumed to cool instantly (compared to the optical pulse width) following the excitation pulse and are then assumed to diffuse away
from the sample surface into the bulk produced qualitative agreement with most of the observed tendencies in the data, as shown in
Fig. 14. The effects of diffusion can be experimentally identified
by repeating the above studies on even thinner samples. For sample
thicknesses on the order of the inverse of the direct absorption
coefficient (0.7 μm), the carrier density can be expected to be
approximately homogeneous at all excitation levels. Such samples
have been produced and the results of such studies will be described below.

Finally, we note that the measurements described above were
performed for lattice temperatures of 35 K, 100 K, 150 K, and 300 K.
The time required for the sample to exhibit its peak transmission
was found to depend only very weakly on lattice temperature. The
lack of a strong temperature dependence would normally be an indication that diffusion is not the mechanism dominating the particular
structure reported in this section. However, since the diffusion
process is not clearly understood at these excitation levels (as
we shall discuss in the next section), we should be careful about
such statements.

The two types of excitation experiments described above, i.e.:
(1) fixed excitation level and variable delay, and
(2) variable excitation level and fixed delay were repeated for a
variety of sample temperatures and for sample thicknesses of 7.0 μm,
5.7 μm, 4.0 μm, 3.0 μm, and 1.8 μm. We emphasize that these are
single crystal wafers not epitaxially grown layers. The 7.0 μm and
5.7 μm samples were prepared by optical polishing and etching,
exactly as described in the introduction. The thinner samples were
produced by ion-milling and etching the 5.7 μm-thick sample. The
sample thicknesses were determined by measuring the Fabry-Perot
fringes in the 2-10 μm wavelength range over a large surface area.
The high visibility of the fringes indicated that removing a few
microns by ion-milling did not measurably degrade the sample surface. Attempts to remove tens of microns by milling did produce a
degradation of the sample surface. All samples were mounted on a
quartz substrate as described in Sec. 2. The dependence of the
rise in probe transmission on sample thickness is shown in Fig. 15
for sample thicknesses of 7.0, 5.7, 4.0, and 1.8 μm. The excitation fluence was held constant at 6 mJ/cm^2 throughout these studies.
(Note that this was the fluence that produced the slowest rise at
a thickness of 5.7 μm in our previous studies). Because the sample
thicknesses differ, we have chosen the ordinate in this case as
the ratio of the probe transmission to the excitation transmission.
The probe transmission rises more slowly and the relative enhancement of the transmission is larger for the thicker samples. For

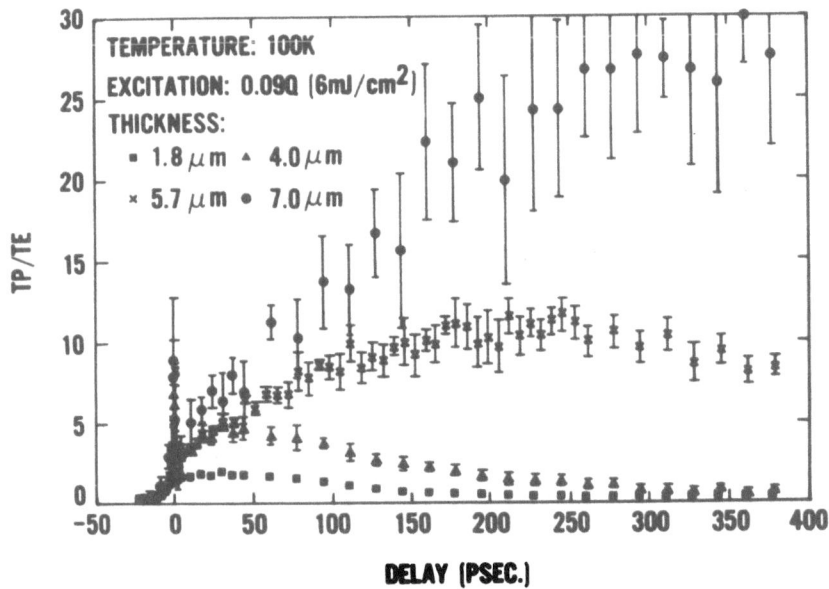

Fig. 15 The dependence of the rise in probe transmission (plotted as the ratio of probe transmission to excitation transmission, T_P/T_E) on sample thickness.

the thinner samples, it appears that a fast decay is dominating the dynamics, prohibiting us from observing the slow rise. We believe that these tendencies are consistent with a high surface recombination velocity at the back surface where the Ge is bonded to the quartz substrate. These effects would be expected to be more pronounced for the thinner samples.

Although surface recombination is a nuisance here, it has allowed us to turn on and off (very inefficiently) an active nonlinearity on a picosecond time scale. These results suggest that perhaps surface recombination could be used to rapidly terminate other nonlinearities induced in thin samples for application in the areas of phase conjugation or optical bistability.

Summarizing the results of this section, we have performed two types of two-pulse excitation-and-probe experiments (fixed excitation-variable delay and variable excitation-fixed delay) for a variety of sample temperatures and sample thicknesses. Only a small portion of the measurements could be presented here. As a result of this parametric investigation of the dynamic Moss-Burstein shift of the absorption edge (band-filling), contributions from direct inter-valence absorption have been identified at the highest excitation levels. In addition, a strong intensity-dependent and weakly tempe-rature-dependent rise in sample transmission is observed that can exceed 200 psec under certain conditions and that is consistent with a slow cooling of the hot carriers (or perhaps diffusion).

4. MEASUREMENT OF NONLINEAR CARRIER DIFFUSION

In the studies to be reviewed in this section, we used a varia-tion of the excitation-and-probe technique to measure the picosecond dynamics of laser-induced transient gratings that are produced in germanium by direct absorption of 35 psec optical pulses at 1.06 µm. By measuring the grating lifetimes, we attempted to investigate the density dependence of the diffusion coefficient and the nonlinear recombination coefficients in germanium. One of the reasons for measuring these coefficients on a picosecond time scale and at high carrier densities (near the damage or melting threshold) is, as we have implied previously, the recent technological interest in the areas of pulsed laser annealing of semiconductors, laser hardening and laser damage. For example, one of the possible effects of the dense free-carrier plasma is the screening of the electron-phonon interaction. The result of such a screening would be a decrease in the carrier cooling rate and an enhancement of the diffusion rate. As we discussed previously, however, the dynamics are further com-plicated by elevated lattice and carrier temperatures as well as by bandgap narrowing. Whether or not the screening is dominant, the diffusion of the hot, dense electron-hole plasma can reduce the rate at which the carriers heat the lattice near the semiconductor surface by increasing the thickness of the region in which energy transfer takes place.

In our application of the transient-grating technique (Fig. 16) two 35 psec (FWHM) pulses at 1.06 µm, separated by an angle 2Θ, were focused onto the germanium sample such that they were both spatially and temporally coincident. The interference between these two (exci-tation) pulses modulated the electric field across the face of the sample. The direct absorption of the excitation pulses produced a spatially-modulated carrier density. Following generation, this optically-created free-carrier grating decayed by bulk recombination and by diffusion of carriers from regions of high concentration to regions of low concentration. The contribution of surface recombi-nation to the grating decay can be neglected for the grating spacings considered here and for etched Ge samples. The grating decay is monitored by measuring the first-order diffracted light from a third (probe) pulse at 1.55 µm as a function of time delay between the ex-citation pulses and the probe pulse. Since the photon energy of the probe at 1.55µm is less than the direct band gap in Ge, we assume that

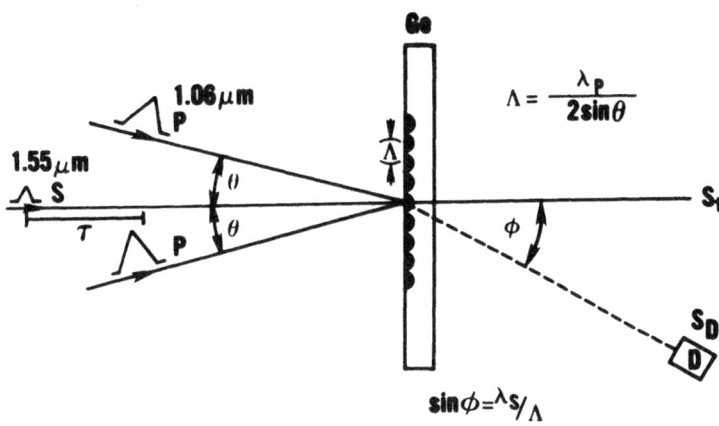

Fig. 16 Schematic of theexperimental technique for producing and measuring the decay of transient gratings.

the probe interrogates principally an index or phase grating.An advan tage of this technique over others is that by controlling the gra- ting spacing (i.e., Θ) we can control which process dominates the grating decay. For small grating spacings,diffusion will dominate; for sufficiently large spacings,recombination will dominate.

The optical excitation source for these studies was a passively mode-locked, Nd:YAG laser system operating at 1.06 μm. The experi- mental apparatus of Fig. 1 could be re-positioned to provide four grating spacings : 22.6 μm, 15.7 μm, 11.5 μm, and 6.8 μm. The peak excitation fluence could be varied, in this case, from approximately 2 mJ/cm^2 to 40 mJ/cm^2 by the use of calibrated attenuators. The probe pulse at 1.55 μm was generated from part of the 1.06 μm light by stimulated Raman scattering in benzene as discussed in Sec. 2. Details of the experimental apparatus, its calibration and its alignment can be found elsewhere [53].

4.1 Moderate excitation levels

Measurements of the picosecond dynamics of these optically- induced gratings were performed first at moderate peak excitation levels of 1.8 mJ/cm^2, for grating spacings of 22.7 μm, 15.7 μm, 11.5 μm and 6.8 μm and for sample temperatures of 295 K and 135 K.

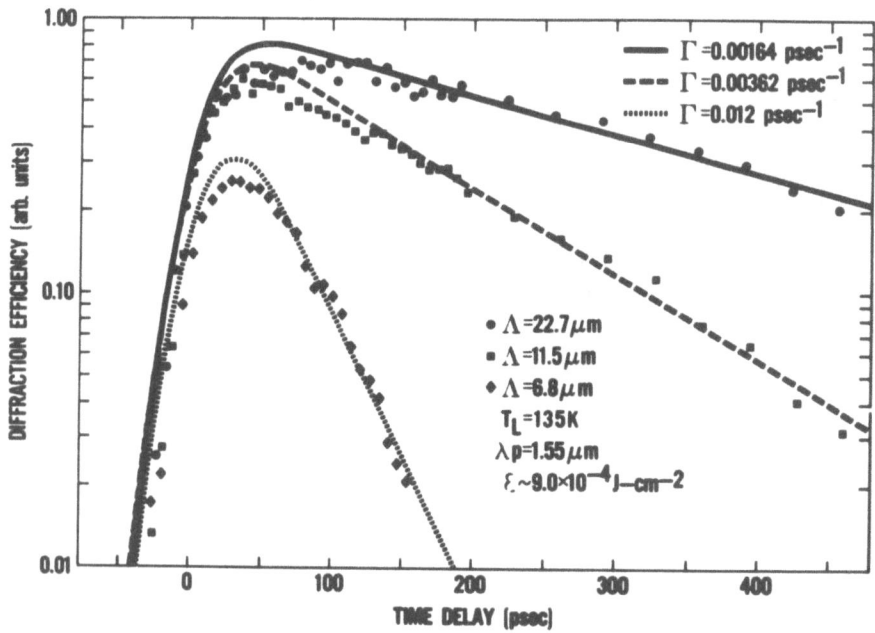

<u>Fig. 17</u> The diffraction efficiency (in arbitrary units) as a function of time delay between the excitation pulses at 1.06 μm and the probe pulse at 1.55 μm for a peak excitation fluence of 1.8 mJ/cm^2, for three grating spacings and for a sample temperature of 135 K. The dotted, dashed and solid curves are the result of numerical calculations described in the text.

The temporal evolutions of the grating diffraction efficiencies for three of these spacings (22.7 μm, 11.5 μm, and 6.8 μm) are shown in Fig. 17 for 135 K. The data for the 15.7 μm grating spacing are omitted for clarity. Each point in Fig. 17 is the average of forty to sixty data points. Note that the fluence labeling this figure is the average fluence; the peak incident fluence is a factor of two higher (1.8 mJ/cm^2). Notice that the grating lifetime decreases with decreasing grating spacing, indicating the increasing importance of diffusion for smaller grating spacings.

Careful examination of the semi-logarithmic plots in Fig. 17 reveals that the decays of the grating diffraction efficiencies are approximately linear for delays much longer than the excitation pulse width. These exponential decays suggest that, for the carrier densities encountered at this excitation level, the dynamics of these gratings can be understood in terms of a simple linear diffusion-recombination model. That is, we can neglect Auger and bimolecular recombination, and we can assume that the diffusion coefficient D is density (and therefore position) independent. In this limit, the time-rate-of-change of the carrier density can be written as;

$$\frac{dn(x,y,t)}{dt} = G(x,y,t) - \gamma_1 n(x,y,t) + D\nabla^2 n(x,y,t) \qquad (6)$$

where γ_1 is the linear recombination coefficient.

Using the Drude model for the optically-created free-carrier plasma, we can show that the free-carrier grating is predominantly an index grating and that the full modulation depth of the phase 2ϕ is proportional to the carrier density at the peak of the grating $n(0,y,t)$ minus that at the minimum $n(\Lambda/2,y,t)$, integrated from the front surface to the back surface of the sample, i.e.:

$$\phi(t) = -\frac{\Pi}{\lambda} \left[\frac{e^2}{2n_o m^* \nu^2 \varepsilon_o}\right] \int_0^\ell dy [n(0,y,t) - n(\Lambda/2,y,t)] \qquad (7)$$

where λ and ν are the wavelength and circular frequency of the probe, respectively, n_o is the index of refraction, ℓ is the sample thickness, e is the elementary charge on the electron, m^* is the reduced effective mass, ε_o is the permittivity constant and Λ is the optically-created grating spacing.

The instantaneous diffraction efficiency, D.E.(t), of a thin phase grating is then proportional to the square of the first order Bessel function $J_1(\phi)$:

$$D.E.(t) \propto J_1^2 [\phi(t)]. \qquad (8)$$

The effect of the finite width of the probe pulse on the time integrated (energy) diffraction efficiency of the grating $\eta(\tau)$ as a function of delay τ between excitation pulses and probe pulse is determined by convoluting the instantaneous diffraction efficiency produced by the excitation (Eq. 8) with the temporal intensity profile of the probe, $I_p(t)$:

$$\eta(\tau) \propto \int_{-\infty}^\infty dt\, I_p(t-\tau) J_1^2 [\phi(t)]. \qquad (9)$$

The expected energy diffraction efficiency of the grating (Eq. 9) can be calculated then if the modulation depth of the phase is known at all times (Eq. 7). Since the peak excitation level of 1.8 mJ/cm^2 is observed to produce a negligible saturation of the direct absorption coefficient, the spatially-modulated generation rate can be assumed to obey Beer's law :

$$G(x,y,t) = 2(1-R) \, \alpha_o I_e(t) \, [\, 1 + \cos 2\Pi x/ \, \Lambda] \, \exp\, [\, - \, \alpha_o y] \, /\hbar\omega_o$$

(10)

Again Λ is the optically-created grating spacing, R the reflection coefficient for Ge, α_o the linear absorption coefficient and $I_e(t)$ is the time dependent intensity profile of a single excitation pulse. (Note : x is taken along the sample surface and y normal to the surface). The linear partial differential equation (Eq. 6) can be readily solved using the generation rate given by Eq. 10 and the result substituted into Eq. 7 for the modulation depth of the phase ϕ to show that :

$$\phi(t) \propto \int_{-\infty}^{t} I_e(t') \, \exp\, [-\Gamma(t-t')] \, dt'$$

(11)

where the characteristic grating decay rate Γ is :

$$\Gamma = 4 \, \Pi^2 D/\Lambda^2 + 1/\tau_R$$

(12)

and where $\tau_R = \gamma_1^{-1}$ is the linear recombination lifetime.

The instantaneous diffraction efficiency of these gratings is then given by Eq. 8, if the grating can be considered thin. Smirl et al [53] show that the gratings can not only be considered thin, justifying the use of Eq. 8, but that the Bessel function $J_1(\phi)$ can be approximated by $\phi/2$ and Eq. 8 re-written as :

$$\text{D.E.}(t) \propto \phi^2(t).$$

(13)

Notice that, for a grating lifetime Γ^{-1} that is long compared to the probe pulse width, for delays that are long compared to the excitation pulse width and for the small modulations ϕ encountered here, the expression for the energy diffraction efficiency of the grating is reduced to :

$$\eta(\tau) \propto \phi^2(\tau) \propto e^{-2\Gamma t}.$$

(14)

That is, the diffraction efficiency decays exponentially at a rate twice the grating decay rate.

The results of fitting the data with the simple model just outlined are shown by the solid, dashed and dotted curves in Fig. 17. The fit is good and, for delays much longer than the excite pulse width, the decays are simple exponentials, as predicted by Eq. 14. The fit was equally good for the 15.7 μm grating spacing (not shown for clarity). The grating decay rate Γ was determined for each grating spacing from the best straight line fit to the data at long delays. These values are listed in Fig. 17.

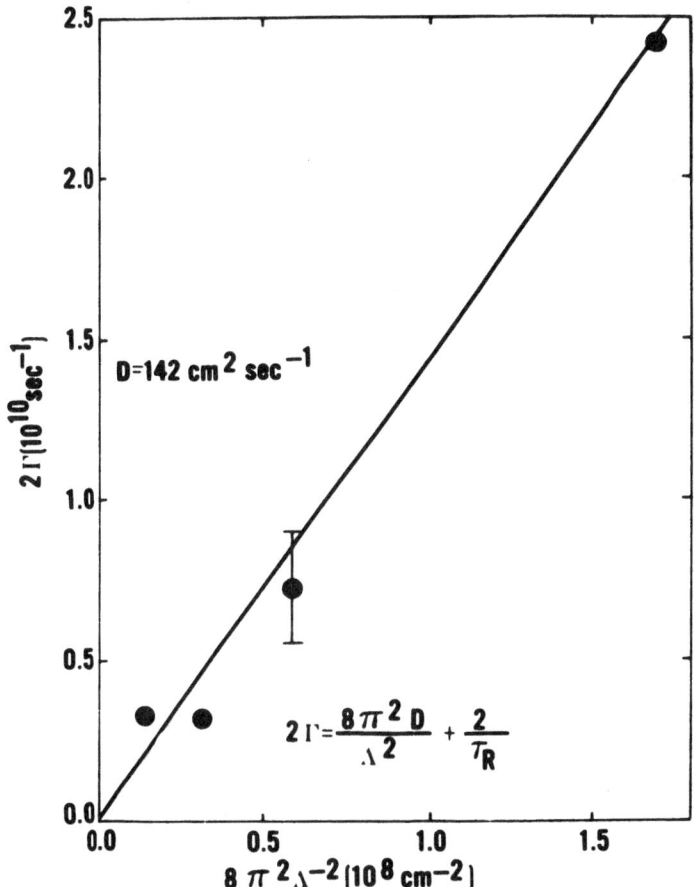

<u>Fig. 18</u> Determination of the diffusion coefficient D from the measured grating decay rates Γ for four grating spacings at a sample temperature of 135 K.

In fact, inspection of Eq. 12 indicates that, if the decay rate for the diffraction efficiency 2Γ is plotted versus $8\Pi^2/\Lambda^2$ for each of the four grating spacings, the slope of such a line will be the diffusion coefficient and the intercept will be proportional to the reciprocal of the recombination lifetime. This procedure, shown in Fig. 18, yields an ambipolar diffusion coefficient of 142 cm^2/sec for Ge at 135 K, a value slightly *smaller* than the accepted value determined from transport measurements at lower densities on doped samples. As a consequence of the error bars associated with these measurements, it is impossible to exactly determine the intercept. Since the recombination coefficient is positive, the lower limit on the intercept must be zero. An estimated upper limit is 2 x $10^9 sec^{-1}$ Consequently, one may safely conclude that at these excitation levels recombination effects are unimportant for times less than a nanosecond. Since the grating decay for each grating spacing was dominated by diffusion, no conclusions can be made as to the exact nature of the dominant recombination process (linear, bimolecular, or Auger). That is, a satisfactory fit to the data can be obtained by entirely omitting recombination from our analysis.

These measurements were repeated for a sample temperature of 295 K. An identical procedure yields an ambipolar diffusion coefficient of 53 cm^2/sec, again a value slightly smaller than the accepted value of 65 cm^2/sec.

At the peak densities encountered here ($\sim$4 x 10^{19} cm^{-3}) and when the carrier temperature is taken to be the same as that of the lattice, the electron-hole plasma is degenerate. Consequently, simple Boltzmann theory would predict that the diffusion coefficient should be higher than its nondegenerate value. In fact, using the simple model to be described below, we can estimate an increase of approximately a factor of 2, if we assume that lattice scattering dominates i.e., let S = - 1/2 in Eq. 17 . To properly investigate the possible influence of such an expected density dependence of the diffusion coefficient on the observed decay of the grating diffraction efficiency, the full density (spatial) dependence of the ambipolar diffusion coefficient must be included as prescribed below. The results of such a study [53] indicate that the effects of the density-dependence of the diffusion coefficient, as given by Eq. 17, on the grating decay should be readily detectable; we apparently observe no such effect.

4.2 High excitation levels

Measurements were also performed by Smirl et al. [53] for peak pump fluences of 36 mJ/cm^2 (1 GW/cm^2). This excitation level is 20 times that reported in the previous section and is within a factor of three of sample damage. The results of measurements of the grating diffraction efficiency as a function of time delay between the excitation and probe pulses at these high excitation levels are

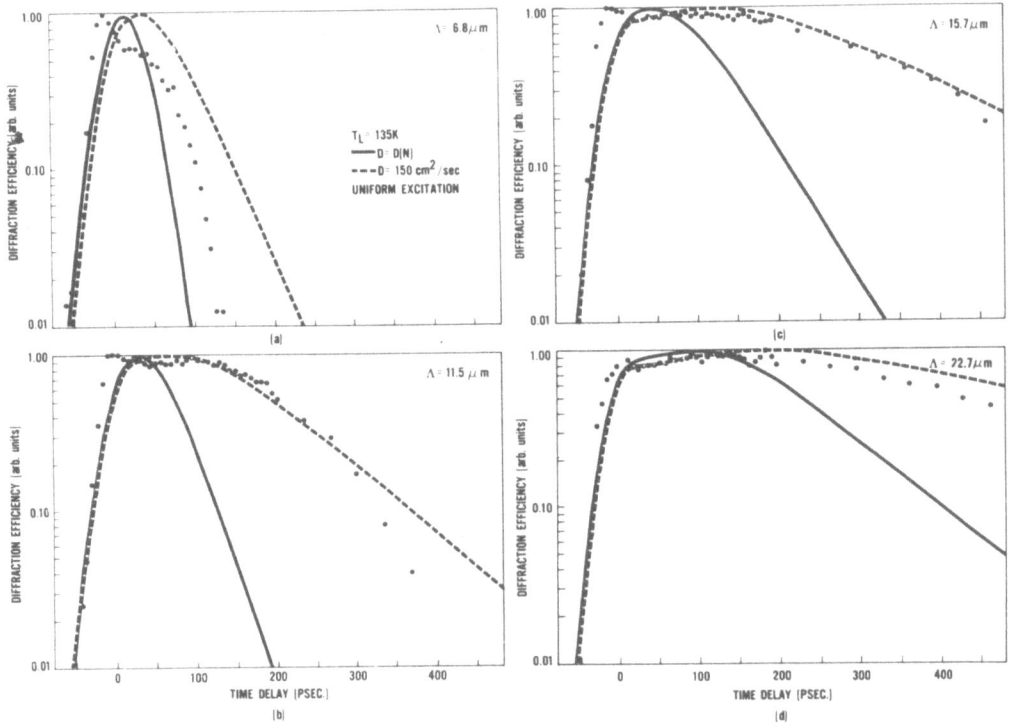

<u>Fig. 19</u> The diffraction efficiency (in arbitrary units) as a func-
tion of time delay between the excitation pulses and the probe pulse
for a peak excitation fluence of 36 mJ/cm^2, for a sample temperature
of 135 K and for four grating spacings : (a) 6.8 µm, (b) 11.5 µm,
(c) 15.7 µm and (d) 22.7 µm. The dashed and solid curves are the
result of numerical calculations described in the text.

shown in Fig. 19 for the same four grating spacings, 6.8 µm, 11.5 µm,
15.7 µm and 22.7 µm, as before and for a sample temperature of
135 K. The data are shown by the solid circles. The dashed and
solid lines are the results of numerical calculations to be discussed
later. Notice that the decays are no longer simple exponentials.

For these higher excitation levels, the instantaneous diffraction efficiency is still given by Eq. 8 and the phase by Eq. 7 but
the expression for the time-rate-of-change of the carrier density
(Eq. 6) must be modified to include higher order recombination processes and we must allow for a density (therefore spatial and temporal) dependence of the diffusion coefficient :

$$\frac{dn(x,y,t)}{dt} = G(x,y,t) - \gamma_1 n(x,y,t) - \gamma_2 n^2(x,y,t)$$
$$- \gamma_3 n^3(x,y,t) + \nabla[D\nabla n(x,y,t)] \qquad (15)$$

where γ_1, γ_2 and γ_3 are the monomolecular, bimolecular and Auger recombination coefficients, respectively, and D is the diffusion coefficient. For excitation levels described in this section, the photogenerated plasma is degenerate, and simple models based on Boltzmann theory predict that the diffusion coefficient differs from its non-degenerate value [54]. If one makes the effective mass and relaxation-time approximations and assumes that the relaxation time τ_{rel} has the form:

$$\tau_{rel} = \tau_o(\underline{r}, k_B T_L) \, E^S \tag{16}$$

where the value of S depends on the scattering process and where E is the electron (hole) energy, k_B is the Boltzmann constant and T_L the temperature of the lattice, then the diffusivity for the electrons D_e (holes D_h) can be written:

$$D_{e,h} = \frac{2\tau_o (k_B T_c)^{S+1}}{3m^*_c} \frac{\Gamma(S + 5/2) \, F_{S+1/2}(\beta)}{\Gamma(1/2) \, F_{-1/2}(\beta)} . \tag{17}$$

In Eq. 17, Γ denotes a Gamma function, T_c is the carrier temperature, β the electron (hole) quasi-Fermi level relative to the band edge in units of $k_B T_c$, m^*_c is the conductivity effective mass for the relevant band and the $F_J(\beta)$ are Fermi integrals defined by:

$$F_j(\beta) = \Gamma^{-1}(j+1) \int_o^\infty dx \, x^j [1 + \exp(x - \beta)]^{-1} , \tag{18}$$

that have been tabulated by Blakemore [55]. Over the range of densities encountered and discussed here, the density-dependent ambipolar diffusion coefficient can now be found from:

$$D = 2 \, D_e D_h / (D_e + D_h). \tag{19}$$

To illustrate the density dependence of Eq. 17 and 19, we have plotted the ambipolar diffusion coefficient as a function of carrier density in Fig. 20, assuming lattice scattering dominates (i.e., $S = -1/2$). Early excitation and probe experiments at intermediate excitation levels [13] initially appeared to corroborate the use of the Boltzmann theory given above at high carrier densities.

It is known from separate studies that at excitation levels of 36 mJ/cm^2 the transmission of the Ge sample is significantly enhanced (Fig. 2). In fact, numerical solutions similar to those described in Sec. 3 show that diffusion and recombination during the carrier generation process and saturation of the direct absorption coefficient produce an almost uniform carrier concentration from the front to the back sample surface in the range of 1 to 2×10^{20} cm^{-3}.

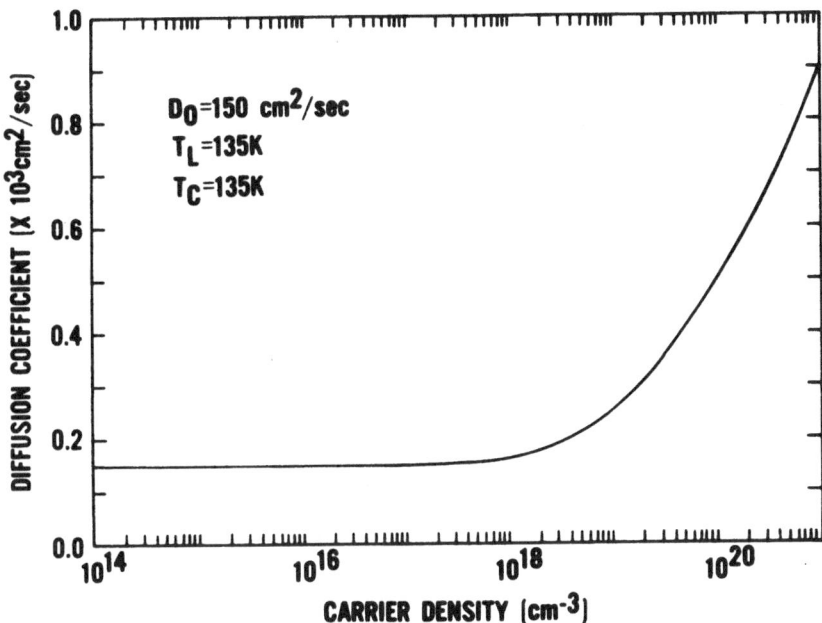

Fig. 20 The density dependence of the ambipolar diffusion coefficient expected from simple Boltzmann theory.

In this case, consequently, the carrier generation rate is taken to be uniform in the y-direction :

$$G(x,y,t) = 2(1-R)\ I_e(t)\ [\ 1 + \cos(2\Pi x/\Lambda)]\ /\hbar\omega_o\ . \qquad (20)$$

At these densities, the dominant recombination process is Auger. Furthermore, the simple Boltzmann theory described above indicates that the ambipolar diffusion coefficient should be highly density dependent. Consequently, the full nonlinear rate equation given by Eq. 15 must be used in any analysis of these results.

The solid and dashed lines in Fig. 19 represent numerical solutions to the coupled set of equations for the carrier density (Eq. 15), the modulation of the phase (Eq. 17) and the diffraction efficiency (Eq. 9). The solid curve is the result of including the full density-dependence of the diffusion coefficient as given by Eq. 17. Thus, the value of the diffusion coefficient should vary with position at any point in time through the spatial dependence of the carrier density. The dashed curve is the result of the same calculation except that the diffusion coefficient was held constant at its accepted low density value of 150 cm^2/sec. The Auger coefficient in both cases was taken as 1 x 10^{-31} cm^6/sec. Notice that the simple Boltzmann theory outlined above produces a grating decay that is much more rapid than the observed decay. Neither of the two models shown in Fig. 19 produces a satisfactory fit for short time delays. In fact, Smirl et al. [53] report the results of extensive numerical studies for various forms for the spatial generation rate (Eq. 10 and 20) and for both purely phase and mixed gratings, always using the standard expression for the degenerate diffusion coefficient given by Eq. 17. The conclusions were the same.

To this point, we have solved the coupled nonlinear partial differential equations given by Eqs. 15 and 9 for various forms of the spatial generation rate and for purely phase gratings, always using the standing expression for the degenerate diffusion coefficient given by Eq. 17. This expression is often used to account for the density dependence of the diffusion coefficient (for a recent example see [13]). Apparently the density dependence contained in this expression for the diffusion coefficient is *not* sufficient to account for the slower decays that we have observed near sample damage. This qualitative result is of fundamental and practical importance. As far as we know, these measurements provide the first experimental suggestion that the evolution of these free-carrier gratings, at these densities, is far more complicated than allowed by the conventional model outlined above. Several other nonlinear diffusion mechanisms that involve carrier screening, bandgap narrowing, and carrier and lattice heating were briefly discussed in Sec. 3. Unfortunately, the inclusion of these processes complicates the quantitative extraction of the exact density dependence of the diffusion coefficient or the identification of the primary mechanism from this single set of experiments.

Briefly summarizing then the results of these studies, a variation of the transient grating technique was used to investigate the picosecond dynamics of laser-induced transient gratings produced in germanium by the direct absorption of 35 psec optical pulses at 1.06 μm. Subnanosecond grating lifetimes were measured for peak excitation levels of 1.8 mJ/cm^2 and 36 mJ/cm^2, for sample temperatures of 295 K and 135 K, and for four grating spacings (6.8 μm, 11.5 μm, 15.7 μm, and 22.7 μm). The studies were performed using a probe wavelength of 1.55 μm. This wavelength was chosen because a quantum of energy at 1.55 μm (0.8 eV) is less than the direct band gap in Ge. Thus, the probe at 1.55 μm should be insensitive to modulations in the direct absorption coefficient and could be expected to interrogate principally a free-carrier index grating, although direct intervalence-band absorption contributions can also be important at the highest excitation levels, as discussed in Sec. 3.

For the lower excitation levels (peak carrier densities of $\sim5 \times 10^{19}$ cm^{-3}), a linear diffusion-recombination model provided a good fit to the experimental data for both sample temperatures and all grating spacings and allowed the extraction of an ambipolar diffusion coefficient of 53 cm^2/sec at a sample temperature of 295K and of 142 cm^2/sec at 135 K. The recombination lifetime was estimated to be much greater than 1 nsec in both cases.

For the higher excitation intensities (near sample damage), the possible density-dependence of the diffusion coefficient was included in the conventional manner (Eq. 17) and nonlinear recombination processes were considered. Numerical solutions to the resulting nonlinear partial differential equation were obtained. Parametric studies of the influence of the degenerate diffusion coefficient and Auger recombination rate on the grating decays did not yield a satisfactory fit to the data, particularly for times less than 200 picoseconds. This indicates that the commonly-used degenerate expression for the diffusion coefficient given by Eq. 17 may be insufficient to account for the observed slow grating decay and that other nonlinear diffusion mechanisms must contribute. This is of fundamental importance in laser-induced damage and laser annealing experiments. Any successful model will have to include the contributions of the complicated relaxation dynamics of the initially hot electron-hole plasma, bandgap narrowing, lattice heating and screening. Further details are available in Moss et al. [56] and Smirl et al. [53]. The presentation in this section is derived from those studies.

5. SUMMARY AND CONCLUSIONS

In this lecture, we have reviewed two separate but related series of experiments that measure the nonlinear, nonequilibrium optical properties of germanium. (1) In the first of these, we have described structure in the picosecond interband saturation of germanium. The excitation-probe technique was employed in an extensive study of thin slices of crystalline germanium as a function of temperature, excitation pulse energy, sample thickness and time delay between the excitation and probe pulses from a Nd-glass laser. The results were interpreted successfully in terms of a dynamic band-filling by a hot carrier distribution with intervalence-band absorption contributing at the highest carrier densities ($\sim10^{20}$ cm^{-3}). Also, a strongly intensity-dependent and weakly temperature dependent rise in sample transmission was observed that can exceed 200 psec under certain conditions and that was consistent with a *slow cooling* of the hot carriers (or perhaps diffusion). (2) In the second, a three-pulse variation of the excitation-and-probe technique was used to measure the picosecond evolution of laser-induced transient gratings that were produced in germanium by the direct absorption of 35 psec optical pulses at 1.06 μm. Grating lifetimes were determined for peak optical excitation levels between 1.8 mJ/cm^2 (~50 MW/cm^2) and 36 mJ/cm^2 (~1 GW/cm^2) as a function of grating spacing and sample temperature. For the lower fluence, a linear diffusion-recombination model for the grating decay provided a good fit to the experimental data and allowed the extraction of the diffusion coefficient and an estimation of the linear recombination lifetime. For the higher excitation level, the usual degenerate (and therefore density-dependent) expression for the diffusion coefficient and nonlinear (Auger) recombination processes were included. Comparison of the numerical solutions of the resulting nonlinear partial differential equation to experiment indicated that the commonly-used degenerate version of the diffusion coefficient (from simple Boltzmann theory) was not sufficient to account for the density dependence of the grating decay and that other *nonlinear diffusion mechanisms* must be considered.

Finally, we point out that in experiments not described here [17-20] , we have observed picosecond self-diffraction from transient orientational gratings produced in germanium by an anisotropic (in k-space) filling of the optically-coupled states by direct absorption of the nearly monochromatic polarized exciting radiation, the first direct observation of anisotropic state-filling (as opposed to isotropic state-filling or to band-filling) in semiconductors. A theoretical model confirming the existence and behavior of anisotropic state-filling has also been presented. Agreement is found with all aspects of the experiment. These results indicate that some portion of the *distribution function* is *non-Fermi-like* at these excitation levels.

These experiments demonstrate that qualitative (and eventually quantitative) information can be obtained at these excitation levels and on picosecond and subpicosecond time scales. Because of the large number of simultaneously active processes that are either not observed at lower carrier densities or modified at high densities, more questions remain unanswered than answered and there is room for much future work. Future progress will depend on our ability to experimentally isolate the contributions of individual processes.

The author gratefully acknowledges the active collaboration of T.F. Boggess, J.R. Lindle, A. Miller, S.C. Moss, G.P. Perryman and B.S. Wherrett during various stages of these studies and many useful conversations wit E.W. Van Stryland. He also wishes to thank R.M. Walser for ion-milling the thinner samples and H.J. Mackey and Dwight Maxson for providing technical support in the form of automated data acquisition facilities. This work was supported by the Office of Naval Research, The Robert A. Welch Foundation and the North Texas State University Faculty Research Fund.

REFERENCES

1. Ulbrich, R., this volume.
2. Dewel, G., this volume.
3. Voos, M., this volume.
4. Kennedy, C.J., Matter, J.C., Smirl, A.L., Weichel, H., Hopf, F.A., Pappu , S.V. and Scully, M.O., Phys. Rev. Lett. 32 (1974) 419.
5. Shank, C.V. and Auston, D.H., Phys. Rev. Lett. 34 (1975) 479.
6. Lindle, J.R., Moss, S.C. and Smirl, A.L., Phys. Rev. B 20 (1979) 2401.
7. Auston, D.H., Shank, C.V. and LeFur, P., Phys. Rev. Lett. 35 (1975) 1022.
8. Smirl, A.L., Lindle, J.R. and Moss, S.C., Phys. Rev. B 18 (1978) 5489.
9. Smirl, A.L., Matter, J.C., Elci, A. and Scully, M.O., Opt. Commun. 16 (1976) 118.
10. Elci, A., Scully, M.O., Smirl, A.L. and Matter, J.C., Phys. Rev. B 16 (1977) 191.
11. Bosacchi, B., Leung, C.Y. and Scully, M.O., Opt. Commun. 27 (1978) 475.
12. Leung, C.Y. and Scully, M.O., Phys. Rev. B 23 (1981) 6797.
13. Auston, D.H. and Shank, C.V., Phys. Rev. Lett. 32 (1974) 1120.
14. Smirl, A.L., in Physics of Nonlinear Transport in Semiconductors, edited by D.K. Ferry, J.R. Barker,and C. Jacoboni (New York, Plenum, 1980) pp. 367-399.
15. Smirl, A.L., in Physics of Nonlinear Transport in Semiconductors, edited by D.K. Ferry, J.R. Barker and C. Jacoboni, (New York, Plenum, 1980), pp. 517-545.
16. Glenn, W.H. and Brienza, M.J., Appl. Phys. Lett. 10 (1967) 221.
17. Smirl, A.L., Boggess, T.F., Wherrett, B.S., Perryman, G.P. and Miller, A., 1982 (to be published).
18. Boggess, T.F., Smirl, A.L. and Wherrett, B.S., 1982 (to be published in Opt. Commun.).
19. Wherrett, B.S., Smirl, A.L. and Boggess, T.F., 1983 (to be published).
20. Smirl, A.L., Boggess, T.F., Wherrett, B.S., Perryman, G.P. and Miller, A., 1983 (to be published).
21. Kane, E.O., J. Phys. Chem. Solids 1 (1956) 82.
22. Smirl, A.L., in Semiconductor Processes Probed by Ultrafast Laser Spectroscopy, edited by R.R. Alfano (New York, Academic, to be published, 1983).
23. Fawcett, W., Proc. Phys. Soc. 85 (1965) 931.
24. Cardona, M., J. Phys. Chem. Solids, 24 (1963) 1543.
25. Cardona, M. and Pollak, F.H., Phys. Rev. 142 (1966) 530.
26. Von Borzeszkoswki, J., Phys. Stat. Sol. (b) 61 (1974) 607.
27. MacFarlane, G.G., McLean, T.P., Quarrington, J.E. and Roberts, V., Phys. Rev. 108 (1957) 1377; Proc. Phys. Soc. Lond. 71 (1958) 863.
28. McLean, T.P., in Progress in Semiconductors, edited by A.F.Gibson

(New York, John Wiley, 1960), Vol. 5, pp. 53-102.

29. Levinger, B.W. and Frankel, D.R., J. Phys. Chem. Solids 20 (1961) 281.

30. Dexter, R.N., Zeiger, H.J. and Lax, B., Phys. Rev. 104 (1956) 637.

31. Lax, B. and Mavroides, J.G., Phys. Rev. 100 (1955) 1650.

32. Van Driel, H.M., Phys. Rev. B 19 (1979) 5928.

33. Conwell, E.M., in Solid State Physics, edited by F. Seitz, D. Turnbull,and Ehrenreich, H. - (New York, Academic, 1967), Suppl. 9, p. 171.

34. Meyer, H.J.G., Phys. Rev. 112 (1958) 298.

35. DeVeer, S.M. and Meyer, H.J.G., in Proc. 6th Intern. Conf. Phys Semicond. (Exeter, 1962), pp. 358-367.

36. Reik, H.S. and Risken, H., Phys. Rev. 126 (1962) 1737.

37. Jorgensen, M.H., Meyer, N.I. and Schmidt-Tiedemann, K.J., in Proceedings 7th Intern. Conf. Phys. Semicond. (Paris, 1964), pp. 457-466.

38. Ito, R., Kawamura, H. and Fukai, M., Phys. Lett. 13 (1964) 26.

39. Jorgensen, M.H., Phys. Rev. 156 (1967) 834.

40. Fawcett, W., Page, E.G.S., J. Phys. C 4 (1971) 1801.

41. Herbert, D.C., Fawcett, W., Lettington, A.H. and Jones, D., in Proceedings 11th Intern. Conf. Phys. Semicond. (Warsaw, 1972), pp. 1221-1226.

42. Seeger, K., Semiconductor Physics (Wien, New-York, Springer-Verlag, 1973).

43. Costato, M., Fontansi, S. and Reggiani, L., J. Phys. Chem. Solids 34 (1973) 547.

44. Yoffa, E.J., Phys. Rev. B 21 (1980) 2415.

45. Wautelet, M. and Van Vechten, J.A., Phys. Rev. B 23 (1981) 5551

46. Van Driel, H.M. and Young, J.F., J. Phys. C 15 (1982) L31.

47. Van Vechten, J.A. and Wautelet, M., Phys. Rev. B 23 (1981) 5543

48. Van Driel, H.M., Preston, J.S. and Gallant, M.I., Appl. Phys. Lett. 40 (1982) 385.

49. Smirl, A.L., Miller, A., Perryman, G.P. and Boggess, T.F., J. Phys. (Paris) C7 (1981) 463.

50. Miller, A., Perryman, G.P. and Smirl, A.L., Opt. Commun. 38 (1981) 289.

51. Ippen, E.P. and Shank, C.V., in Ultrashort Light Pulses, edited by Shapiro, S.L. (New York, Springer-Verlag, 1977) p. 110.

52. Latham, W.P., Smirl, A.L., Elci, A., Solid State Electron, 21 (1978) 159.

53. Smirl, A.L., Moss, S.C. and Lindle, J.R., Phys. Rev. B 25 (1982) 2645.

54. Smith, R.A., Semiconductors (London, Cambridge, 1978).

55. Blakemore, J.S., Semiconductor Statistics, (New York, Pergamon, 1972).

56. Moss, S.C., Lindle, J.R., Mackey, H.J. and Smirl, A.L., Appl. Phys. Lett. 39 (1981) 227.

TRANSIENT OPTICAL PROPERTIES OF LASER-EXCITED Si

A. Compaan

Department of Physics, Kansas State University,
Manhattan, Kansas 66506 U.S.A.

ABSTRACT

The optical transmission characteristics of the ~100 nsec duration, high reflectivity phase induced by intense pulsed laser excitation have been studied for photon energies from 0.79 eV to 3.1 eV.
The transmission through ~1μm thick silicon-on-sapphire (SOS) samples
is $\lesssim 1\%$ throughout this range. On ion-implantation-amorphized SOS
the rapid rise in transmission which occurs in the 50-100 nsec following the high reflectivity phase can be used to estimate the regrowth
velocity. Beyond 200 nsec the transmission recovery at 633 and 514nm
can be used to estimate the Si temperature and cooling rate. From
the differences observed between amorphous and crystalline starting
material we estimate the latent heat of recrystallization of the a-
morphous phase.

Most of the direct information concerning the nature of the
transient annealing state in semiconductors produced by pulsed beams
has been obtained from optical probes involving lasers. Time-resolved reflectivity is widely used as an indicator of the annealing
state in silicon [1] . Time-resolved transmission through silicon-
on-sapphire (SOS) has been used across the visible spectrum to infer
the optical constants of the high reflectivity, annealing state
[2] and optical multiple interference effects in reflectivity have
been used to infer conditions in the SOS *after* the high reflectivity
state [3] . Finally, pulsed-laser Raman scattering has been used to
obtain lattice temperatures immediately after the high reflectivity
phase [4] .

The observation of a high reflectivity state during annealing has been widely taken to indicate the presence of a molten silicon phase at the surface, although other explanations of the high reflectivity have been offered [5,2] . However, the Raman measurements consistently indicate lattice temperatures $\lesssim 400$ C, well below the 1412 C melting point of Si [4] . This unusually low lattice temperature raises questions about the nature of the high reflectivity phase and the period of time immediately after. For example, if the Si absorption coefficient at the Raman probe frequency were to decrease substantially, then the Raman experiment might be probing the relatively unheated material below the laser-excited region. This could account for the low temperatures observed. On the other hand, the temperature shift of the band gap in Si normally leads to an increase in absorption with increasing temperature. Thus, if conditions are near thermal equilibrium, one may be able to use the transmission behavior itself to infer an average crystal temperature. Such a procedure is most meaningful in SOS samples where the silicon is epitaxial, yet the heat flow into the sapphire is relatively slow.

We focus our attention here on three types of transmission measurements in SOS samples : 1) a measurement of the spectral dependence of the absorption during the ~100 nsec high reflectivity phase [2], 2) a study of the recrystallization rate in ion-implantation amorphized SOS [6], and 3) an inference of crystal temperature beyond 200 nsec via the transmission behavior [7] . By comparing the results of this last measurement on amorphized SOS with that on crystalline SOS, we are able to infer the latent heat of recrystallization of the amorphous to crystalline transition.

1. EXPERIMENTAL

Most of the experiments were performed using a broad-band N_2-pumped-dye laser at $\lambda = 485$ nm ($\Delta\lambda = 9$nm) as the excitation source focused to a spot size of ~200μm with an 8 ns pulse duration [2] . For the measurement of the spectral dependence of the absorption during the high reflectivity phase, the N_2 laser beam was split to pump a second tunable dye laser (Molectron DL-200) which was used over nearly its full tuning range. For data points in the infrared a frequency doubled Nd : YAG laser (Quanta-Ray DCR-1) was used with part of the fundamental ($\lambda = 1.06$μm) used for a probe or 1.58μm light generated by third-order stimulated Raman scattering in an H_2 pressure cell. This configuration is shown in Figure 1. For the pulsed probes the pulses were optically delayed 25 nsec in a spherical mirror delay line. Transmitted signals in the IR were detected with a Ge p/n diode and part of the undelayed probe was also incident directly on the diode to serve as a monitor to normalize for probe power fluctuations. Signals in the visible were detected either by RCA 7102 or 1P28 photomultipliers. All signals were photographed on

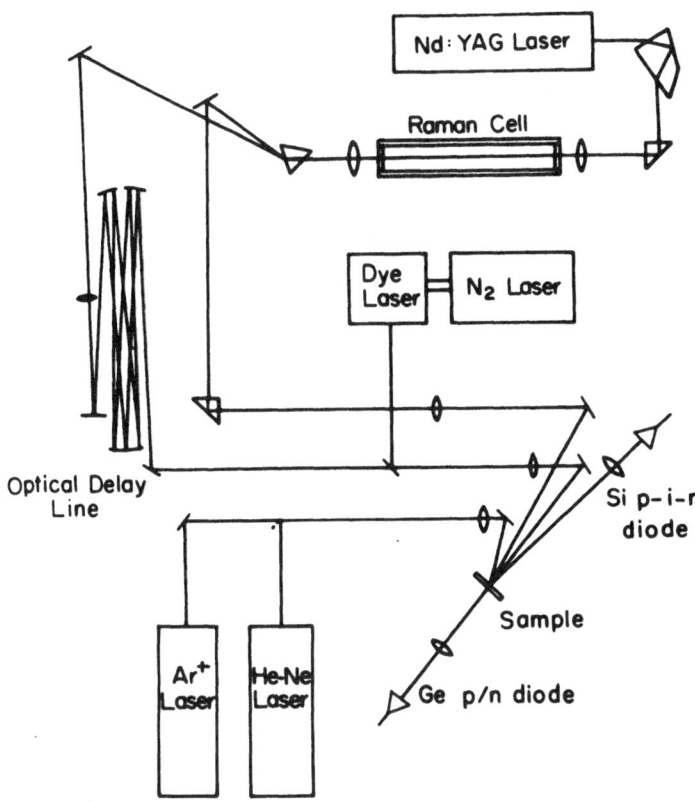

Fig. 1 Experimental apparatus used for time-resolved transmission measurements showing both the pulsed and cw probe beam lines.

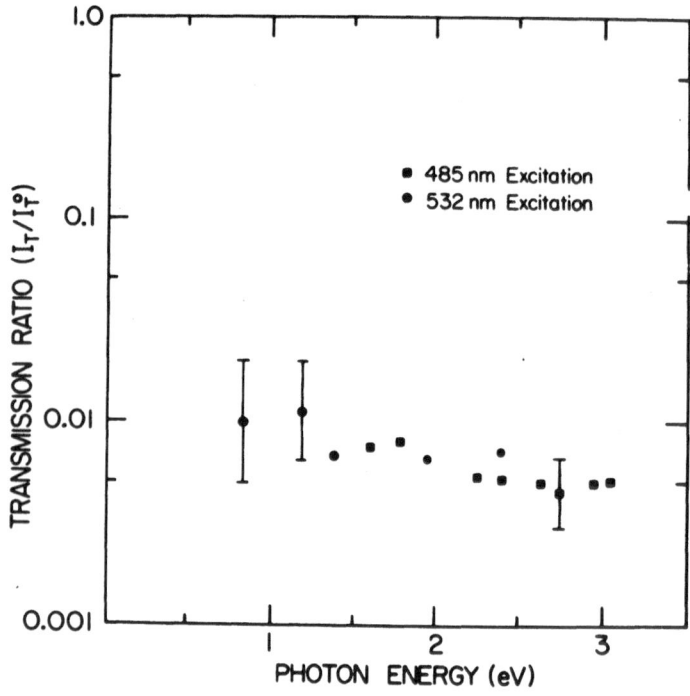

<u>Fig. 2</u> Transmission of SOS sample 25 nsec after pulsed excitation, normalized to the transmission without excitation pulse.

a Tektronix 7904 oscilloscope.

Other experiments used the 633 nm and 514 nm He-Ne and Ar laser lines as a continuous monitor of transmission to measure the recrystallization rate and temperature vs. time after excitation. These cw beam paths are also shown in fig. 1. The focal diameters of the pulsed and cw probe beams varied from 20 μm to 50 μm depending on which excitation laser (485 nm dye or 532 nm YAG) was used. No dependence on probe size was observed in this range. Signals could generally be reproduced for several consecutive laser shots on the same sample position when crystalline SOS was used although usually a fresh sample spot was used for each trace. The general features of both the reflectivity and transmission signals were independent of the various SOS thickness which ranged from 0.6 μm to 2.0 μm and also included a polished wedge with thickness as low as 0.05 μm.

2. HIGH REFLECTIVITY PHASE

The spectral dependence of the transmission during the high reflectivity phase (probe delay 25 nsec peak-to-peak) is shown in Fig. 2. Of the total drop in transmission the well-known transient reflectivity change should account for about a factor of two. At the shorter probe wavelengths this change is somewhat less. The measurements of Fig. 2 were performed with 0.8 J/cm^2 of excitation power at 485 nm on a silicon-on-sapphire (SOS) sample 0.6 μm thick although a variety of SOS thickness ranging from 2.0 μm to 0.07 μm gave identical transmission results. Since the transmission drop during the high reflectivity phase was independent of silicon thickness down to .07 μm we conclude that the absorbing layer must be no deeper than .07 μm. This result was confirmed by calculating reflectivity from a Kramers-Kronig analysis of the spectral dependence of the transmission. Thus, if one takes the transmission drop due to absorption as 10^{-2} and a thickness of .07 μm, the induced absorption is $\alpha'=6\times10^5$ cm^{-1}. If this is taken constant from 3.1 eV to 0 eV (see fig. 3) the K-K analysis yields reflectivity changes which are close to those we observed at $\lambda=1152$, 633, and 514 nm [4] .

The value for the absorption coefficient given above is somewhat below that measured for the molten Si phase (uppermost curve in Fig. 3). However, considering the systematic uncertainties, mainly in the depth of the absorbing region, we believe the transmission data do not provide *independent* evidence against the existence of a normal Si molten phase in this time period. The data do show that there is no *increase* in the absorption length in this spectral region which might explain the low observed Raman temperatures in the time period immediately after the high reflectivity phase.

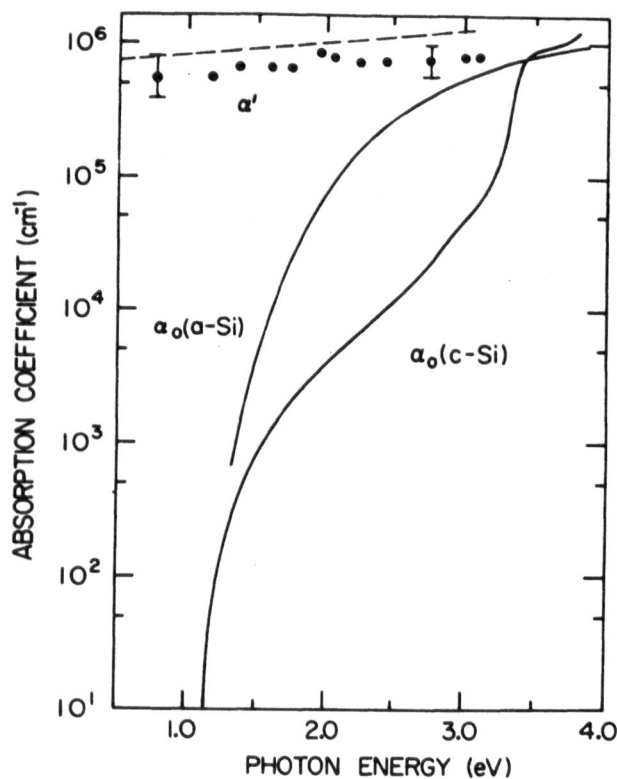

<u>Fig. 3</u> Induced absorption coefficient, α', measured 25 nsec after
pulsed laser excitation, top curve is for molten Si, crystalline
Si and amorphous Si curves are from refs. 8 and 10.

3. RECRYSTALLIZATION KINETICS

As seen above, time-resolved transmission studies on crystalline SOS suffer from two complications : 1) The transmission through the laser-excited silicon during the high reflectivity phase drops to a value 10^{-2} to 10^{-3} of the surrounding, unexcited crystalline material. This creates potential problems of light leakage through the relatively transparent material surrounding the exciting laser spot. 2) Additionally it is difficult to determine the maximum depth of the transient phase change in the silicon (e.g. the maximum depth of melting, if it occurs). The use of ion-implanted amorphous SOS (a-SOS) solves both of these problems. The absorption at 633 nm and at 514 nm in amorphous SOS is much higher than in crystalline SOS so potential problems of light leakage are greatly reduced. Also the complete annealing of the amorphous layer indicates that any transient phase change, e.g., melting, must have penetrated at least to the amorphous-crystalline interface.

Time-resolved transmission measurements on ion-implantation-amorphized SOS were performed using cw lasers at 633 nm and 514 nm to probe during pulsed excitation at 485 nm or 532 nm [6] . Kotani et al.[9] have recently reported similar transmission studies using the self-transmission of a 250 nsec annealing pulse at 590 nm. In the time domain of overlap our results appear to be in general agreement with those of reference 8.

The 0.6 μm SOS sample was implanted with 180 keV Si^{+} to a dose of 3×10^{15} cm^{2}. The projected range, R_p, and range straggling, ΔR_p, are calculated to be .246 μm and .061 μm respectively. Amorphization is generally believed to extend at least to a depth equal to $R_p + \Delta R_p \approx 0.3$ μm. Our cw Raman studies on the implanted surface show no evidence of a residual crystalline-like Raman signal, and from the sapphire side the 520 cm^{-1} Raman line is approximately 1/3 of the strength of that seen from the unimplanted region. Thus, amorphization appears to extend 0.3 to 0.4 μm below the silicon-air interface.

Most of the data were obtained with the 532 nm Nd : YAG pulse with an energy density of $1.1 J/cm^2$ and a 10 nsec pulse duration. Complete epitaxial anneal occurred over at least a 300 μm diameter region with the YAG pulse and 100 μm to 200 μm with the dye pulse. CW Raman polarization studies showed the recrystallized material in both cases to be oriented the same as the substrate; this is our check for epitaxial regrowth.

The time-resolved transmission of the cw lasers at 633 nm and 514 nm was monitored with an RCA 1P28 photomultiplier. The system rise time was less than 5 nsec. A fast photomultiplier was used rather than a photodiode because of the strong absorption of the 514 nm probe and the need to minimize self-heating from the cw probe

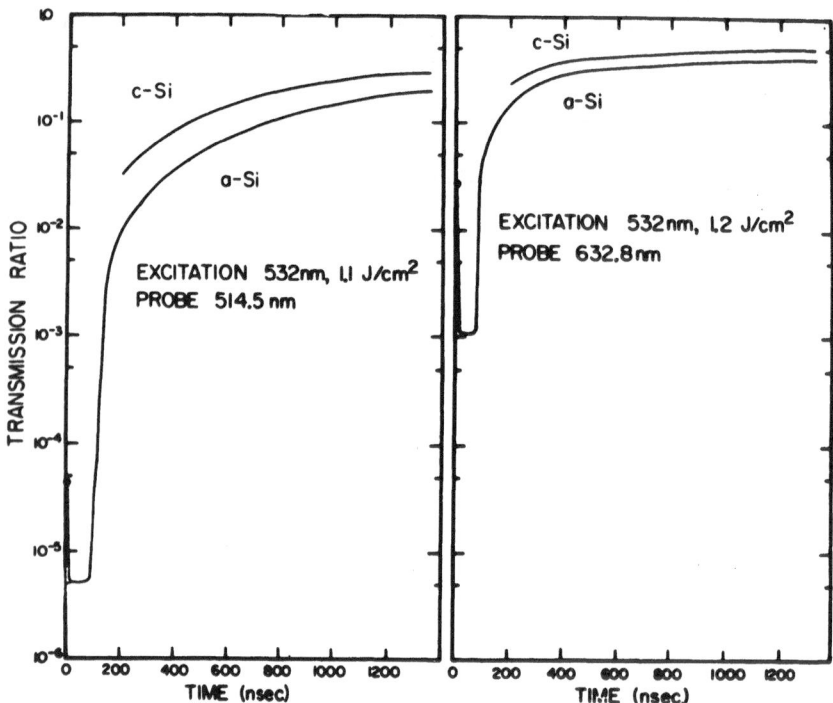

<u>Fig. 4</u> Time-resolved transmission of a 0.35 μm ion-implantation-amorphized layer on SOS 1.83 μm thick following a 1.1 J/cm² pulse at 532 nm.

forced the use of laser powers below 300 mW. The transmission data of Fig. 4 were compiled from sequences of three or four photographs of oscilloscope traces taken with a variety of neutral density filters in front of the detector to prevent saturation. Before each trace the sample was moved to a virgin, unannealed spot. The initial transmission levels are consistent with a .3 to .4 μm thick amorphous layer [10], in agreement with the LSS range calculations [11] and the CW Raman data described above. The data shown were obtained with the YAG laser as the excitation source; data obtained with the pulsed dye laser for excitation were similar.

The data divide naturally into three qualitatively distinct time regimes. The first 80 nsec is a period of very low transmission which coincides with the high reflectivity period. During the next 50 to 100 nsec an extremely rapid rise in transmission occurs at both 633 nm and 514 nm as recrystallization sets in. Finally, beyond ~200 nsec the transmission rises slowly and at a rate consistent with the cooling of the hot Si layer into the sapphire substrate. We discuss each regime more fully below.

The period of minimum transmission coincides well with the period of enhanced reflectivity which was simultaneously monitored using a silicon p-i-n photodiode and a Tektronix 585A oscilloscope. These minima at both wavelengths correspond to a reduction by approximately a factor of ten from the amorphous values. At 633 nm the absolute value of this minimum (relative to the final crystalline SOS level) corresponds reasonably well with the minimum transmission observed from unimplanted SOS [2]. However, at 514 nm the transmission from 10 to 80 nsec is 5×10^{-6} relative to the final value of crystalline SOS. This is far lower than the 3×10^{-3} observed from crystalline SOS during the high reflectivity phase. We cannot completely exclude the possibility that the minimum transmission might be experimentally limited by light leakage around the excited spot. However, we have used a system and technique identical with that for observations on c-SOS which allowed us to observe transmission drops of as such as 1×10^{-3}. Also at 514 nm we cannot exclude the possibility of light leaking through the excited spot itself due to possible tiny "pinholes" in the silicon layer. Checks for pinholes were made in the *amorphous region* by probing the transmission at 457.9 nm where the room temperature absorption in a-Si is much higher [9]. We found the transmission to be less than 6×10^{-6} at this wavelength. In addition optical microscopic examination of the annealed spots showed no evidence of pinholes after laser excitation. Thus the data show that the minimum transmission at 633 nm is about a factor of 100 higher than at 514 nm and in both cases the minimum is only a factor of ten less than the surrounding unexcited a-SOS region.

In the period from 80 to 200 nsec an extremely rapid rise in transmission occurs for both probes. At 514 nm, for example, the signal increases by 2×10^3 during this period. We believe this is a signature of the progress of recrystallisation since by 200 nsec the transmission for both wavelengths is consistent with a hot (~500°C) silicon layer 1.8 μm thick (see below). Essentially all of this rapid rise in transmission occurs in 50-60 nsec. Since the amorphous layer is known to be ~0.35 μm thick, the transmission data imply a regrowth rate of ~6m/sec. This value is near the top of the range of melt front velocities calculated [12] for an infinite Si sample based on a melting model. It is also considerably higher

than the 2.7 m/s inferred by Thompson, et al. [13] for unimplanted
0.4 μm SOS in their electrical conductance measurements using a
pulsed ruby laser (λ= 694 nm).

4. COOLING RATE BEYOND 200 NSEC

The temporal behavior of the transmission beyond 200 nsec
shows a much slower increase which we believe is characteristic of
the cooling of a hot crystalline Si layer. With a laser spot size
of ~300 μm and a Si layer thickness of 1.8 μm the heat flow is essentially one dimensional into the sapphire. However, since the thermal conductivity of sapphire is only 1/6 that of crystalline Si,
the Si layer should rapidly reach a fairly homogeneous temperature
followed by a cooling rate limited by the sapphire thermal conductivity. The transmission rise is simply a response to the decreasing optical absorption coefficient of Si as it cools.

We have calibrated this optical response by measuring the
transmission of 633 and 514 nm beams through oven-heated SOS 2.0μm
thick. Scaling the transmission to 1.8 μm allows one to infer a
silicon crystal temperature for the points beyond 200 nsec. The
inferred temperatures are shown in Fig. 5 for both wavelengths.
The fact that the temperatures show good agreement for the two wavelengths strengthens our assumption that thermalization has occurred
in the Si layer beyond 200 nsec.

The second set of curves in Fig. 5 show the thermal response
of the SOS to a second identical laser pulse on the same spot of
the sample. I.e., the sample in this case starts out entirely in
the crystalline state. Although the laser pulse energy density is
the same to within ±5% the temperatures are clearly lower than if
the sample begins in the amorphous state. The difference in temperature of the 1.8 μm Si layer is approximately 60°C at 200 nsec.
Since the originally amorphous layer thickness was 0.35 ±.05 μm,
we can infer the latent heat of crystallization of the amorphous
state. The difference in the enthalpies of melting of the crystalline and amorphous states is then calculated as :

$$\Delta H_{\ell c} - H_{\ell a} = c\Delta T \left(\frac{1.8\mu m}{0.35\mu m}\right) = 280 \text{ J/gm}$$

where we have taken a constant specific heat of c =.90 J/g-K for
crystalline Si and scaled by the ratio of final Si thickness to the
original amorphous layer thickness. The difference is considerably
smaller than the 570 J/g difference recently estimated by Baeri, et
al., [14] from an indirect method based on recrystallization features observed after pulsed electron beam annealing. However, this
number is only slightly below the result of 339 J/g obtained by Fan
and Anderson [15] from calorimetric data. The estimate we have

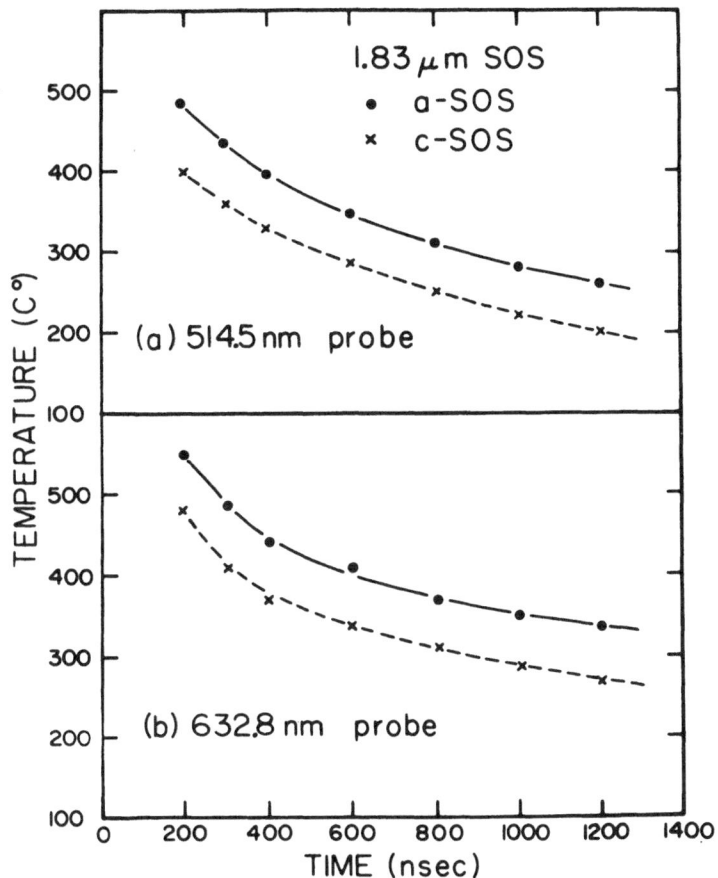

<u>Fig. 5</u> Temperature as inferred from the transmission data on amor-
phous Si similar to Fig. 4 and from additional data taken for a second identical pulse on the same sample spot (i.e. previously recrystallized).

given above should be considered preliminary since we have neglected heat flow into the sapphire, differences in absorbed laser energy due to the different reflectivity and absorption between the crystalline and amorphous states, and also the fact that the amorphous state may enter the high reflectivity phase sooner. Further work with thinner Si layers is in progress to improve this estimate.

5. SUMMARY

Studies of time-resolved optical transmission and reflectivity provide a powerful technique for *in situ* determination of the transient conditions present during the laser annealing process. The optical probe has the important advantages of being non-intrusive and capable of extremely fast time-response. For example, transmission and reflectivity have recently been studied with picosecond time resolution to examine the onset of the high reflectivity phase [16] . In addition to the work described above, Murakami, et al., [3] and also our group [7] have examined multiple interference effects which occur in the cw-reflected beam from SOS samples. These oscillations can also be used to infer a crystal temperature for times greater than approximately 200 nsec. Our results [7] for the temperature inferred from reflectivity oscillations are in close agreement with those of Fig. 5 from the transmission behavior.

The time-resolved transmission studies described here have provided valuable information on the properties of the high reflectivity phase. They have provided an estimate of the recrystallization growth velocity in ion-implanted SOS and allowed an estimate of the latent heat of recrystallization of the amporphous phase.

The support of the Office of Naval Research (contract no. N00014-80-C-0419) is gratefully acknowledged.

REFERENCES

1. Auston, D.H, Surko, C.M., Venkatesan, T.N.C., Slusher, R.E. and Golovchenko, J.A., Appl. Phys. Lett. 33 (1978) 437; Murakami, K., Kawabe, M., Gamo, K., Namba, S. and Aoyagi, Y., Phys. Lett. 70 A (1979) 332.
2. Aydinli, A., Lo, H.W., Lee, M.C. and Compaan, A., Phys. Rev. Lett. 46 (1981) 1640; Yamada, M., Kotani, H., Yamamoto, K. and Abe, K., Phys. Lett. 85A (1981) 191.
3. Murakami, K., Takita, K. and Masuda, K., Jpn, J. Appl. Phys. 20 (1981) L867.
4. Lo, H.W. and Compaan, A., Phys. Rev. Lett. 44 (1980) 1604 ; Compaan, A., Aydinli, A., Lee, M.C. and Lo, H.W., in Laser and

Electron-Beam Interactions with Solids, ed. by Appleton, B.R. and Celler, G.K. (New York, Elsevier, 1983), p. 43.

5. Van Vechten, J.A. and Wautelet, M., Phys. Rev. B 23 (1981) 5543; Wautelet, M. and Van Vechten, J.A., Phys. Rev. B 23 (1981) 5551.

6. Aydinli, A., Lo, H.W., Lee, M.C. and Compaan, A., Bull. Am. Phys. Soc. 27 (1982) 236.

7. Lee, M.C., Lo, H.W., Aydinli, A. and Compaan, A., Bull. Am. Phys. Soc. 27 (1982) 236.

8. Philipp, H.R. and Taft, E.A., Phys. Rev. 120 (1960) 37; Shuarev, K.M., Baum, B.A. and Gel'd, P.B., Sov. Phys.-Solid State 16 (1975) 2111; Lampert, M.O., Koebel, J.M. and Siffert, P., J. Appl. Phys. 52 (1981) 4975.

9. Kotani, H., Yamada, M., Yamamoto, K. and Abe, K., Sol. St. Commun. 41 (1982) 461.

10. Pierce, D.T. and Spicer, W.E., Phys. Rev. B 5 (1972) 3017.

11. Johnson, W.S. and Gibbons, J.F., Projected Range Statistics in Semiconductors (Stanford, Stanford Univ., 1969).

12. See for example Baeri, P., Campisano,S.U.,Foti, G. and Rimini, E., J. Appl. Phys. 50 (1979) 788 and Wood, R.F. and Giles, G.E. Phys. Rev. B 23 (1981) 2923.

13. Thompson, M.O., Galvin, G.J., Mayer, J.W., Hammond, R.B., Paulter, N., Peercy, P.S. in Laser and Electron-Beam Interactions with Solids, ed. by Appleton, B.R. and Celler, G.K. (New York, Elsevier, 1982) p. 209; Galvin, G.J., Thompson, M.O., Mayer, J.W., Hammond, R.B., Paulter, N. and Peercy, P.S., Phys. Rev. Lett 48 (1982) 33.

14. Baeri, P., Foti, G., Poate, J.M. and Cullis, A.G., Phys. Rev. Lett. 45 (1980) 2036.

15. Fan, J.C.C. and Anderson, C.H., J. Appl. Phys. 52 (1981) 4003.

16. Yen, R., Liu, J.M., Kurz, H. and Bloembergen, N., in Laser and Electron-Beam Interactions with Solids, ed. by Appleton, B.R. and Celler, G.K. (New York, Elsevier, 1982) p. 37; Kim, D.M., Shah, R.R.,van der Linde, D. and Crosthwait, D., ibid, p. 85.

TIME-RESOLVED RAMAN STUDIES OF LASER-EXCITED SEMICONDUCTORS

A. Compaan

Department of Physics, Kansas State University,
Manhattan, Kansas 66506

ABSTRACT

The phonon Raman spectrum of Si has been studied as a function
of time during and after the ~100 nsec high reflectivity phase induced by an intense ($\sim$ 1 J/cm^2) laser pulse. In ion-implantation-
amorphized Si the onset of recrystallization is clearly characterized
by the appearance of a sharp Raman feature near 520 cm^{-1} well within
the first 50 nsec. We find no Raman line during the laser-induced
high reflectivity period in either crystalline or amorphous Si and
find immediately thereafter that the optic phonon population is characteristic of a temperature no greater than 450 C. We show also
that the correction factors necessary for evaluating a temperature
from the Stokes/anti-Stokes ratio may be empirically obtained rigorously from the time-reversal invariance of the Raman cross section.

In the few years since its discovery, pulsed laser annealing of
semiconductors has stimulated unusual excitement concerning its potential for reconstruction of ion-implanted surfaces, amorphous deposited layers and for growth of unusual surface alloys. The technique has also generated intense controversy concerning the physical mechanisms responsible for the ultra-rapid regrowth. Khaibullin,
et al. [1] one of two Soviet groups largely responsible for demonstrating the technique, have suggested that a non-thermal-equilibrium
mechanism must be involved. However, the theoretical work of several groups has been predicated on an assumption that normal thermal
melting occurs following the pulsed laser excitation [2] . The laser energy is initially deposited in the electronic system of the
semiconductor due to band-to-band and free-carrier absorption, but

rapid electron-lattice energy relaxation is presumed to channel the laser energy into the lattice on a sub-nanosecond time scale. Such an assumption seems reasonable since picosecond electron lattice relaxation times are normally observed under low carrier density and low temperature conditions. The advocates of a thermal melting model have shown that most of the characteristics of the annealed material have a reasonable explanation in terms of this model. This includes the high diffusion coefficients of impurities and the segregation and precipitation of impurities. The striking observation of cell formation in heavily implanted silicon clearly suggest a moving phase boundary such as that between molten and solid silicon, although segregation coefficients need to be adjusted in a somewhat *ad hoc* fashion to obtain quantitative agreement with observations.

On the other hand, Van Vechten and Wautelet [3] have argued that electron-lattice relaxation may be inhibited under conditions of intense laser excitation and furthermore certain other physical characteristics of the molten phase seem inconsistent with the melting hypothesis [4]. It is argued that a high density plasma may be created by the laser pulse which will substantially alter the properties of the crystalline semi-conductor. It is suggested that the dense plasma may lead to a phase change, possibly a fluid phase, and thus lead to all the properties so characteristic of a moving phase boundary during rapid regrowth.

The motivation for the experiments reported here was to obtain information in as direct a way as possible on the characteristics of the transient annealing state induced by the laser pulse. In many ways Raman scattering with pulsed lasers is an ideal tool for probing these transient conditions in the semiconductor. Preliminary conclusions have been reported earlier [5,6]; here we present a complete description of the experiments and emphasize the extensive checks which have been made to validate the results. Our early Raman studies [5] utilized two simultaneously pumped dye lasers with somewhat limited power. These have now been improved [6] by using a frequency-doubled Nd:YAG laser as the excitation source with a spot area more than ten times greater than the original dye laser pulse. We have placed a major emphasis on understanding and measuring independently the corrections necessary for extracting a phonon population from a raw Stokes/anti-Stokes ratio. We describe here a method which allows the unknown corrections to be obtained experimentally by exploiting the time-reversal invariance of the Raman cross section [7] and does not require the assumption of thermal equilibrium or steady-state conditions in the semiconductor. Our results confirm our original interpretation that for energy densities below 1.5 J/cm^2 the phonon population remains far below that required if silicon were to melt at its normal 1412 C melting temperature.

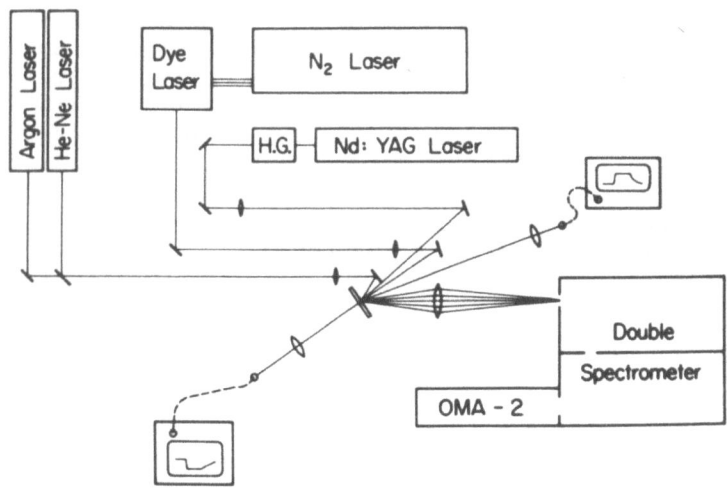

Fig. 1 Experimental apparatus used for time-resolved Raman studies.

OMA-2 DETECTION SYSTEM

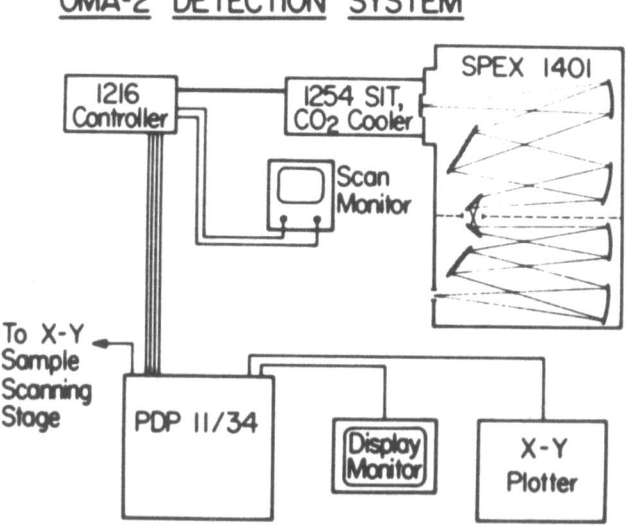

Fig. 2 Vidicon detection system used for Raman data acquisition.

1. EXPERIMENT

The measurement of lattice temperature was obtained by Raman scattering with a two-beam system. An intense first pulse was used to excite the semiconductor followed by a weak, delayed pulse which generated the Raman signal. The first measurements used a Molectron UV-1000 pulsed nitrogen laser simultaneously pumping a coumarin dye (7DTMC) ($\Delta\lambda$ = 9 nm at $\lambda \simeq$ 485 nm) and a Molectron DL-200 dye laser operating at λ = 405 nm. The laser pulse durations were typically 7-8 nsec and the 405 nm pulse was optically delayed thru a 2.55 m long confocal spherical mirror delay line with a high reflectivity multilayer dielectric coating. These first experiments [5] used an ITT-FW 130 photomultiplier with gated photon counting for signal acquisition. The most recent system [6] (see Fig. 1) employs a frequency-doubled Nd:YAG laser (Quanta Ray DCR-1) with 10 nsec duration for the excitation pulse and an electronic delay to trigger the N_2-laser-pumped dye laser. Signal is integrated for typically 5 to 10 minutes on a PARC OMA-2 vidicon cooled to dry ice temperature. The Spex 1401 spectrometer is operated with no intermediate slit and no exit slit. The use of 1800 ℓ/mm holographic gratings provides a dispersion of 21 cm^{-1}/mm at the face plate of the model 1256 SIT. The model 1216 controller of the OMA-2 was interfaced to a PDP 11/34 computer for scan programming and data acquisition. (See Fig. 2) Most of the pulsed Raman data was obtained by combining 10 vidicon channels before digitizing in the 1216 controller. Optimum background subtraction was obtained by acquiring an identical spectrum with only the Raman excitation beam blocked.

Typical amplitude fluctuations of the dye laser pulses were less than ±5% and ±10% for the frequency-doubled pulse at 532 nm. Timing jitter with the electronic delay was ±5 nsec and arose largely from variations in the N_2 laser thyratron firing. Pulse energy densities of the various beams were determined by using a 50 or 200 µm calibrated pinhole placed at the sample position and monitoring the transmitted energy at 10 pulses per second with a Scientech disc calorimeter – the overall size of the various beams was determined from the fraction of energy transmitted through the 50 µm pinhole and by optical microscopic examination of the size of the anneal spots produced on an implanted sample. To serve as an additional and continuous monitor of spot location and excitation pulse power while the Raman data was acquired, a cw neon or argon laser was focussed to the center of the pulsed laser spots on the sample. The duration and shape of the characteristic reflectivity rise [8] associated with the annealing state served as a sensitive monitor of beam stability and energy density.

The use of a two-beam system in an excite/probe configuration provided several advantages; 1) With a 1 mm diameter exciting beam and a 200 µm diameter probe beam, the transverse temperature gradients become insignificant; 2) The absorption length in room temperature, crystalline Si [9] at the probe wavelength of 405 nm is 1.4×10^{-5} cm thus the effective e^{-1} probe depth in a backscatte-ring geometry is only 700 Å since the Raman light is absorbed on the way out at essentially the same rate; 3) The separate, time-dela-yed probe pulse allows a study of lattice temperature as a function of time after the excitation pulse; 4) Finally, flexibility in probe wavelength allows the selection of a wavelength region where resonance enhancement of the Raman cross section could be exploited but where the slope of the cross section variation with photon fre-quency is not extreme *(vide infra)*.

2. THEORY

For Raman scattering with laser frequencies, ν_L, above the band gap in a semiconductor and using the back-scattering geometry, the total Stokes photon generation rate is given by :

$$R_S = \frac{\nu_S^3}{\alpha_L + \alpha_S} \sigma(\nu_L, \nu_S) (1 + n(\Delta\nu)) . \qquad (1)$$

The anti-Stokes generation rate is similarly:

$$R_{AS} = \frac{\nu_{AS}^3}{\alpha_L + \alpha_{AS}} \sigma'(\nu_L, \nu_{AS}) n(\Delta\nu) \qquad (2)$$

where α_L, α_S and ν_{AS} are the absorption coefficients at the laser (ν_L) Stokes (ν_S), and anti-Stokes (ν_{AS}) frequencies, respectively. σ and σ' are the Stokes and anti-Stokes Raman cross-sections, respec-tively, and $n(\Delta\nu)$ is the phonon occupation factor for the optic pho-non at $\overline{\Delta\nu} \simeq 520$ cm^{-1} in silicon.

Strictly, $n(\Delta\nu)$ is the occupation factor at the phonon wave vector $q \simeq 2k$ where k is the photon wave vector in the crystal; however, for convenience, we shall assume that the phonons have equilibrated among themselves (see conclusion section for further discussion) so that the occupation factor may be expressed in terms of the usual Planck distribution characterized by a lattice tempe-rature T_L. In this case, the ratio of phonon generation rates uni-quely determines the lattice temperature from :

$$\frac{R_S}{R_{AS}} = (\frac{\nu_S}{\nu_{AS}})^3 \frac{\alpha_L + \alpha_{AS}}{\alpha_L + \alpha_S} \frac{\sigma(\nu_L,\nu_S)}{\sigma'(\nu_L,\nu_{AS})} e^{h \Delta\nu/k T_L} . \qquad (3)$$

The observed ratio of counting rates may be related to this genera-tion rate by correcting for the spectrometer throughput and detec-tor efficiency at the Stokes and anti-Stokes frequencies. These were determined using a tungsten-halogen standard lamp traceable to NBS standards.

There are two non-trivial corrections to the Raman ratio that appear in Eq. 3. The absorption correction : $(\alpha_L + \alpha_{AS})/(\alpha_L + \alpha_S)$ has the effect of suppressing the anti-Stokes signal because of slightly stronger absorption at the anti-Stokes frequency than at the Stokes [9]. However, Eq. 3 shows that the Raman cross section correction $\sigma(\nu_L, \nu_S)/\sigma'(\nu_L, \nu_{AS})$; has the effect of *enhancing* the anti-Stokes peak because the Raman cross section is increasing as one nears the direct gap in Si [10]. Thus, the two important corrections act in opposite directions on the Raman ratio.

The variation of the absorption coefficient with frequency is available in the literature for several temperatures [9] and has recently been studied systematically by Jellison and Modine [11] up to T=1000 C in Si. The spectral dependence of the Stokes Raman cross section has also previously been measured over the range of interest here [10]. Using these results, we have chosen the Raman probe wavelength to lie near 3.2 eV because there appears to be a relative plateau in the Raman cross section here which would minimize the corresponding correction term in Eq. 3. It should be emphasized that this correction actually involves also the anti-Stokes cross-section, $\sigma'(\nu_L, \nu_{AS})$, which was not measured by Renucci, et al. [10]. However, a simple time reversal argument can be used to obtain the anti-Stokes cross-section if the Stokes cross-section is known over the nearby region [12]. Thus, all of the correction factors in Eq. 3 are known at room temperature. We have readily validated their accuracy by de-focussing the probe pulse to a diameter of 500 µm, removing the first pulse and calculating the temperature from the Stokes/anti-Stokes ratio. With the sample at room temperature (295 K) we obtained a Raman temperature of 300 ± 25°K.

For higher lattice temperature, such as may occur when the first pulse is turned on, the correction $(\alpha_L + \alpha_{AS})/(\alpha_L + \alpha_S)$ is expected to increase since the slope $d\alpha/d\nu$ itself is increasing [9]. Heating narrows the band gap in Si; thus a temperature rise is roughly equivalent to an increase in the laser frequency ν_L. However, the recent data of Jellison and Modine [11] show the effect to be quite small with the correction peaking at 300°C, (see Fig.4) this is apparently the result of the broadening of spectral features which occurs at high temperature. Thus, the spectral dependence of α should have little effect on our inferred lattice temperature, contrary to the claim of Wood, et al. [13].

In addition, however, the corrections due to the spectral dependence of the Raman cross-section must be considered. Again the slope $d\sigma/d\nu$ is increasing near λ=405 nm [10] so that with increasing temperature the size of this correction should also *increase*. But its effect is to further *lower* the inferred temperature, just opposite to the effect of the absorption correction. We have directly checked these effects by heating the substrate to 700K; again with the first pulse off and the probe defocussed a Raman temperature of 680 K was obtained using the *same* corrections as at room temperature. This result confirms our expectation that the product of the two corrections is not likely to be strongly temperature dependent.

3. TIME REVERSAL INVARIANCE

However, it is very important to recognize that all of the above discussion pre-supposes that the behavior of $\alpha(\nu)$ and $\sigma(\nu)$ are the same under intense laser irradiation as under oven-heated conditions. We have, moreover, no direct evidence that conditions are close to thermal equilibrium throughout the time scale of our interest or even that a well-defined lattice temperature exists. Fortunately, the α and σ correction factors may be rigorously obtained without relying on any assumption of thermal equilibrium. We now show how one may obtain these corrections *under the identical conditions of the original Stokes/anti-Stokes ratio determination* by performing one additional Raman measurement at new laser frequency ν_L, chosen to be the same as the original anti-Stokes frequency ν_{AS}. The ratio of the two Stokes rates is then formed :

$$\frac{R_S}{R_{S'}} = \frac{\nu_S^3 R_L \sigma_S(\nu_L, \nu_S)(\alpha_{L'} + \alpha_{S'})}{\nu_{S'}^3 R_{L'} \sigma_S(\nu_{L'}, \nu_{S'})(\alpha_L + \alpha_S)} \tag{5}$$

where, since $\nu_{L'} = \nu_{AS}$ and $\nu_{S'} = \nu_L$, necessarily $\alpha_{L'} + \alpha_{S'} = \alpha_{AS} + \alpha_L$.

We now make use of the fact that the new Stokes cross-section $\sigma_S(\nu_{L'}, \nu_{S'})$ is identical to the original anti-Stokes cross-section $\sigma_{AS}(\nu_L, \nu_{AS})$ by time-reversal invariance. This is readily verified by reference to the diagrams of Fig. 3. (The time reversal symmetry of the Raman cross-section has been discussed by Loudon [12] and experimentally verified under strong resonance conditions by Compaan et al.) [14]. Consequently, if the incident laser flux is held constant, the ratio of Eq. 4 gives exactly the product of correction factors required in Eq. 3 with no knowledge of lattice temperature required nor even that thermal equilibrium or steady state conditions be present. The results of this measurement of the product of the two corrections is shown in Fig. 4 where the temperature scale was obtained from an experiment described later. A correction of less than unity tends to reduce the inferred lattice temperature. Apparently the downward correction from the σ term is greater than the upward correction from the α term. This is not unexpected since, as resonance is approached in most semiconductors, one typically finds the slope $\frac{d\sigma}{d\nu}$ to be larger than $\frac{d\alpha}{d\nu}$ because σ involves the product of two electron-radiation field operators (one for virtual absorption, one for emission) whereas α

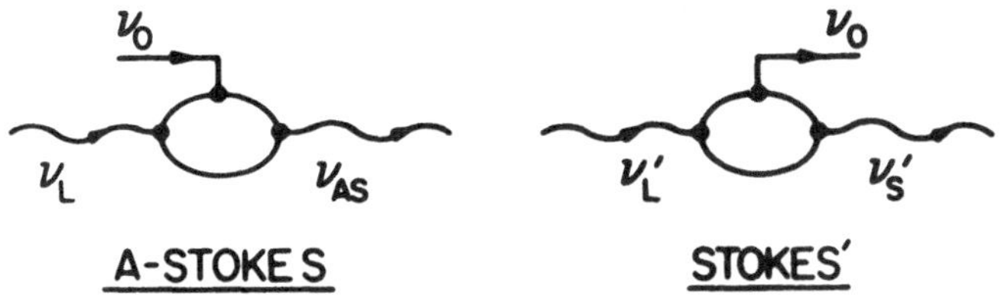

ν_0 ν_L ν_{AS} A-STOKES

ν_0 ν_L' ν_S' STOKES'

<u>Fig. 3</u> Diagrams illustrating the time-reversal symmetry of the original anti-Stokes and the frequency-shifted Stokes processes.

involves only one [10,14,15] .

4. TEMPERATURE MEASUREMENTS

The experimental arrangement of Fig. 1 was used for the time-reversal studies shown in Fig. 4 and also for two studies of Stokes/anti-Stokes ratios (1) as a function of 532 nm laser power at 30 nsec delay and (2) as a function of Raman probe delay with constant 532 nm excitation power.

In Fig. 5a we show the same total correction factor plotted as a function of 532 nm excitation pulse power. We have then applied the correction to the observed Raman ratio and expressed the resulting phonon occupation factor in terms of a lattice temperature in Fig. 5b even though a lattice temperature is not necessarily well-defined under these conditions. The maximum phonon population observed was n=.58. At excitation powers greater than $0.4 J/cm^2$ no Raman signal could be observed at the 30 nsec probe delay of this experiment. The amplitude of the Stokes Raman signal as a function of 532 nm laser energy density is shown in Fig. 6. One observes a rapid decrease in signal for powers beyond $.3 J/cm^2$. This drop in signal occurs just as one observes in the reflectivity monitor that the duration of the high reflectivity phase is beginning to reach the 30 nsec probe delay. Thus, this drop in Raman signal appears to be due to the presence of a Raman-silent phase, no doubt the phase responsible for the enhanced reflectivity also.

The temporal dependence of the Raman temperature following a 532 nm pulse of energy $0.8 J/cm^2$ is shown in Fig. 7. For this energy density the high reflectivity phase lasts for approximately 80 nsec as shown in the inset. Again, no Raman signal was observed for probe delays of 80 nsec or less, i.e., any discrete peak in the

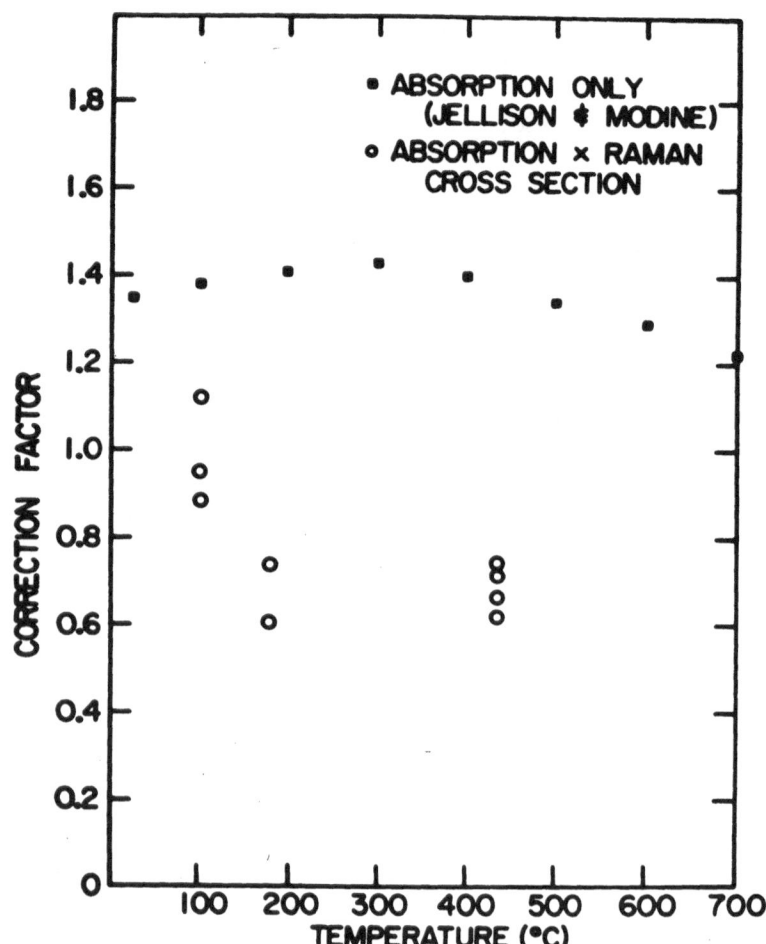

Fig. 4 Upper curve : absorption correction as a function of temperature from Ref. 11, and lower curve : product of both corrections
at 405 nm as measured by the time reversal Raman method. Temperature is inferred from Fig. 5.

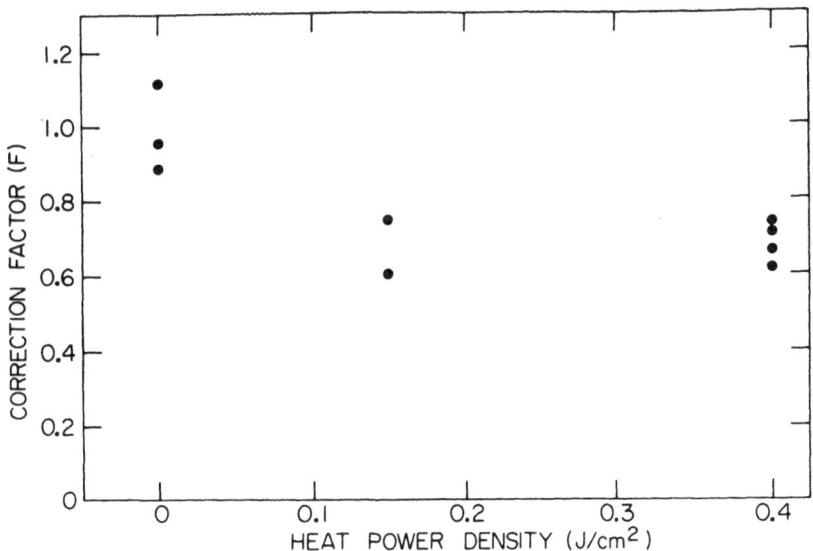

Fig. 5a Product of correction factors at 405 nm as a function of
532 nm laser power.

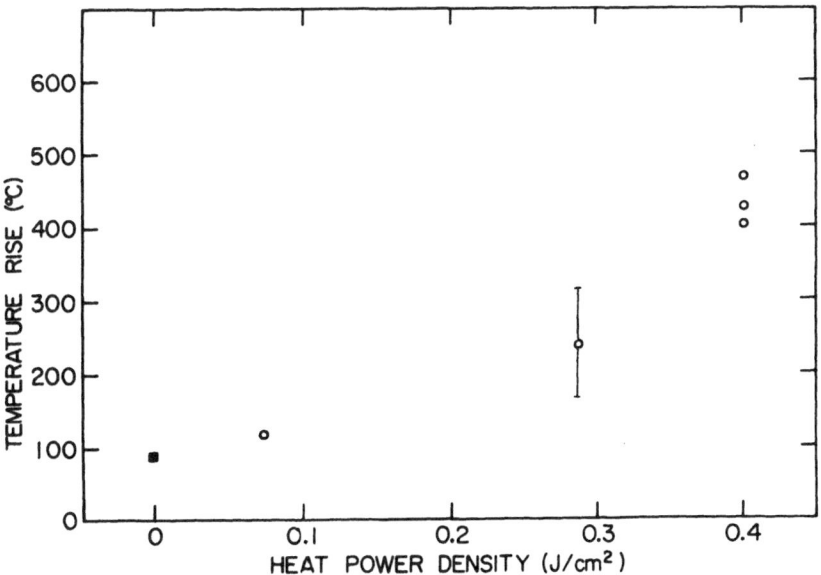

Fig. 5b Temperature rise for the same range of laser powers using
the correction factors of Fig. 5(a). Probe delay 30 nsec.

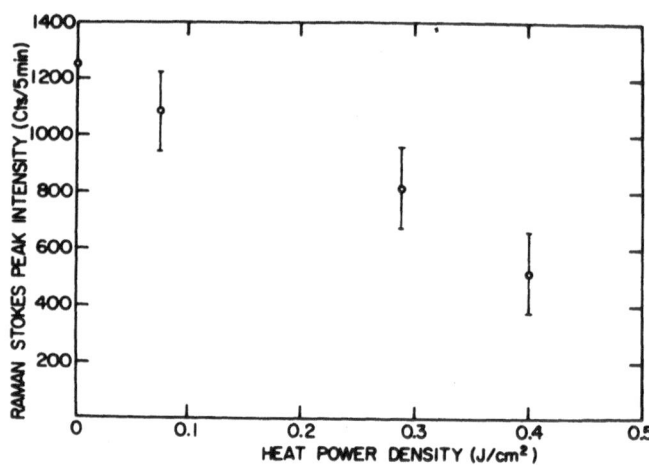

Fig. 6   Integrated Stokes Raman signal near 500 cm^{-1} for the same range of powers as Fig. 5.

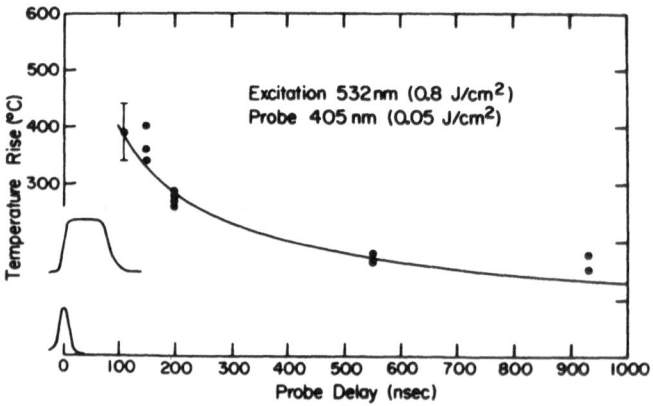

Fig. 7   Phonon population as expressed in a lattice temperature vs. probe delay for excitation power of 0.8J/cm^2.

range from 200 to 600 cm^{-1} must be less than 5% of the normal 520 cm^{-1} Raman line strength. The peak Raman temperature, observed less than 20 nsec after the reflectivity returns to normal, is less than 400°C.

The solid curve in Fig. 7 is a calculation of the surface temperature as a function of time for an instantaneous heat input Q at t=0 using a value for the thermal diffusivity averaged between 300K and 700K and adjusted to fit near 100 nsec. The calculation clearly shows the cooling rate to be consistent with normal crystalline Si values for thermal transport. However, the peak temperature again is far below that expected if the high reflectivity is due to the normal molten phase of Si.

5. RAMAN LINE SHIFT

The multispectral detection capability of the vidicon lends itself naturally to a study of the shift of the normal Raman line. This provides additional evidence on the conditions present in the semiconductor during the 532 nm laser excitation. Figure 8 shows both Stokes and anti-Stokes peaks for very low laser power and for 0.4 J/cm^2. Probe delay was 30 nsec and all other parameters were the same as for the data of Figures 5 and 6. A distinct shift of the peak is clearly noticeable and a slight increase in the width. The observed shift of 13 cm^{-1} is consistent with a crystal temperature of 470 C *if* one assumes a shift equal to that of oven-heated silicon [16] . This is probably not a valid assumption for several reasons including the fact that the laser-excited region is clamped in the transverse direction by the surrounding unexcited material. Furthermore, there may be phonon frequency changes due to the high electronic excitation. These effects have recently been calculated by Biswas and Ambegaokar [17] . Nevertheless, the shift of the Raman line appears to be consistent with the Stokes/anti-Stokes ratios in indicating a surprisingly low lattice temperature.

6. CONCLUSIONS

The use of the time reversal invariance of the Raman cross-section has allowed us to remove any ambiguity regarding the correction factors which are needed to infer a phonon population from the Raman ratio. The combined corrections for absorption and Raman cross-section appear to be slowly decreasing over the range of power densities of this experiment.

We have chosen to express the optic phonon population in terms of a lattice "temperature" although the Raman data themselves provide

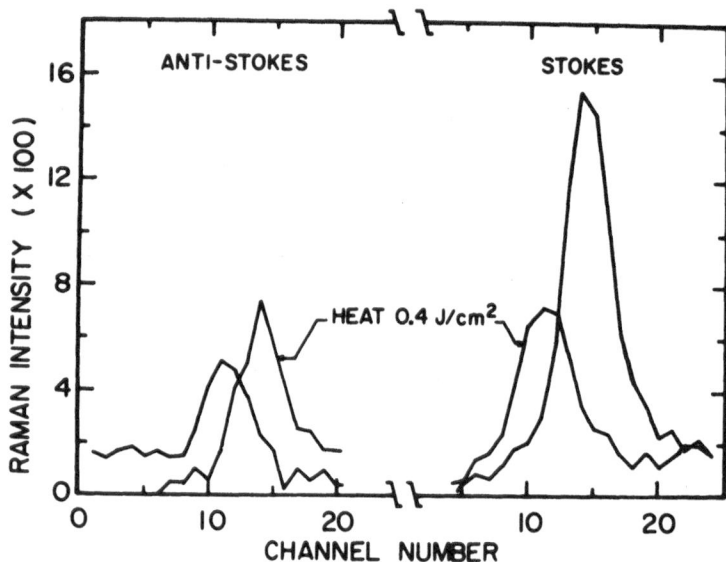

Fig. 8 Raman spectra at 405 nm for zero and high excitation powers at 532 nm.

no strong evidence that a well defined lattice temperature exists. The Raman measurement is sensitive only to optic branch phonons of wave vector $k = \frac{4\pi n}{\lambda} \cong 1.5 \times 10^6$ cm^{-1} where n = 5.0 is the real part of the index of refraction at λ = 405 nm. The optic phonon populations further from the zone center have not been measured nor have populations on the acoustic branches. With this *caveat* in mind, it must be emphasized that the measurement of a very small phonon population (relative to that at the melting point) at only one point on the optic branch is quite sufficient to invalidate the assumptions built into the normal thermal melting model of laser annealing. *viz.* that there is complete thermalization of electron and phonon systems.

It appears that either the laser-deposited energy does not reach the lattice in the subnanosecond time scale normally assumed or that the energy diffuses over much longer distances than normally assumed before reaching the lattice. This would require, e.g., that the laser energy, absorbed by band-to-band transitions and by free-carrier absorption into the carrier system within 1 μm of the surface, be spread over a much larger depth before the energy ultimately comes to rest in the lattice modes. The Raman data

do not directly indicate whether either of these possibilities
actually obtains under strong pulsed laser excitation conditions.

The author is deeply indebted to H.W. Lo, A. Aydinli and M.C.
Lee for their collaboration and especially for their patience and
meticulous attention to experimental detail in the course of this
work.

The author acknowledges many stimulating conversations with
J.A. Van Vechten during the course of this work. He expresses
appreciation also to G.B. Jellison, R. Tsu, M. Balkanski and E.
Haro for discussions of their work prior to publication. The
support of the U.S. Office of Naval Research (contract no. NOOO 14-
80-C-0419) is gratefully acknowledged.

REFERENCES

1. Khaibullin, I.B., Shtyrkov, B.I., Zaripov, M.M., Bayazitov,
   R.M. and Galjautdinov, M.F., Radiation Effects 36 (1978) 225.
2. Baeri, P., Campisano, S.U., Foti, G. and Rimini, E., J. Appl.
   Phys., 50 (1979) 788; Wood, R.F. and Giles, G.E., Phys. Rev.
   B. 23 (1981) 2923.
3. Van Vechten, J.A. and Wautelet, M., Phys. Rev. B 23 (1981)
   5543; Wautelet, M. and Van Vechten, J.A., Phys. Rev. B 23
   (1981) 5551.
4. Van Vechten, J.A., Tsu, R., Saris, F.W. and Hoonhout, D.,
   Phys. Lett. 74a (1979) 417.
5. Lo, H.W. and Compaan, A., Phys. Rev. Lett. 44 (1980) 1604.
6. Compaan, A, Aydinli, A., Lee, M.C. and Lo, H.W., in Laser and
   Electron-Beam Interactions with Solids, ed. by B.R. Appleton
   and G.K. Celler (New York, Elsevier, 1982), p. 43, and references therein.
7. Compaan, A., Lo, H.W., Lee, M.C. and Aydinli, A., Phys. Rev.
   B, Rapid Communication (in press, July 1982).
8. Auston, D.H., Surko, C.M., Venkatesan, T.N.C., Slusher, R.E.
   and Golovchenko, J.A., Appl. Phys. Lett. 33 (1978) 437.
9. Philipp, H.R. and Taft, E.A., Phys. Rev. 120 (1960) 37.
10. Renucci, J.B., Tyte, R.N. and Cardona, M., Phys. Rev. B 11
    (1975) 3885.
11. Jellison, G.E. and Modine, F. (private communication and to be
    published).
12. Loudon, R., Proc. Roy. Soc. London A 275 (1963) 218; Hayes, W.
    and Loudon, R., Scattering of Light by Crystals (New York,
    Wiley, 1978) p.31.
13. Wood, R.F., Rasolt, M. and Jellison, G.E., in Laser and Elec-
    tron-Beam Interactions with Solids, ed. by B.R. Appleton and
    G.K. Celler, (New York, Elsevier, 1982) p. 65.

14. Compaan, A., Genack, A.Z., Cummins, H.Z. and Washington, M., in Light Scattering in Solids, ed. by M. Balkanski, R.C.C. Leite and S.P.S. Porto (Paris, Flammarion, 1975) p. 39.
15. Scott, J.F., Leite, R.C.C. and Damen, T.C., Phys. Rev. 188 (1969) 1285; Richter, W., in Solid State Physics (vol. 78 of Springer Tracts in Modern Physics, Höhler, G., ed) Berlin, Springer-Verlag (1976); Williams, P.F. and Porto, S.P.S., Phys. Rev. B 8 (1973) 1782.
    Birman, J.L., Theory of Crystal Space Groups and Infrared and Raman Processes in Insulating Crystals, in Handbuch der Physik edited by S. Flügge (New York, Springer, 1974) Vol. 25; Genack, A.Z., Cummins, H.Z., Washington, M.A. and Compaan, A., Phys. Rev. B 12 (1975) 2478.
16. Hart, T.R., Aggarwal, R.L. and Lax, B., Phys. Rev. B 1,(1970) 638; Balkanski, M., Wallis, R.F. and Haro, E., (private communication; Tsu, R. (private communication).
17. Biswas, R. and Ambegaokar, V. (private communication) and Bull. Am. Phys. Soc. 27 (1982) 322.

ULTRAFAST PHASE TRANSITIONS IN SILICON INDUCED BY PICOSECOND LASER
INTERACTION

H. Kurz, J.M. Liu and N. Bloembergen

Gordon McKay Laboratory, Division of Applied Sciences,
Harvard University, Cambridge, Massachusetts 02138, USA.

ABSTRACT

Picosecond time-resolved reflectivity and transmission measurements of silicon provide evidence for an ultrafast phase transition
which occurs within the duration of a single picosecond pulse ($\sim$25ps).
Complementary measurements of electron emission and charged particle
emission demonstrate that the lattice is heated to the melting point
even within the range of picoseconds. The phase transition can be
entirely explained by simple thermal melting.

A large amount of experimental work done on laser annealing of
ion-implanted silicon with ns-pulses (PLA) demonstrated conclusively
the thermal nature of the corresponding phase transition [1-6]. It
is clear that the fundamental mechanism of PLA is governed by the
energy transfer from the electron-hole plasma to the lattice. An
important issue is the time scale of this transfer and the distribution of energy among the different phonon modes. Another important
question is at which time scale the simple thermal melting mode may
become inapplicable, or whether melting requires an established thermal equilibrium between O- and A- phonons.
The intention of this paper is to summarize the results of picosecond laser irradiation experiments with respect to the question
whether melting can occur on a picosecond time scale.
The phase transition at the surface of single crystal silicon is
manifested by morphology changes after the irradiation with single
picosecond pulses exceeding certain energy fluences [7-9]. At loca-
tions where the incident picosecond laser pulse fluence lies between
$0.2J/cm^2$ and $0.26J/cm^2$ the surface is transformed into the amorphous
state. At region where the fluence at λ = 532nm lies above $0.26J/cm^2$

transformation back into the original single crystal structure occurs.

The formation of amorphous and recrystallized layers depends strictly on the energy of the laser pulse and is nearly independent of the pulse duration and intensity as the comparison with data of ns-pulse experiments show. Three orders of magnitude changes of the laser intensity does not affect significantly the energy level required for the phase transition. According to the rate equations for plasma formation, this implies that the phase transition is not governed by a critical *density* of electron-hole pairs. It has to be explained by accumulative processes such as increases of phonon occupation numbers.

The thickness of the amorphous layer is of the order of several hundred angstroems and depends on the crystal orientation. The formation of amorphous and recrystallized structures can only be explained by the formation of a liquid surface layer. Following the pulse, such a thin layer is cooled so rapidly by thermal conduction to the underlying substrate that the liquid-amorphous transition can take place. For fluences larger than 0.26 J/cm^2 the melt depth and the thermal gradient are sufficient for the surface to re-solidify at a slower rate, so that epitaxial regrowth of the original crystalline structure occurs. It is known that critical cooling rates for liquid-amorphous transitions depend on the crystal orientation consistent with this observation.

Direct information about the energy transfer between plasma and lattice can be obtained from several distinct photo-induced emission processes:i)Thermionic emission of electrons under normal heating conditions, in which the lattice and the plasma are in thermal equilibrium, is described by the well-known Richardson-Dushman equation:ii)Thermionic emission under anomalous heating conditions in which the plasma retains the energy and the lattice remains cold $(T_e{>}T_l)$. Due to the low value of the specific heat of the plasma, small energy differences between plasma and phonon system would result in highly disparate temperatures. Amplified by the exponential temperature dependence, the thermionic process appears extremely sensitive to the splitting of energy levels between plasma and lattice. If the energy relaxation time is longer than the laser pulse duration, considerable thermionic emission of electrons should be observed.

In addition to these processes, photoelectric emission occurs following simple power laws and dependent on the crystal orientation. Therefore, as soon as the photoelectric emission appears to be independent of the crystal orientation, the formation of a liquid layer has to be assumed.

A positive indicator for high phonon temperatures is the evaporation of atoms from the surface, which occurs nearly simultaneously at the phase transition. As optical reflectivity measurements show, the phase transition at the surface is accompanied by a drastic increase of the optical absorption. This results in a rapid heating

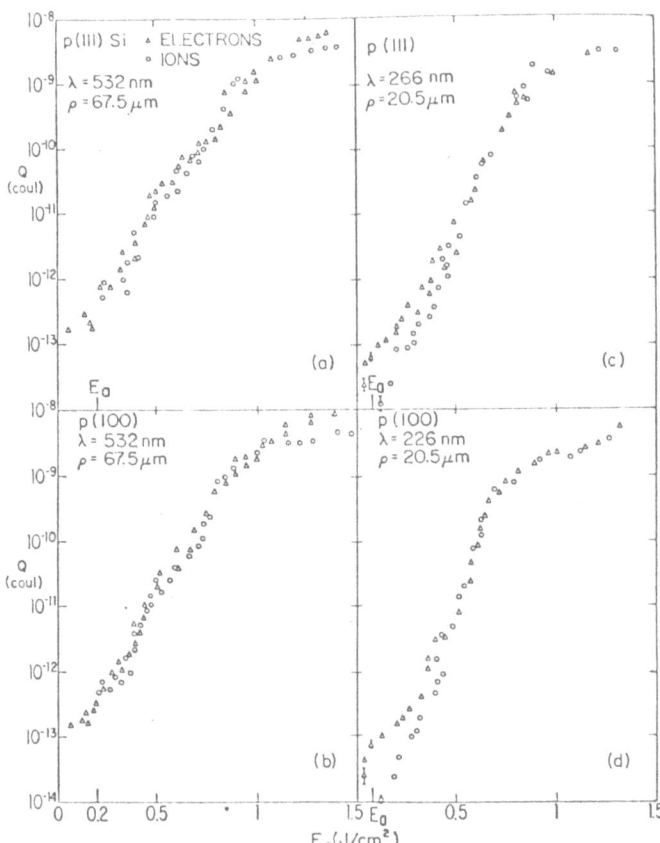

Fig. 1 Integrated positive and negative charges emitted from (111)
and (100) surfaces during and following laser pulses at 532 nm and
at 266 nm. Below E_a, the critical threshold for surface amorphization, no positively charged particles are emitted. The density
differences between positive and negative charges decreases continuously above E_a. Above 0.5 J/cm² in all cases the same amount of
electrons and positive ions are collected. Above 1J/cm² space
charge limitation saturates the emission process. The quantum yields
of the particle emissions above E_a is nearly independent of the
wavelength , indicating that most of the ions are emitted by irradiation of a liquid surface layer formed at the threshold value.

of the new presumably liquid phase·to the boiling point. Recent experiments on the velocity distribution of atoms evaporated by nanosecond pulses of the same fluence as are used to induce phase transitions demonstrate clearly the thermal nature of the particle emission [10]. Evaporation by picosecond pulse is certainly a more complex process whose qualitative understanding needs more experimental and theoretical experimentation. However the observation of positive particle emission in this experiment requires a kinetic energy of the atoms increased significantly above room temperature conditions.

In Figure 1 a typical set of data on electron and ion emission from silicon surfaces under picosecond laser irradiations are shown. Experimental details are published elsewhere [11]. They are scaled to the same energy fluences on axis. There are four cases to be considered, combining different sample orientations and different laser wavelengths. For both orientations the ion emission starts exactly at the laser fluence E_0 where phase changes at the surface can be induced. Below these values ($E_0 = 0.2$ J/cm^2 at $\lambda = 532$ nm and $E_0 = 0.08$ J/cm^2 at $\lambda = 266$ nm) electron emission remains observable. This charge inequality is more pronounced in the case of UV-irradiation. Above E_0 the deviation of charge densities decreases. Above $E_0 = 0.5$ J/cm^2 and $\lambda = 266$ nm a nearly equal amount of negative and positive charges is detected.

Application of the Richardson-Dushman equation for charge densities observed around the phase transition energy E_0 sets an upper limit of the plasma temperature with $T_e < 5000$ K. This indicates that most of the energy is transferred to the lattice during the laser pulse. Numerical calculations of the energy transfer in silicon under conditions described here show that this upper limit corresponds to an electron-phonon relaxation time in the order of one picosecond. This fast energy relaxation process is consistent with the generally accepted theory of deformation potential scattering. Based on this figure, numerical heat flow calculations predict a temporal temperature development as displayed in Figure 2. Taking into consideration the temperature dependence of optical and thermal properties of silicon, the calculation demonstrates that at a fluence level of 0.2 J/cm^2 the surface (Z=0) is heated to the melting point and sufficient energy is delivered to the lattice to supply the required latent heat for melting. Figure 2 also illustrates the high temperature gradients involved in the picosecond heating process.

Optical experiments have played a decisive role during the investigation of laser-induced phase transition. Time-resolved measurements of reflectance and transmittance during illumination with ns-pulses reveal abrupt changes at the phase transition to values consistent with the optical properties of liquid silicon [13,14]. After some careful measurements, refuting earlier erroneous results, there is little doubt that during irradiation of ns-pulses the observed phase transition is due to thermal melting.

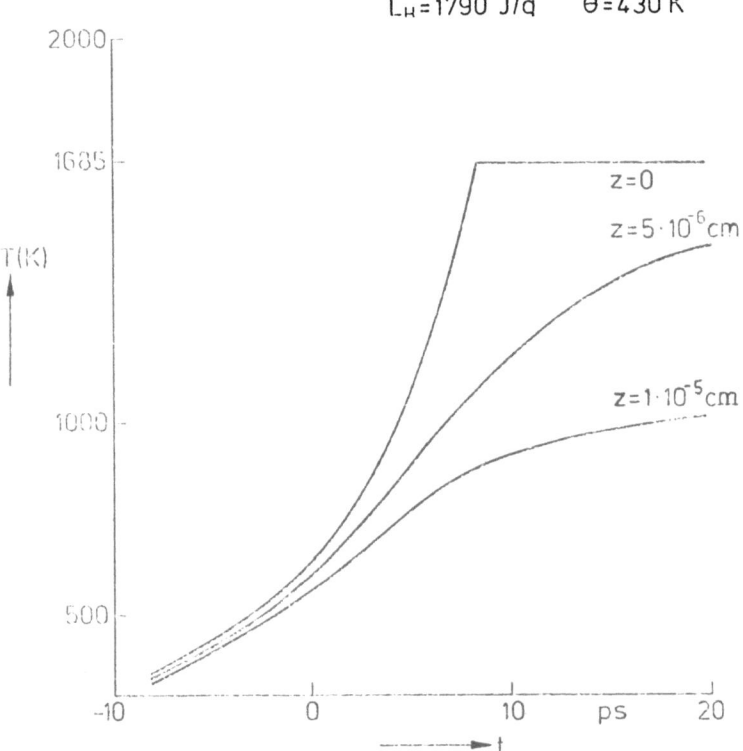

"thermal model" $\tau_e \sim 1\,ps$
temporal temperature profile during
vis ps irradiation $E_0 = 0.2\ J/cm^2$
$\alpha = \alpha_0\ exp\ T\!/_\theta,\ \alpha_0 = 5000\ cm^{-1}$
$L_H = 1790\ J/g$ $\theta = 430\ K$

Fig. 2 Numerical calculation of the temporal temperature profile immediately ($\Delta T = 25$ ps) after irradiation with a single picosecond pulse at 532 nm. At an incident energy fluence of $E_0 = 0.2\ J/cm^2$ the melting temperature at the surface (Z=0) is reached within the Gaussian shaped temporal profile of the laser pulse whose center lies at t = 0. An exponential increase of optical absorption with increasing temperature assumed and the temperature dependence of the thermal conductivity is taken into account.

It is clear that experiments on a picosecond time scale provide more insights and more stringent tests for the validity of the thermal model, as well as for other proposals.

Using probe-and-excite configurations with picosecond laser pulses, the changes in the complex index of refraction induced by heating pulses are determined with a time resolution of 30 ps above

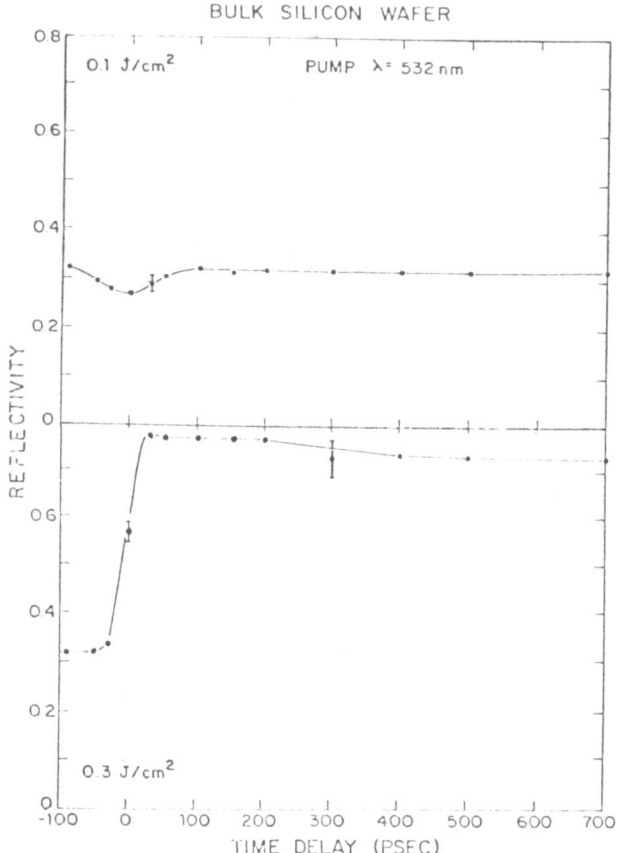

Fig. 3 Reflectivity changes at the surface of a bulk silicon wafer as a function of the probing time delay at two different fluences at 532 nm, 0.1 and 0.3 J/cm^2. The wavelength of the probing pulse is 1064 nm. The observed reflectivity decrease is due to the formation of a dense electron-hole plasma which lasts for approximately 100 ps. At 0.3 J/cm^2, above the fluence threshold for the phase transition, the reflectivity increases rapidly to the liquid state value, indicating melting of the surface within the pulse duration. The plasma contribution is completely masked by the drastic changes of the optical properties encountered in the solid-to-liquid transition.

and below the fluence levels for phase transitions [15-17]. By
using thin silicon films on sapphire the detection sensitivity for

286

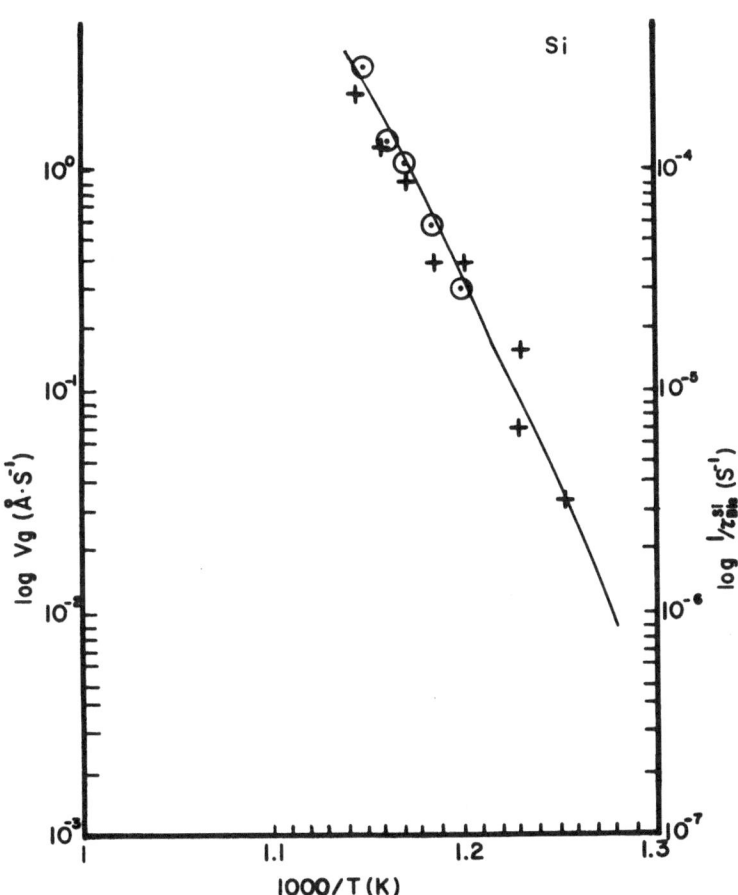

Fig. 3 Arrhenius plot of the growth rate v_g of crystallization in
amorphous silicon layers : + data of Zellama et al [3]. Arrhenius
plot of $1/\tau_{BIC}^{Si}$: $\odot$ present study.

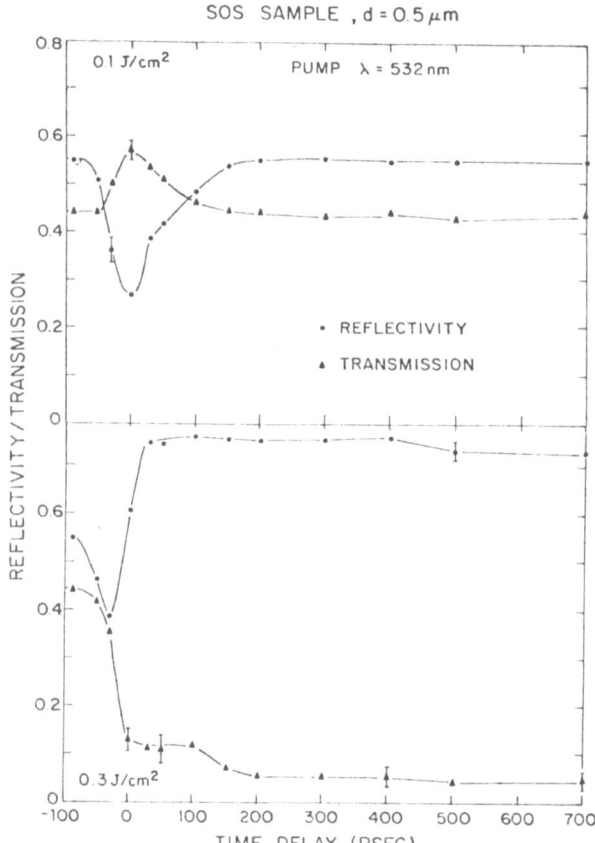

Fig. 4 Reflectivity and transmission changes of a silicon on sapphire sample under the same conditions as in Figure 3. Due to the multiple interferences within the thin silicon film the reflectivity in the unexposed state is higher as compared to the bulk sample. At 0.1 J/cm^2 the signatures are mainly defined by the changes of the real and imaginary part of the refractive index induced by the formation of the plasma. At 0.3 J/cm^2 the transmission drops suddenly to the detection limit and the multiple interferences are suppressed, indicating conclusively the formation of a highly absorbing liquid layer of metallic silicon.

time the transmission drops abruptly. Experimentally we observe
a remanent transmission of 3-5 per cent, even after a long delay.
We believe this data cannot be used to deduce the thickness of the
liquid layer, because most of the transmitted light arises from
scattering by evaporated material in front of the irradiated surfaces. This background can be reduced by placing a pinhole smaller
than the pump spot just behind the sample. The signal-to-noise
ratio is, however, low.

In conclusion, the specific details of surface morphology after
irradiation, emission of electrons and positive ions as well as
changes in the optical properties demonstrate independently that
the phase transition induced by picosecond pulses is caused by melting of the surface as predicted by simple heat-flow calculations.
The energy stored in the plasma is transferred to the lattice in
less than 10 picoseconds. Even without sincere knowledge whether
thermal equilibrium is established between the different phonon
modes it is clear that the energy transferred from the plasma is
sufficient to melt the surface. Lattice heating, melting and vaporization are the dominant processes which occur during the irradiation at these high fluence levels.

ACKNOWLEDGMENT

This research was supported by the Joint Services Electronics
Program under Contract N00014-75-C-0648.

REFERENCES

1. Laser and Electron Beam Processing of Materials, ed. by C.W.
   White and P.S. Peercy (New York, Academic Press, 1980).
2. Laser and Electron Beam Solid Interaction and Material Processing, ed. by J.F. Gibbons, L.D. Hess and T.W. Sigmon (Amsterdam, North-Holland, 1981).
3. Laser and Electron Beam Interactions with Solids, ed. by B.R.
   Appleton and G.K. Celler (Amsterdam, North-Holland, 1982).
4. Baeri, P., Campisano, S.U., Forti, G. and Rimini, E., J. Appl.
   Phys. 50 (1979) 788.
5. Wood, R.F. and Giles, G.E., Phys. Rev. B23 (1981) 2923.
6. Bell, A.E., RCA Review 40 (1979) 295.
7. Liu, P.L., Yen, R., Bloembergen, N., Hodgson, R.T., Appl. Phys.
   Lett. 34 (1979) 864.
8. Liu, J.M., Yen, R., Kurz,H., Bloembergen, N., Appl. Phys. Lett.
   39 (1981) 755.
9. Yen, R., Liu, J.M., Kurz, H., Bloembergen, N., Appl. Phys. Lett.
   27 (1982) 153-160.
10. Stritzker, B., Pospieszczyk, B., Tagle, J.A., Phys. Rev. Lett.

47, (1981) 356.

11. Liu, J.M., Yen, R., Kurz, H., Bloembergen, N., Laser and Electron Beam Interactions with Solids, ed. by B.R. Appleton and G.K. Celler (New York, North-Holland, 1982), p. 29.

12. Dumke, H.P., Phys. Lett. A78 (1980) 477.

13. Auston, D.H., Golovchenko, J.A., Simons, A.L., Surko, C.M., Venkatesan, T.N.C., Appl. Phys. Lett. 34 (1979) 777.

14. Lowndes, D.H., Phys. Rev. Lett. 48 (1982) 267.

15. Liu, J.M., Kurz, H., Bloembergen, N., Appl. Phys. Lett. 41 (1982) 643.

16. Liu, J.M., Kurz, H., Bloembergen, N., in Picosecond Phenomena III, ed. by K.B. Eisenthal, R.M. Hochstrasser, W. Kaiser and A. Lauberau (Springer-Verlag, Berlin Heidelberg 1982), pp. 332-335.

17. von der Linde, D. and Fabriscus, N. in Picosecond III, ed. by K.B. Eisenthal, R.M. Hochstrasser, W. Kaiser and A. Lauberau (Berlin Heidelberg, Springer Verlag, 1982), pp. 336-340.

PLASMA ANNEALING AND LASER SPUTTERING; ROLE OF THE FRENKEL EXCITON

J.A. Van Vechten

IBM Thomas J. Watson Research Center
Yorktown Heights, New York 10598 U.S.A.

ABSTRACT

For sputtering with ion or electron beams it is well established that the contribution of thermal energy in the lattice to
the rate of material desorption is quite negligible. The same can
be shown to be true for pulsed laser sputtering. By considering
the non-thermal mechanisms by which laser irradiation causes atoms
and ions to be sputtered from the surface, we may gain a better
understanding of pulsed beam annealing. In particular, it is noted
that a fraction vastly exceeding the thermal equilibrium value of
the material sputtered (by ions, electrons or photons) is in highly
excited electronic states. The contribution of these excited electronic states to the sputtering and laser annealing processes is
indicated. They are also related to the "Frenkel" or "continuum"
or "hyperbolic" excitons that were previously invoked to explain
interband optical spectra and to the zero-crossings of the real
part of the dielectric constant, $\varepsilon_1(q,\omega)$. It is also shown that
both the established theory of internal photoemission and the time
resolved thermal radiation measurements of Hanabusa and Suzuki
show that the effective heating depth for laser irradiated molten
Si of order 1 μm, far greater than the optical absorption depth of
9 nm. This result contradicts recent speculations about superheating molten Si with the much lower energy pulses used for laser
annealing. Finally, additional evidence for the Bose condensation
theory of pulsed beam annealing is found in a sharp peak reported
by Nakayama et al. of the laser sputtering yield of P from GaP at
photon energies near the band gap and the observation that GaP
cannot be driven into the high reflectivity phase characteristic
of pulsed laser annealing when photon energies are as low as the
peak yield value.

1. INTRODUCTION

When one rates various forms of energy according to their degree of order or coherence, heat is rated absolute lowest and laser radiation amongst the absolute highest. The word "heat" connotates total disorder and equipartition of energy amongst all the modes of the system. Laser radiation, and other forms of directed energy, may be monoenergetic, unidirectional, and tightly localized spatially, within the limits of the Uncertainty Principle. Consequently, if there is some specific task (excitation of a particular electronic level, hopping of an interstitial atom to the next equivalent site, activation of some chemical reaction, or whatever) to be accomplished, some mode of radiation can generally be found to do this with maximal efficiency. Furthermore, *any radiation* that excites the target to energy levels significantly greater than ambient thermal energies, kT, may generally be expected to be much more efficacious than simple heat at driving processes which have an activation energy (enthalpy), ΔH, much greater than kT, provided only that appreciable matrix elements couple the radiation to the initial, intermediate and final states of the system. This radiation enhancement of the rate will persist for as long a period of time as the excitation of the system retains its high energy, nonthermal character, i.e., until the energy is scattered down into all the low-energy modes of the thermal distribution. This requires not one, but a great many scattering events [1] (For examples, Tsai and Trevino find [1] that 200 (40 ps in liquid Ar) are not enough to thermalize the phonons of a shock wave).

The degree to which the rate of the radiation induced process exceeds the strictly thermal rate may be correlated to the magnitude of ΔH. Let us write the strictly thermal rate, R_y, in the elementary form (neglecting entropy factors and the like):

$$R_y(T) = \nu_0 \exp(-\Delta H/kT) \tag{1}$$

where ν_0 is the "attempt frequency", i.e., the minimal time in which the system could accomplish the task. Suppose the system is irradiated with photons of energy $h\nu$. If :

$$h\nu > \Delta H \gg kT , \tag{2}$$

the maximal radiation induced rate of the process, $R_\varrho(I)$, where I is the intensity of the radiation, may be written for the simplest (linear) case :

$$R_\varrho(I) = \nu_0 f(I) \tag{3}$$

where :

$$f(I) = 1.0 \quad \text{if} \quad I > I_0$$

$$f(I) = I/I_0 \quad \text{if} \quad I < I_0 \tag{4}$$

and I_0 is a saturation intensity at which the radiation source is so intense that the system never lacks a quantum to drive the process. In this case, the rate of the process may be enhanced over the strictly thermal value, $R_T(T)$, by the factor exp $(\Delta H/kT)$. If $\Delta H = 2.6$ eV, a value reported [2] for the thermal crystallization of amorphous Si, the maximal enhancement (estimated in this very crude manner) over $R_T(1200K)$ would be a factor of 8×10^{10}; a process that would require 20 minutes at 1200 K would then require only 14 ns even at T = OK.

Examples of radiation-enhanced processes that are not so marred by controversy as pulsed beam annealing include the migration of point defects [3-6] , gliding of dislocations [7-9] and sputtering [10-12] . (A typical value for the radiation induced enhancement rate for point defect migration [6] is 10^5 with a typical $\Delta H \simeq 1.3$ eV. Enhancement factors for dislocation glide have not been accurately determined but are known [8] to be at least a factor of 10^3). This author has previously discussed [13-14] the relation between pulsed beam annealing and the radiation-enhanced rates of point and line defect migration. Particularly because much of the recent controversy regarding pulsed beam annealing has stemmed from varying interpretations [13-19] of laser sputtering data [16-30] it may be useful to compare the very rapid annealing effect that may be had with pulsed laser, electron or ion beams to the sputtering processes produced by the same types of radiation. Also, as the ΔH's for sputtering are relatively large (the heat of vaporization of Si is 4.72 eV), radiation enhancements of the rates are relatively large and unmistakable. Therefore, in Sec. 2, we shall consider sputtering in general and laser sputtering in particular.

Hanabusa and Suzuki measured [21-24] the thermal (black-body) radiation transient during their study of very high energy density laser sputtering of Si. Their targets were clearly melted, [24] , but the maximum T was not reached until 20 ns after the peak of a 7 ns FWHM pulse. Moreover, this maximal T is only about 2100 K, much less than estimated by thermal model calculations that neglect the migration of photoexcited carriers. We find, also in Sec. 2, that this is a natural consequence of the long mean free path between inelastic scattering events in metals, such as molten Si, and the tendency of photoexcited carriers to cross the Schottky barrier into the crystalline Si, where they cause more Si to melt rather than to produce any superheating of the surface layer, despite the short optical absorption length. (This effect is known as internal photoemission.)

One of the more striking features of the sputtering data, for all forms of sputtering, is the very large fraction of the material that leaves the surface in highly excited electronic states [30,31]. Moreover, for the case of laser sputtering from molten Si, atoms continue to leave the surface in these highly excited states for more than 100 ns after the end of the laser pulse [23] which indicates the corresponding highly excited states in the (condensed phase) target have lifetimes at least this long. These excitations are degenerate with band states of the target so they can be only metastable. Such metastable excitations have previously been invoked to describe interband optical spectra [32-39] and local field effects [40-42]. They have been denoted by several terms, including "continuum exciton", "small exciton", "hyperbolic exciton", and "Frenkel exciton"; the term Frenkel exciton is adopted here as it was also in the most developed theory of local field effects [42] and in this author's contribution to the Bose condensation theory [14,43-45] of pulsed beam annealing. These "Frenkel excitons" and their relation to the values of frequency, ω, and of wavelength, q, at which the real part of the dielectric function, $\varepsilon_1(q,\omega)$, crosses zero are discussed in Sec. 3.

While it is rather obvious that radiation ought to, and does, dramatically increase rates for some period of time, it is not so obvious or well established how long that period, before the absorbed energy becomes thermalized, ought to be. The sputtering data also gives us some information on that subject. In Sec. 4 we (again) consider the rate of thermalization in the light of the most recent data and insight.

Conclusions are drawn in Sec. 5. In particular, it is noted that a sharp peak [20] in the yield, $Y_p(h\nu)$, of P atoms laser-sputtered from GaP as a function of photon energy at values of $h\nu$ near the band gap, i.e., the fact that $Y_p(h\nu)$ *decreases* abruptly for *increasing* $h\nu$ somewhat above the band gap (Fig. 1), can be understood in terms of the Bose condensation theory when this is coupled with consideration of the requirements for self-confinement [46,47] of the photoexcited plasma of the Frenkel excitons. Only if the absorption constant for the incident light is sufficient that the initial gradient of plasma reaches a critical value will self-confinement occur. It is here proposed that there is an inverse relation between laser sputtering and pulsed beam annealing; if the Frenkel excitons undergo Bose condensation, the fraction in the superfluid phase is decoupled from the lattice and allows the annealing to proceed. The fraction of Frenkel excitons remaining in the normal fluid phase (See Table 1) when there is Bose condensation, and all of the Frenkel excitons produced when there is no condensation may contribute to the damage of the crystal that is observed

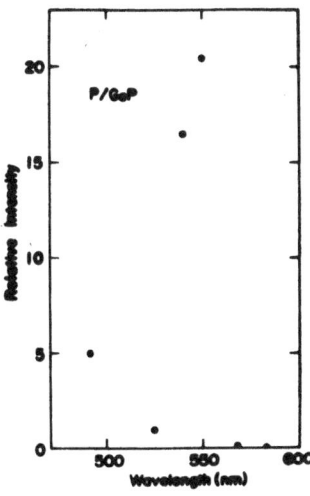

Fig. 1 Laser sputter yield of P from GaP as a function of photon wavelength at *constant incident energy density* 0.1 J/cm^2 in a 10ns dye-laser pulse, as reported by Nakayama et al., [20] . Note the sharp peak in the yield for photon energies near the room temperature band gap of this indirect gap semiconductor, where the optical absorption constant is very low and the energy is absorbed with only a small gradient. The sputter yield is less for $\lambda < 545$ nm even though more energy is absorbed nearer the surface and each photon imparts a higher excitation. It is also known (see I.B. Khaibullin et al., Radiat. Effects 36 (1978) 225 and K. Murakami et al., Phys. Lett. 70A, (1979) 332)that GaP can be laser-annealed at the peak wavelength and at longer wavelengths without ever entering the characteristic high reflectivity phase that this authors ascribes to the self-confined, Bose condensation of Frenkel excitons [14, 43-47] . A.D. Compaan (private communication) has shown that the laser power at $\lambda = 545$ nm can be raised to the point the sample surface is clearly damaged without producing the characteristic high reflectivity phase. As a critical gradient in the absorbed energy density is required for the self-confinement phenomenon according to the theory of Wautelet and Van Vechten, [46]and [47], this and the Bose condensation are expected to occur only for wavelengths shorter than a critical value where the absorption is sufficient to produce the critical gradient. Thus, this yield curve is explicable in terms of the requirements for self-confinement leading to Bose condensation, which would in turn decouple the superfluid fraction of the excited carriers from the sputtering processes and reduce the yield. The obvious consequence that the high reflectivity phase should appear for λ shorter than the peak seems to be confirmed by preliminary observations.

Table 1

T_L	N_{fc} 10^{21}	σ_{fc} 10^{-17}	α_{fc} 10^{4}	σ_{fc} 10^{-17}	α_{fc} 10^{4}	σ_{fc} 10^{-18}	α_{fc} 10^{4}
°C	cm^{-3}	cm^{2}	cm^{-1}	cm^{2}	cm^{-1}	cm^{2}	cm^{-1}
		(3.39µm)		(1.06µm)		(0.514µm)	
300	1.1	3.3	3.6	0.97	1.1	3.3	0.36
400	2.1	3.9	8.2	1.1	2.4	3.9	0.81
500	3.7	4.4	16.	1.3	4.9	4.4	1.6
600	6.0	5.0	30.	1.5	8.9	5.0	3.0
700	9.3	5.6	52.	1.6	15.	5.6	5.2

For various surface lattice temperatures, T_L, that have been measured during pulsed laser annealing by Raman scattering [54,55], the fraction N_{fc} of the total density of excited electrons and holes, 4×10^{22} cm^{-3}, estimated to be in the self-confined plasma layer [46,47 and 54] are calculated [14] to be in the normal fluid, or free carrier state while the remainder are in the Bose condensed, superfluid state. Also shown is the free carrier absorption, α_{fc}, due to these carriers at various wavelengths calculated according to the empirical formula of D.E. Ackley and J. Tauc, Appl. Optics 16 (1977) 2806 and using the cross section, σ_{fc}, measured at 1.06 µm by K.G. Svantesson and N.G. Nilsson, J. Phys. C : Solid State, 12 (1977) 3837.

to extend to depths 10 µm below the irradiated surface [48] and to the sputtering. This hypothesis may be tested by determining if the value of hν at which GaP enters the high reflectivity phase characteristic of pulsed beam annealing corresponds to the hν at which Y_p (hν) drops. Preliminary data seem to support this prediction [49].

2. SPUTTERING

It is well established [10-12] that sputtering with ion or electron beams are essentially non-thermal processes. Although the target may be warmed by the radiation, the thermal part of the energy in the lattice makes a very minor contribution to the net rate of sputtering. A convincing demonstration of this fact can be made from measurement of the distribution of total excitation energy (electronic plus kinetic) of the sputtered species. Data for Al sputtered [31] with 30 keV Ar^+ ions, with an RF plasma at low density (0.3 Pa) and at intermediate density (3 Pa) are shown in Fig.2.

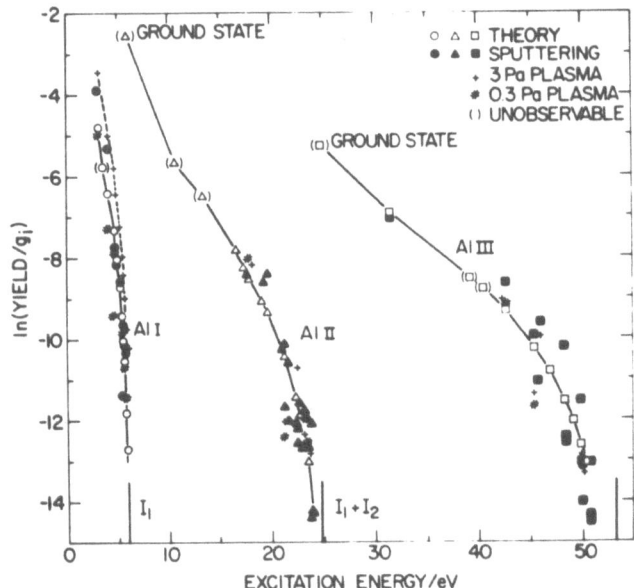

<u>Fig. 2</u> The fraction of sputtered Al atoms and ions (Al neutral ≡
AlI, Al$^+$ ≡ AlII, and Al^{+2} ≡ AlIII) that are in excited states 4 eV
or more above the ground state of each species is many orders of
magnitude above that consistent with thermal equilibrium. The kinetic energy of the species is also far higher than thermal values.
But most strikingly, there are as equally as many species with 5 eV
total excitation energy (kinetic plus electronic) as there are with
20 eV and as many of each as there are with 45 eV; there is no re-
semblance to any thermal distribution for which the number must decrease as the energy increases. From Kelly et al., [31] .

(The theoretical curves are from [11].) For all three types of
radiation the distributions are indistinguishable from type to type.
Note that there are equally as many Al^{+2} ions (denoted AlIII) with
45 eV excitation energy as there are Al$^+$ ions (denoted AlII) with
20 eV excitation energy and this number is also equal to that of
neutral Al atoms with 5 eV excitation energy. There is no resem-
blance to any Maxwell-Boltzmann distribution !

     Moreover, the velocity distribution, f(v), of the various spe-
cies [10-12,18-21] is found not to be of the Maxwellian form characteristic of thermal desorption (evaporation), f_m(v) :

$$f_m(v)dv \propto v^3 \exp(-Mv_X^2/2kT)dv \qquad (5)$$

where M and T are the mass and characteristic temperature of the

sputtered species. Instead, sputtering with ions or electrons usually produces [9-10] a velocity distribution of the form:

$$f(v)dv \propto M^2 v^3 / (Mv^2/2 + U)^3 \, dv \, . \tag{6}$$

The parameter U is taken to be the surface binding energy of the
species. This form may be derived [50,51] within the "linear
collision cascade" mode [52] of sputtering. Note that the probability for large v, high energy, excitation is exponentially greater
for (6) than for (5). (Despite its general success at predicting
the f(v), the linear collision cascade model does not accurately
account for the absolute yield nor for the large fractions of electronic excited states [11,18,30]).

The reason that the thermal energy of the lattice should make
negligible contribution to the net rate of sputtering by ion or electron beams is not difficult to understand. The energy required to
desorb an atom is so high that there is very little probability that
so much energy will ever be localized on any one atom if the distribution is thermal, i.e., according to the equipartion of all the
modes of the solid. However, when the incident sputtering particle,
say a 30 keV Ar$^+$, first strikes the target the energy transferred
to the first several generations of displaced atoms is quite large
compared with U. These atoms have a good probability to leave the
surface with substantial excess energy. The heat produced as the
energy of the energetic collisions dribles down into the thermal
modes and warms the target simply cannot compete with these collisions themselves in concentrating several eV of energy onto single
atoms. Part of the reason for this is that the quantum of lattice
heat, the phonon, has a typical energy of only about 10 meV, or
10^{-3} of what is involved in sputtering.

As a further illustration of this point we show in Fig. 3 a
color photograph of the thermal radiation coming from a resistively
heated Au sample (an effusion furnace) and from a Au sample being
sputtered by an ordinary (10 keV) CW e-beam evaporator [53]. The
two sources have been carefully adjusted to produce equal total
yields through a matched pair of well columnated orifices localized
well within the center region of both sources and measured by a
pair of quartz crystal oscillator thickness monitors. It is evident that the e-beam (sputter) source is cooler than the (thermal)
resistively heated source. (An optical pyrometer indicated a 300K
difference.) However, the e-beam has heated the Au target sufficiently that normal thermal melting has occurred. When the two
sources are at the same temperature, the difference in the rates is
a factor of 13. For a Si target at its melting point the e-beam
induced rate of sputtering was found to be more than 100 times the
thermal rate of 6×10^{16} cm^{-2} s^{-1}. A similar effect, but with even
larger magnitude, is found for the desorption of oxides and organic
species from various metal surfaces.

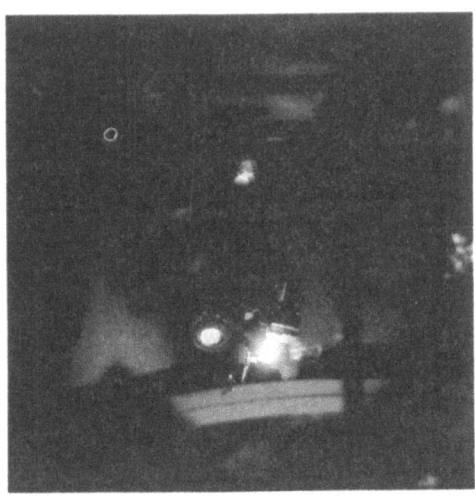

<u>Fig. 3</u> Color photograph showing surface of resistively heated effu-
sion source of Au and 10 keV, CW e-beam "evaporation" (sputtering)
source, also Au, adjusted to give equal rates. Note that e-beam
source is much cooler indicating that the electronic excitation due
to the impacting electrons is much more important than simple thermal excitation at producing the "evaporation" yield.

Let us begin the subject of pulsed laser sputtering with the case of
very high energy density, so that clear evidence of surface heating
and formation of the molten phase is found. Such data is available
from the work of Hanabusa et al. on molten Si [21-24] . With 7 ns
FWHM pulses containing [24] 2 J/cm^2 of 532 nm plus 6 J/cm^2 of 1.06
light (well above the "laser annealing" regime and into the laser
damage regime), they made time-resolved measurements of thermal
(black-body) radiation near 656 nm. (See Fig. 4.) The ratio of
the peak intensity (which does not occur until 30 ns after the beginning of the pulse) to that after 120 ns, when the intensity is
essentially time-invariant and T must be near the melting point
(1685K), is only 11 to 1. We note that this ratio and the radiation law :

$$I_p(\nu,T) {}^+\!\propto \nu^5 \left[\exp(h\nu/kT) - 1\right]^{-1} \tag{7}$$

imply a maximal T(t) at t = 30 ns of only about 2100 K.

Why should the maximal T be reached so late after the peak of
the laser pulse and why should the degree of superheating of the

438

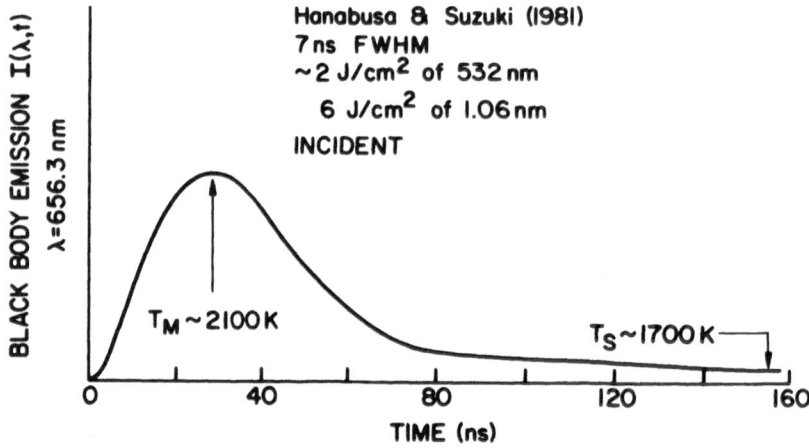

Fig. 4 Time resolved thermal emission transient measured by Hanabusa et al., [21 to 24] , during their laser sputtering study with 7 ns FWHM pulses having an order of magnitude more energy density than those used for pulsed laser annealing. These pulses severely damage the target and drive its surface temperature above the melting point, 1685 K, but the degree of super heating, to about 2100K is much less than simple thermal models predict and the maximal temperature is only reached 20 ns after the peak of the pulse.

molten layer, where the light is absorbed,be so slight ? With an order of magnitude less energy incident and absorbed (once the high reflectivity state of pulsed laser annealing is reached, the 1.06μm light is absorbed with about the same reflectivity and absorption constants as the 532 nm light [54,55]), Yen et al. estimated [17] the surface temperature would be driven to 3500 K within the duration of a 20 ps laser pulse ! Simple thermal models, which assume the energy of the photon is delivered as heat in the lattice at the point where photon is absorbed [56] , predict that the surface temperature in Hanabusa et al.'s experiment would have gone to 10^4 K 2ns past the peak of its 7ns FWHM pulse were it not for the effect of surface evaporation limiting the maximum possible temperature to about 3500 K [15,17] . Thus, these calculations, that neglect the motion of the photoexcited carriers, underestimate the delay to the maximal temperature by a factor of 10 and underestimate the degree of superheating (above the thermal melting point at 1685 K) by a factor between 4 and 20.

The long delay in reaching the maximal T and the low degree of superheating can be understood if we take account of the fact that the energy of the photon is not delivered to the lattice at the point where the photon is absorbed. The photoexcited carriers move about emitting phonons into some volume that depends on their group velocity, mean free path and phonon emission rate. These phonons will then scatter into a thermal distribution and become heat in

the lattice after about 10^3 phonon-phonon scattering events [1] .
In her treatment of this problem [57] , Yoffa concluded that the
"effective heating depth", δ', over which the atoms receive energy
of photons absorbed with absorption length, δ , may be written as:

$$\delta' = \delta (\alpha/ \delta + 1) \qquad (8)$$

where α is a parameter characterizing the range of the photoexcited
carriers. For *crystalline* Si at its melting point with $5 \times 10^{19} cm^{-3}$
carrier pairs excited, Yoffa estimated $\alpha = 250$ nm. In *molten* Si
the corresponding values would be significantly greater for several
reasons. First, because molten Si is metallic and nearly free-elec-
tron like, the group velocity :

$$v = \hbar^{-1} dE/dk \qquad (9)$$

is essentially the Fermi velocity and far larger than the corresponding value in a semiconductor at such a moderate level of excitation; obviously, $v = 0$ for a carrier at either band edge, or any-
where else that $dE/dk = 0$. The Fermi velocity in molten Si is
2.2×10^8 cm/s, corresponding to a Fermi energy of 13.9 eV for the 4
valence electrons per atom. Second, because molten Si is liquid,
there are no transverse phonon modes, which make up 2/3 of the phonons in the solid. Third, because scattering cross sections [58]
generally decrease with increasing v at least as v^{-2}, the much
greater value of v in the melt implies that the mean free paths increase at least in proportion to v. (We return to this point in
Sec. 4). Fourth, because the melting temperature is large compared
with the energy of the remaining (LA) phonons of the liquid, the
phonon occupation number, $\mathcal{N}$, is large and the ratio $(\mathcal{N} + 1)/\mathcal{N} \approx 1.0$;
the carrier is almost as likely to gain energy by absorbing a phonon
as it is to loose it by emitting one.

For fast (hot) carriers the dominant effect of the phonon scattering is to change their direction of motion and make their paths
diffusive rather than ballistic. In fact, phonon scattering in metals is commonly called "elastic" because so little energy is lost
this way; the dominant mode by which fast carrier loose energy is
by "inelastic" scattering of other carriers [59-62] . This is similar to, but more rapid,than impact-ionization in semiconductors
because there is no band gap setting a large minimal value for the
excitation. Both inelastic scattering and impact-ionization may be
viewed as the inverse of Auger recombination; none of the three
transfer energy to the lattice.

This subject was rather thoroughly studied in the field of
"internal photoemission" several years ago[59-63] . Internal photoemission refers to the injection of photoexcited carriers from a
metal surface layer into a semiconductor substrate and is one method
of measuring Schottky barrier heights. This can only succeed if the

metal layer is not so thick that the carriers loose too much of their energy before reaching the Schottky barrier to cross it. For metals, it follows from simple consideration of phase space [59-62] that the inelastic mean free path, Λ , is :

$$\Lambda = L/(E-E_f)^2 \tag{10}$$

where E is the carrier's energy, E_f is the Fermi energy, and L is a length parameter. Seah and Dench showed [62] that:

$$L = 538 \text{ nm(eV)}^2 \tag{11}$$

for almost all crystalline, elemental metals. (The discussion above indicates it may be somewhat larger for molten metals.) At room temperature, where the band gap of Si is 1.1 eV, a typical value for the Schottky barrier height is 0.6 eV, at which one would calculate from Eqs 10 and 11 that Λ (T=300 K) = 1.5 μm. At the melting point, the band gap is [63] 0.6 eV, so we may estimate the Schottky barrier height to be 0.3 eV at that temperature. This would then imply Λ (T=1685 K) = 6 μm. Given that the carrier velocity within the molten layer is about the Fermi velocity, 2.2×10^8 cm/s and estimating the time between phonon scattering events that randomize the carrier's direction to be $\delta t = 1 \times 10^{-13}$ s (corresponding to a mean phonon energy of 41 meV in the melt and assuming one scattering event per period), the carrier goes about 220 nm in a given direction between "elastic" phonon scattering events. Thus, the 6 μm inelastic mean free path corresponds to an average of about 27 elastic scattering events, and one may estimate the diffusion length α to be :

$$\alpha = (\Lambda/v\delta t)^{1/2} \times (v\delta t) = 1.1 \ \mu m. \tag{12}$$

As long as the thickness of the layer of molten Si in Hanabusa et al.'s experiment is much thinner than this, most of the photoexcited carriers (both electrons and holes) absorbed with $\delta \approx 9$ nm [64] will simply cross the Schottky barrier (the interface recombination states are totally saturated at these power levels) into the crystalline Si and cause more of that to melt. The electronic energy will not be confined to the molten Si layer causing it to superheat to the maximal degree until the layer has melted to a depth greater than α(T = 1685K).

Is the time delay between laser peak and maximal T consistent with this picture ? The rate of melting, and of crystal growth, $\mathcal{V}$ for Si near 1685 K is [65,66] :

$$\mathcal{V} = 60 \ \Delta T \ \text{cm/s} \tag{13}$$

where ΔT is the degree of superheating or supercooling. If we take the mean value of ΔT to be 200 K, then the extrapolated rate is $\mathcal{V} = 1.2 \times 10^4$ cm/s $= 0.12 \, \mu m/ns$, so that during the observed delay of 20 ns, $2.4 \, \mu m$ of Si would melt. The discrepancy between these two estimates, 1.1 versus $2.4 \, \mu m$, seems to be accountable within the accuracy to be expected from the methods used. Actually the $1.1 \, \mu m$ ought to be an underestimate because it neglects the diffusion that occurs while the carriers have an energy more than 0.3 eV from E_f. A more accurate estimate would be a finite series, limited by the maximal excitation of the carriers :

$$\alpha = 1.1 + 0.55 + 0.27 + .. \simeq 2 \mu m . \tag{14}$$

A second check on the validity of this explanation of the thermal radiation transient of Hanabusa et al. can be made from observation of the characteristic thickness of the lamellae of molten and solid Si produced by Bösch and Lemons by CW laser irradiation of Si on Al_2O_3 and on SiO_2 substrates[67]. (See Fig. 5.) As the solid Si lamellae are facetless, while the polycrystalline grains that form as they cool have the normal distinct facets, this author concluded [68] the solid is amorphous Si. For films 0.4 to $2.0 \, \mu m$ thick, these lamellae are seen to have a characteristic width of order $5 \, \mu m$. This width ought to be, and is, approximately twice α. Photons are much more strongly absorbed in the molten regions than in the surrounding solid. Carriers photoexcited in the melt will enter the solid and cause it to melt as long as they can cross the Schottky barrier. Diffusion in the vertical direction is limited by the insulating substrate.

Other aspects of the laser sputtering data[21-24] of Hanabusa et al. also give clear indication the laser sputtering mechanism is essentially non-thermal. In particular, the sputter yield, $Y \sim 10^{24}/cm^2$ s, and the fraction of excited states around 5 eV above the ground state of the Si atom far exceed that expected for a thermal distribution of the sputtered atoms and ions at the surface temperatures determined from the thermal radiation transient. These values are $Y_T(2100 \text{ K}) = 1.4 \times 10^{19}/cm^2$ s, a factor of 10^4 less than the observed rate (compare with the e-beam evaporator result), and (using "*" to indicate electronic excited states) $Y(Si^*)/Y(Si) = 10^{-12}$, $Y(Si^+)/Y(Si) = 10^{-7}$, $Y(Si^{+*})/Y(Si^+) = 10^{-31}$. The velocity distribution of the excited neutral atoms, $f(v_{Si}^*)$, was measured [21] and found to be distinctly non-thermal, with a most probable velocity normal to the target surface corresponding to 9 eV of kinetic energy. For a thermal distribution this would require T = 10^5K. Finally, depositions patterns showed that all Si species had their velocity vector nearly perpendicular to the target surface rather than in the cosine distribution that is characteristic of thermal evaporation.

In the work of Liu, et al.[16], laser sputtering was carried out at even higher levels of I, but lower E_0. These data are

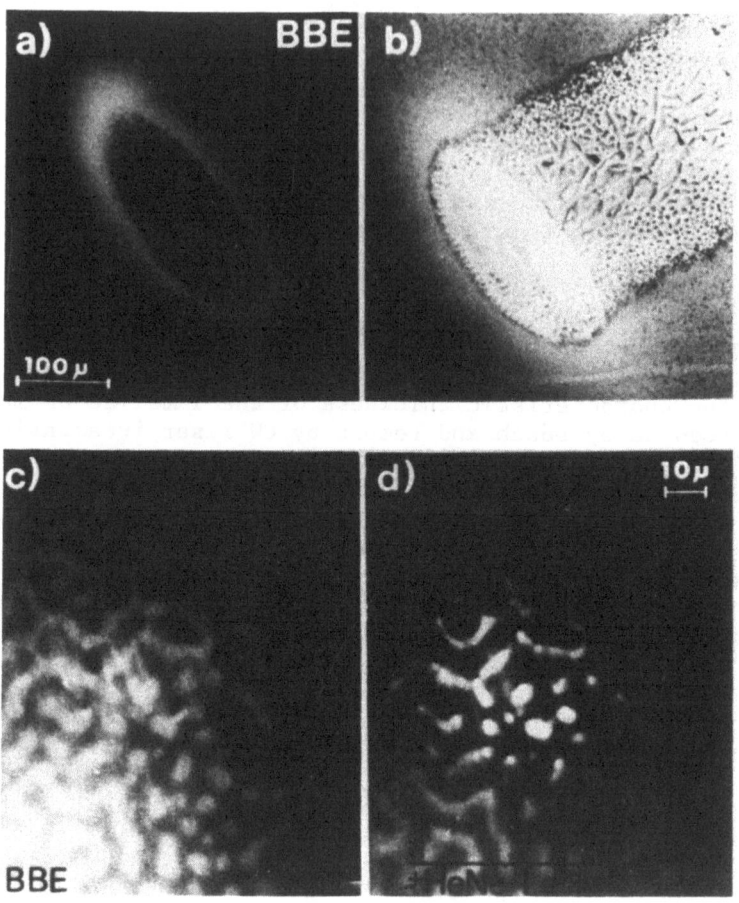

Fig. 5 Black body emission (BBE) (parts a and c), reflectivity
(part b) and transmission (part d) microscopy of a μm layer of Si
on an SiO$_2$ substrate heated and melted by a CW laser by Bösch and
Lemons [67]. Note the region near the melting point comprises
lamellae of molten and of facetless solid Si (believed to be amorphous Si) with a characteristic width of about 5 μm. It is here
argued the characteristic width is related to the range of photoexcited carriers in molten Si, α ≈ 2 μm, that retain enough energy
to cross the Schottky barrier to solid Si. (Courtesy of M.A. Bösch,
Bell Laboratories, Holmdel.)

obviously not related to thermal evaporation for several reasons
[15] :
i) the sputter yields are exponential rather than of the Arrhenius
form;
ii) the Y ratios are non-thermal; and
iii) direct measurements of T for comparable laser irradiation [55]
show that T < 600 C.

For neutrals emitted from GaP [20] and ZnO [18]:

$$f(v) \propto v^{-2}, \qquad (15)$$

rather than having a Maxwellian form, down to 0.01 eV. Typical velocity distributions are given in Fig. 6 and 7. For O^+ ions from
ZnO, Nakayama et al. [18] found the most probable velocity to
correspond to 1 eV of kinetic energy perpendicular to the target
surface. Of course, this also is much too large to correspond to
the target T and the distribution was found to be non-thermal.

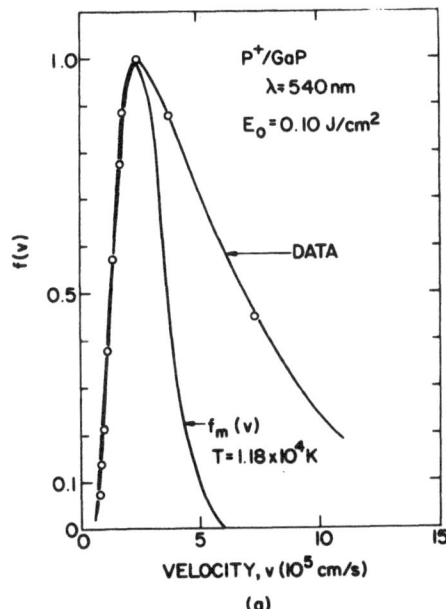

(a)

Fig. 6 Velocity distribution, $f(v)$, of P^+ ions laser sputtered
from GaP by Nakayama, et al. [20] . The distribution is distinctly
non-thermal, as can be seen by comparing with Maxwellian function,
$f_m(v)$, Eq. 2, which has T = 11,800 K fitted to match the most probable velocity. Of course, this T could not possibly represent the
true surface temperature for these conditions (E_0 = 0.10 J/cm^2 of
540 nm in 10 ns).

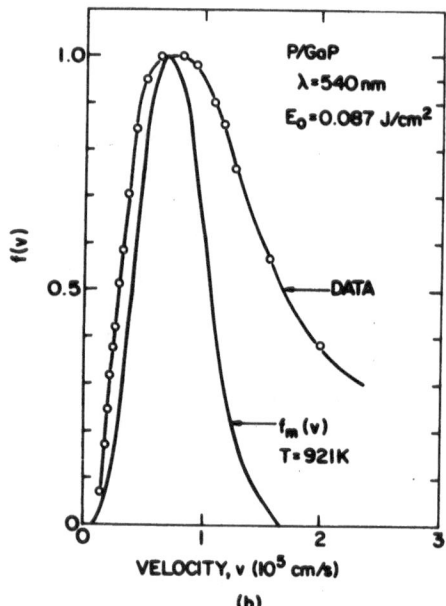

(b)

Fig. 7 Velocity distribution, f(v), for neutral P laser-sputtered from GaP in the same experiment as Fig. 6. Note that this distribution is no closer to being Maxwellian than that of the P^+ ions, but the most probable velocity and the fitting temperature, T = 921 K, are much lower. However, this T also is far larger than any lattice temperature possible for these conditions.

3. FRENKEL EXCITONS AND THE ZERO-CROSSING OF THE DIELECTRIC FUNCTION

The well known, stable Frenkel exciton may be viewed as an excited atom, or molecule, imbedded in a (wide band gap insulating) solid. It is usually said that the spatial extent of a Frenkel exciton is only through one (or perhaps two) atomic sites of the host solid. This would seem to imply that its effective radius :

$$R_F \approx b \tag{16}$$

where b is the bond length. One might immediately object to this notion on the ground that the spatial extent of the electronic wave function of an excited state of the same atom in free space, R_S, (as obtained, e.g., from the Herman-Skillman tabulated calculations [69]), is much larger than b. Why should :

$$R_F \ll R_S \ ? \tag{17}$$

To resolve this issue we must consider the dielectric response of the host for the case of the Frenkel exciton.

In our discussion of Frenkel excitons, the wave functions of interest will be spread throughout most of the Brillouin zone rather than restricted near the bottom of the conduction band, E_c, and the top of the valence band, E_v. For simplicity we shall adopt approximations that are accurate for this case. We shall take the mass of both e^- and h^+ to be m so that the reduced mass of the two in the exciton, $M_r = m/2$. Therefore, from the Uncertainty Principle: $\Delta p \, \Delta x = \hbar$ with $\Delta x = 2R = \Delta y = \Delta z$, we have, for the total (kinetic) localization energy in three dimensions :

$$E_\ell = 3\hbar^2/4mR^2 .\tag{18}$$

It follows then that the orbital frequency of the $e^- - h^+$ pair of the Frenkel exciton, ω_F, is :

$$\omega_F = 3\hbar/4mR_F^2 ,\tag{19}$$

corresponding to the orbital frequencies in free atoms (or ions) excited to analogous electronic states, and far larger than any ω_F or any phonon frequency. Again we write the net coulomb attraction as :

$$E_C(R_F) = e^2/ \, \varepsilon^*(R_F,\omega)R\tag{20}$$

which in fact serves to define $\varepsilon^*(R,\omega)$, the dielectric function for the screening of the interaction.

The existent theoretical machinery to calculate the dielectric response of a solid [40-42,70-76] generally does not allow a direct evaluation of $\varepsilon^*(R,\omega)$. (The review by Hanke and Sham [42], which treats Si as an example in great detail, probably represents the current "state of the art" of theory of dielectric response.) Instead, one first calculates a "proper response function" matrix, the inverse of which gives the dielectric matrix $\varepsilon(q+K,q'+K',\omega,\omega')$, where K and K' are reciprocal lattice vectors of the host crystal lattice, and (q+K) and ω are the wave vector and frequency of one Fourier component of the displacement induced by a Fourier component of the external field with wave vector (q'+K') and frequency ω'. In ordered cyrstals : q = q', but in general K $\neq$ K' and $\omega \neq \omega'$. This object is usually calculated in the "random phase approxima- tion" [83], RPA, i.e., $\omega = \omega'$, but many studies [32-42] have shown this approximation neglects effects that are needed to account for the interband optical spectra of all types of solids and of the "local field effects" in nonmetals [40-42]. In order to give a trac- table description of the problem at hand, we shall drop the "local field effects" from the present qualitative discussion, although there is reason to believe they will have significant quantitative effect [42]. This means that we take: q+K = q'+K' as well $\omega = \omega'$. The object that is obtained from the one-electron band structure in this approximation is denoted $\varepsilon(q,\omega)$. The relation between $\varepsilon(q,\omega)$ and $\varepsilon(R,\omega)$ is [76]:

$$1/\varepsilon^*(R,\omega) = RE_c(R,\omega)/e^2 = R/2\pi^2 \int (e^{-iq\cdot R}/q^2\varepsilon(q,\omega))d^3q .$$

(21)

Suppose now that the host material is a wide gap insulator with a low dielectric constant, i.e., that $\Delta E_{cv} > \Delta E^*$ for several free atom excitation levels, $\varepsilon_\infty \approx 1$ and $\varepsilon(q,\omega) \approx 1$ for all q and for the ω such that $h\omega \approx \Delta E^*$. Eq. 21 will then give $\varepsilon(R,\omega) \approx 1$ for the same ω. Evidently, the excitation levels of the same atom in the host lattice corresponding to those with ΔE^* in free space that are less than ΔE_{cv} will have about the same excitation energy, $\Delta E^{*'}$, and radius, R_F, in the solid as it has in vacuum if these conditions are met. (Note that we here discuss the dielectric response at the large ω corresponding to $\Delta E^{*'}$ and to the electromagnetic field that creates the exciton rather than the much lower ω corresponding to E_T. In the language of Feynman diagrams, this is the distinction between "vertex corrections" which are here discussed and "self-energy corrections" which are appropriate, e.g., to a discussion of the attraction of a free e^- and free h^+ to form a Wannier exciton. The initial polarization field induced by radiation that would produce the exciton by direct excitation and the final polarization field as the exciton decays in a radiative transition clearly have the large ω that we discuss. Therefore, for our discussion of the lifetimes of metastable continuum excitons, the response at the large ω is the more important, and less developed, issue.) As $\Delta E^{*'}$ is then less than ΔE_{cv}, the electron thus excited from the valence band edge E_v does not have enough energy to reach E_c. If it could reach the free electron states at E_c, it would dissociate from the h^+ left in the valence band after some time that would depend upon matrix elements between the localized and band wave functions. (It is the polarization field of that interaction that has the low ω corresponding to E_T and is relevant to such considerations.) Consequently, we have an excitation that may be associated with a single atom or molecule. Such an excitation is the traditional, stable Frenkel exciton with a two particle energy level that is by convention placed within the one electron band gap, ΔE_{cv}, at an energy $\Delta E^{*'}$ above E_v. Note that the excitation will retain its atomic character even though its radius may encompass several other atoms if the solid (those atoms) cannot react to screen the Coulomb attraction between the e^- and the h^+ of the excitation at their kinetic frequency, ω_F.

Now let us consider the cases of a smaller band gap insulator or a semiconductor for which $\Delta E^* > \Delta E_{cv}$ or of a metal. Consider the

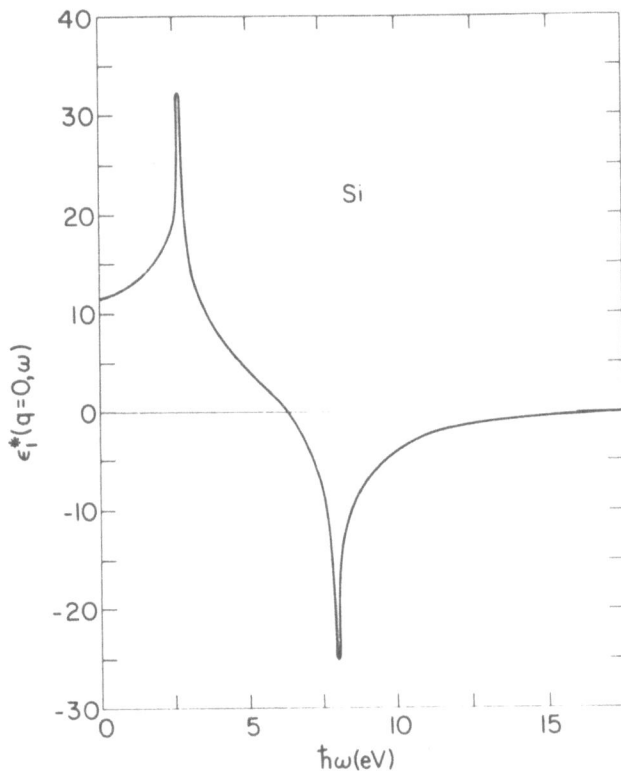

Fig. 8 The real part of the empirical dielectric function of Si as a function of energy for photons (q ≈ 0) by Kramers-Krönig transform of reflectivity data by Philipp and Ehrenreich, [78]. (This empirical ε_1 is shown with an "*" to distinguish it form the calculated object, $\varepsilon_1(q, \omega)$ for which significant local field and excitonic contributions are virtually impossible to calculate with good accuracy. See text.)

particular case of Si. Empirical values for $\varepsilon(q,\omega)$, $\varepsilon^*(q, \omega)$, can be obtained from analysis of the reflectivity spectrum [77,78]. The distinction between the $\varepsilon(q, \omega)$ that is calculated in lowest approximation and the empirical $\varepsilon^*(q,\omega)$ is significant because contains the local field, dynamical and many body effects which are so difficult to calculate rigorously [40-42]. The real part of $\varepsilon^*(q=0,\omega)$, $\varepsilon_1^*(q=0,\omega)$, for Si is shown in Fig. 8. In Fig. 9, we show the values of $\varepsilon_1(q,\omega)$ for a few ω''s as calculated according to the tight binding RPA model of Chadi and White (Eq. 25 of [75] using the empirical $\varepsilon_1^*(q=0,\omega)$ as input parameters). Hanke and Sham discuss in great detail [42] the consequence of effects neglected in such a calculation, but these curves should suffice to make a qualitative point. We immediately note in Fig. 8 that $\varepsilon_1(q=0,\omega) < 0$ for $\hbar\omega > 6.4$ eV. Also $\varepsilon_1(q=0,\omega)$ gets much larger than ε_0 as $\hbar\omega$ approa-

448

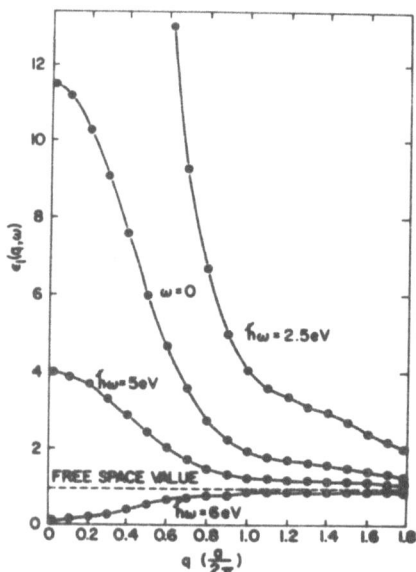

Fig. 9 Real part of dielectric function, $\varepsilon_1(q,\omega)$, calculated for Si from Eq. 25 of [75] , using the empirical values of $\varepsilon_1^*(q=0,\omega)$ as fitting parameters in that model of Chadi and White.

ches 2.5 eV from below (anomalous dispersion), but as ℏ ω increases above 2.5 eV, $\varepsilon_1(q=0,\omega)$ drops sharply again to values lower than ε_0 above 3.5 eV and to values less than 1.0, the value in free space, at about 6.0 eV. In Fig. 9 we see that the excursion of $\varepsilon_1(q,\omega)$ with ω becomes progressively less pronounced as q becomes larger. (See also the $\varepsilon(q,\omega)$ plots of Sramek and Cohen, [74]). For $q > 1.0\ 2\pi/a$, where a is the lattice constant (0.543nm), there is no longer a range of negative values. When q is larger than this, $\varepsilon_1(q,\omega) \approx 1$ for all ω. This last fact is just the point that the host medium cannot screen the coulomb interaction at distances much less than the interatomic spacing. From the tight binding or linear combination of atomic orbitals, LCAO, view point for the electronic structure of solids, which seems most appropriate for a discussion of the highly localized processes that lead to sputtering, let us now consider what happens if one Si atom is excited, as by impact with a sputter ion or electron, instantaneously into a state characteristic of the free Si atom. (As sputtering results from interaction that may accurately be regarded as two-body, as

these may then excite one atom and as the continuum of band states
of the perfect crystal are derived from LCAO's of the free atoms,
there seems no doubt that this is a meaningful dynamic process to
consider). Again we denote the free space excitation energy of
this excitation as ΔE^*. As the host responds to the excitation
and intermediates the interaction between the excited e^- and h^+,
for the finite period that the excitation remains associated with
the single atom, this excitation energy shifts to $\Delta E^{*,}(t)$ in time
t. If $\Delta E^* < 4$ eV in Si so that $\varepsilon_1(q,\omega)$ is large compared with
1.0 (free space), Eq. 21 shows that $1/\varepsilon^*(R,\omega)$, with R and ω initially corresponding to R_s and ΔE^*, receives a much smaller incre-
ment to the integral from the region near q = 0 (corresponding to
regions of the host that are far from the excited atom) than it
would in free space. Contributions to the integral from regions of
larger q (nearer to the center of the excited atom) are not so different from those in free space. As the total integral is thus
much reduced, $\varepsilon^*(R,\omega)$ is increased to a value that, to *first approximation*, might be estimated to be $\varepsilon_1(q=1/R,\omega)$. Thus the $\Delta E^{*,}(t)$
shifts to lower values as the medium screens the Coulomb attraction
between the excited e^- and h^+ and R increases, causing a further
increase in $\varepsilon(R,\omega)$. The time required for this expansion of R, the
motion of the e^- away from the h^+, can be estimated from the orbital velocity of the e^- (1 eV = 6×10^7 cm/s) to be of order 10^{-15} s.
Thus, an initially atom-like excitation with ΔE^* in this range will
expand in a time of this order and the carriers will either form a
Wannier exciton or become free carriers.

Now suppose $\Delta E^* \approx 6$ eV so that $\varepsilon_1(q,\omega) < 1.0$ for all q. Then
the integral in Eq. 21 will be greater than in free space, implying,
within the RPA and neglect of local field approximations, that the
e^- and the h^+ of the initial atomic excitation feel a stronger coulombic binding than they would in free space, due to the dielectric
response of the sourrounding host. I assert that the dynamical,
many body and local field corrections to this approximate treatment
will not alter this conclusion qualitatively. I base this assertion
on three considerations. First, we have made use of the empirical
[77,78] $\varepsilon(q=0,\omega)$, which in fact has had these corrections to theory
put into it by nature. Second, as noted above, the more basic the-
oretical object is the proper response function, $\varepsilon^{-1}(q+K, q+K', \omega, \omega')$
and the zeros of $\varepsilon(q,\omega)$ correspond to terms on the diagonal of that
tensor tending to infinity; the effect of the Umklapp terms, $K \neq K'$,
off the diagonal of the proper response function are not going to
budge those infinities on the diagonal very significantly. Third,
the general effect of the dynamical corrections is to strengthen,
not weaken, the $e^- - h^+$ binding [40,42]. If this is correct, then
for such values of ΔE^*, R_w will not expand rapidly from R_s but
may even contract which would have the effect of further reducing
$\varepsilon^*(R,\omega)$ below 1.0, further increasing the coulombic attraction of
the excited e^- and the h^+ and further stabilizing and localizing
this excitation. Of course, this excitation can be only metastable

because there are-one electron band states into which both the e^- and the h^+ can make transitions and thereby become free carriers. However, the transition rate for this process may be much smaller than the rate of expansion to a Wannier exciton when $\Delta E^* < 4$ eV. We denote this metastable excitation the "metastable Frenkel exciton" and note an analogy to the "Frenkel exciton"[42] or "continuum exciton" or "extra exciton" or "hyperbolic exciton" or "contact exciton" as the effect of Coulomb attraction between excited carriers has previously been denoted in discussions of optical interband spectra, derivative spectroscopy and other electronic properties of all classes of solids [32-42].

Finally, let us suppose that $\Delta E^* > 6.4$ eV so that $\varepsilon_1(q,\omega) < 0$ for small q. This will greatly reduce the integral in Eq. 21 by giving negative contributions from the region near q = 0 to partially cancel the positive one from large q. Thus, $\varepsilon^*(R,\omega)$ will be quite large and the excitation will expand rapidly, because E_0 will be large compared with E_C. However, this expansion of R_W will cause both E_1 and ω to decrease. If the excited pair remains correlated through the point that ε_1 passes through zero to small positive values, the excitation will again be stabilized. Alternatively, the e^- or the h^+ or both may become detached from the initially excited atom. However, in this case both the e^- and the h^+ are then left in energetic states and may form metastable pairs with other, less energetic h^+'s and e^-'s leading to excitations with ΔE^*'s in the range of stability around 6 eV (for the case of Si). Thus, these excitations also may lead to long lived, metastable Frenkel excitons.

We note that the photons observed to be emitted [21-23] from sputtered neutral Si atoms are generally somewhat less than 6 eV in energy. They have been observed to be laser-sputtered off the surface of molten Si in this excited state for as long as 100 ns after the end of a 7 ns (FWHM) laser pulse [23]. We are compelled to conclude that the lifetime of the metastable Frenkel exciton in metallic, molten Si is at least this long. Also, Eberhardt et al. observed in photoemission [79] a $1.2 \times 10^{21} cm^{-3}$ density of laser excited states near the surface of Si which had a measured lifetime between 1 and 5 μs and caused a 2 eV shift in the 2p core emission line; this author interpreted [14] excitations to be metastable Frenkel excitons which live even longer at a free surface than in the bulk.

4. LIFE TIME OF ELECTRONIC EXCITATION

As was noted in the previous section, Hanabusa and Suzuki observe [23] neutral Si atoms excited to states more than 5 eV above their ground state to be emitted from the molten Si surface with 9 eV of kinetic energy for more than 100 ns after a 7 ns high energy

density laser pulse and Eberhardt et al., observe [79] 10^{21} cm^{-3}
excitations at the surface of a crystalline Si sample persisting
between 1 and 5 μs after a low energy density laser pulse. Both
of these observations contradict the assertion that the energy of
a laser pulse is completely transferred to the lattice and thermalized in less than 20 ps [16,17]. An understanding of the phenomenon of pulsed beam annealing that is consistent with direct measurements of the surface temperature [54,55] and a great deal of
other evidence [80-86] that the phenomenon is non-thermal seems
to require [14] that the energy stay in electronic excitations during that process for a period at least as long as the duration of
the high reflectivity phase, as long as 0.6 μs.

This life time for the electronic excitations is much longer
than many would estimate and constitutes a major theoretical problem. This author has addressed that problem before [13,14] as hav
others [87]. Rather than reviewing points made before, here we
shall consider two new points. These relate to the metastability
of the Frenkel excitons relative to the band states and the effect
of the velocity dependence of deformation potential scattering cros
sections.

Consider the total energy E_T of an exciton of effective radius
R as a function of R :

$$E_T = E_\ell - E_C \tag{22}$$

where E_ℓ is the (positive) energy associated with the localization
of the e$^-$ and h$^+$ into a restriction region of real space and E_C
is the energy regained by the Coulomb attraction between the e$^-$
and the h$^+$. We use Eq. 20 for E_C which in effect defines $\varepsilon(R,\omega)$,
the dielectric function for the screening of the interaction. As
was noted in the previous section, a first principles, quantum
mechanical calculation of $\varepsilon(R,\omega)$ would necessarily begin by calculating the proper response function, then calculate $\varepsilon(q,\omega)$, then
finally calculate $\varepsilon(R,\omega)$ from Eq. 21 and the rigor with which this
could be done in the present state of development of that theory
[42] is not as great as one might hope. Therefore, I shall here
adopt a more empirical approach in order to get an analytic expression and, hopefully, some useful insight.

The $\varepsilon_1(R, \omega = 0)$ as calculated by Srinivasan [76] can be approximated for small R in the form :

$$\varepsilon(R) = 1 + (R/R_0)^2 \tag{23}$$

where R_0 is a parameter of order the radius of the atom. From the
calculation of Chadi and White [75] for the case of Si we obtain
the value $R_0 \approx 1.0$ Å. Using these approximations, the calculated

values of E_C, E_ℓ and E_T are shown in Fig. 10. If we solve for
$dR = 0$, we find that one local minimum of the total excitation
energy generally occurs where :

$$R = 0.79 \ \overset{\bullet}{A} \ [\ 1 \ + \ (R/R_0)^2]^2 \ / \ [\ 1 \ + \ 3R^2/R_0{}^2] \ . \qquad (24)$$

Thus, as in the particular case of Fig. 10, there is a range of
(meta)stability for excitons with small radii, generally not far
from $R = 0.8$ Å and there will be so in other types of host, metals,
for which Eq. 23 is a reasonable approximation. These may be associated with the continuum "Frenkel excitons".

In general, the function $E_T(R)$ for semiconducting and insulating hosts will have a second local minimum at much larger R. Of
course, this corresponds to the Wannier excitons. In insulators
with narrow band widths and low values of ε for large R, the Frenkel excitons will be more stable than the Wannier excitons but
in many insulators and probably in all semiconductors, the Wannier
exciton (i.e., the larger R minimum) will be more stable. On the
other hand, for metal hosts, the local minimum responsible for the
Wannier excitons is apt to be entirely missing, so that only Frenkel excitons will be metastable.

In any case, once a Frenkel exciton is formed by whatever
means, with the excited e^- and h^+ at an effective radius less than
1 Å, it is evident in Fig. 10 that there is a barrier of order 2eV
preventing them from separating. If the temperature characterizing the excited e^- and h^+ is comparable with the lattice temperature, measured to be [54,55] about 300 C, and if the attempt frequency $\nu_0 = 10^{13} s^{-1}$, from Eq. 1 we estimate it would take 160 μs
for the pair to dissociate. This is then more than enough to account for the observations of Eberhardt et al. [79] .

Now let us consider the how the group velocity, v, of a free
carrier (recall Eq. 9) affects the rate of phonon emission through
deformation potential scattering. Of course, this rate is also
affected by dielectric screening [88,89] and significant slowing
of the rate of phonon emission have been observed [90-92] at excitation densities far below those involved in pulsed beam annealing.
The deformation potential can, and normally is, written as a product of structure factors, which characterize the deviation of the
nuclei from the perfect lattice positions, times form factors,
which characterize the scattering potential of each atom individually. Scattering potentials are in general velocity dependent [58]
so the form factors are velocity dependent and so to the same
degree is the product, the deformation potential itself.

For a Coulomb potential, the scattering cross section, σ ,
varies as v^{-4}. Potentials of finite range generally vary as v^{-2}

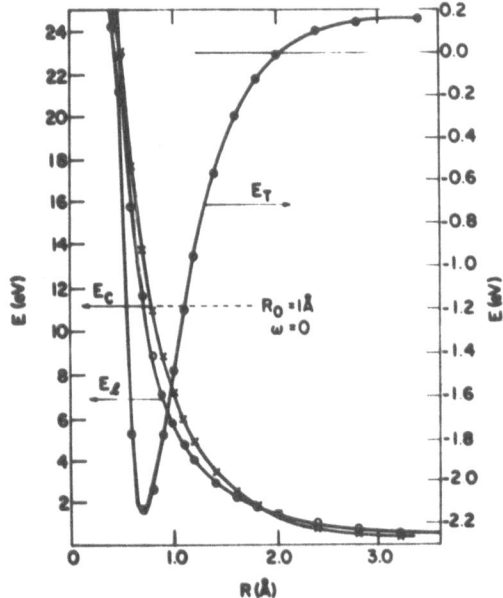

Fig. 10 Screened Coulomb attraction energy, E_C, localization energy, E_ℓ, and total energy, E_T of a continuum "Frenkel" exciton in crystalline Si as estimated in text. Note the very large barrier between the minimum and the free carrier states, or the Wannier exciton. This may help to explain the very long lifetime of the excitations observed by Hanabusa, et al., more than 100 ns in [23] (with molten Si), and by Eberhardt et al. 1 to 5 μs in [79].

for large v and are nearly constant for small v. Schiff solves explicitly the case of the square well potential [58] which is the opposite extreme of the Coulomb potential. (The scattering of fast carriers in solids is intermediate between these cases.) If a is the radius of the square well, the cross section to scatter at angle θ, σ(θ), varies as:

$$\sigma(\theta) \propto g(2ka \, \sin(\theta/2)) \qquad (25)$$

where $k = mv/\hbar$ is the carrier wave length and

$$g(x) = (\sin(x) - x \cos(x))^2/x^6 . \qquad (26)$$

454

This begins to fall off as v^{-2} when $2ka \sin(\theta) \approx 1.0$. ($\sigma(x=1) = 0.8\ \sigma(x=0)$; $\sigma(x=3) = 0.1\ \sigma(x=0)$). For a free carrier with unit effective mass and 1 eV of kinetic energy, $k = 0.5/\text{Å}$, so if a = 1.18 Å, the covalent radius of Si, this threshold will be exceeded for $\theta = \Pi$ scattering already at that energy.

5. CONCLUSIONS

Pulsed beam annealing is a remarkably gentle phenomenon [14] considering the rate at which energy passes in and out of the material treated. It is far more gentle than one could expect from a quenching process, the processes conventionnally considered to be the opposite of annealing. However, the laser, electron and ion beam sources similar to those producing the annealing effect can also produce sputtering, which is a rather violent process, in which material is ablated with kinetic energies much greater than thermal values [10-12,31] . They also tend to produce deep level defects up to 10 μm into the substrate behind the annealed layer [48] which indicates that violent processes take place there.

In trying to obtain a consistent understanding of all these phenomena, we find compelling evidence for a crucial role of electronic excitations. These may weaken bonds both to allow the material to relax out strain and defects, i.e., to anneal, to promote the expulsion of atoms or ions that have acquired sufficient momentum to carry them through the surface, i.e., to sputter, or to dislodge atoms from their perfect crystal sites in the bulk i.e., to create deep level defects. The most important electronic excitations are highly localized into what may be called continuum Frenkel excitons. A more detailed discussion of the role of localized electronic excitations in sputtering is given in [30].

The wave length dependence of the sputter yield and the onset of the high reflectivity phase of pulsed laser annealing noted in Fig. 1 seems to show a nice dichotomy between the gentle and the violent behavior of Frenkel excitons. The Bose condensation theory of these excitons [14,43-45] seems to offer the only explanation of this dichotomy yet proposed.

REFERENCES

1. Tsai, D.H. and Trevino, S.F., Phys. Rev. A 24 (1981) 2743 and references therein.
2. Kokorowski, S.A., Olson, G.L. and Hess, L.D., J. Appl. Phys. 53 (1982) 921.
3. Bourgoin, J.C. and Corbett, J.W., Rad. Eff. 36 (1978) 157 and

therein.

4. Stoneham, A.M., Adv. Phys. 28 (1979) 457.

5. Lang, D.V. and Kimerling, L.C., Phys. Rev. Lett. 33 (1974) 489.

6. Troxell, J.R., Chatterjee, A.P., Watkins, G.D. and Kimerling, L.C., Phys. Rev. B 19 (1979) 5336.

7. Iwamoto, M. and Kasami, A., Appl. Phys. Lett. 28 (1976) 591.

8. Monemar, B., Potemski, R.W., Small, M.B., Van Vechten, J.A. and Woolhouse, G.D., Phys. Rev. Lett. 41 (1978) 260; see also Basson, J.H. and Van Vechten, J.A., Phys. Rev. B 23 (1981) 2032.

9. Porter, W., Parker, D.L., Richardson, R.T. and Swenson, J.E., Appl. Phys. Lett. 33 (1978) 896.

10. Behrisch, R., Sputtering by Particle Bombardment 1, Physical Sputtering of Single-Element Solids (Springer Verlag, Berlin, 1981).

11. Kelly, R., Phys. Rev. B 25 (1982) 700.

12. Kelly, R., Rad. Eff. (1982) (in press).

13. Van Vechten, J.A., J. de Phys. 41 (1980) C4-15.

14. Van Vechten, J.A., in Laser and Electron-Beam Interactions with Solids edited B.R. Appleton and Celler, G.K. (New York, Elsevier Science, 1982), p. 49.

15. Van Vechten, J.A., J. Appl. Phys., to be published (1982).

16. Liu, J.M., Yen, R., Kurz, H. and Bloembergen, N., Appl. Phys. Lett. 39 (1981) 755.

17. Yen, R., Liu, J.M., Kurz, H. and Bloembergen, N., Appl. Phys. A 27 (1982) 153.

18. Nakayama, T., Itoh, N., Kawai, T., Hashimoto, K. and Sakata, T., Radiation Effects Lett. 67 (1982) 129.

19. Itoh, N. and Nakayama, T., to be published in Radiation Effects.

20. Nakayama, T., Ichikawa, H., Itoh, N., Kawai, T., Hashimoto, K. and Sakata, T. (to be published in Phys. Lett. A).

21. Hanabusa, M., Suzuki, M. and Nishigaki, S., Appl. Phys. Lett. 38 (1981) 385.

22. Hanabusa, M. and Suzuki, M., Appl. Phys. Lett. 39 (1981) 431.

23. Hanabusa, M. and Suzuki, M. in Laser and Electron-Beam Interactions with Solids, Op. Cit. Ref. 14 p. 559.

24. Hanabusa, M., private communication of 2/22/82.

25. Kawai, T. and Sakata, T., Chem. Phys. Lett. 69 (1980) 33.

26. Gauther, R. and Guttard, C., Phys. Stat. Sol a38 (1976) 477.

27. Donaldson, T.P., Lädrach, P. and Wägli, P., Phys. Lett. 70A (1979) 419.

28. Moison, J.M. and Bensoussan, M., J. Vac. Sci. Techn. 21 (1982).

29. Williams, R.T., Kabler, M.N., Long, J.P., Rife, J.C. and Royt, T.R in Op. Cit. Ref. 8, p. 97 and to be published.

30. Van Vechten, J.A., Itoh, N. and Kelly, R., to be published.

31. Kelly, R., Shivashankar, S.A. and Cuomo, J.J., J. Vac. Sci. Techn. to be published (1982).

32. Cardona, M. and Harbeke, G., Phys. Rev. Lett. 8 (1962) 90.

33. Philipps, J.C., Phys. Rev. Lett. 10 (1963) 329.

34. Philipps, J.C., Phys. Rev. 136 (1964) A1705.

35. Mayer, H. and El Naby, M.H., Z. Physik 174 (1963) 280.
36. Hamakawa, Y., Germano, F. and Handler, P., J. Phys. Soc. Japan, Suppl. 21 (1966) 111.
37. Mueller, F.M., Phys. Rev. 153 (1967) 659.
38. Hermanson, J., Phys. Rev. 166 (1968) 893.
39. Rowe, J.E. and Aspnes, D.E., Phys. Rev. Lett. 25 (1970) 162.
40. Van Vechten, J.A. and Martin, R.M., Phys. Rev. Lett. 28 (1972) 446.
41. Martin, R.M., Van Vechten, J.A., Rowe, J.E. and Aspnes, D.E., Phys. Rev. B 6 (1972) 2500.
42. Hanke, W. and Sham, L.J., Phys. Rev. B 21 (1980) 4656.
43. Nagy , M. and Noga, M., Czech. J.Phys. B 31 (1981) 1358.
44. Van Vechten, J.A. and Compaan, A.D., Solid State Comm. 39 (1981) 867.
45. Van Vechten, J.A., Solid State Comm. 39 (1981) 1285.
46. Wautelet, M. and Van Vechten, J.A., Phys. Rev. B 23 (1981) 5551.
47. Wautelet, M., J. Phys. C : Solid State Phys. 14 (1981) 4303.
48. Mooney, P.M., Young, R.T., Karins, J., Lee, Y.H. and Corbett, J.W., Phys. Stat. Sol. (a) 48, (1978) K31.
49. Compaan, A.D. and Itoh, N., private communication.
50. Thompson, M.W., Philos. Mag. 18 (1968) 377.
51. Politiek, J. and Kistemaker, J., Radiat. Eff. 2 (1969) 129.
52. Sigmund, P., Phys. Rev. 184 (1968) 383.
53. Guarnieri, C.R. and Van Vechten, J.A., to be published.
54. Aydinli, A., Lo, H.W., Lee, M.C. and Compaan, A., Phys. Rev. Lett. 46 (1981) 1640.
55. Compaan, A., Aydinli, A., Lee, M.C. and Lo, H.W., Op. Cit., Ref. 14, p. 43 and to be published.
56. I have used a program supplied by R.A. Ghez that is a version of that described by Ghez, A and Laff, R.A., J. Appl. Phys. 46 (1974) 2103, modified to treat Si substrates. It gives results essentially the same as those described by Yen et al., Ref. 17, or Wood, R.F. and Giles, G.E., Phys. Rev. B 23 (1981) 5543.
57. Yoffa, E.J., Appl. Phys. Lett. 36 (1980) 37.
58. Schiff, L.I., Quantum Mechanics (New York, Mc-Graw-Hill,1955) pp. 168.
59. Crowell, C.R. and Sze, S.M. in "Physics of Thin Films" (Academic Press, New York, 1967) Vol. 4 pp. 325.
60. Krolikowski, W.F. and Spicer, W.F., Phys. Rev. 185 (1969) 882 and B 1 (1970) 478.
61. Kanter, H., Phys. Rev. B 1 (1970) 2357.
62. Seah, M.P. and Dench, W.A., Surf. Interf. Anal. 1 (1979) 2.
63. Thurmond, C.D., J. Electrochem. Soc. 122 (1975) 1133 and therein.
64. Sharev, K.M., Baum, B.A. and Geld, P.V., Sov. Phys. Solid State 16 (1975) 2111.
65. Voronkov, V.V., Kristallografiya, 17 (1972) 909.
66. Chernov, A.A., J. Crystal Growth 24/25 (1974) 13.

67. Bösch, M.A. and Lemons, R.A., Phys. Rev. Lett. 47 (1981) 1151.
68. Van Vechten, J.A., Solid State Comm., to be published (1982).
69. Herman, F. and Skillman, S., Atomic Structure Calculations, (Prentice-Hall, Englewood-Cliffs, New Jersey, 1963).
70. Ehrenreich, H. and Cohen, M.H., Phys. Rev. 115 (1959) 786.
71. Pick, R.M., Cohen, M.H. and Martin, R.M., Phys. Rev. B 1 (1970) 910.
72. Pick, R.M., Advan. Phys. 19 (1970) 269.
73. Walter, J.P. and Cohen, M.L., Phys. Rev. B 5 (1972) 3101.
74. Sramek, S.J. and Cohen, M.L., Phys. Rev. B 5 (1972) 3800.
75. Chadi, D.J. and White, R.M., Phys. Rev. B 11 (1975) 5077.
76. Srinivasan, G., Phys. Rev. 178 (1969) 1244.
77. Ehrenreich, H. and Philipp, H.R., Phys. Rev. 128 (1962) 1622.
78. Philipp, H.R. and Ehrenreich, H., Phys. Rev. 129 (1963) 1550.
79. Eberhardt, W, Brickman, R. and Kaldor, A., Solid State Comm. 42 (1982) 169.
80. Khaibullin, I.B., Shtyrkov, E.I., Zaripov, M.M., Bayazitov, R.M. and Galjautdinov. M.F., Radiat. Eff. 36 (1978) 225.
81. Hoonhout, D. and Saris, F.W., Phys. Lett. 74 A (1979) 253.
82. Van Vechten, J.A., Tsu, R., Saris, F.W. and Hoonhout, D., Phys. Lett. 74A (1979) 417.
83. Yamada, M., Kotani, H., Yamazaki, K., Yamamoto, K. and Abe, K., J. Phys. Soc. Japan 49, Suppl.1980 A 1299.
84. Dvurechenskii, A.V., Mustafin, T.N., Smirnov, L.S., Geiler, H.D., Götz, G. and Jahn, U., Phys. Stat. Sol. 63a (1981) K203.
85. Bensoussan, M. and Moison, J.M., J. de Phys. 42 (1981) C7-149.
86. Kotani, H., Yamada, M., Yamamoto, K. and Abe, K., Solid State Comm. 41 (1982) 461.
87. Tsu, R. and Jha, S.S., J. de Phys. 41 (1980) C4-25.
88. Yoffa, E.J., Phys. Rev. B 21 (1980) 2415.
89. Yoffa, E.J., Phys. Rev. B 23 (1981) 1909.
90. Nurmikko, A.V. and Schwartz, B.D., J. Vac. Sci. and Techn. 21 (1982) 229.
91. Seymour, R.J., Junnarkar, M.R. and Alfano, R.R., 41 (1982) 657.
92. Yao, S.S., Buchert, J. and Alfano, R.R., Phys. Rev. B 25 (1982) 6534.

MULTI-ELECTRON DEFECTS IN THE ELEMENTAL SEMICONDUCTORS

Wolfgang Schröter

IV. Physikalisches Institut der Universität Göttingen
und Sonderforschungsbereich 126 Göttingen/Clausthal,
3400 Göttingen, Bunsenstrasse, 11-15, Germany.

ABSTRACT

This lecture aims at two aspects : (1) to give an introduction
to the properties of dislocations in semiconductors, and (2) to work
out some of those concepts which are needed to describe the electric
properties of many-electron defects.

The first aspect is pursued mainly in chapter 1 where a brief
summary of the electric and dynamic behaviour of dislocations and
of their interaction with point defects is given, and chapter 3,
where recent results of the core structure are discussed.

The treatment of the second aspect is confined mainly to some
defects which, in a simple ball and spokes model, have many dangling
bonds : the vacancy in silicon with four dangling bonds (chapter 4.
1) and the dislocation with a linear arrangement of dangling bonds
in its core (chapters 2, 3.2, 4.2 - 4.4).

1. INTRODUCTION

1.1 Definition and some basic properties of dislocations

Two basic parameters characterizing a dislocation may be easily derived from a gedanken-experiment : by performing a cut parallel
to the slip plane of the crystal up to a line with line element ds,
translating the two shores of the cut relative to each other by
the vector $\underline{b}$ and then reconnecting the bonds across the cut, one
obtains a dislocation with Burgers vector $\underline{b}$ and line element ds.
(Fig. 1a).

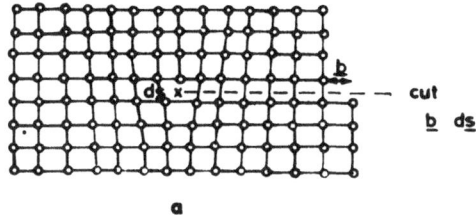

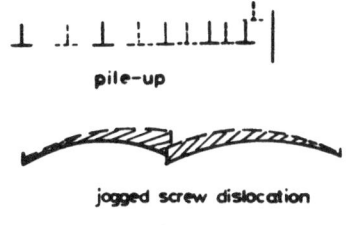

Fig. 1 a A dislocation may be thought to originate in a cut parallel to the slip plane. The end line ds of the cut and the vector b, by which the shores of the cut have been translated relative
to each other, define the line element and Burgers vector of the
dislocation respectively.
Fig. 1 b Two-region model of a dislocation : core ($r < r_c$) and region of the long-range field ($r > r_c$).
Fig. 1 c A dislocation arrangement of minimum strain energy is the
small-angle grain boundary.
Fig. 1 d Stress-concentrating configurations are the pile-up of
dislocations and the jog-dragging screw dislocation.

The description of the properties of dislocations uses the concept of the inner cut-off radius r_c, separating the core $(r < r_c)$ which comprises a topological disturbance and a region of unusual physical properties, from the outer region $(r > r_c)$ where the dislocation gives rise to elastic and electric long-range fields (Fig.1b). The cut-off radius has to be carefully defined for every property. It is of the order of the lattice constant.

In the crystalline solid under atmospheric pressure, dislocations are non-equilibrium defects, since their self energy being of the order of 10 eV/(period along the core) is much too large to be counterbalanced by the entropy gain associated with the introduction of dislocations into the crystal. However, outer stresses, which are quite small compared to the theoretical shear strength of the material, lead to an appreciable plastic deformation if, at the temperature considered, the dislocations are mobile and if stress concentrating configurations -like small particles or scratches, edges and holes at the surface etc. - are present, at which first dislocations can be generated. Also the stress field associated with concentration gradients of impurities, temperature gradients or interfacial misfits and the chemical force of a super- or undersaturated system may relax by generation of dislocations. The limiting velocity with which dislocations may follow any perturbance is the sound velocity.

Let us consider a crystal with many dislocations. In a single-dislocation approximation one describes the interaction between the dislocations by a mean internal stress τ_i, acting on a single dislocation due to its interaction with all other dislocations. For a random distribution of parallel dislocations the applied stress τ_a is replaced by an effective stress : $\tau_e = \tau_a - \tau_i$. An often used empirical ansatz for τ_i is [1] : $\tau_i = A\sqrt{N_d}$

A : a constant; N_d : dislocation density in cm $\times$ cm^{-3}

The self energy of a set of many dislocations depends also on their distribution in space. One configuration of low interaction energy is that of the small angle boundary (Fig. 1c), which also represents a rather stable configuration in the relaxation of a deformed crystal.

Rather important for the role of dislocations in the production of annihilation of point defects are two stress-concentrating configurations : the pile-up of n dislocations and the dislocation with jogs (height d, mean distance l), which can follow its motion only by emitting or absorbing vacancies or interstitials (Fig. 1d). The effective stress acting on the leading dislocation of the pile-up is of the order of $\tau_e \approx n\tau_a$ and that acting on the jog is: $\tau_e \approx (1/d)\tau_a$

1.2 Dislocations in semiconductors

In Fig. 2a, a 60°-dislocation ($\angle(\underline{b},d\underline{s})$=60°) in a crystal with diamond or sphalerite structure is shown. In the core, a row of atoms can be seen which are missing one neighbour and which there-fore have one dangling bond. If the semiconductor is a III-V or II-VI compound, this row is composed of atoms of one kind, so that one could, due to the ionicity of the crystal, expect an intrinsic line charge. However, at present there are neither theoretical cal-culations nor experimental results which would allow to determine the magnitude of this line charge.

The role of dangling bonds for surfaces and amorphous semi-conductors is also discussed by Gaspard in his contribution to this summer school. The best studied and understood defect with dang-ling bonds is the vacancy in silicon due to the thorough works of Watkins and coworkers [2] and to the theoretical work of various groups, especially Baraff, Kane and Schlüter[3] . It is now well established that the vacancy with its charge states V^{++}, V^+ and V° is the first defect which has been proven to be an Anderson negati-ve-U center. We shall discuss the important interactions and ex-periments which manifest this result in some detail later (see chapter 4.1). Considering dislocations we have to ask if the ato-mic and bond structure as shown in Fig. 2a is already a realistic one. That there are electrically active states associated with the dislocation core has been demonstrated by many experiments [4] . As an example we show in Fig. 3 the temperature dependence of the hole density, as determined from the Hall effect, for p-type germa-nium before and after plastic deformation[5] . A dislocation may be taken as an example of an extended many-electron center and we shall outline various consequences of this fact in the main part of the lecture.

To conclude this introduction let us briefly mention two basic properties which we shall not describe and discuss in detail here : the dynamic behaviour in which dislocations in semiconductors differ significantly from those in metals and the role of disloca-tions as sinks and sources for point defects, into which new in-sights have been and will be obtained by application of some expe-rimental methods which have been developped recently for semicon-ductor research.

The dynamic behaviour of dislocations and the plasticity in the elemental semiconductors is governed by an exponential tempera-ture dependence and weak stress dependence of the dislocation velo-city : $v(\tau,T) \simeq \tau^m \exp(-Q/kT)$ [1] . Appreciable plasticity is found for temperatures above $0.6\ T_m$ (T_m melting temperature). This beha-viour has been ascribed to a large Peierls potential, i.e. a large variation of the core energy with position between two stable posi-tions of the dislocation. The movement of a dislocation under

462

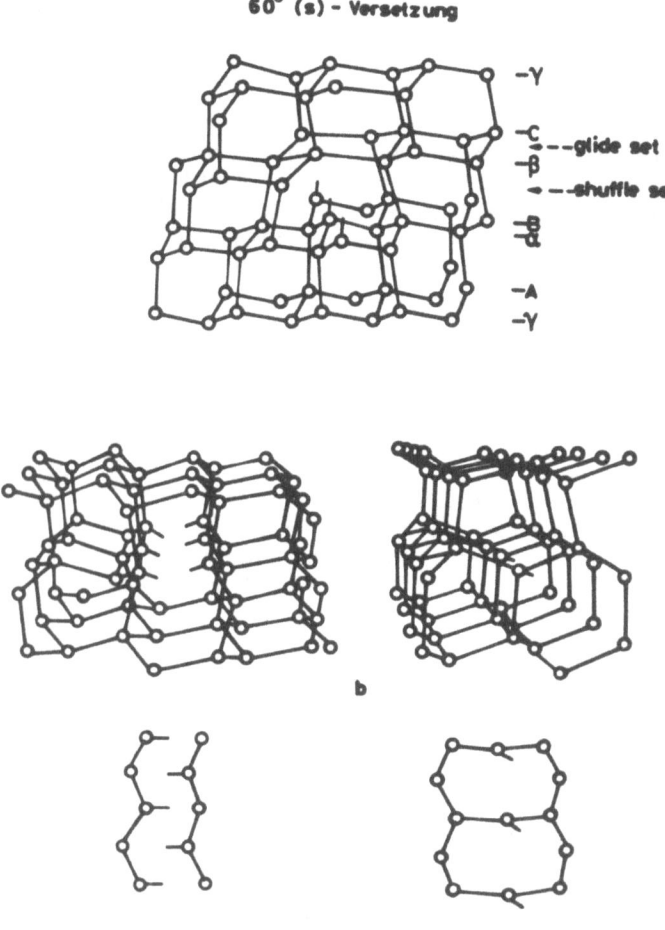

60° (s) - Versetzung

−γ
−C −−glide set
−β
−−shuffle set
−B
−A
−γ

b

Fig. 2a 60°-dislocation of the shuffle set in a crystal with diamond structure.
Fig. 2b 60°-dislocation of the glide set which is dissociated into
a 30°- and a 90°-partial with a stacking fault in-between.

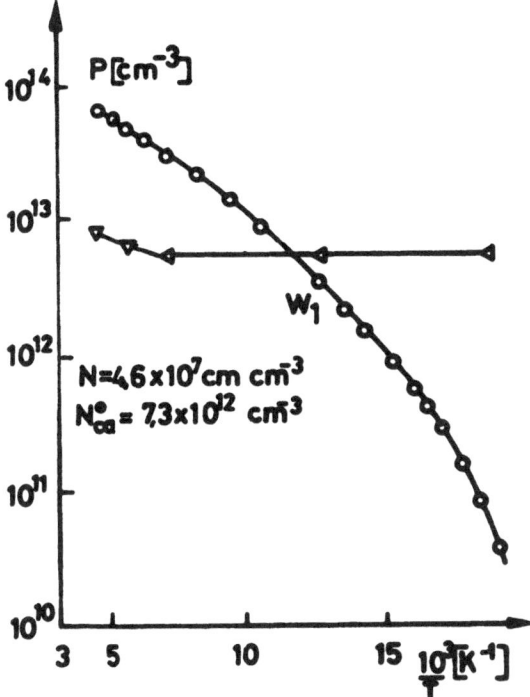

<u>Fig. 3</u> Hole density in p-type germanium as a function of inverse
temperature before and after plastic deformation.

the influence of a force then does not proceed as a free string like
in many metals, but is resolved into the generation of double kinks,
the movement and mutual annihilation of kinks (see Fig. 4). Each
of these steps may be thermally activated.

In 1966 Patel et al found [6] a second basic difference in the
plasticity of semiconductors as compared to that of metals. The
plasticity and dislocation velocity in semiconductors is dependent
on the doping concentration if this exceeds the density of intrinsic electrons or holes. There is strong support now that this effect
is due to a coupling between some of the elementary processes of
dislocation motion and the electrons and holes of the crystal, although there is no clear theoretical understanding of the relevant
mechanism. Meanwhile it has been demonstrated by numerous investigations that also light, electric field, chemical surrounding etc...
are means to vary the plasticity in semiconducting and isolating
materials [7-9] .

The interaction between dislocations and point defects can be
nicely demonstrated by a diffusion experiment : long tails in the
concentration profiles indicate a significantly enhanced transport
in the dislocation core (see Fig. 5) [10] . If the isotope studied
does not interact with the stress or electric field of the dislocation, the enhanced transport must be due to a larger concentration
and mobility of those point defects which are responsible for
diffusion within a dislocation-core region ($r \lesssim r_c$). Point defect
clouds around dislocations that extend much farther into the bulk
have been detected recently by DLTS [11], EBIC [12], CL [13] and
local conductivity [14] measurements in crystals which have been
deformed at temperatures below about 0.6 T_m. Apparently these
clouds strongly influence the electric and plastic behaviour of the
material but they cannot be stable at the deformation temperature,
since the interaction with dislocation is much too short-ranged to
keep them together [15] . To all experimental evidence,these clouds
are difficult to avoid in the III-V and II-VI compounds but the
research on this field is just at its beginning so that more definite statements can be expected soon.

2. SIMPLE MODELS FOR DISLOCATIONS IN THE ELEMENTAL SEMICONDUCTORS

2.1 Energy spectra

The model that claims for a dangling-bond array in the dislocation core has been introduced by Shockley in 1953 [16] . He speculated that the associated electron states may form a one-dimensional band which, partially filled, causes each dislocation to become
a one-dimensional degenerate electron-gas conductor.

One year later, Read [17],on the grounds of first experimental data, associated acceptor centers with the dangling bonds at
dislocations. At first sight this proposal seems to be basically

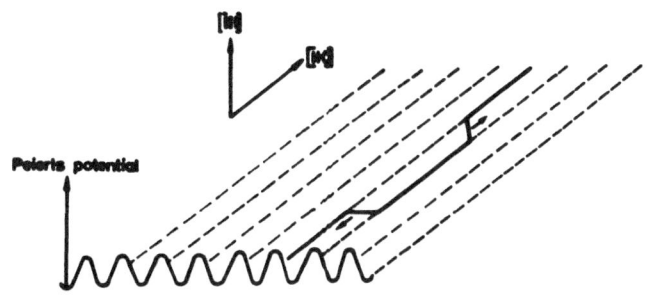

Fig. 4 Due to a large Peierls potential in the <110> - directions, the movement of dislocation is believed to proceed of the generation of double kinks, the motion of kinks, their interaction with obstacles and their initial annihilation.

different from Shockley's, but indeed it is only the other limiting case of the same concept.

To see the underlying concept we describe a semiconductor with dislocations by an Anderson-type Hamiltonian [18] :

$$H = H_b + H_{db} + H_{int}$$

where the first term on the right side represents the two bulk bands, H_{db} the dangling-bond states and H_{int} the interaction between bulk states and dangling-bond states.

For the non-magnetic case we write :

$$H_{db} = \sum_{i,\sigma} (\varepsilon_i - \mu) n_{i\sigma} + \sum_i U n_{i\uparrow} n_{i\downarrow} + V_c \qquad (1)$$

where i numerates the core sites with a dangling bond, $n_{i\sigma}$ the number of electrons occupying the dangling bond at site i with spin σ, μ is the Fermi level.
ε_i are the one-electron energies of the dangling bond at site i.
If the dangling bonds form an ideal one-dimensional array, one obtains a one-dimensional continuum of states (see Fig. 6) :

$$E_d(k_z) = E_{do} + 2 J \cos (k_z s). \qquad (2)$$

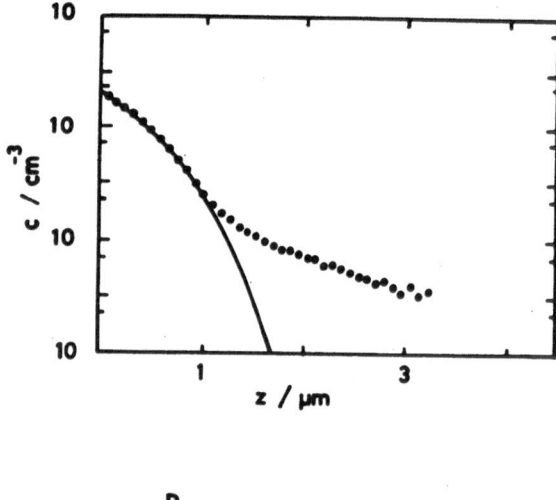

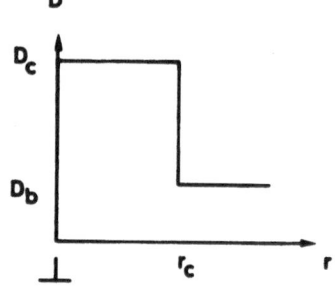

Fig. 5 a   Enhanced diffusion in the dislocation core gives rise to
tails in the concentration profile, here for the diffusion of
gallium in deformed germanium.
Fig. 5 b   The profiles have been analyzed in terms of the parameters
of a simple two-region model. Within the pipe ($r \leqslant a$) the diffusion
coefficient D_p is assumed to be constant   but considerably larger
than that in the volume D_v. The interaction of the diffusing atom
with the long-range elastic and electric fields of the dislocation
for $r > a$ is weak in the case considered and has been neglected.

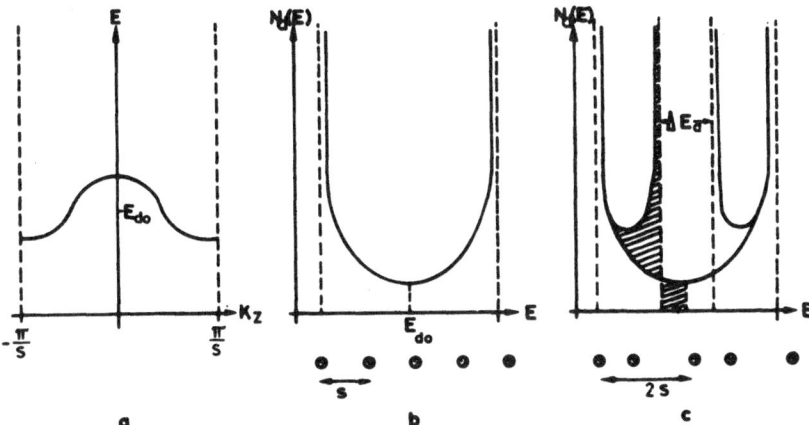

<u>Fig. 6 a</u> The dispersion $E_d(k_z)$ of a one-dimensional periodic system. For $60°$-dislocations in germanium the maximum of $E_d(k_z)$ is according to results of spectral photo-conductivity expected at $k_z = 0$.
<u>Fig. 6 b</u> The density-of-states of a one-dimensional system has singularities at the band edges.
<u>Fig. 6c</u> As a consequence a dislocation with a half-filled one-dimensional band should undergo a Peierls transition, leading to a doubling of the periodicity length and a splitting of the half-filled band into a full and an empty band.

E_{do} is the center of the band, J is the matrix element of a direct or indirect overlap between neighbouring dangling bonds and has been estimated to be of the order of 0.05 to 0.1 eV. The width of the band is $\Delta\varepsilon = 4 |J|$.

The second term in equation (1) describes the Coulomb interaction U between two particles (electrons or holes) in the same dangling bond, V_c the interaction between particles in different dangling bonds. For a value of U, which is large compared to $\Delta\varepsilon$, a half-filled band would split off into a full band and an empty band (Hubbard splitting), which could be the situation faced by Read, while a value of U smaller than $\Delta\varepsilon$ would yield Shockley's concept of a partially filled band.

We shall discuss the important role of U for the behaviour of dangling bond arrays in some detail in chapter 4. For a free atom, U is of the order of 10 eV, but we shall also see that for the dangling bond configuration of the vacancy in silicon the effective

U_{eff} has been shown to be negative.

The experimental proof that a onedimensional band structure has evolved from a Hubbard splitting has to verify the existence of a gap between full and empty states and of a dangling bond array with the lattice periodicity in the direction of the dislocation line. This criterion delineates the Hubbard splitting from two others. One, which we call reconstruction and shall discuss further in chapter 3.2, is the regrouping of two dangling bonds to form a bond which results into the disappearance of dangling bonds. The second results from the Peierls instability of a one-dimensional system and is due to the $1/E^{1/2}$ singularity in the density-of-states at the band edges. By doubling the periodicity length, which needs some strain energy, the system gains electronic energy (see Fig. 6) and a half-filled band splits off into an empty band and a full band. In this case, a dangling bond array with modified periodicity and a transition has to be verified experimentally. Up to now, a Peierls transition for dislocation has not been identified unambiguously [14,19] .

2.2 Occupation statistics

We shall briefly describe simple calculations of those part of our Hamiltonian which contain the self energies of the line charge of the dislocation V_c (part of H_{db}), of the screening cloud surrounding it V_{sc} (part of H_b) and of the Coulomb interaction energy H_c (part of H_{int}) between line charge and screening cloud.

The electrostatic energy of line and screening charge :

$$E_{el} = V_c + V_{sc} + 2 H_c$$

has been first calculated by Read [7] , a more general treatment is given in [20] and a summarizing description is given in [21] . E_{el} together with an entropy part enters the free energy of the dislocation, which is coupled to the reservoir of the electrons and holes of the bulk and which has to be varied with respect to the occupation f_{db} to obtain the equilibrium occupation. The result of the calculation contains a part :

$$\frac{\partial E_e}{\partial f_{db}} = E_e = \frac{e^2 f_{db}}{2\pi\epsilon \ S} \ [\ln \frac{\lambda}{r_c} \ - \ C] \approx \alpha^* f_{db} \ , \qquad (3)$$

ϵ static dielectric constant, λ screening radius, r_c the inner cut-off radius and C a constant between 0 and 1.
E_e has been interpreted as a rigid shift of the energy spectrum at a neutral dislocation on the energy scale. But note that for a given occupation f_{db} the shift E_e depends on the screening radius, e.g. for screening by free electrons, i.e. $\lambda = (\epsilon K_B T/e^2(n+p))^{1/2}$

on temperature and on the electron density and on the inner cut-off radius r_c, defining the transition between the short range part and the long range part of the potential (see chapter 4.2).

3. CORE STRUCTURE OF DISLOCATIONS

3.1 Atomic arrangement

The diamond structure consists of a face-centered cubic space lattice with a basis of two atoms. There exist two sets of {111} - slip planes and also two sets of dislocations with different core structure, called the shuffle and the glide set.

Common to the face-centered cubic and the diamond structure is an instability of dislocations which have the shortest Burgers vector $\underline{b} = \frac{a}{2} <110>$ (a lattice constant). These dislocations lower their strain energy by dissociation into two partial dislocations with Burgers vector $\underline{b_p} = a/6 <112>$ (note : $b^2 > (b_{p1})^2 + (b_{p2})^2$), which cannot exist as isolated species but are bordering a stacking fault. The partials repel each other, so that the width d of this planar defect results from the balance between the interaction energy of the two partials and the formation energy of the stacking fault.

Since in semiconductors also the bonding energies have to be taken into account, the questions as to what set and what configuration - unsplit or split - is realized, could not be answered from simple arguments. These questions are also not accessible to present calculation techniques and have to be decided experimentally.

To differentiate between an unsplit and split dislocation, the resolution of transmission electron microscopy (TEM) had to be brought below about 15 Å. This aim was reached around 1970 with the development of the weak beam mode of TEM [22] which, using a beam with larger indices and therefore weaker intensity forms an image of the defect by a contrast with larger Bragg angle, i.e. by a region around the defect of larger strain, as compared to conventional TEM. It has been found that under most conditions of plastic deformation experiments the dislocation are dissociated over most of their length with constrictions or segments of unsplit dislocation in-between [23,24].

From the width of the stacking fault its energy has been derived. The values are $\gamma = 58 \pm 8$ mJ/mm^2 for silicon and $\gamma = 68 \pm 6$ mJ/mm^2 for germanium [25]. For the cubic III-V compounds a linear decrease of γ with increasing ionicity f_I has been found, where f_I is measured in the scale proposed by Philips [26].

To differentiate between the two possible core structures needs an image with a spatial resolution of about 3 Å which has been achieved recently with the development of the lattice imaging mode of TEM. The dislocation is viewed end-on in a foil of about 50 Å thickness and is imaged by a set of beams outside the Gaussian image plane, so that optimum contrast is obtained with the slight phase shift induced by the atoms of the lattice and with the image aberrations inherent to TEM (Scherzer focus) [27] .

The interpretation of the pictures obtained by this method needs the computer simulation on the basis of a supposed defect structure and the careful comparison with a series of pictures recorded as a function of the defocusing distance.

Till now the 30°-partial in silicon has been investigated by two groups and both arrive at the conclusion that this type is of the glide set [28,29] .

3.2 Bond arrangements

In how far do the findings of the last section modify the simple model of the dislocations core which we have developed for an undissociated dislocation of the shuffle set?

We show in fig. 2b the bond configuration which one would expect from a ball-and-spokes model for a dissociated 60°-dislocation which has split off into a 30°-partial and a 90°-partial. A screw dislocation dissociates into two 30°-partials. The line elements of both partials as well as that of the undissociated dislocation follow the open channels of the diamond structure in the $\langle 110 \rangle$ - directions, so that dangling bond arrays with small periodicity length result. The major difference between an undissociated dislocation of the shuffle set and partial dislocation of the glide set is the direction of dangling bonds with respect to the glide plane, which is perpendicular in the first and almost parallel in the second case.

According to cluster calculations on the basis of a one-electron tight-binding approximation the dangling bonds, lying in the glide plane, tend to reconstruct whereby a half-filled band is transformed into an empty band within the band gap, while the filled states disappear with the valence band [30,31] . Let us denote by $\Delta E_r = E_{rec} - E_{unrec}$ the total energy difference between the reconstructed and unreconstructed core. At present the calculations schemes are not precise enough to establish for a given dislocation type the sign of ΔE_r. Also a stability sequence which has been extracted from the calculations, viz. $\Delta E_r^{30} < \Delta E_r^{90} < \Delta E_r^{60}$ [32,33] has to be considered to be preliminary because important interactions have been neglected.

4. MANY-ELECTRON DEFECTS

An isolated dangling bond is expected to be a donor or an acceptor and therefore to realize three charge states $(db)^+$, $(db)^\circ$ and $(db)^-$. In this chapter we describe defects which have more than one dangling bond and can be occupied by more than one electron or hole. As examples we take the vacancy in silicon and dislocations in semiconductors, which are many-electron centers and show rather peculiar characteristics.

4.1 Vacancy in silicon

To have an appropriate description of the various electron transitions between defect and bulk bands, we introduce the concept of occupancy level [3] by considering the energy variation of a system consisting of the defect and the crystal as a reservoir with chemical potential μ. When the defect occupancy changes from $(n+1)$ to n, we find the energy change $E(n+1/n) - \mu$ and associate the "level" $E(n+1/n)$ with this transition. If two electrons are involved in one transition, so that the occupancy of the defect changes from $(n+2)$ to n, we find the energy change $2(E(n+2/n) - \mu)$ and associate a "level" $E(n+2/n)$ with this transition. If we may neglect modifications due to spin and entropy, the so-defined "level" denotes the value of μ for which the occupancy changes from $(n+1)$ to n in the first case and from $(n+2)$ to n in the second case.

In a one-electron approximation the orbitals at the vacancy are constructed from linear combinations of the four dangling bond orbitals. There are four different combinations : the totally symmetric one A_1, whose energy is lowest (no nodes) and lies according to various calculations about 1 eV below the upper valence band edge, and three combinations T_2, which have the same energy of 0.7 eV above the upper valence band edge [34, 3] .

The T_2-states determine the electric behaviour of the vacancy and are also of great importance for many deep impurities on lattice sites whose localized states result from a weak interaction of the orbitals of the free atom with the dangling bond orbitals of the vacancy.

The degeneracy of the T_2-states is lifted by a Jahn-Teller distortion when these states are occupied by electrons [35] . The energy needed for the distorsion is compensated by a gain in electron energy. The A_1-state is always occupied with two electrons corresponding to the charge state V^{2+} of the vacancy. When the T_2-states are occupied, an appropriate distorsion reduces the symmetry of the surrounding atomic configuration and thereby lowers the energy of all occupied states, but leaving the center of gravity of the T_2-states unchanged (Fig. 7).

472

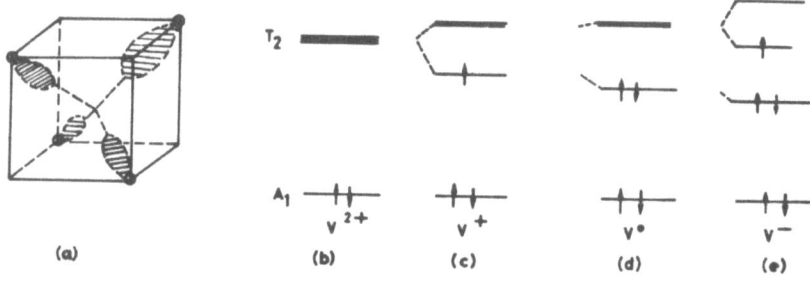

(a)   (b)   (c)   (d)   (e)

Fig. 7 Lifting of the degeneracy of the T_2-states by Jahn-Teller distorsion for the different charge states of the vacancy in silicon.

According to the experimental evidence the vacancy realizes five charge states V^{2+}, V^+, V°, V^-, V^{2-}, when the Fermi level μ is moved from the valence band edge to the conduction band edge.

Taking now also into account the electron-electron correlation, a simple approach would add to the matrix element of an LCAO treatment charge dependent terms which take into account intra- and interatomic Coulomb energies(see chapter 2.1). To achieve a partial self consistency, charge transfer between bonds, polarization of the bonds leading to a variation of the potential and finally of the matrix elements are admitted to lower the total energy. Using such a procedure,Lannoo [36] has shown that the intra-atomic Coulomb energy U of a dangling bond at the vacancy is reduced from a value of about 7eV for the isolated dangling bond to about $U_{eff} \approx 0.25$ eV by delocalization of added electrons among the four dangling bonds and by a polarization of the back bonds, which also corresponds to a delocalization of charge.

Baraff, Kane and Schlüter[3] have incorporated the static Jahn-Teller distorsion, the electron-electron correlation (neglecting correlation induced spin-multiplets, but see [37]) and static (outward) relaxations of the atomic neighbours (breathing modes) in a total energy functional. Using a self-consistent Green's function formalism to calculate electronic states, they arrive at the conclusion that the charge states V^{2+}, V^+, V° form an Anderson negative-U center. This means that actually $U_{eff} < 0$ so that the V^+ charge state is unstable and the lowest level for the vacancy is E(2+/0).

A possible over-compensation of U by lattice distorsions has been originally proposed by Anderson [38] to account for the diamagnetic behaviour of amorphous semiconductors.

The experimental proof that the vacancy in silicon is a negative-U center has been given experimentally by Watkins et al. [39] . They compared, using DLTS, the maximum number of holes that are emitted from the lowest lying level of the vacancy into the valence band with the number of vacancies in their specimen that can be bound to tin impurities to form tin-vacancy pairs. The ratio of these numbers is two which demonstrates that two holes are involved in the single emission process : $V^{2+} \rightarrow V^{\circ} + 2h$. Meanwhile also interstitial boron in silicon has been shown to be a negative-U center [40].

4.2 Dislocations in the elemental semiconductors

As an example of an extended dangling-bond defect we consider a dislocation with a linear array of dangling bonds in its core (density N_{db} per unit volume). If we denote by n_{db}^- the number of electrons transferred to the dislocation, we define the occupation ratio $f_{db} = n_{db}^- / N_{db}$ and the line charge $q = -e n_{db}^- / N_{db}$. The localized states at dislocations can be presented by a dispersion relation $E_\alpha(k_z)$, where k_z is the wave vector in direction of the dislocation line. To determine the energy spectrum at a dislocation from experimental data, we would have to define carefully a core radius r_c which separates the short range part of the potential ($r < r_c$), giving rise to bound states, from the long range part $V(r) = (q/2\Pi\epsilon)\ln(r/r_c)$ ($r > r_c$) responsible for the energy shift with occupation : $E_e(q, \lambda, r_c)$ (see chapter 2.2). At present, a physical argument which allows to quantify r_c cannot be given so that for the analysis of experimental data the assumption that r_c = const. has been made.

Can we save this simple concept if we now also take into account correlation effect ? The most important contribution is the intraatomic Coulomb interaction U, i.e. the term $\Sigma U n_{i\uparrow} n_{i\downarrow}$ in equation (1). Its magnitude will be lowered to a value U_{eff} by delocalisation of electrons along the dislocation core determined by the hopping strength J and the polarization of the surrounding, mainly of the back bonds, determined by the hopping strength Δ between dangling bond states and bulk states (part of H_{int}).

For the vacancy, Lannoo [36] has shown that one can separate the range of strong correlation with $8|J|/U < 1$ from the range $4|J|/U = \Delta\epsilon/U > 1$, within which the one-electron picture should be appropriate and U can be obtained from a Hartree-Fock calculation. In the limit of strong correlation we expect a finite energy separation between the states with one electron and those with two electrons in the dangling bond. One could then incorporate correlation effects in the energy band structure of dislocations.

In the other limit, U_{eff} adds to the electrostatic shift E_e of the dangling bond states. A calculation with this limit, using an Anderson-type Hamiltonian (see chapter 2.1), has shown that it gives a good description of the electric properties of 60°-dislocation in germanium [18] .

To screw-dislocations in germanium [42] and to 60°-dislocation in silicon [15,41] a full and an empty bands, separated by an energy gap, have been associated, but at the moment it is undecided by which effect the energy separation is caused.

Summarizing we state that (i) the energy spectrum of extended dangling bond defect is expected to be sensitively dependent on the arrangement of dangling bonds and that(ii) there is a strong long-range potential due to the charge on the defect, giving rise to a continuous shift E_e of the dangling bond states on the energy scale with occupation. Note that E_e is not only dependent on the line charge q but also on the screening length λ and on the cut-off radius r_c. The variation of the occupation ratio,which is realized when the Fermi level is shifted through the band gap, is in general small $|\Delta f_{db}| \ll 1$ and dependent on the screening.

4.3 Dislocations, experimental results

The energy spectra of dislocation in silicon and germanium have been investigated by many authors with a variety of experimental methods : Hall effect, transient and stationary photoconductivity, luminescence, capacitive methods like DLTS etc., EPR, EBIC etc.

It is important to state that a consistent analysis of experimental data and interpretation in terms of the core states is available only for dislocations which have been introduced or annealed at temperatures T_d above 0.6 T_m (T_m melting temperature). If T_d lies below 0.6 T_m, the production of point defects during plastic deformation and their interaction with dislocations may lead to considerable variation of the measured properties that are associated with the introduction of dislocations [18] . Point defect clouds surrounding the dislocations [15] and ,in one case ,structural changes in the dislocation core [43] have been made responsible for these variations.

For $T_d > 0.6$ T_m ,the energy spectra associated with dislocation core states in silicon and germanium are shown in Fig. 8. In Fig.6 a schematic energy dispersion curve for 60°-dislocations in germanium, as derived from spectral photoconductivity data, is presented. The electrostatic shift E_e of dislocations states with occupation, which has been derived from Hall effect and capacity data, is shown in Fig. 9.

For $T_d < 0.6$ T_m measured properties are not any more related to the dislocation density N_d for a given deformation mode but in general depend on deformation and annealing parameters (Fig. 10).

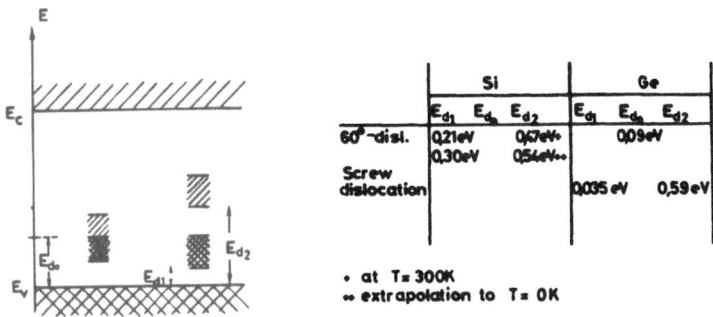

Fig. 8 Energy spectra of dislocation in silicon and germanium, as derived from Hall effect data. Note that actually 60°-dislocations are dissociated into a 30°- and 90°- partial, screw dislocations into two 30° -partials over most of their length.

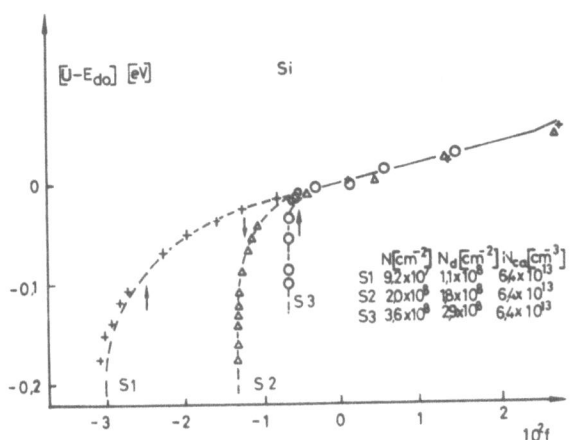

Fig. 9 Electrostatic shift E_e as a function of the occupation ratio for three silicon specimen with different doping and dislocation density, as derived from Hall effect data. The tails in the low-temperature range have been attributed to a lack of screening, so that neighbouring dislocation exert an electrostatic interaction on each other (the full lines present a quantitative description of this effect).

476

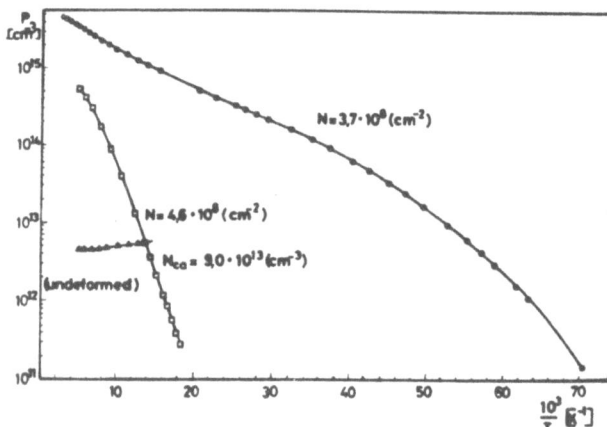

Fig. 10 Hole density P as a function of inverse temperature 1/T
for an undeformed and deformed germanium. When the deformation temperature is lowered below 0.6 T_m point defect clouds surrounding
the dislocation lead to significant variations of the P(1/T)-curves.

For silicon an EPR-signal has been found to be associated with
the introduction of dislocations at $T_d < 0.6\ T_m$. It consists of a
broad line [44] which represents a spin density of the order of the
possible dangling bond density in the dislocation core, and - superimposed - a set of lines [45] , whose associated spin density is
about two orders of magnitude smaller and which are believed to be
due to special sites or segments at the dislocation core. From an
analysis of the g-tensor and the hyperfine splitting of these fine
lines (Si K-1) the s- and p-contributions of the wave function have
been found to be 0.12 and 0.88, respectively, while the fractional
amount of the total wave function localized at the site is 0.54 [45] .

Both the broad line and the set of fine lines vanish on annealing around 0.6 T_m, although the dislocation density does not change
significantly. From the fact that together with these lines also
a DLTS-signal vanishes, it has been proposed that structural changes
in the dislocation core occur around 0.6 T_m [43] .

From the few data available for III-V and II-VI compounds a
consistent model of the core states could not be derived up to now.
There are several indications that point defect clouds play a significant role and are difficult to avoid.

4.4 Capture and emission processes at dislocations

Transition rates from defect states to the bulk bands have been widely investigated during the last few years by capacitance transient spectroscopy (DLTS). While the analysis of the data obtained with these methods to determine cross sections, activation enthalpies etc. is straightforward for point defects [46], it is up to now not established for extended defects. We therefore adopt in what follows a comparative description of the properties of point defects and of extended defects. The definition of those parameters which can be used to characterize a defect outside of electronic equilibrium, are given in Fig. 11. Note that we consider only the exchange of electrons with the conduction band, which means that we confine our description to electron traps.

Assuming a rigid shift of dislocation states with occupation, the balance between capture and emission rates may be written :

$$\frac{df_T}{dt} = c_n \, n(1-f_T) - e_n f_T \qquad \text{for point defects,} \qquad (4)$$

$$\frac{df_{db}}{dt} = c_n \, n(1-f_{db})\exp(-\frac{E_e}{K_B T}) - e_n f_{db} \quad \text{for dislocations,} \quad (5)$$

where the first term on the right side represents the capture, the second the emission rate. Writing $f_T = f_T^o + \Delta f_T$ and $f_{db} = f_{db}^o + \Delta f_{db}$, where the index o denotes the occupation in the stationary state, one sees immediately that the first equation is linear in Δf_T, while the second is non-linear in Δf_{db}. Following the lines of chapter 2.2, we write $E_e(f_{db}) = E_e(f_{db}^o) + \alpha \Delta f_{db}$.

Note that in a photoconductivity experiment, which in the past has been often used to study capture and emission at deep centers, the variation of the free-carrier density $\Delta n = n - n_o$ is measured so that equ. 4 becomes non-linear through terms proportional $\Delta n \Delta f_T$ and thereby also coupled, if more than one center is involved.

In DLTS it is the variation of the charge on the trap Δn_T^-, which determines the capacitive transient, while the free charge carrier density is kept constant ($n_o \approx N_{cd} \gg N_T$ during capture and $n \approx 0$ during emission). Since the emission rate is exponentially dependent on $\Delta H_{T,C}/k_B T$, where $\Delta H_{T,C}$ is the enthalpy of this process, it is at a given temperature the emission from one trap that determines $\Delta C(t) \sim \Delta n_T^-(t)$. Consequently only the solutions of one linear rate equ. 4 are needed to analyze DLTS data in terms of the capture cross section and energy level of the point defect. To determine also the defect concentration N_T needs a relation between Δn_T^- and Δf_T, i.e. an estimation of the effective volume. An appropriate evaluation for point defects has been given by Pons [47] .

478

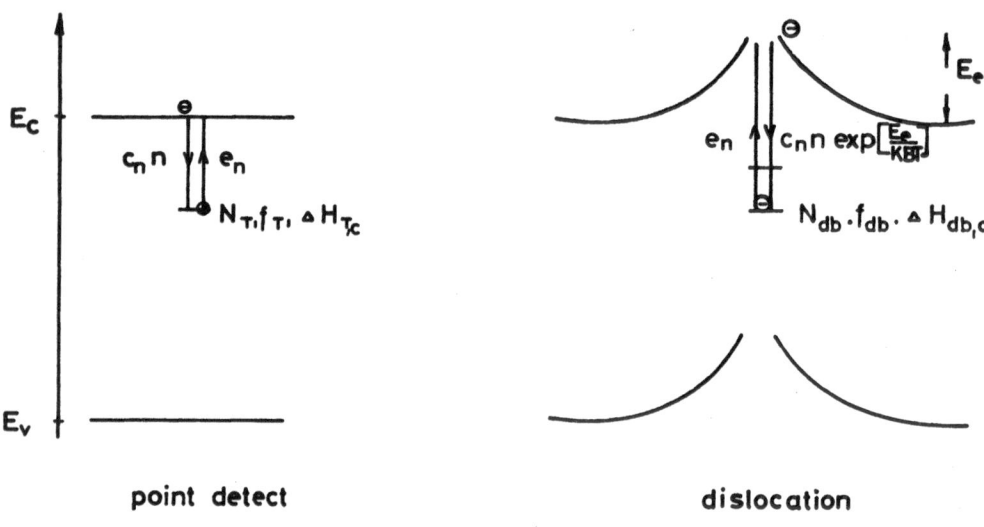

point detect   dislocation

Fig. 11 Parameters characterizing electron emission and capture at point defects and dislocations.

For extended defects an equivalent analytic procedure has not been developed up to now. By the loss of linearity of equ. 5 and due to the fact that the charge at the defect gives rise to a long range potential, the solutions are rather complex. We therefore will confine our description to those experimentally established criteria which allow for a differentiation between point defect and dislocation.

By periodic variation of the bias voltage V_b, the space charge region of a Schottky contact oscillates in width so that within a certain volume the value of the stationary occupation f_T of the traps, whose density has to be small compared to the doping concentration, also varies periodically. To adjust their occupation to the stationary value, electrons are trapped or emitted with a time law given as the solution of equ. 4.

In the usual DLTS the transient $\Delta C(t)$ during the emission period is reduced to one value by means of a filter (lock-in amplifier; $\sin 2\Pi v_L t$, or box-car detector $\delta(t-t_1)$ $\delta(t-t_2)$) and recorded as a function of temperature. In a $\Delta C/C_0$ (T) plot a point defect is represented by an almost symmetric line. From the maximum position, one easily derives the thermal emission rate at T_{max} : $e_n(T_{max}) \approx 0.4/v_L$ if the lock-in amplifier is used. Varying v_L allows to establish $e_n(T)$ within a certain temperature range.

Capture and emission processes are related by the principle of detailed balance :

$$e_n = c_n x N_c \exp\left(\frac{\Delta S_{T,C}}{k_B}\right) \exp\left(-\frac{\Delta H_{T,C}}{k_B T}\right) , \qquad (6)$$

with $c_n = \sigma_n \langle v \rangle$, σ_n the electron capture cross section of the defect, $\langle v \rangle \sim T^{1/2}$ the thermal velocity of electrons in the conduction band, $N_C \sim T^{3/2}$ the effective density-of-states of the conduction band, and $\Delta H_{T,C}$ $\Delta S_{T,C}$ enthalpy and entropy changes due to the emission of an electron from the defect to the conduction band.

If σ_n and $\Delta S_{T,C}$ are independent of temperature, one derives $\Delta H_{T,C}(T)$ from an Arrhenius plot of e_n/T^2 versus $1/T$. By a more refined measuring procedure also a separate determination of $\Delta H_{T,C}$ and $\Delta S_{T,C}$ and a check of their temperature dependence is possible.

From the capture characteristics $\Delta C(t_p)$, where t_p is the duration of the filling pulse, one also obtains the trap density N_T.

Where do DLTS data for dislocations differ from those for point defects ? The differences that we report here do not necessarily establish general criteria, since they have been derived from a few experimental investigations [48] :

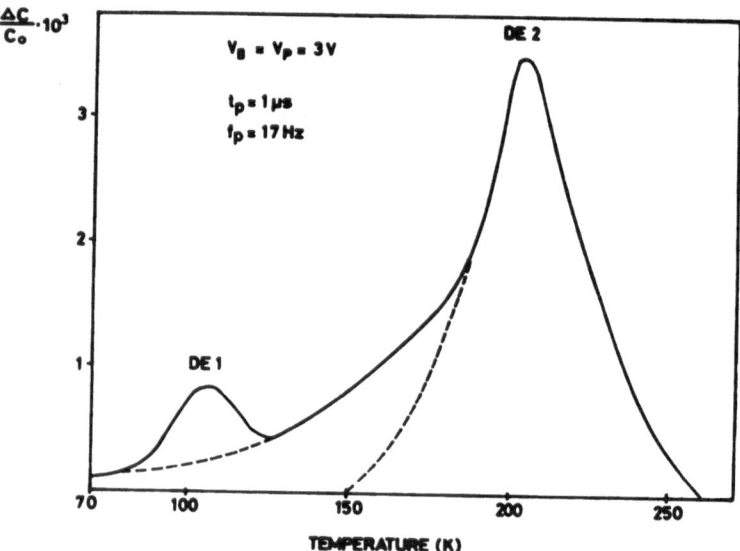

<u>Fig. 12</u> Capacitance-transient spectrum $\Delta C/C_0$ (T) of n-type silicon (doping concentration 2.4×10^{15} cm^{-3}) after plastic deformation. The small peak at low temperature is associated with a point defect. If the large peak would be associated with a point center, one expects the dashed line form. The strongly asymmetric line-shape is characteristic of extended defects.

1. The line shape is asymmetric with a long low-temperature tails (see Fig. 12). The tail is exponential with 1/T;
2. Within a certain t_p-range (about two orders of magnitude), one finds $\Delta C \sim \ln t_p$;
3. Within a certain range of T around the maximum of the DLTS line, the emission rate is exponential with time so that e_n(T) may be extracted in the same way as for point defects.

Note that the capture characteristics for dislocations should be definitely different for dislocations from that of point defects, because it is the capture rate which is affected most by the long range potential, while the term describing the emission rate is identical in the rather simple view of equations 4 and 5.

Although the density of localized states associated with a dislocation cannot be derived from a total filling of those states by electrons during a sufficiently long filling pulse, it should be deductable from an analysis since $\Delta C \simeq \Delta n_{db}^-$ while $\Delta E_e \simeq \Delta f_{db} = n_{db}^-/N_{db}$.

ACKNOWLEDGEMENT

I am grateful to M. Lannoo (Lille) for critical discussions and to H. Bock for the preparation of the figures.

REFERENCES

1. Alexander, H. and Haasen, P., Solid State Physics, 22 (1968) 27.
2. See for example : Watkins, G.D., in Point Defects in Solids, Vol. 2, ed. J.H. Crawford Jr. and Slittin, M.L., Plenum Press, New York, 1975, p. 333.
3. Baraff, G.A., Kane, E.O. and Schlüter, M., Phys. Rev. B21, (1980), 5662.
4. Schröter, W., Inst. Phys. Conf. Ser. 46 (1978) 114.
5. Schröter, W., Phys. Stat. Solidi 21 (1967) 211.
6. Patel, J.R. and Chaudhari, A.R., Phys. Rev. 143 (1966) 601.
7. For a summary see : Surface Effects in Crystal Platicity, ed. by R.M. Latanision and J.T. Fouric, Noordhoff, Leyden 1977.
8. Ossipyan, Y.A. and Petrenko, V.R., J. Physique C6 (1979) 161.
9. Schröter W. in Electronic Structure of Crystal Defects and Disordered Systems ed. by F. Gautier, M. Gerl and P. Guyout, (Paris, Les Eds. de Physique, 1980), p. 129.
10. Ahlborn, K., J. Physique, C6 (1979) 185.
11. Baumann, F. and Schröter, W., Phys. Stat. Sol. to be published
12. Ourmard, A., Weber, E., Gottschalk, H., Booker, G.R. and Alexander, H., Inst. Phys. Conf. Ser. 60 (1981) 63.
13. Böhm, K. and Fischer, B., J. Appl. Phys. 50 (1979) 5453.
14. Döding, G. and Labusch, R., Inst. Phys. Conf. Ser. 60 (1981) 57.
15. Schröter, W., Scheibe, E. and Schoen, H., J. Microscopy 118 (1980) 23.
16. Shockley, W., Phys. Rev. 91 (1953) 228.
17. Read, W.T., Pil. Mag. 45 (1954) 775 and 1119.
18. Usadel, K.D. and Schröter, W., Phil. Mag. B37 (1978) 217.
19. Elbaum, C., Phys. Rev. Letters 32 (1974) 376.
20. Schröter, W. and Labusch, R. Phys. Stat. Sol. 36 (1969) 539.
21. Labusch, R. and Schröter, W. in Dislocation in Solids, ed. by F.N.R. Nabarro, Amsterdam, North Holland Publ. Company, 1980, vol. 5, 127.

22. Cockayne, D.J.H., Ray, I.L.F. and Whelan, M.J., Phil. Mag., 20 (1969) 1265.
23. Cockayne, D.J.H. and Hous, A., J. Physique C-6 (1979) 2.
24. Packeier, Cr. and Haasen, P., Phil. Mag. 35 (1977) 821.
25. Alexander, H., J. Physique C-6 (1979) 1.
26. Alexander, H., Proc. of the 6th European Congress on Electron Microscopy (1976), page 208.
27. See the book by John C.H. Spence, Experimental High Resolution Electron Microscopy, Oxford, Clarendon-Press, 1981.
28. Bouret, A., Desseaux, J. and D'Anterroches C.Inst. Phys. Conf. Ser. 60 (1981) 9.
29. Austis, G.R., Hirsch, P.B., Humphreys, C.J., Hutchison, J.L. and Ourmazd, A., Inst. Phys. Conf. Ser. 60 (1981) 15.
30. Marklund, S., Phys. Stat. Sol. (6) 92 (1979) 83.
31. Jones, R., J. Phys. C-6 (1979) 33.
32. Marklund, S., Phys. Stat. Sol. (6) 100 (1980) 77.
33. Jones, R., Inst. Phys. Conf. Ser. 60 (1981) 45.
34. See the article by P. Pécheur in Electronic Structure of Crystal Defects and Disordered Systems, ed by F. Gautier, M. Gerl, P. Guyot,(Paris, Les Editions de Physique, 1981),p. 93.
35. Watkins, G.D., Inst. Phys. Conf. Ser. 23 (1974) 1.
36. Lannoo, M., Inst. Phys. Conf. Ser. 46 (1978) 1.
37. Lannoo, M., Baraff, G.A. and Schlüter, M., Phys. Rev. B24 (1981) 955.
38. Anderson, P.W., Phys. Rev. Lett. 34 (1975) 953.
39. Watkins, G.D. and Troxell, J.R., Phys. Rev. Lett. 44 (1980) 593.
40. Watkins, G.D., Cattergec, A.P. and Harris, R.D., Inst. Phys. Conf. Ser. 59 (1980) 199.
41. Grazhulis, V.A., J. Physique, C-6 (1979) 59.
42. Wagner, R. and Haasen, P., Inst. Phys. Conf. Ser. 23 (1975) 56.
43. Kveder, V.,Ossipyan, Yu.A., Schröter, W. and Zoth, G., Phys. Stat. Sol., in Press.
44. Grazhulis, V.A. and Ossipyan, Yu.A., Sov. Phys. JETP 33 (1971) 623.
45. Weber, E. and Alexander, H., J. Physique C-6 (1979) 101.
46. Miller, G.L., Lang, D.V. and Kimerling, L.C., Ann. Rev. Mater. Sci. 7 (1977) 377.
47. Pons, D., Thesis Paris 1979.
48. Seibt, M., Diploma thesis Göttingen 1982.

OPTICALLY EXCITED DEFECTS

Michel Wautelet

Université de l'Etat, Faculté des Sciences, 7000 Mons, Belgium.

1. INTRODUCTION

Laser irradiation of a solid is known to result in the excitation of electrons from states located below the Fermi level, E_F, up to states located above E_F.
It is generally believed that delocalized (initial and final) states play the major role during laser processing of materials. However, localized (defects) excited states may also play an important role, as evidenced by a lot of experimental facts. Among others, let us mention :
1) the optical absorption oscillations observed in a-GeSe$_2$ and a - As$_2$Se$_3$ films for laser powers below the crystallization threshold P_c [1,2] ;
2) the metastable amorphous state of Si and Ge, evidenced by conductivity changes well below P_c [3,4] ;
3) the laser and electron-induced enhanced crystallization rate of Si [5,6] ;
4) the enhanced kinetics of As diffusion after laser irradiation [7] ;
5) the creation of new-metastable donor states after laser annealing of ion-implanted Si [8,9] .
The aim of this lecture is to present the physical mechanisms responsible for the optical excitation of defects and their properties. The prerequisite to such a study is a knowledge of the properties of the fundamental and excited defects. This is treated in Section 2, with particular emphasis on dangling bonds in Si. In Section 3, the mechanisms of optical excitation and the various decay paths (radiative, non-radiative, Auger) are discussed. Atomic diffusion is described in Section 4. First, one discusses the factors influencing diffusion. Then, it is shown how defect excitation is

able to modify diffusion in different ways.

2. DEFECT STATES

In semiconductors, defects and impurity levels are classified in two groups : shallow and deep levels. The shallow levels lie within a few 10^{-2} eV from the band edges and can be ionized at room temperature. In Si and Ge, impurities of columns III and V of the Mendeleev periodic table are from this category. The deep levels category includes polyvalent metals, transition metals as well as vacancies, dislocations and surfaces. The corresponding levels are located "deep" in the gap. In the following, one will be mainly concerned with these states.

Impurity and defect electronic states may be separated into four classes [10] , depending on the spatial arrangement of the atomic orbitals involved : isolated, semi-isolated, molecular and periodically distributed. The typical isolated orbital is most probably the dangling bond (DB) in amorphous Si or Ge. It is characterized by the fact that it points inside a cavity where all other bonds are satisfied. Semi-isolated and molecular orbitals are present in the cavities of crystals. Mono-vacancies consist exclusively of molecular ones, while poly-vacancies show semi-isolated orbitals at the extremities. Semi-isolated orbitals are only weakly coupled to other ones, while molecular ones are strongly coupled to each other. Periodically arranged orbitals are located at the free surface of a crystal and along dislocations.

2.1 Fundamental states

Excellent reviews about theoretical and experimental determination of defect states [11,12] exist in the literature, so that one will briefly discuss one typical case, namely the dangling bonds in Si, since it is probably the most studied defects.

Si and Ge have four electrons (s and p) outside their closed shells. They can then hybridize into the so-called sp^3 hybrid states, where atoms have the tetrahedral symmetry with four unpaired orbitals. They will combine with other similar atoms by covalent bonding. In the amorphous phase, it is generally recognized that the four-fold coordination is retained for all atoms, except at cavities, impurity sites, dangling bonds and the surface. Under these circumstances, one observes a variation of bond angles and lengths, which induce simultaneous changes in the electronic structure associated with these bonds [10,12,13] .

The variation of the electronic structure with the deformation of the bonds is evaluated theoretically by using the tight-binding method. We refer to other reviews for a description of the calcu-

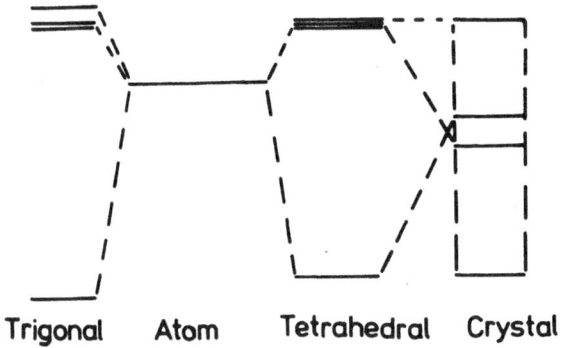

Trigonal   Atom   Tetrahedral   Crystal

Fig. 1 Schematic energy level diagram for Si and Ge, as a function of their hybridization state.

lations [10,12] . Let us summarize the results obtained for the isolated, molecular and periodically arranged DBs in Si. The symmetry
of the isolated DB configuration is trigonal. As a result, the two
electronic atomic-like levels of the sp^3 configuration(one three-fold
degenerate and one non-degenerate) are split into one doublet
and two singlets, as shown in figure 1. The most interesting fact
is that the (sp^3) three-fold degenerate level, associated with the
top of the valence band is split into one doublet and one singlet,
whose level is higher than the sp^3-triplet, i.e. lies within the band
gap of the semiconductor. This is generally believed to be the origin of band gap states in amorphous silicon.

The electronic levels associated with periodically arranged DBs
may be calculated starting from the isolated one, plus some shift
and broadening due to orbital overlap, like for the calculation of
the electronic structure of a solid starting from atomic orbitals
in the tight-binding approach. This is schematized in figure 2.
It is worth noting that the fundamental state of a periodic array
does not necessarily correspond to an arrangement of neutral, equivalent DBs, but may be associated with a periodical arrangement of
alternatively raised and lowered ions. In this case, the partially
filled DB band is split into one full and one empty bands. This

situation is well established in the case of the Si(111) 2 x 1 surface structure [14] , and is probably also present in some kinds of
dislocations in Si and Ge [15] . The typical molecular DB association is the mono-vacancy in Si. The calculations are similar to
the calculations of molecular levels, starting from the atomic properties. The four DB orbitals levels combine into two molecular
levels (one triplet and one singlet) in the tetrahedral symmetry.
However, this state is unstable against a distorsion of the vacancy
towards a tetragonal distorsion, and the upper triplet is split
into one singlet (b) and one doublet (e). This is summarized in
figure 2.

2.2 Excited states

The mono-vacancy in Si possesses four electrons and the last
occupied level is the b level. The excited state is then the e level. The excited state of the isolated DB in a-Si is not unambiguously determined theoretically at the present time. Two models
have been proposed to account for its properties : the self-trapped
exciton and the dehybridized DB.

The so-called self-trapped exciton [16] is a state made of a
pair of spatially separated charged centres, DB^+ and DB^-. The formation energy of this pair is lowered by the Coulomb interaction between them. The resulting density of gap states is shown in figure 3. This is similar to the density of states of the Si(111) 2 x
1 surface. Let us note that the DB^+ most likely exhibit sp^2 bonding
with a bond angle near 120°, while DB^- are expected to exhibit p
bonding with bond angle in the 10°-100° range.

In the dehybridized DB model, the excited state corresponds
to a neutral DB one of $sp^2 - p_z$ electronic configuration. The DB
state is assumed to lie within the band gap [10,17,18] .

The self-trapped exciton is also believed to explain the origin
of the metastable states observed in a-Se [19] and a-GeSe$_2$ [1] .
As noted before, at the Si (111) 2 x 1 surface and for some kinds
of dislocations, the electronic structure of DBs is characterized
by one filled band (associated with negatively ionized DBs) and one
empty band (associated with positively ionized DBs). When the system is excited, (either thermally or electronically), the relative
populations of the initially filled and empty bands vary, i.e. the
total electronic energy increases. Due to the strong electron-
deformation coupling, it turns out that the relaxation of DB sites
is modified, which in turn modifies the resulting electronic level
structure. In order to see this, let us consider the total (electronic + vibrational) energy vs. some normal coordinate, q :

$$E_{tot} = Cq^2 - Aq \qquad (1)$$

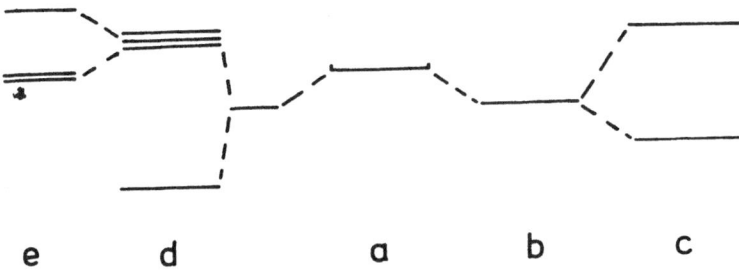

Fig. 2 Schematic energy level diagram of the dangling bond state; a) unreconstructed; b) relaxed; c) reconstructed; d) mono-vacancy (un-reconstructed); e) mono-vacancy (Jahn-Teller distorted). Cases a, b, c may be associated with the isolated dangling bond in a-Si, a-Ge dislocations, or surface.

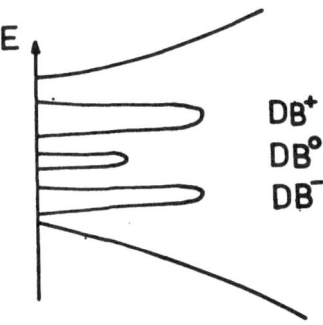

Fig. 3 Density of states corresponding to the dangling bond ionic configurations in a-Si.

where C is an elastic constant, and A is related to the separation between the electronic filled and empty bands. The minimum of E_{tot} occcurs at $q_0 = A/2C$. When a large number of electrons are excited, the electronic energy is modified, and the minimum of E_{tot} occurs at q_1, such that [20] :

$$q_1 = q_0 B = q_0 \frac{\sinh(x) + 0.5 \sinh(2x) + x \cosh(x) + x}{[1 + \cosh(x)]^2} \quad (2)$$

where $x = Aq_1/2kT$. T is the temperature of the system, related to the number of excited electrons, n, by the law of mass action. B is shown in figure 4.

Under thermal equilibrium conditions, T is the normal temperature of the solid. Under laser or electron excitation, n may exceed the equilibrium value. Then T is much larger than the lattice temperature, and the distortion evolves rapidly to $q_1 = 0$, i.e. there is no more reconstruction. When applied to the Si (111) 2 x 1 surface, this induces a transition to a Si(111) 1 x 1 structure. A similar situation takes place for the reconstructed dislocation. Simultaneously, the two DB bands approach each other until they mix exactly. Given a mean separation between the bands of 0.3 eV and a width of 0.5 eV, the 2 x 1 to 1 x 1 transition occurs for T≈3000K. However, it is not clear whether the laser-induced reconstruction of the Si(111) surface [21,22] is due to this effect or to a normal heating effect.

3. OPTICAL EXCITATION AND DECAY

Absorption of photons in a solid results in the transitions of electrons up to excited states. The optical transition is so short $(10^{-14}s)$ that nuclei have no time enough to move during the transition. Indeed, the characteristic period of vibration of nuclei in solids is given by the inverse of the Debye frequency, i.e. about $10^{-12}s$. This is the origin of the Franck-Condon principle, stating that atomic nuclei do not move during optical excitation.

When one electron goes from a delocalized valence band state to a delocalized conduction state, all interatomic distances remain unchanged, so that the transition occurs between pure electronic states.

If the optical transition involves one or two localized states, the situation may be different. Although here one is dealing with defect and impurity states, the "localized" situation applies also to the (localized-like) band edges of amorphous Si and Ge and to the surface states. In all these cases, one has to consider the total (electronic plus lattice) energy vs. some configurational co-ordinate

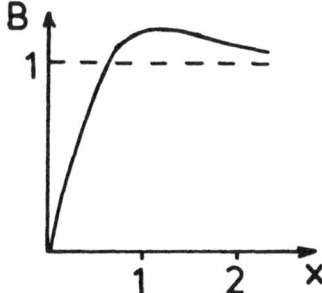

Fig. 4 Variation of the Jahn-Teller distortion parameter, B, with x (see text).

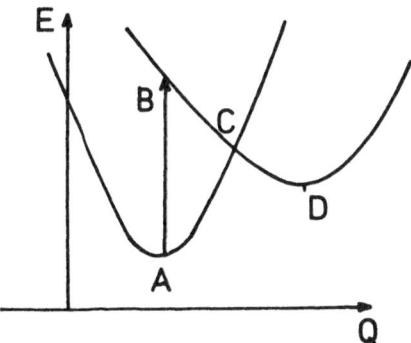

Fig. 5 Configurational coordinate diagram of the total energy (E) of the fundamental and excited states of a given defect.

diagram, as illustrated in figure 5. This arises from the strong connection between the electronic level of the "defect" and the distortion of the lattice, the equilibrium configuration being quite different for a ground state and for an excited state. This is a general situation which occurs every time the wave function is localized so that its corresponding energy is sensitive to small displacements of the neighboring nuclei. In figure 5, the equilibrium fundamental and excited electronic states are in A and D, respectively.

3.1 Optical excitation

Due to the Franck-Condon principle , it is impossible to excite directly the atom from state A to D by a photon of energy (E_D-E_A), since this transition is not vertical on the diagram of figure 5. The only way is to excite optically an electron from A to B, i.e. $h\nu > (E_D-E_A)$. However, the excitation is not exactly monochromatic, since one has to consider the vibrations. They introduce discrete energy levels in the bonds, corresponding to different vibrational quantum numbers m. At 0 K, only the m=0 vibrational level is occupied and transitions occur only from this level, from any position consistent with the spatial extent of the m=0 vibrational wave functions. Moreover, the probability of any transition occuring from a particular value of the configurational co-ordinate is a function of this co-ordinate. An absorption band results from this effect. As the temperature increases, different vibrational levels become occupied, which have different probability distribution functions.

When the optical excitation couples two localized states, the absorption band is a Gaussian [23] , with a half-width which varies with temperature proportionally to coth $(h\nu/2kT)$.

In most cases of interest in semiconductors, optical excitation couples one localized and one delocalized states, i.e. electrons are excited either from a valence band state to a defect state, or from a defect state to a conduction band state. Then, the optical cross section is calculated to be proportional to [24] :

$$\sigma_T(h\nu) = \frac{1}{h\nu} \int_0^\infty dE\, \rho(E) \left| \frac{(1\pm\eta)\, E^{1/2}}{|E_{io}|+E} + \frac{(1\mp\eta)\, E_F^{1/2}}{|E_{io}|-E-(Eg+Ep)/2} \right|^2$$

$$\times \exp\left[-\,(h\nu-[|E_{io}|+E])^2\,/\,4kT\,d_{FC}\right] \qquad (3)$$

where $\eta(E) \approx \exp(-2E/E_A)$ and $\rho(E)$ represents the density of electron states. The other parameters entering equation 3 are given in figure 6. $\sigma_T(h\nu)$ presents a maximum when $h\nu$ increases, and then decreases. $\sigma_T(h\nu)$ depends on temperature, due to the phonon-depending contribution [24] . It is also worth noting that the defect-conduction (or valence) band separation measured optically is *not*

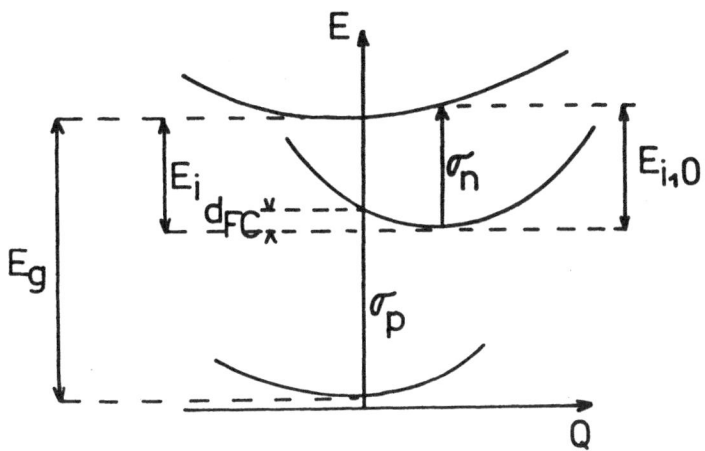

<u>Fig. 6</u> Configurational coordinate diagram of a defect. E_i is the binding energy. The valence and conduction bands are separated by Eg. d_{FC} is the Franck-Condon parameter, and $E_{i,0}$ is the optical ionisation energy. σn and σp represent the transitions to the conduction band and from the valence band, respectively.

equal to the separation given by electrical transport measurements, due to the strong electron-phonon coupling.

3.2 Mechanisms of decay

The optically excited state is obviously unstable and the system tends towards its fundamental state. Various mechanisms are involved in this decay, depending on the kind of configurational coordinate vs energy diagram which applies to any particular defect. Let us consider the following mechanisms : emission of phonons, luminescence and Auger de-excitation.

3.2.1 Emission of phonons

Within the configurational coordinate diagram of figure 5, immediately after state B has been reached, the lattice relaxes around the defect up to state D, by emission of an appropriate number of phonons. It is worth noting that these phonons are emitted in the immediate vicinity of the excited site which behaves, therefore, as

492

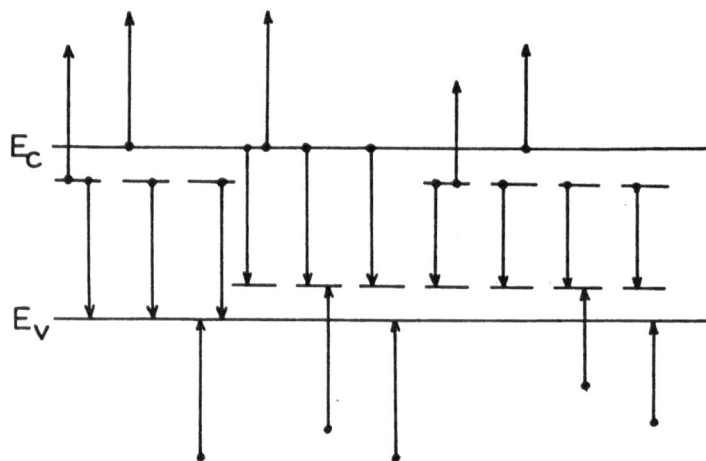

<u>Fig. 7</u> Examples of possible Auger recombination processes involving defect states.

a localized source of "heat". This is different from the conventional electron-phonon recombination mechanism in which the vibrational energy is released over a relatively large volume.

When the excited state corresponds to an optically ionized defect (i.e. the electrons are in a delocalized state), the vibrational energy release is in part localized at the defect site, in part delocalized via the de-excitation of the electron within the conduction band.

3.2.2 Luminescence

Once in state D, the system may de-excite radiatively to the ground state by emitting a photon of energy (E_D-E_c). It is apparent that there is a shift in the relative positions of the absorption and emission bands : this is usually referred to as the Stokes shift. The difference in energy between the absorption and emission bands appears in the form of vibrational energy during the cyclic process.

3.2.3 Auger de-excitation

In the Auger effect, the energy lost by the excited to fundamental state transition excites a nearby carrier in the solid. Various possible Auger recombination processes involving defect states are shown in figure 7. Let us note that the energy exchange is not always limited to similar states. The hole decay rate is given by [25] :

$$\frac{dp}{dt} = - \gamma \, p \, n_c \, n \tag{4}$$

where p, n and n_c are the densities of holes, recombination centers for holes and conduction band electrons, respectively, as schematized in figure 8. Theoretically, it is found that the Auger rate constant, γ, is of the order of $10^{-26} \mathrm{cm}^6 \mathrm{s}^{-1}$ [26] to be compared with $10^{-31} \mathrm{cm}^6 \mathrm{s}^{-1}$ for the band-to-band Auger recombination process [27].

3.2.4 Metastable states

Up to now, we considered configurational coordinate diagrams like illustrated in figure 5. Other situations are also seen, as shown in figure 9, where the minima corresponding to the fundamental (A) and excited (D) states are such that it is not possible to go "vertically" from D to any point of the non-excited curve. This is believed to be the case of the dangling bonds in a-Si [16-18] and the metastable configuration of a-As$_2$S$_3$ and a-GeSe$_2$ - like materials [1, 19,28] .
The optical excitation process is the same as previously described, Immediately after state B has been reached, the system relaxes to point C by emission of phonons. From C, there is some probability to go to A and D, again by emission of phonons. If the system relaxes in state D, it may remain there for a long time, since, returning to A would necessitate a distortion of the defect site. The recombination mechanism needs to pass through C and requires an activation energy, E_a. This can be achieved by vibrational (mechanical and/or thermal) activation. In this mechanism, it is possible to produce some damage in the vicinity of the defect similarly to what is assumed in halides [29] .
It is also important to note that the system cannot decay radiatively from D to the fundamental curve, due to the Franck-Condon principle. The decay mechanisms are therefore non-radiative and involve capture of phonons and/or electrons or Auger transitions (see above).

Such a situation explains very well the long-lived metastable state observed after low power laser irradiation of glow-discharge(GD) a-Si [3] and a-Ge [4] films. In GD a-Si, E_a has been deduced to be 1.52 eV, by measuring the temperature dependence of the recovering

494

time. It is worth noting that E_a is larger than the fundamental band gap and much larger than the DB level-to-band separation. This confirms the role of phonons in the relaxation process. The problem remains to associate the diagrams with physical situations (pairs of ionized defects, dehybridized dangling bonds).

Let us also mention that the previous discussion applies to surface states, since the configurational coordinate diagram is also valid. The surface state curve is not symmetric, due to the presence of vacuum, as shown in Fig. 10.

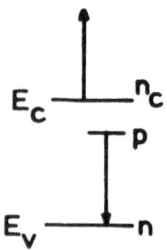

Fig. 8 One possible Auger recombination process. p, n and n_c are the populations of the defect, hole and electron states, respectively.

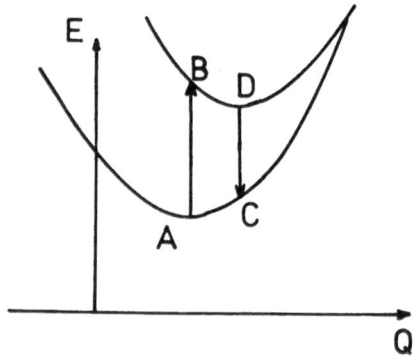

Fig. 9 Configurational coordinate diagram of a defect, with a long-lived metastable excited state.

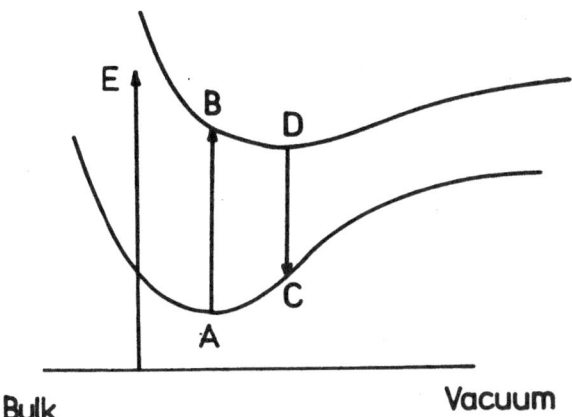

E
B
D
C
A

Bulk   Vacuum

Fig. 10 Configurational coordinate diagram of a surface state.

4. ATOMIC DIFFUSION

4.1 General

An atom in a crystal or an amorphous material may jump form
one site to another provided :
1) the final site is vacant, and
2) it acquires an energy sufficient to allow it to pass a "potential barrier". This process may be described by using the configurational coordinate diagram shown in figure 11.
The jump frequency is given by [30,31] :

$$\Gamma = X\omega ,$$
(5)

where ω is the frequency at which an atom exchanges position.
Γ depends on the probability that the final site is vacant, so that
ω is multiplied by the ratio of vacant to occupied sites, X, expressed by :

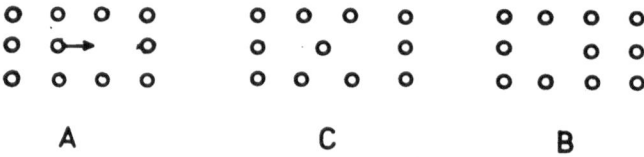

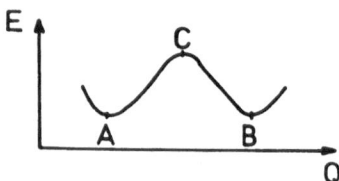

<u>Fig. 11</u> The vacancy diffusion mechanism, with the corresponding total energy (E) vs. displacement (Q) diagram.

$$X = \exp (\Delta S_f/k) \exp (-\Delta H_f/kT) \tag{6}$$

where ΔS_f and ΔH_f are the entropy and enthalpy of formation. It is obvious that $X = 1$, in the case of interstitial diffusion, and that $\Delta H_f = 0$ when the vacant site concentration is independent of temperature (for instance, when the material is non-stoichiometric, so that the intrinsic vacancy concentration is much lower than the concentration of "missing" atoms).

The frequency ω is not easy to calculate from first principles. It is generally expressed like :

$$\omega = \omega_o \exp (\Delta S_m/k) \exp (- \Delta H_m/kT) \tag{7}$$

where ΔS_m and ΔH_m are the entropy and enthalpy of migration.

ω_o is often taken as the Debye frequency, ω_D. However, it has to be recognized that ω_D is an approximate upper bound to the values of ω_o [32] . This value should be approached only if the mechanism of migration requires no correlated motion of the neighboring atoms. If it requires such a correlated motion, ω_o is smaller than ω_D. Then, $\omega_o = a \ \omega_D$, where $a \approx 1$ for simple mechanism, and $a \ll 1$ for a complex one.

Combining equations 6 and 7, one obtains :

$$\Gamma = a \; \omega_D \; \exp \; (\frac{\Delta S_f + \Delta S_m}{k}) \; \exp \; - (\frac{\Delta H_f + \Delta H_m}{k_T}) \; . \qquad (8)$$

Let us now examine the parameters affecting the entropy and enthalpy terms.
The vibrational entropy of a system is given by [33] :

$$\Delta S/k = 3 \; N \; [1 - \ln(h\omega/kT) - 1/2 \sum_j \ln \; \omega_j^2 + o(1/T2)] \qquad (9)$$

where 3 N is the number of vibrational modes and ω_j are the vibrational frequencies of the system.
At the saddle point position for migration (point C, in figure 11), some of the ω_j are replaced by ω'_j. Since the first term at the righhand part of equation 9 is constant, the change in the entropy is expressed under the form :

$$\frac{\Delta S}{k} = - 1/2 \sum_j \ln \; \omega'_j^2 + 1/2 \sum_j \ln \; \omega_j^2 \; . \qquad (10)$$

The ω_j^2 are known to be proportional, in first approximation, to the interatomic force constants, C_{1m}. Then, if some "reconstruction" of defects occur, it turns out that ΔS may be modified. For instance, in the case of the mono-vacancy in Si, Lannoo and Bourgoin [34] have shown that a softening of the C_{1m} (by elongation of the bonds) leads to an increase of $(\Delta S_f + \Delta Sm)$. The enthalpy of motion $(\Delta H_f + \Delta H_m)$ is related to three parameters : the surface energy of the vacant site (in ΔH_f); the change in bonding energy between the saddle point and the equilibrium configurations (in ΔH_f and ΔH_m); and a ballistic effect(in ΔH_m).
Let us discuss this for the case of a monovacancy in Si : other cases are detailed, for instance, by Van Vechten [35] .

To a first approximation, the enthalpy of formation of the vacancy is the surface energy of the corresponding cavity [35 and ref. therein] :

$$\Delta H = E_S + nU \qquad (11)$$

where E_s is due to a free-electron-like effect, proportional to the 5/6 th power of the electron density [36] ; and nU is due to the number, n, of broken bonds (dangling bonds) present in the cavity. U is the heat of bonding.

Before a given atom may jump into another site, a number of bonds have to be broken [32,35] . This means that ΔH_m is, at least, equal to the product of the number of broken bonds, n' (not n !) by the binding energy of one bond. Theoretical estimates of this energy

498

<u>Fig. 12</u> The vacancy diffusion mechanism, showing the effect of high
temperature. Dashed circles : extreme motion of neighboring atoms
at high temperature.

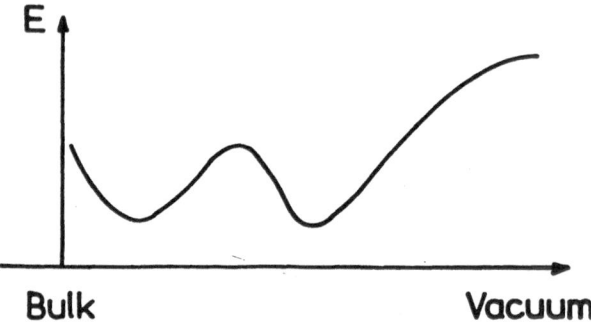

<u>Fig. 13</u> Energy (E) of a given atom in the bulk and surface regions.

are in good agreement with the empirical values [35] .

      However, at high temperature ΔH_m of the single-vacancy in Si
and Ge are much larger than those measured at low temperature. This
may be explained by a simple ballistic model [37] illustrated in
figure 12. At low temperature, the vibration of atoms B and C, perpendicular to the motion of the vacancy, is not large. Then, the
moving atom has only to break n' bonds before jumping. On the contrary, at high temperature, atoms B and C vibrate much more. Then

the energy required to move the vacancy is modulated by the displacement, d, of atoms B and C. During part of the period of d, the energy of the saddle point is decreased, while it is increased during another part of the period. The net result is an increase of ΔH_m upon increasing temperature, proportionally to ω_D^2 [37] .

The previous reasoning may be extended to other kinds of defects [35] and to desorption of atoms or ions. The configurational coordinate diagram for desorption is given in figure 13. From the previous discussion, it turns out that surface reconstruction would affect the rate of desorption, via the variations of ΔS and ΔH.

4.2 Excited defects

When a defect is excited (thermally, by stress, electronically or optically), one may affect one or more quantities discussed before (a, ω_d, ΔS, ΔH), so that one may reduce or enhance Γ. One is generally interested by enhanced diffusion, but it has to be borne in mind that the opposite may (and presumably do) occur in some cases. One may also conceive that there exist situations where processes with higher ΔH have higher Γ, due to the modification of a, E_s and/or ΔS. Also, it is worth noting that diffusion mechanisms are often complex and, thus, difficult to be predicted theroretically. Both the decay mechanisms (dynamic) and the metastable states may affect the diffusion properties of a given defect.

4.2.1 Diffusion during decay.

As described in Section 3.2.1, defects may desexcite by emission of phonons. If these phonons remain localized near the defect site, the vibrational energy of the system is increased by E_v, and the thermal activation energy is reduced from ΔH to $\Delta H - c\,E_v$, where c is a number less than unity. However, this will not often be very effective, since :
1) the excited phonon modes have to be normal modes of the *localized* motion, and;
2) the localized motion has to survive long enough. In most cases, the vibrational energy is not localized but is spread over the whole system in delocalized modes. The vibrational energy would most probably remain localized for very light masses, like H in Si or Ge, where local modes or resonances are present. Moreover, the configurational coordinate of the excitation may differ from the one necessary for motion. As an example, the excited state may be displaced perpendicular to the jump direction, as illustrated in figure 14. This would help the migration event only if some transfer of movement is possible to some vibrational modes parallel to the migration direction.
When effective, this mechanism of enhanced diffusion is called the recombination-enhanced diffusion mechanism. Altogether, it turns out that Γ (equation 8) is replaced by :

500

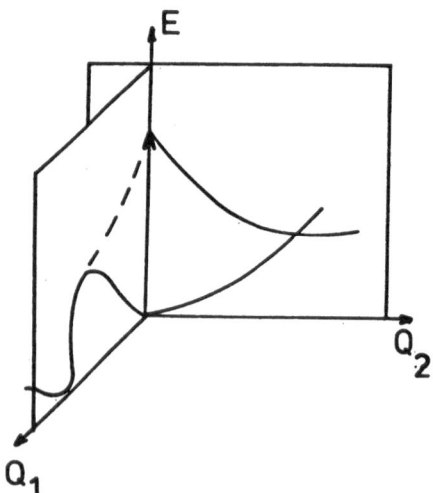

<u>Fig. 14</u> Configurational coordinate diagram, showing the effect of optical excitation of normal modes (parallel to Q_2), which are not modes for diffusion (parallel to Q_1).

$$\Gamma' = K_o \exp \left[- \frac{(\Delta H_f + \Delta H_m - c\ E_V)}{kT} \right] \tag{12}$$

where K_o and c depend mainly upon two physical properties of the defect [38,39] : the number of effective oscillators of the defect and the rate of dissipation of local vibrational energy to the lattice. K_o is generally some orders of magnitude less than the pre-exponential term in equation 8 . However, the activation energy in equation 12 is less than the correponding one in equation 8, so that, in some case, it appears that $\Gamma' > \Gamma$, like interstitial Al in Si. It also may occur that $c\ E_V > \Delta H_f + \Delta H_m$; then, the diffusion process is *a-thermal*, since no activation energy is involved. Local heating, i.e. local increase of T, is also possible when phonons remain localized for a time sufficient to help jumping the barrier.

4.2.2 Excited state diffusion

Once in the (metastable) excited state, the defect may have different diffusion properties.

The normal, so-called charge state, mechanism occurs when the equilibrium spatial configurations of the fundamental, F, and excited, E, states are not very different, but the ΔH corresponding to E is less than the one of F. As a result, excitation enhances the mobility of the defect. This is well established for the As- and P- vacancy pairs in Si [30], for instance. The fact that As$^+$ migrates less easily than As in Si is also due in part to a decrease of the entropy by a variation of bond lengths [40], like for the single vacancy in Si.

The saddle-point mechanism takes place when the equilibrium configuration of the excited state is the saddle point for migration of the fundamental state (or vice-versa). Therefore, upon desexcitation, the defect has a 50% probability to jump either in the initial or in the adjacent vacant site.

During non-radiative transition from the metastable to the fundamental state, a diffusion similar to the recombination-enhanced mechanism is possible.

Let us also mention the mechanism proposed by Monemar et al. [41,42] to account for optically induced dislocation glide in GaAs/AlGaAs double heterostructures. The authors assume that a large density of electron-hole pairs exist in the immediate vicinity of the dislocation. This weakens the frictional forces of the lattice against movement. This reasoning is the basis of the model of "plasma annealing" [43].

4.2.3 Amorphous to crystal transition of Si and Ge

The enhancement of the crystal growth rate of Si and Ge under electrons [5] and laser [6] irradiation may be understood easily in the framework of the diffusion theory presented before.

As a starting point, it has to be noted that amorphous Si and Ge are characterized by the presence of a large concentration (10^{19}-10^{20} cm^{-3}) of isolated DBs, i.e. such that all bonds around one DB are satisfied. This does not occur in the crystalline phase, where DBs are always associated, either in multiplets (like in vacancies), rows (in dislocations), planes (at surfaces). Moreover, in the amorphous phase, each isolated DB is connected to a lattice which is not topologically equivalent to the crystal one. This implies that the amorphous to crystal transition requires the breaking of bonds. Also, in a simplified model, crystallization may be seen as the result of an ad-hoc migration of DBs, which either annihilate

or agglomerate into dislocations, voids, ... The growth velocity
is then proportional to Γ.

In the self-trapped exciton model, spatially separated charged
centers, DB^+-DB^-, are created. The migration properties of these
differ from the neutral ones. Indeed, the force constants corresponding to the back bonds increase by going from DB^- to DB^+, since
the bond lengths are known to be larger in the DB^- than in the DB^+
configuration. As discussed before, this results in a modification
of the entropy of migration of the associated vacancy, and the
simultaneous change of the pre-exponential term in Γ. This pre-exponential term is larger for DB^- centres than for DB° than for DB^+
ones. If one assumes that the enthalpy of migration of DBs is not
affected by their excitation, it turns out that the growth velocity
under laser irradiation is proportional to :

$$\Gamma_n \div (N-n) + \frac{n}{2} \exp\left(\frac{S^-}{k}\right) + \frac{n}{2} \exp\left(-\frac{S^+}{k}\right) , \qquad (13)$$

where $(N-n)$ and $\underline{n}$ are the numbers of non-excited and excited DBs,
respectively. S^- and S^+ are positive entropies associated with DB^-
and DB^+ centres, respectively. From this relation, it is obvious
that the effects of the DB^- and DB^+ are different, due to the different signs in the exponential terms. Provided S^-/k and S^+/k are
equal or larger than unity, it is obvious from equation (13) that
optical excitation will increase Γ_n.

Γ_n may also be modified by doping, since this is known to
change the relative concentrations of DB^+ and DB^- centres. This is
in agreement with the results presented by Germain et al. at this
meeting, and may serve as the basis for an interpretation of their
data.

4.3 Other effects

Up to now, only effects attributed to the diffusing defect and
its immediate vicinity have been considered. However, diffusion
results from gradients of *entropy* in the solid, so that gradients
of temperature and/or density of electron-hole pairs, for instance,
are expected to modify the diffusion of defects, and vice versa.
Without any gradient, the mean displacement of defects in the solid
is zero. The gradients will add a non-zero component to their
mean displacement. So, gradients of stress or temperature enhance
the diffusion properties. It is also well known that diffusion of
ionized species is enhanced when an electric field is applied to the
sample [30] , and that oxidation reactions are functions of the
intensity of the electric field which appears due to oxygen

chemisorption [44] .

5. CONCLUDING REMARKS

From this lecture, it turns out that most properties of defects are easily understood phenomenologically by use of the so-called configurational coordinate diagram. However, the difficulty is to predict a priori the behaviour of a given defect.
Moreover, although important in the field of laser or electron beam induced effects, the basic characteristics of excited defects are much less known theoretically than for the corresponding fundamental states. In particular, the nature of the excited dangling bond states in amorphous silicon is not unambiguously determined, despite the large number of works actually performed on this particular subject. This is of primary importance for an understanding of, for instance, enhanced crystallization rates [45] .

The previous discussion has been examplified on defects in group IV semiconductors. It is obvious that it is also valid for other materials (III-V, II-VI, Se, Te, ...).

Finally, it is worth noting that the effects described here would be seen under low or medium laser power irradiation, but may also persist after high power irradiation.

ACKNOWLEDMENTS

The author thanks Prof. L.D. Laude, Drs. R. Andrew and M. Failly-Lovato for helpful discussions and comments. Part of this work was supported by Project IRIS of the Belgian Ministry of Science Policy.

REFERENCE

1. Hajto, J., J. de Physique, 41 (1980) C4-63.
2. Tanaka, K., J. Non-Cryst. Solids, 35/36 (1980) 1023.
3. Staebler, D.L. and Wronski, C.R., Appl. Phys. Lett, 31 (1977) 292.
4. Lovato, M., Wautelet, M. and Laude, L.D., Appl. Phys. Lett, 34 (1979) 160.
5. Germain, P., Doctor Thesis, Paris VII (1977).
6. Lietoila, A., Gold, R.B. and Gibbons, J.F., Appl. Phys. Lett, 39 (1981) 810.
7. Chu, W.K., Appl. Phys. Lett, 36 (1980) 273.
8. Kimerling, L.C. and Benton, J.L., in : Laser and Electron Beam Processing of Materials, edited by White, C.W. and Peercy, P.S. (New York, Academic Press, 1980) p. 385.
9. Mesli, A., Muller, J.C., Salles, D. and Siffert, P., Appl. Phys. Lett, 39 (1981) 159.
10. Wautelet, M., Failly-Lovato, M. and Laude, L.D., J. Phys. C (Solid St. Phys.), 13 (1980) 5505.
11. Corbett, J.W. and Bourgoin, J.C., in : Point Defects in Solids Semiconductors and Molecular Solids, Vol. 2, edited by Crawford, J.H. Jr. and Slifkim, L.M. (New York, Plenum Press, 1975) p. 1.
12. Pécheur, P., in : Electronic Structure of Crystal Defects and of Disordered Systems, edited by Gautier, F., Gerl, M. and Guyot, P. (Les Ulis, Les Editions de Physique, 1981), p. 93.
13. Joannopoulos, J.D., Phys. Rev. B, 16 (1977) 2764.
14. Appelbaum, J.A. and Hamann, D.R., Rev. Mod. Phys., 48 (1976) 479.
15. Schröter, W. (ibid. ref. 12), p. 129.
16. Adler, D., J. de Physique, 42 (1981) C4-3.
17. Wautelet, M., Laude, L.D. and Andrew, R., Phys. Lett. 77A (1980) 274.
18. Wautelet, M., Laude, L.D. and Failly-Lovato, M., Sol. St. Comm., 39 (1981) 979.
19. Street, R.A., Phys. Rev. B, 17 (1978) 3984.
20. Wautelet, M., Phys. Stat. Sol. (b), 103 (1981) 703.
21. Zehner, D.M., White, C.W. and Ownby, G.W., Appl. Phys. Lett. 36 (1980) 56.
22. Zehner, D.M., White, C.W. and Ownby, G.W., Surf. Sc. 92, L 67 (1980).
23. Klick, C.C., in : Point Defects in Solids : General and Ionic Crystals, Vol. 1, edited by Crawford, J.H., Jr and Slifkin, L.M. (New York, Plenum Press, 1972), p. 291.
24. Jaros, M., Phys. Rev. B, 16 (1977) 3694.
25. Bräunlich, P., Kelly, P. and Fillard, J.P., in : Thermally Stimulated Relaxation in Solids, edited by P. Bräunlich (Berlin, Springer-Verlag, 1979), p. 35.
26. Haug, A., Phys. Stat. Sol. (b), 108 (1981) 443.
27. Voos, M., this Summer School.

28. Wautelet, M. and Failly-Lovato, M., in : Physical Processes in Laser-Material Interaction, edited by M. Bertolotti (New York, Plenum Press) in press.
29. Itoh, N., J. de Physique, 37 (1976) C7-27.
30. Casey, H.C.,Jr. and Pearson, G.L., (ibid. ref. 11), p. 163.
31. Le Claire, A.D., in : Treatise on Solid State Chemistry, Vol 4 : Reactivity of Solids, edited by N.B. Hannay (New York, Plenum Press, 1976), p. 1.
32. Van Vechten, J.A., Phys. Rev. B, 10 (1974) 1482.
33. Landau, L. and Lifchitz, E., Physique Statistique (Moscow, Mir, 1967).
34. Lannoo, M. and Bourgoin, J.C., Sol. St. Comm, 32 (1979) 913.
35. Van Vechten, J.A., in : Handbook of Semiconductors, edited by S.P. Keller (Amsterdam, North Holland, 1980), p. 1.
36. Schmit, J. and Lucas, A.A., Sol. St. Comm, 11 (1972) 415.
37. Van Vechten, J.A., Phys. Rev. B, 12 (1975) 1247.
38. Weeks, J.D., Tully, J.C. and Kimerling, L.C., Phys. Rev. B, 12 (1975) 3286.
39. Troxell, J.R., Chatterjee, A.P., Watkins, G.D. and Kimerling, L.C., Phys. Rev. B, 19 (1979) 5336.
40. Wautelet, M., Phys. Lett., 84A (1981) 263.
41. Monemar, B., Potemski, R.M., Small, M.B., Van Vechten, J.A. and Woolhouse, G.R., Phys. Rev. Lett, 41 (1978) 260.
42. Basson, J.H. and Van Vechten, J.A., Phys. Rev. B, 23 (1981) 2032.
43. Van Vechten, J.A., this volume.
44. Hauffe, K. (ibid. ref. 31), p 389.
45. Bourgoin, J.C. and Germain, P., Phys. Lett., 54 A (1975) 444.

THE ROLE OF IONIZED DEFECTS IN Ge AND Si CRYSTALLIZATION

P.J. Germain,[+] M.A. Paesler
Department of Physics, North Carolina State University
Raleigh, North Carolina 27650, USA.
K. Zellama
Groupe de Physique des Solides de l'ENS
Tour 23, 2, Place Jussieu, 75221 Paris Cédex 05, France.

ABSTRACT

   Crystallization of amorphous germanium has been studied as a function of temperature and the flux of ionizing radiation ϕ. The crystallization growth rate v_g takes on the form :

$$v_g = v_o(\phi) \exp (-E_{Ge}/kT) ,$$

where E_{Ge} is constant and v_o is an increasing function of ϕ.

   Experimental studies of crystallization of doped and undoped amorphous silicon show that v_g roughly follows :

$$v_g = v_o (C_d) \exp (-E_{Si}/kT) ,$$

where v is an increasing function of the doping level C_d at $C_d \sim 10^{20} cm^{-3}$.

   On the basis of these observations we propose the following model of crystallization in amorphous group IV semiconductors : a concentration of mobile dangling bonds (DBs) exists in the bulk and near the amorphous-crystalline (a-c) interface. Ionization and doping induce transitions from the uncharged state D° to the charged states D^+ and D^-. The origin of the process controlling crystallization resulting in the activation energies in the above equations is discussed. Only certain sites on the a-side of the interface are available for crystallization and these sites are those which have captured DBs. The charged D^+ and D^- states have a larger capture cross section than the uncharged D° state. This is due to the coulombic interaction between the charged DBs and the interface states.

Increased concentrations of charged DBs result in an enhancement of the pre-exponential factor in the above equations. A study of oxygenated a-Si showing similar E_{Si} but decreased v_0 is described in terms of a reduction in the total concentration of DBs.

1. INTRODUCTION

The issue of energy flow during pulsed laser annealing experiments performed on ion-bombarded silicon has lain at the heart of a fundamental controversy [1]. One school of thought suggests that a photo-generated electron-hole plasma may carry a large fraction of the energy of a short (~10 ns) laser pulse and that the network temperature is increased by only about 300°C at 10 ns after the pulse [2]. Authors emphasizing this channel have interpreted experiments in terms of a relatively slow thermalization of carriers in the bands followed by recombination controlled by Auger processes which conserve the energy. Many other authors have interpreted their results assuming that the energy of the laser pulse is effectively immediately transferred to the network. They report network temperature rises of ~1000°C. We propose that there is a third channel into which energy may flow and that consideration of this channel may be crucial to the understanding of crystallization phenomena as well as many atomic transport effects. This channel is provided by the large number of localized states in the band gap of the amorphous (a-) silicon prior to laser irradiation.

In the following we discuss experiments in terms of a model (presented in the framework of the thermal model) where photo-generated electron-hole pairs are rapidly thermalized such that their temperature is close to that of the network. Thus electron-hole pairs produce a small effect in the conduction and valence bands. Then, however, we submit that an appreciable fraction of the thermalized electrons and holes are trapped in gap states producing an ionization. Such a model of band-gap ionization has been used to interpret permanent photoconduction [3]. Our model may also impact on the plasma annealing model. One can easily predict the temperature rise in *purely* thermal model (PTM) and the band-gap ionization model (BGIM). Let us consider the particular case of a layer of a-Si containing N_{DB} = 3 x 10^{19} cm^{-3}. In a typical experiment described below, ionization is produced by a high energy electron beam, but we need only propose here that a homogeneous concentration of carriers is produced. Let E be the total energy transferred to the material per unit volume. In the PTM, the network temperature rise is T_1 where $E = NkT_1$, and N is the total number of atoms per unit volume (5×10^{23} cm^{-3}). In the BGIM, we submit that $E = NkT_2 + N_{DB}\Delta E_{DB}$ where T_2 is the network temperature and ΔE_{DB} is the energy of excitation of a dangling bond (~0.3 eV). Solving, we find $T_1 - T_2$ = 0.1°C. Thus as far as the network temperature is concerned, consideration of the ionization of band-gap states doe not present a major departure from the PTM model. We demonstrate below, however, that the crystallization process can be demonstrably affected by changes in the charge state of band-gap states. We discuss results of experiments on crystallization in group IV semiconductors and propose that the results can best be understood in terms of the BGIM.

Several experiments have resulted in measurements of the crystallization growth rate. Thermal crystallization enhanced by ionizing radiation has been studied in undoped a-Ge [4]. Other thermal crystallization studies of doped and undoped a-Si have revealed an enhancement in crystallization brought on by doping [5]. Studies of self-implanted [6] and As-doped [7] silicon show respectively an enhancement by a factor of 100 and no difference in the growth rate of laser-annealed samples compared with respect to the thermally annealed ones. One aim of our arguments below is to present a model of crystallization which allows one to account for all of these results. Crucial to our model is inclusion of an often neglected channel of energy flow : the ionization of gap states.

2. DETERMINATION OF CRYSTALLIZATION GROWTH VELOCITIES

2.1 Ionization-enhanced crystallization in a-Ge

Barna et al. [8] using electron microscopy, and Germain et al [9] using conductivity measurements, have obtained the crystallization growth velocity v_g for a-Ge over a large range of temperatures. These results are summarized in Figure 1a where we have plotted log v_g versus 1000/T. The activation energy of v_g is found to be ~1.5 eV. This value is in agreement with photoemission results obtained by Laude et al [10].

Germain et al. [4] have observed that ionizing radiation enhances v_g in a-Ge. For a flux $\phi_1 = 0.9$ μA.cm^{-2} of 1 MeV electrons, this enhancement is shown in Figure 1a. In addition, these authors measured v_g at a fixed temperature, varying the electron flux. These results are shown in Figure 2a. From these results, Germain et al. [11] have written v_g in the form :

$$v_g = v_{g_o} (\phi) \exp(-E/kT) , \qquad (1)$$

where v_g (ϕ) does not depend on T and is an increasing function of ϕ, and E is independent of T and ϕ.

2.2 Thermal crystallization of doped and un-doped a-Si

Csepregi et al.[5] have used channeling to observe the recrystallization of ion-implanted (100) Si. These results are plotted in Figure 1b. The variation of v_g at a constant temperature (475°C) as a function of the concentration of implanted dopant (phosphorus) is shown in Figure 2b. These authors observe that v_g increases with P concentration up to $C_P \sim 2 \times 10^{20} cm^{-3}$. Above this concentration, v_g is not strongly dependent on C_P [5]. For no doping and for P and As doping, the activation energies are equal to 2.4 eV within the range of the error. This result is in agreement with reference 12 which shows that the crystallization of P doped a-Si:H and the H exodiffusion stays constant as a function of P concentration for $C_P \sim 2 \times 10^{20} cm^{-3}$. For doping with B, a slightly lower (1.9eV) activation energy is observed and for O doping a slightly higher activation energy was determined. From the data of reference 5 we estimate the error in the activation energies to be ± 0.2 eV [13]. Thus within the range of the error, all dopants result in nearly equal activation energies with O having a slightly higher and B perhaps a slightly lower value. The doping concentrations for P, As, B and O were 2.0, 2.4, 2.5 and 4.0x10^{20}cm^{-3}.

The results of Csepregi et al. may be expressed using the form:

$$v_g = v_{g_o} (C_d) \exp (-E/kT), \tag{2}$$

where v_{g_o} (C_d) does not depend on T and is an increasing function of the doping concentration C_d. E is independent of T and may vary slightly with doping.

2.3 Cw laser induced crystallization

Leitoila et al. [6] have measured the rate of cw laser-induced solid phase epitaxy in self-implanted amorphized silicon by determining the dwell time required to regrow the entire amorphous layer at the center of a scanned laser beam. The measurement was performed in the annealing temperature range of 800 to 900°C. The measured regrowth rates were about two orders of magnitude higher than those extrapolated from low temperature furnace annealing data [5].

Kokorowski et al. [7] and Roth et al. [14] have made measurements of the solid phase epitaxial regrowth rate (v_g in our notation) of As implanted a-Si using time-resolved reflectivity measurements. The temperature rise inducing the epitaxial regrowth was induced by laser heating. The computation of the temperature rise required the determination of a phenomenological factor γ.

510

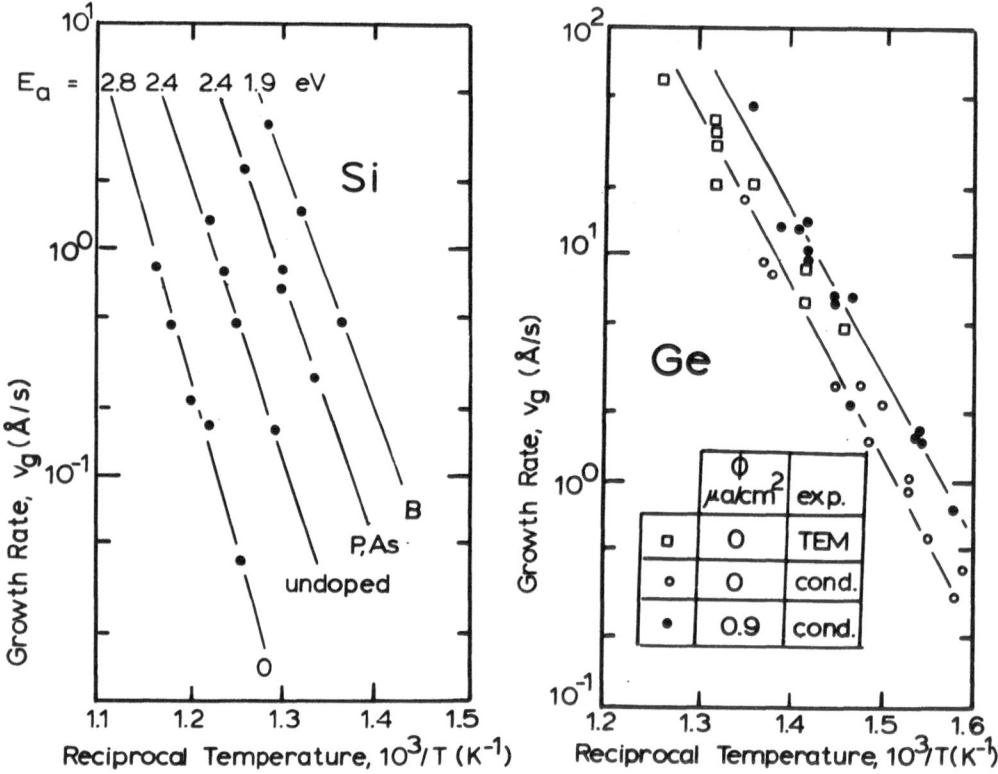

Fig. 1a Crystallization growth rate v_g for a–Ge as a function of reciprocal temperature for samples with and without electron ionization. Data are from reference 8 (square and references 4,9,11) (circle).

Fig. 1b Crystallization growth rate v_g for a–Si as a function of reciprocal temperature for several doping condition. Data are from reference 5.

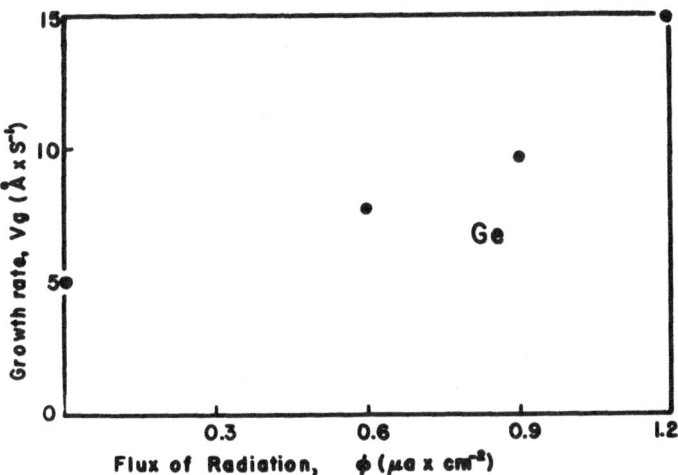

Fig. 2a Plot of v_g for a-Ge as a function of the flux of ionizing electron radiation. Sample temperature was 430°C. Data are from reference 11.

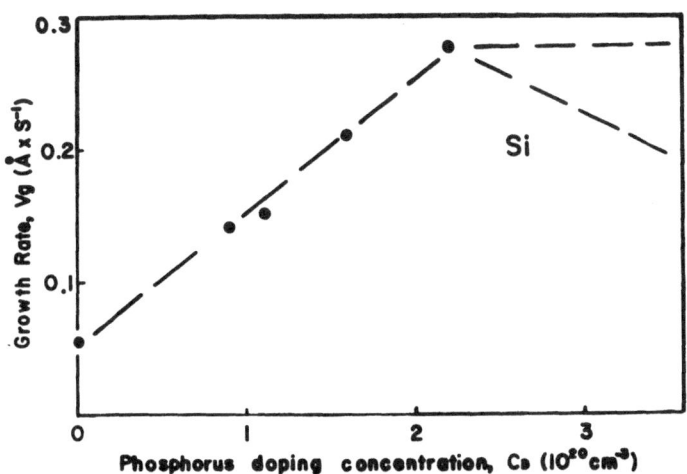

Fig. 2b Plot of v_g for a-Si as a function of P doping concentration. Sample temperature was 475°C. Data are from reference 5.

In order to determine γ these authors assume that v_g at a given temperature is the same for furnace and laser crystallization. Then

286

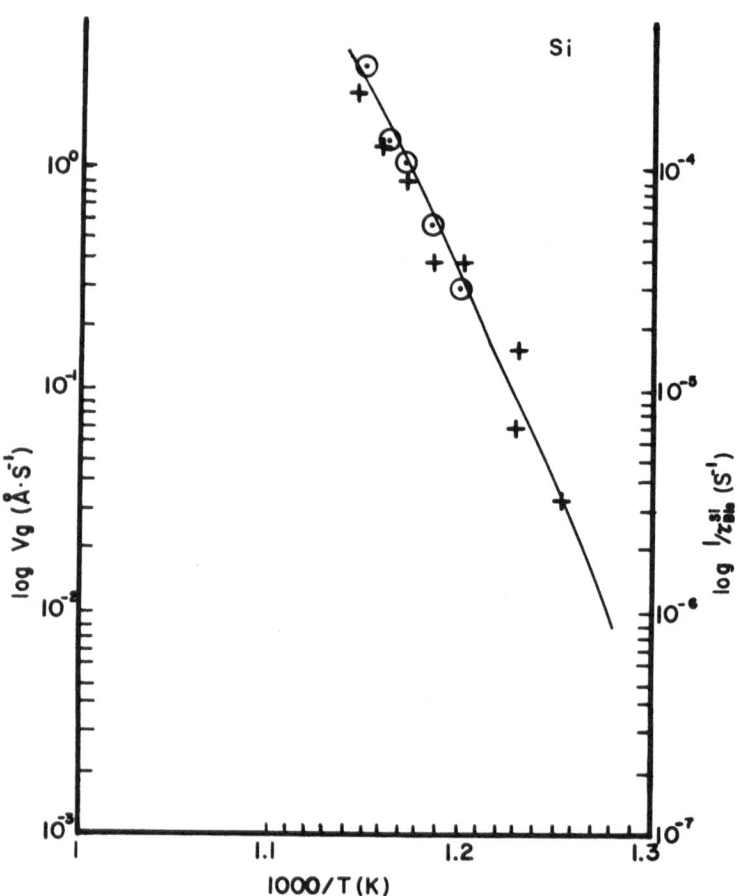

Fig. 3 Arrhenius plot of the growth rate v_g of crystallization in amorphous silicon layers : + data of Zellama et al [3]. Arrhenius plot of $1/\tau_{BIC}^{Si}$: ⊙ present study.

negative defects states D°, D^+ and D^- respectively, we can describe this interaction as $2D^\circ \rightarrow D^+ + D^-$.

For the laser experiments on As-implanted Si [7,14] we must consider the effects of irradiation with a laser beam on an a-Si sample with C_D =2 to 4 x 10^{19}cm^{-3} ($\sim C_{DB}$). We show below that for such doping most of the dangling bonds *near* the a-c interface will be in the D^- state prior to irradiation. If such a sample is irradiated at a low level ϕ_1, then some DBs will participate in the reaction $D^- \rightarrow D^\circ$ and the growth rate $v_g(C_D, \phi = \phi_1)$ should be smaller than $v_g(C_D, \phi=0)$, i.e. v_g would be a *decreasing* function of the flux. If, on the other hand, high level radiation $\phi = \phi_2$ is present then in the limit where the effect of doping is negligible, most DBs will be in charged states and the growth rate $v_g(C_D, \phi=\phi_2)$ should be an *increasing* function of the flux. The Kokorowski experiments correspond to a situation for which the effects of doping and photo-ionization or "photo-doping" roughly cancel [17].

3.2 Diffusion of dangling bonds

Brodsky et al. [15] found 10^{20}cm^{-3} dangling bonds in a-Si samples deposited in a system with a residual pressure of $\sim 10^{-6}$ Torr and have interpreted their results in terms of spins on internal surfaces. Thomas et al. [18] prepared a-Si under ultra-high vacuum conditions ($\sim 10^{-10}$ Torr) and have shown for samples deposited at room temperature that $\sim 7 \times 10^{19}$cm^{-3} randomly distributed and isolated spins in the bulk were inferred. These authors were able to reproduce the previous results by artificially increasing the residual pressure, thus they called their high purity samples "intrinsic" a-Si. They have further shown that the spin concentration decreases with increasing annealing temperature, from which one is led to the conclusion that DBs diffuse from the bulk toward the surface as a result of annealing [19].

The experimental results obtained by Csepregi on the crystallization of ion-implanted a-Si cannot be correlated with the results obtained by Brodsky et al. [18] but may be compared to the results of Thomas et al. [18] because of the purity of the bombarded samples. Indeed, Mayer [20] has shown that the samples evaporated by Thomas [18] and studied during crystallization by Zellama [21] have the same behavior as the Csepregi samples.

3.3 The amorphous-crystalline interface

The c-Si surface has been the object of many experiments. In order to theoretically describe these results, Harrison [22] and Chadi [23] have performed calculations of the surface energy taking into account the reconstruction of the surface. Although the structures proposed by these authors are different, they agree on one point : the dangling bonds of the surface atoms are not neutral, rather there is an alternation of doubly occupied and empty states, i.e. D^- and D^+. We know of no calculations of the surface energy of the external surface of a-Si but find it reasonable to transpose the crystalline results to the amorphous surface.

The growth of crystallinity has been reviewed by Turnbull [24] and Christian [25]. During crystallization, the a-c boundary moves perpendicular to itself at a rate v_g. Crystallization occurs from the a- to the c- phase and the interface is necessarily incoherent. We propose that the surface dangling bonds on both sides of the interface present an alternation of doubly occupied and empty configurations, D^- and D^+.

3.4 The microscopic mechanism at the a-c interface

We suggest that as crystallization proceeds, charged dangling bonds diffuse towards the a-c interface. A non-fully co-ordinated charged atom at the interface would capture the diffusing DBs resulting in a decrease of the co-ordination number of the atom on the a-side of the interface, allowing it to jump to the c-side. An atom that has captured a dangling bond shall be defined as being "available" for crystallization. The exact natures of the diffusion, captures and jump are not proposed. Presumably a similar mechanism involving neutral DBs, D°, exists, but the efficiency for such a mechanism is lower. In addition, several dangling bonds might be captured at one site. Crucial to our model are only : i) the diffusion of charged dangling bonds, ii) capture at the a-c interface resulting in a reduction of the co-ordination number of the a-interfacial atom, and iii) jump of this atom to the c-side.

3.5 Origin of the growth rate activation energy

Assuming a steady state of crystallization, the growth rate v_g takes the form :

$$v_g = \alpha \delta v^1, \text{ where :} \tag{3}$$

$$v^1 = v \exp \left\{ - \frac{\Delta g^*}{kT} \right\}, \tag{4}$$

where δ is the distance of the jump from the a- to the c-side of the interface; v is an atomic frequency $v = kT/h$; Δg^* is the free energy barrier; and α is the fraction of atoms on the a-side available for cyrstallization.

If there are N_S surface atoms, N_a of which are available for crystallization, then $\alpha = N_a/N_S$ and the rate of change of N_a is given by :

$$\frac{dN_a}{dt} = (N_s - N_a) \, \nu_{DB} - N_a \nu' .$$ (5)

In the above, $(N_s - N_a)$ is the number of sites which may capture a DB and ν_{DB} is a frequency characteristic of the capture. ν_{DB} should be thermally activated. The frequence ν_{DB} which globally describes the phenomenon should be a function of : i) the concentration of DBs in the space charge of the bulk, ii) the diffusion of DBs in the bulk, and iii) the magnitude of the electric field at the interface. The first term on the right hand side of Equ. 5 is related to the creation of available sites. The second term represents an annihilation due to crystallization. Assuming a steady state, $dN_a/dt = 0$ and :

$$\frac{N_S - N_a}{N_a} = \frac{\nu'}{\nu_{DB}} ,$$ (6)

thus :

$$\alpha = \frac{\nu_{DB}}{\nu' + \nu_{DB}} ; \text{ so that } v_g \text{ becomes :}$$ (7)

$$v_g = \left(\frac{\nu' \nu_{DB}}{\nu' + \nu_{DB}}\right) \delta .$$ (8)

We consider three limiting cases of Equ. 8 : i) $\nu' \ll \nu_{DB}$ which implies $v_g = \nu' \delta$: in this case nothing in the expression for v_g depends on the DB concentration; ii) $\nu' \gg \nu_{DB}$ which implies $v_g = \delta \nu_{DB}$: in this case the growth rate is controlled by the jump of a DB at the a-c interface; iii) ν' and ν_{DB} have the same order of magnitude and roughly the same activation energy.

Finally, we apply these three limiting cases to the data of Figures 1 and 2 :

i) For $\nu' \ll \nu_{DB}$, the expression for v_g depends only on ν', the frequency of the jump of an atom from the a- to the c- side of the interface. Thus v_g is independent of the DB concentration, and the assumption that $\nu' \ll \nu_{DB}$ does not fit the data of Figures 1 and 2. ii) For $\nu' \gg \nu_{DB}$ it must be the case that $v_g = \nu_{DB} \delta$ and v_g takes the form $v_g = \delta \nu_{DB_0} \exp(\frac{-E_{DB}}{kT})$. This may be compared to Eqs. 1 and 2 and the data of Figures 1 and 2 where E_{DB} represents the activation energy for the diffusion of DBs and/or the electrostatic barrier of their capture at the interface. In order to allow comparison with Equ. 1, E_{DB} must be independent of ϕ and should depend slightly on

C_D. Furthermore, in order to describe the data, ν_{DB_0} should be an increasing function of charged DB concentration which is reasonable. iii) If ν' and ν_{DB} are roughly the same order of magnitude and have roughly the same activation energy, we may write :

$$\nu' = \nu'_o \exp\left(\frac{-E'}{kT}\right) = \nu'_o \exp\left(\frac{-E_{DB}}{kT}\right) \tag{9}$$

and equ. 11 becomes :

$$v_g = \left[\frac{\nu_{DB_o} \nu'_o}{\nu'_o + \nu_{DB_o}} \delta\right] \exp\left(\frac{-E'}{kT}\right), \tag{10}$$

where the factor in square brackets represents the prefactors of Equations 1 and 2 and the intercepts in the Figures. If E' and E_{DB} are only approximately equal, the activation energy E' may vary slightly with doping concentration.

In conclusion, we feel that cases ii) or iii) may describe the data.

REFERENCES

1. See, for example, articles in Materials Research Society Symposia Proceedings, 1980 and 1981, North Holland, NY.
2. Lo, H.W. and Compaan, A., Phys. Rev. Letters, 44 (1980) 1604.
3. Wautelet, M., Laude, L.D. and Andrew, R., Physics Letters 77A (1980) 274.
4. Germain, P., Squelard, S. and Bourgoin, J., J. Non-Cryst. Solids 23 (1977) 159; and
   Germain, P., Squelard, S. and Bourgoin, J., Radiation Effects in Semiconductors, Dubrovnik 1976, N. Urli and J. Corbett eds.
5. Csepregi, L., Kennedy, E., Gallagher, T., Mayer, J. and Sigmon, T., J. Appl. Phys. 48 (1977) 4234.
6. Leitoila,A., Gold, R., and Gibbons, J., Appl. Phys. Lett. 39 (10) (1981) 810.
7. Kokorowski, S., Olson, G. and Hess, L., J. Appl. Phys. 53 (1982) 921.
8. Barna, A., Barna, P. and Pocza, J., J. Non-Cryst. Solids 8 (1972) 36.
9. Germain, P., Zellama, K., Squelard, S., Bourgoin, J. and Gheorghiu, A., J. Appl. Phys. 50 (1979) 6986.
10. Laude, L.D. and Willis, R.F., AIP Conf. Proc. No. 20, p. 65 (1975).
11. Germain, P. and Squelard, S., to be published.

12. Squelard, S., Zellama, K., Germain, P. and Bourdon, B., Revue de Phys. Appliquée 16 (1981) 123.
13. The error in activation energies form reference 5 stems principally from the RBS instrument resolution of 20keV (434 Å in depth), which results in an error in the determination of the activation energy of approximately 0.2 eV based on a least squares fit to a straight line on the Arrhenius plot.
14. Roth, J., Olson, G., Kokorowski, S. and Hess, L., Proc. Materials Research Society, vol. 1, J. Gibbons, L. Hess and T. Sigmon, eds (New York, North Holland) p. 143.
15. Brodsky, M., Title, R., Weiser, K. and Petit, G., Phys Rev B1 (1970) 2632.
16. Mott, N. and Davis, E., Electron Processes in Non-Crystalline Materials, 2nd edition (Oxford, Clarendon Press, 1979).
17. Applying the error bars of Fig. 3 in ref. 14 to the earlier work of the Hughes group [7], one cannot at any rate make firm conclusions about the combined effects of doping and irradiation.
18. Thomas, P., Brodsky, M., Kaplan, D. and Lepine, D., Phys. Rev. B 18 (1978) 3059.
19. Kaplan, D., private communication.
20. Mayer, J. in Thin Films, Preparation and Properties, K. Rosenberg, ed. Pasadena, CA , 1981.
21. Zellama, K., Germain, P., Squelard, S., Bourgoin, J. and Thomas, P., J. Appl. Phys. 50 (1979) 6995.
22. Harrison, W.A., Surface Science 55 (1976) 1.
23. Chadi D., Phys. Rev. Letters, 43 (1979) 43.
24. Turnbull, D., Sol. State Phys. vol. III, p. 226 (1954).
25. Christian, J., Phase Transformations in Physical Metallurgy, R.W. Cahn, ed., North Holland, NY, 1970.

+ on leave from Université Paris VII, Groupe de Physique des Solides de L'E.N.S., Tour 23, 2, Place Jussieu, 75221 Paris Cédex 05,France.

One of the authors (MAP) acknowledges the support of the General Electric Corporation.

INTERFACES UNDER LASER IRRADIATION

Martin F. von Allmen

University of Bern, Switzerland

1. INTRODUCTION

This lecture deals with structural and chemical changes in binary samples induced by short laser pulses, typically pulses in the ns regime.

The reaction of a composite sample to irradiation by a short laser pulse is, conceptually and experimentally, much more complex than that of an elemental one, say, a Si wafer. Examples of composite samples are ion-implanted semiconductor crystals, or vapor-deposited films or multilayers on top of suitable substrates. Phenomena observed upon irradiation of such structures include the production of supersaturated solid solutions, the occurence of growth instabilities, as well as the formation of metastable compounds and glassy phases. As diverse as these structures are, they may all be understood on the basis of a three-step process : i) heating and melting due to absorption of laser light; ii) redistribution of the elements in the molten state; and iii) rapid solidification of the liquid mixture. The last step clearly is the crucial one.

The principles of crystal nucleation and growth from elemental melts have been treated in previous lectures and are not repeated here. The presence of more than one chemical species introduces several new aspects :
- There is, in general, a compositional mismatch between the liquid and the solid phase. As a result, crystal growth involves mass transport;
- There tend to be gradients in melt composition. Concomitant is the presence of gradients in equilibrium melting point, and hence in the chemical driving force for crystal growth;

- There are chemical reactions, associated with liberation of heats of reaction (mixing, compound formation).

Let me, before discussing some of these points in more detail, give a brief (and somewhat schematized) overview of the experimental situation. I shall use metal-silicon systems as an example.

2. EXPERIMENTAL

The structures resulting in composite laser-irradiated structures are conveniently categorized according to the "impurity" content of the melt. Pure melts are able to crystallize very rapidly and tend to form single-crystals, if there is a suitable seed (such as a single crystal substrate) available. Single-crystallinity may persist up to several atomic % of ad-mixture of suited impurities (such as group III or V dopants in Si). This is the regime where solute trapping by the rapidly advancing melt-crystal interface is observed which leads to the formation of single-crystalline super-saturated solid solutions [1] .

If the impurity content of the melt exceeds a few atomic %, then single-crystal growth tends to be suppressed. Crystal nucleation from the melt typically results in a polycrystalline mixture of phases, depending on the local impurity concentration and on the type of phase diagram. Alternatively, the melt freezes into a glassy state, provided it cools too rapidly for nucleation to occur (the critical cooling rate is observed to be a function of melt composition). This property has recently been used to obtain metallic and semiconducting glassy phases in a number of Si-metal binary systems [2] . The method turns out to be a far more powerful means for glass formation than splat cooling. In the mentioned work amorphous phases were obtained at all compositions except those close to that of a pure element or a congruently melting compound.

Finally, irradiation of structures consisting of a single thin layer, say of material A, on top of a substrate B, results in a melt composition varying from pure B to pure A within the melt depth, typically a few 100 nm. Rather complicated structures are observed to solidify from such inhomogeneous melts. They may contain cells, dendrites, even bubbles and tend to consist of a mixture of stable or metastable phases. Many of these phenomena have been known from conventional cast metallurgy, if on a totally different temporal and spatial scale [3] .

These experimental findings can be understood on the basis of the following mechanisms.

3. SEGREGATION AND "SOLUTE TRAPPING"

In a binary system A-B the composition of a crystal is generally different from that of its melt (an exception are congruently melting phases). Upon growth of the crystal, atoms of one species are thus accumulated in the melt. This is characterized by the interfacial segregation coefficient:

$$k_0 = (x_s/x_l) \qquad (1)$$

where x_s and x_l denote the crystal and melt composition (expressed in atomic % of the solute A), respectively. In the limit of very diluted mixtures the segregation coefficient is given by the ratio of the slopes of the solidus- and liquidus-lines near the "pure A end" of the A-B phase diagram. In the same limit segregation is the basis of the process of zone refining (Figure 1). The definition of k_0 in terms of the equilibrium phase diagram is, of course, valid only at small growth velocities. Further, in concentrated mixtures, k_0 depends on the composition as well.

Experiments with ion-implanted Si annealed by short (ns) laser pulses show the incorporation of substantially more impurity atoms than expected from the equilibrium segregation constant. This may be understood from the following argument.

The segregation of impurity atoms requires enough mobility of the impurity atoms in the interfacial region to allow them to diffuse by at least one molecular diameter a within the time a/v it takes the crystal to grow by the same amount. Otherwise they are overtaken and "trapped" by the advancing crystal boundary. The minimum time required for diffusion by a distance a is equal to a^2/D_i, D_i being the diffusivity at the interface. The interface velocity u^i at which trapping should be expected is estimated as $a^2/D_i = a/u$, or $u = D_i/a$. Rough estimates give 0.1 m/s for the critical interface velocity (Figure 2). This coincides by order-of-magnitude with the typical velocities at which impurity trapping is observed; see, e.g., White et al [1].

4. INTERFACE INSTABILITY

Segregation, as discussed above, results in the presence of a gradient in composition within the melt ahead of the l-s interface. Even apart from segregation, compositional gradients are present whenever short pulses are used for mixing or alloying of layered samples. A gradient in composition means a gradient in equilibrium melting point and thus, in general, a gradient in undercooling. The magnitude and slope of the latter depends on the actual temperature distribution in the region of the interface, as illustrated

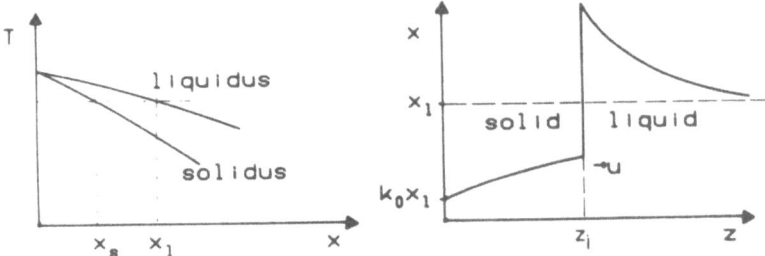

Fig. 1 Left : Definition of the interfacial segregation coefficient k_0 from the equilibrium phase diagram. Right : solute concentration profile during crystal growth for constant k_0.

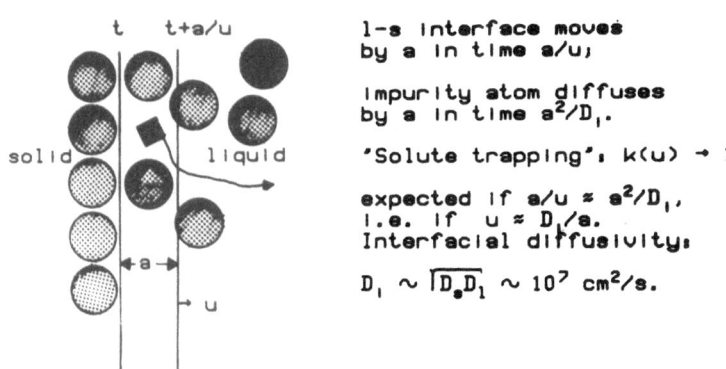

1-s interface moves by a in time a/u;

Impurity atom diffuses by a in time a^2/D_l.

'Solute trapping': $k(u) \to 1$

expected if $a/u \approx a^2/D_l$, i.e. if $u \approx D_l/a$. Interfacial diffusivity:

$$D_l \sim \sqrt{D_s D_l} \sim 10^7 \, cm^2/s.$$

Fig. 2 Illustration of the solute trapping in the case of very rapid crystal growth; a is the molecular diameter and D_s and D_l denote the solute diffusivities in the solid and liquid phase, respectively.

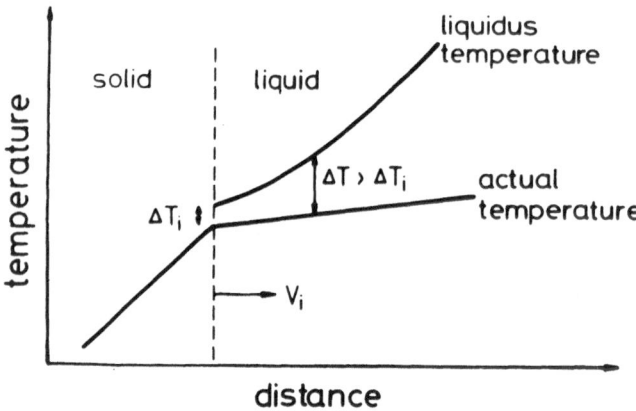

Fig. 3 Schematics of temperature profiles leading to Constitutional Supercooling.

in Figure 3. Under the conditions shown in the Figure (liquidus temperature decreasing with impurity content; gradient in liquidus temperature larger than gradient in actual temperature) the under-cooling, and thus the chemical driving force for growth, increases with distance away grom the l-s interface. This situation is known as Constitutional Supercooling (CSC).

Crystal growth in the presence of CSC is inherently unstable Random protusions in the crystal boundary, due to fluctuations, experience at their tip a larger undercooling and thus grow faster than their surroundings. They grow into columns, the shape of which depends on the details of the microscopic mass and heat flow. The remaining melt is enriched in the impurity and trapped between ad-jacent columns ; it eventually freezes out in the form of cell walls, surrounding the columns of relatively pure material.

This basic cellular pattern, which has been observed in many variations in laser irradiated samples (see, e.g., [4]), is well known from cast metallurgy [3] . However, whereas casting results in cell sizes of tens of μm to several mm, cells made by Q-switch laser pulses range between a few 10 and a few 100 nm. The inverse relationship between the feature size and the growth velocity re-flects the tendency of rapid solidification to suppress long-range mass transport. Ultimately growth of a compound crystal (which re-quires long-range mass transport for all except congruently melting phases) becomes altogether impossible; a glass forms. Conceptually and physically, there is not a large step from a structure with a few 10 nm feature size and a featureless, or glassy, structure.

5. GLASS FORMATION

Glass formation requires cooling a melt while by-passing crystal nucleation and growth. Therefore, the conditions for glass formation are to be derived from nucleation and growth theory. The growth velocity of a crystal into its undercooled melt, according to classical theory, is given by :

$$u(T) = u_o e^{-Q/kT} [1 - \exp(-s_m(T_m-T) /kT] . \qquad (2)$$

Here u_o is essentially a constant which depends on the detailed structure of the crystal-melt boundary, s_m is the entropy of melting per particle and Q is the activation energy for molecular re-arrangement at the boundary (assumed here to be a simple thermally activated process). Q may be estimated from the activation energy for viscous flow in the undercooled melt [5] . The growth velocity increases rapidly with increasing undercooling just below the melting point, but, due to the presence of a thermally activated process, it approaches zero at large undercooling. Thus there is a maximum velocity at which the crystal can grow.

The rate of crystal nucleation, if regarded as a function of temperature, behaves qualitatively similarly to the growth velocity, in that there too is an optimum temperature range, determined by a compromise between chemical driving force and particle mobility. The nucleation rate depends explicitly also on time. If a melt is suddenly undercooled, then a finite time-lag is observed before measurable nucleation sets in. The time-lag t_N may be interpreted as the time required to establish an equilibrium population of clusters corresponding to the new temperature [6] . The rate of homogeneous nucleation may be expressed in a somewhat simplified form as :

$$I(T,t) = I_\infty (T) \times [1 + 2 \sum_{n=0}^{\infty} (-1)^n \exp(-n^2 t/t_N)] ,$$

$$I_\infty(T) = I_o e^{-Q/kT} \exp - \sigma^3/N^2 s_m^2 (T_m-T)^2 kT . \qquad (3)$$

Here I_∞ is the steady-state nucleation rate, σ is the crystal-melt interface energy, N is the particle density and I is a constant. Note the very strong dependence of I_∞ on undercooling; unlike growth, homogeneous nucleation sets in only when the undercooling exceeds a certain finite amount, even in the steady-state case. The time-lag should be minimum in the temperature range where I_∞ is at

524

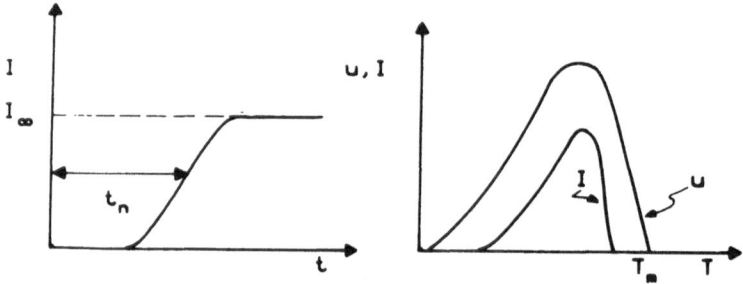

<u>Fig.</u> 4 Schematics of the dependence of the growth velocity u and
the homogeneous nucleation rate I on temperature or time.

its maximum; a lower limit to the minimum time-lag is given by the
ratio of the number of atoms in a critical nucleus to the molecular
re-arrangement frequency.This yields values of the minimum time-lag
of the order of ns for metallic melts. At large undercooling the
nucleation rate approaches zero, in a similar way as the growth velocity. At the same time the time-lag, which is expected to scale
approximately as exp $(-Q/kT)$, becomes very large. The general behavior of u and I as a function of t and T is sketched in Figure 4.

Once the nucleation rate and the growth velocity are known,
one may formulate a condition for glass formation by requiring that
the crystallized volume fraction X, given for small X by:

$$X(T,t) \sim I(T,t) \ u^3(T) \ t^4, \tag{4}$$

stays below some arbitrary limit, e.g., 10^{-6} [7] .

What equations (2) and (3) not allow for is the fact that the
compositions of the crystal and the melt may differ. There is, unfortunately, no theory for this important case. It may be expected
that the general behavior sketched in Figure 4 still holds; however,
it is clear that a change in composition must slow down nucleation
and growth, the more so the larger the difference in composition is.
Growth must, in addition, slow down with time as one species in the
melt is being depleted.

Perhaps the simplest model is to describe nucleation and growth
by (2) and (3) but with u_o and I_o proportional to the concentration
in the melt of that atomic species of which the crystal is richer

than the melt. If x_s and x_1 denote the crystal and melt concentrations of this particular species, then the growth velocity can be shown to decrease with the distance grown approximately like:

$$u(z) = u_o y/ [1 + (1-y)zu_o/D] \qquad (5)$$

where $y = (x_1/x_s) < 1$; D denotes the melt diffusivity and u_o is the growth velocity for $x_s = x_1$. For $u_o = 1$ m/s, $D = 10^{-5}$ cm^2/s and $y = 0.8$, the initial velocity is 0.8 m/s and decreases to 0.4 m/s after 50 Å.

Crystal nucleation and growth from a binary melt is thus severely impaired if the melt composition does not coincide with that of a congruently melting compound (for which $y = 1$). This is in accordance with experimental trends of glass forming ability in Si-metal systems by ns pulse irradiation [2] .

6. MODELING ULTRA-RAPID SOLIDIFICATION

In the following we demonstrate the impact of a reduction of intrinsic growth rate on the structure of the re-solidified melt. We do this with the help of a numerical model which combines the equations governing heat flow with those describing the molecular re-arrangement kinetics [8] . Solidification is governed by an interplay of heat flow and molecular kinetics. This becomes evident if we write the heat flow equation in the following generalized form :

$$(\partial H/\partial t) = (\partial/\partial z) (K\partial T/\partial z) + A(z,t) + (\partial L/\partial t). \qquad (6)$$

Here $H(T)$ is the enthalpy per unit volume of the material, A is the volumetric rate of heat production by light absorption, and L is the latent heat content of the material which varies between 0 and L_m, the heat of melting. The rate $(\partial L/\partial t)$ depends on temperature as well as on time and is determined by nucleation and growth kinetics. A computer code was designed to solve (6) with allowance for the dynamical processes during melting and solidification. In short, the program is based on the following model :
i) Radiation is absorbed in a thin surface layer and converted into heat instantaneously. This is a valid approximation for irradiation of metals ;
ii) The material is divided into discrete volume elements which are heated by the laser and exchange heat according to (6). Volume elements absorb or liberate latent heat at a rate that depends on local temperature as well as on time. The quantity $(\partial L/\partial t)$ is determined by a rate function, a generalized and simplified combination of equations 2 and 3. The temperature-dependent part of the rate function is schematized in Figure 5; positive values lead to liberation, negative ones to absorption of latent heat. There is

526

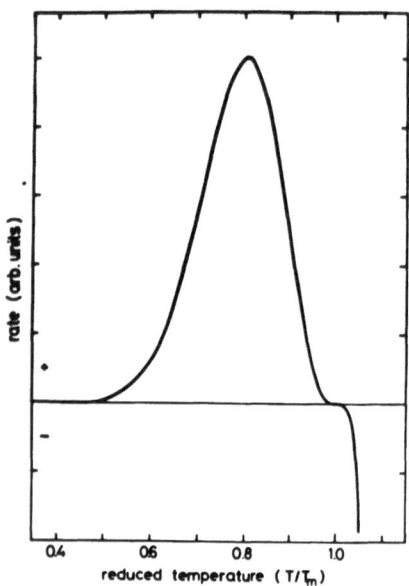

Fig. 5 Temperature-dependent part of the rate function for melting
and crystallization, used in the calculation shown in Figures 6 and
7.

admittedly some arbitrariness in the parameters of this curve; however, it turns out that its detailed shape (apart from the width and
height) is of little influence on the results of the calculation.
The time-dependent part of the rate function allows latent heat to
be absorbed or liberated only after a specified time has elapsed
since the volume element was first superheated or undercooled. This
accounts for the presence of a nucleation time lag;
iii) The l-s interface is defined at the position of that volume
element which has absorbed or liberated 50 % of the latent heat of
melting. This interface does not have a physical meaning in all
cases (see below).

A more detailed account of the calculation will be published
elsewhere. As an illustration, we present some results for the
case of a 30 ns laser pulse of 10 MW/cm^2 of absorbed fluence, incident on a crystalline bulk metal specimen. Typical values for the
thermal data were chosen and kept constant during the process, in
order to facilitate interpretation of the results. Figure 6 shows
the position of the liquid-solid boundary as a function of time.
The three curves shown differ only by a scale factor in the crystallization rate; this simulates three different material compositions :

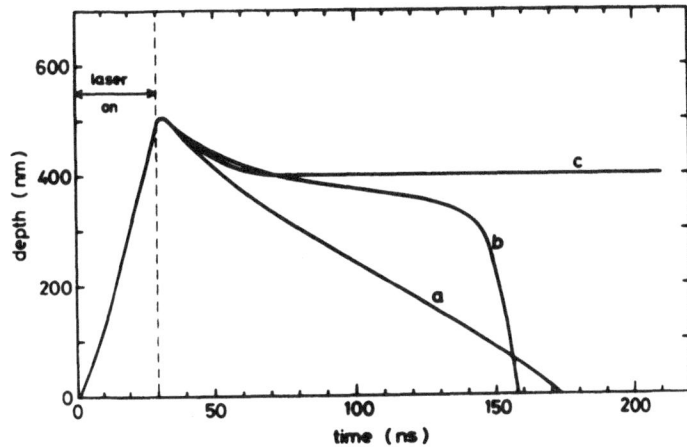

<u>Fig. 6</u> Calculated position of the crystal boundary as a function of time in a bulk metal sample, irradiated by a 30 ns laser pulse. Curves a) – c) are obtained with different scale factors for the rate function of Figure 5. Cases a) and b) correspond to crystal growth, case c) to glass formation.

For curve a the peak rate (assumed to occur at reduced temperature $T/T_m = 0.81$ in all cases) is 3×10^9 s^{-1}, for curve b it is 6×10^6 s^{-1} and for curve c is is 3×10^6 s^{-1}.

The largest rate (curve a) results in crystal growth at moderate undercooling ($T/T_m = 0.98$) and a nearly constant velocity of 3.2 m/s, which is limited by heat flow rather than by the intrinsic growth kinetics. Curves of this kind are known from numerous laser annealing studies (see, e.g. [9]); only in this case does the l-s interface have a clear physical meaning : it represents the interface at which latent heat is liberated. Curve b presents a case close to the limit of crystal growth : after an initial period of slow growth (limited by the rate function), the undercooling grows larger and the interface accelerates; eventually it moves at a velocity close to maximum ($T/T_m = 0.85$). During the last stage the whole remaining melt is strongly undercooled. The interface tends to loose its physical meaning, since the crystallization process is of a volume nature. Curve c is just beyond the limit for crystal growth : the undercooling at the interface exceeds the critical value ($T/T_m = 0.81$) after a few tens of ns and the

528

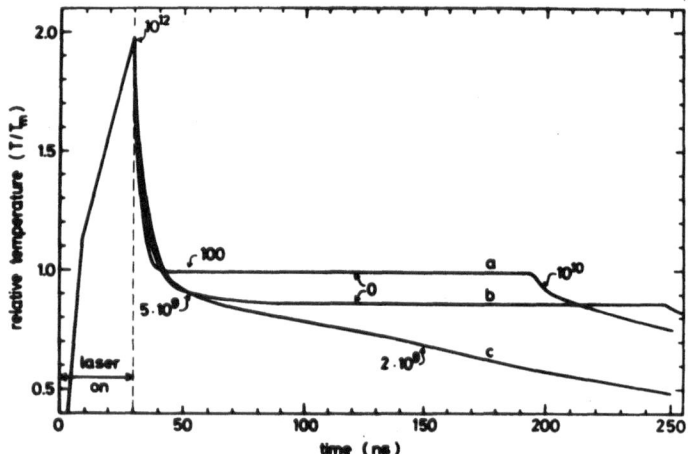

<u>Fig. 7</u> Surface temperature as a function of time for the cases of
Figure 6. Inserted numbers indicate the instantaneous cooling rate
at the crystal boundary.

"interface" stops; the liquid cools down without a phase transition.
The latent heat essentially remains stored in the structure which
may thus be called a glass. (Of course the model accounts only for
the energy content, not for the structure of the glass.)

Figure 7 shows the surface temperature as a function of time
for the cases of Figure 6. The numbers in the figure indicate the
instantaneous value of the cooling rate at the interface (which may
slightly differ from that at the surface). In all three cases the
cooling rate is of the order of 10^{12} °K/s just after the pulse has
ended. However, this cooling rate is of little relevance since it
occurs in the superheated melt. The cooling rate first decreases
by heat flow and then, in addition, by liberation of latent heat.
It stays close to zero until crystallization is completed in cases
a and b . In case c liberation of latent heat is suppressed.
The cooling rate stabilizes around 2×10^9 °K/s until the temperature
has dropped well below the range where the rate function is appreciably different from zero : the glass is stable. The calculation
shows, however, that even in this case some ten percent of the latent heat have been emitted which may be interpreted as partial
nucleation having taken place. Only with a ten times smaller scale factor for the rate function is nucleation found to be completely

suppressed. The cooling rate then remains around 10^{10} °K/s until solidification is completed.

The sample calculation presented here emphasizes the influence of the melt properties. A number of further conclusions regarding the influence of other experimental parameters can be drawn from the model. E.g., it turns out that material b becomes amorphous if irradiated with a ten times shorter pulse of the same energy. Pulse energy also matters : a 30 ns pulse of twice the intensity used in Figures 6 and 7 melts a layer 1000 nm instead of 500 nm thick; cooling is slowed down sufficiently even for material c to crystallize.

These examples demonstrate how critically the solidification process is influenced by the intrinsic crystallization rate if cooling is very rapid. Long-range mass transport at the l-s interface is the decisive (if not the only) factor influencing the intrinsic crystallization rate in non-elemental melts and, therefore, the tendency for glass formation upon fast cooling.

REFERENCES

1. White, C.W., Wilson, S.R., Appleton, B.R., Young, F.W., J. Appl. Phys. 51 (1980) 738-000.
2. von Allmen, M., "Laser Quenching", in "Glassy Metals II" (H. Beck, H.J. Guentherodt, eds), Springer Verlag, Chapter 11 (1982).
3. Chalmers, B. "Principles of Solidification", R.E. Krieger Publ. Co. (1964).
4. von Allmen, M., Lau, S.S., Sheng, T.T., Wittmer, M., in "Laser and Electron Beam Processing of Materials" (C.W. White, P.S. Peercy, eds) (1980) 524-529.
5. Turnbull, D. Contemp. Physics 10 (1969) 473-488.
6. Kashiev, D. Surface Science 14 (1969) 209-220.
7. Uhlmann, D.R. J. Noncryst. Solids 7 (1972) 337-348.
8. von Allmen, M. to be published.
9. Wang, J.C., Wood, R.F., Pronko, P.P. Appl. Phys. Lett. 33 (1978) 455-458.

LASER INDUCED OHMIC CONDUCTION IN GALLIUM ARSENIDE

Yves I. Nissim

Centre National d'Etudes des Télécommunications
196, rue de Paris 92220 BAGNEUX, France

1. INTRODUCTION

Compound semiconductors, and in particular GaAs, have from the start been potential candidates for the application of beam processing techniques. The conventional thermal annealing of GaAs often gives unsatisfactory results in both activation of dopants and removal of damage induced by ion implantation. Furthermore, encapsulation or controlled arsenic overpressure are required to prevent the substrate from surface decomposition (As evaporation) at elevated temperature. The short and localized heat treatment under beam irradiation can be expected to suppress decomposition and thereby improve the crystalline and electrical properties of annealed layers. However the results to date on beam annealing of implanted layers in GaAs have not shown substantial improvements as compared to thermal annealing. Major difficulties arising from the fragility and dissociation of the material under laser irradiation are responsible for the limited success achieved in early attempts. However the beam processing technique has made a significant contribution in the field of ohmic contacts on III-V compounds.

Both alloyed and non-alloyed ohmic contacts to GaAs have been obtained from irradiated high dose implants or deposited thin film on GaAs. An ohmic contact between a metal and a semiconductor is defined as an interface that exhibits current voltage characteristics which is linear for both directions of current flow. The performance and reliability of a number of GaAs microwave, logic and optoelectronic devices are determined in large part by the properties of their ohmic contacts. In microwave MESFET's, for example, parasitic source resistance due to the ohmic contact is a major contributor to noise in state-of-the-art devices. The device performance of a common FET (1 μm gate-length at 10 GHz) degrades significantly for value of specific contact resistance (ρ_c) above the mid 10^{-6} $\Omega\,cm^2$ range. As device geometry becomes smaller, the demands will be even more stringent. Traditional GaAs devices have standardized the use of Au-Ge based contacts [1,2]. For these contacts a eutectic composition of Au-Ge is evaporated onto a n-GaAs substrate and then heated above the eutectic melting point (356°C). In the melt, part of the GaAs is dissolved and Ge is incorporated into Ga sites upon cooling. It is commonly believed that a degenerate n^+ layer is then formed which lowers the barrier between the metal and the semiconductor [3] resulting in ohmic conduction. These contacts have a relatively low specific contact resistance (in the low 10^{-6} $\Omega\,cm^2$ range) but need improvement in both morphology and thermal stability. The major problem associated with this technique is the lack of uniform wetting of the molten Au-Ge to the GaAs (Ni or Pt are often used as overlayers to improve the wetting uniformity). This results in : (i) an evident surface roughness, (ii) a large thickness variation in the heavily doped surface layer and (iii) the formation of micro-precipitates of uneven composition [4]. In addition to these non-uniformities, the Au-Ge contact resistance increases with temperature and time [5,6]. In an attempt to solve these difficulties, new techniques have been recently developed to form ohmic contacts. Three different approach using laser or electron beam irradiation are presented here.

2. THEORETICAL BACKGROUND AND EXPERIMENTAL TECHNIQUE

An ohmic contact is a metal-semiconductor contact that has a negligible contact resistance relative to the bulk or spreading resistance of the semiconductor. In other words, an ohmic contact should supply current to the semiconductor with a voltage drop across the junction that is sufficiently small compared with the drop across the active region of the device. The figure-of-merit of an ohmic contact is mainly contained in its specific contact resistance value, defined as the reciprocal of the derivative of current density with respect to voltage at zero bias :

$$\rho_c = \left(\frac{\partial J}{\partial V}\right)^{-1}_{v=0} \quad (\Omega\,cm^2).$$

When the doping level of the semiconductor is low, the thermionic-emission current dominates the current transport and the specific contact resistance is then [7] :

$$\rho_c = \frac{k}{qA^*T} \exp\left(\frac{q\phi_{Bn}}{kT}\right).$$

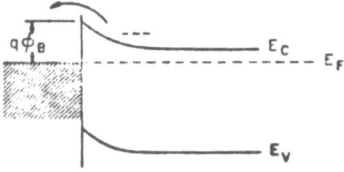

(a) LOW BARRIER HEIGHT

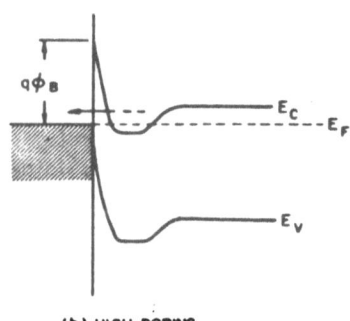

(b) HIGH DOPING

Fig. 1   Low barrier height and/or high doping for ohmic contacts.

Small values of ρ_c are thus obtained for low barrier height (ϕ_{Bn}).
When the doping of the semiconductor is increased, the barrier becomes
thinner and the carrier can tunnel thru it. When the tunneling
current dominates, the specific contact resistance is proportional
to :

$$\rho_c \sim \exp\{\frac{2}{h} (\frac{\varepsilon_s m^*}{N_D})^{1/2} \phi_{Bn} \}$$

Low barrier height and high doping concentration are then necessary
to obtain low values of ρ_c. These two criteria illustrated in
Fig. 1 are the approaches used for ohmic conduction. When tunneling
is predominant, the doping level dependence of the specific contact
resistance in the case of GaAs is illustrated in Fig. 2 [3].   It
should be noted that the barrier height is practically independent
of the metal choice ($\phi_B \simeq 0.7eV$ in GaAs) but is fixed by interface
states. However, as it is shown in a later section of this

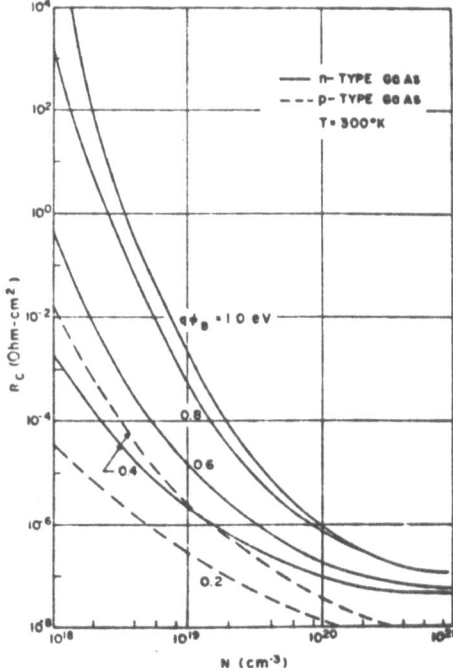

Fig. 2 Specific contact resistance at 300°K for n-type and p-type GaAs as a function of doping concentration (after [3]).

presentation, the barrier can be lowered via a thin interface compound formed by a chemical reaction between the semiconductor and the metal constituants. Finally it is worth mentioning that damages on the surface of a semiconductor, that should be avoided for the performances of a device, can reduce the specific contact resistance of an ohmic contact by forming recombination centers in the metal-semiconductor interface region.

The measurement of the specific contact resistance of an ohmic contact is a very delicate task when the value is small. It requires the extraction of a small contribution from a measured value. Among the different techniques, the Transmission Line Model (TLM) [8] to be described here is the more accurate one. It requires a planar test structure with variable contact spacing where the current is confined in a narrow n channel of an insulating substrate. When the résistance between two contacts in such a planar configuration is measured, one obtains :

534

$$R_+ = 2R_c + \rho_e \frac{L}{7}$$

286

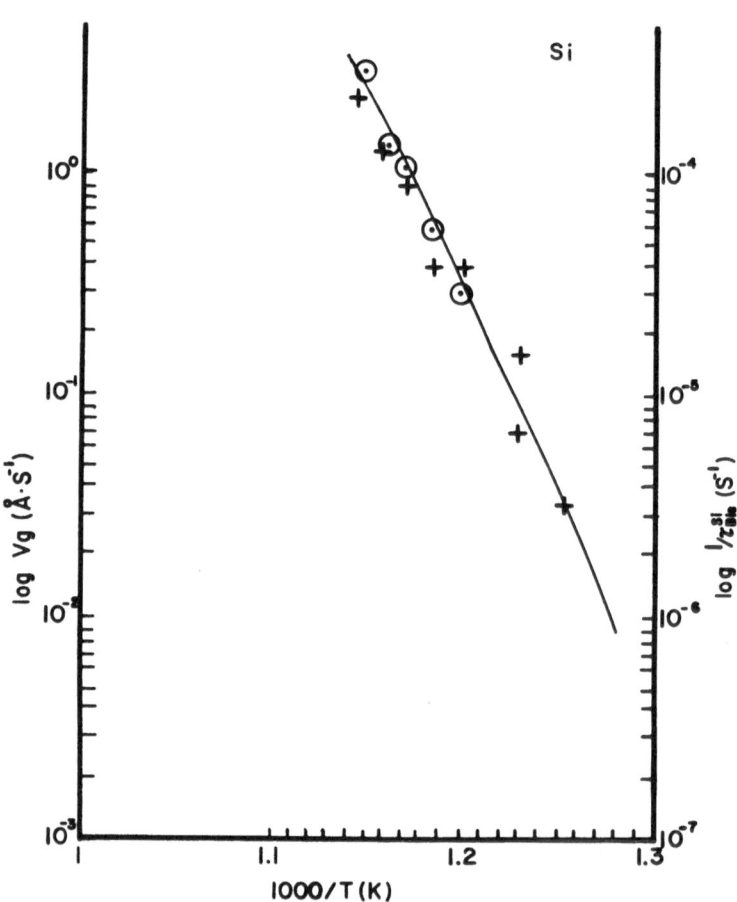

Fig. 3 Arrhenius plot of the growth rate v_g of crystallization in amorphous silicon layers : + data of Zellama et al [3]. Arrhenius plot of $1/\tau_{BIC}^{Si}$: ⊙ present study.

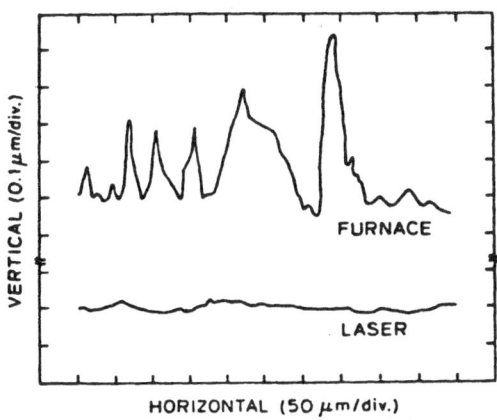

Fig. 3 Surface profilometer characterization of pulsed alloyed Ge-Au/GaAs ohmic contacts. Laser energy was $15J/cm^2$, furnace alloy was 450°C, 10 sec (after [12]).

TABLE 1

Best Au:Ge based ohmic contacts formed on GaAs by CW Ar-ion laser alloying

Contact metal	Specific contact resistance $\Omega\,cm^2$	laser parameters	
		Power (W)	Scan velocity (cm/sec)
Au:Ge-Ni-Au	4.8×10^{-6}	4	0.43
Au:Ge-Pt-Au	1.5×10^{-5}	3.8	0.43
Au:Ge-Ag-Au	2.0×10^{-4}	4.1	0.43
Au:Ge-Ti-Au	1.8×10^{-5}	3.8	0.2
In-Au:Ge	1.3×10^{-6}	3.5	0.43

prevent the metal layers from pulling back during the scan. When the ruby laser was employed the Ni layer was unnecessary, the morphology and definition of the contact was excellent in all cases and the best value of specific contact resistance reported was 2×10^{-6} Ωcm^{-2} at incident energies in the vicinity of $151 J/cm^2$.

An extensive comparison of different beam irradiation technique to alloy Au:Ge based contacts on GaAs was reported by Eckhardt [13]. The best results (lowest ρ_c, firmest adhesion between contact and GaAs and highest reproducibility) were obtained with a single mode (TEMoo) CW Ar-ion laser. The summary of the results and experimental conditions is presented in Table 1. All the samples used in this study consisted of semi-insulating GaAs, implanted with Si ion to produce $n \sim 10^{17}/cm^3$ at the surface. A number of 1 µm-gate GaAs MESFET's were prepared with Au:Ge based contacts and compared with furnace alloyed contacts from the same wafer. The laser alloyed contacts were found to be superior or at least as good in every respect. Their resistance was lower by 10-20 % and the dc characteristics were excellent. Average values of gain and noise measured at 14 GHz were somewhat better for the laser alloyed devices. In all cases, the surface morphology and edge definition of these contacts was far superior to that of their thermally alloyed counterparts.

A number of other laboratories have reported the formation of low contact resistance alloyed ohmic contact using pulsed laser or electron beam, but in all cases reported or mentionned here no reliability or lifetime were tested. If the beam alloying technique has brought an important improvement in the formation of ohmic contacts to GaAs, the degradation of these contacts at elevated temperatures stays a limiting factor to the lifetime of GaAs devices.

The mechanism of laser alloyed ohmic contact formation is similar to the furnace alloy mechanism described in the introduction. But because of the shortness of the laser irradiation, interdiffusion of semiconductor and metal constituants is reduced as well as the out-diffusion of As. This is illustrated in the Auger depth profiles of Fig. 4. In the furnace alloying, the interdiffusion is several hundred angstroms and less than one hundred angstroms in laser alloying case. The instability at elevated temperatures can be explained by instable carrier concentration in the regrown layer, and/or the low melting points of the eutectic and of the Au-Ga compound that has been recently reported to form during the alloy cycle [14].

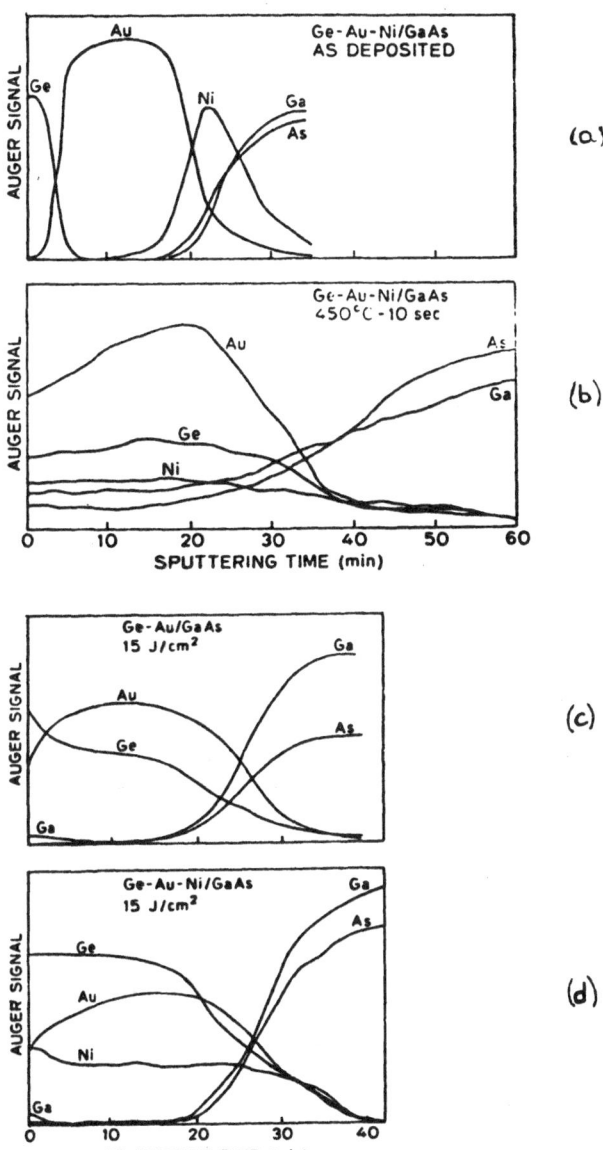

Fig. 4 Auger depth profiles of Au–Ge contacts on GaAs. a) as deposited Ge–Au–Ni; b) Furnace alloyed, 450°C, 10sec, Ge–Au–Ni; c) Pulsed ruby laser alloyed Ge–Au; d) Pulsed ruby laser alloyed Ge–Au–Ni (after [12]).

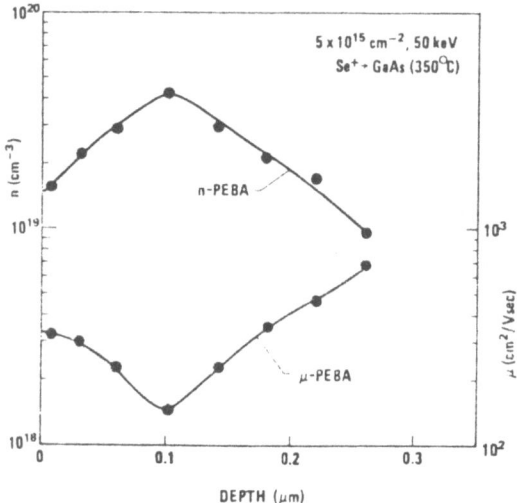

Fig. 5 Carrier concentration and mobility profiles for pulsed e-
beam annealed (0.9J/cm^2) ion implanted GaAs (5x10^{15}, 50keV Se)
(after [18]).

4. PULSED BEAM ANNEALING OF HIGH DOSE IMPLANTS IN GaAs

Degenerate n^{++} layers can be formed in GaAs using controllable
techniques such as ion implantation and laser annealing. When such
layers are formed it is then possible to form ohmic contacts by a
simple metallization. The alloying cycle is thus removed and the
formation of contacts can be achieved at any time of the process of
a device. One of the very few success of beam annealing of ion
implanted GaAs has been obtained with pulsed laser or electron beam
annealing of high dose implants. It has been reported that doses
in the 10^{15} to 10^{16} ions/cm^2 range could be annealed using pulsed
beam sources resulting in 1 to 10 % dopant activation on substitu-
tionnal sites and good recovery of the cristalline structure [15].
Low values of electron mobility have been observed and postulated to
be due to the presence of defects created by rapid quenching during
the transient anneal. But this anomalous behaviour should not affect
the ohmic contact formation. Several authors [16,17,18] have performed Se or Te implants in GaAs at doses reported above, followed
by a pulsed beam annealing. Peak carrier concentrations above 10^{19}
cm^{-3} and up to 4x10^{19} cm^{-3}[18] have been measured as can be seen in
Fig. 5. The anneal is responsible for an As evaporation detected
by an excess of Ga at the surface that can be removed in an HCl deep.
Non-alloyed ohmic contacts made on these layers were measured and
values of specific contact resistance in the mid 10^{-6} cm^2 range
were obtained.

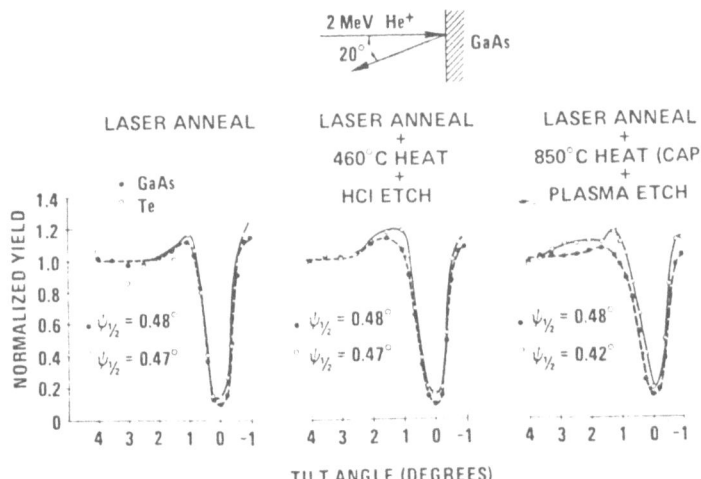

Fig. 6 Angular distribution of 2MeV He$^+$ ions back-scattering yield from GaAs and Te along the <100> axis. The GaAs samples were implanted with 250 keV, 5×10^{15} Te ions/cm^2, laser irradiated with 0.8J/cm^2 and post-laser anneal at 450 and 850°C (after [19]).

The stability at elevated temperature of these contacts has been studied. If the samples are heated above 250°C, the carrier concentration decreases rapidly to a saturation value of 3×10^{14}cm^{-3} above 350°C. Isothermal annealing curves show that this saturation level is reached as fast as 15sec above 350°C. Another decrease in carrier concentration starts at 650°C and reaches 5×10^{13} cm^{-3} at 850°C. The major loss in carrier concentration occurs within the first 1000 Å of the surface. Further studies of this behaviour has been made using angular distribution of He$^+$ ions backscattering yield [19]. In the case of Te implants the angular distribution was observed after each stage of the post-laser-anneal heat treatment. After the first state (450°C) no change were observed in the minimum yield of GaAs, the substitutional Te fraction or the channeling half angle of Te. From these observations, the significant reduction in carrier concentration is proposed to come from a Te vacancy complex formation. The vacancies formed during the rapid quenching following the irradiation are mobile at 450°C and could be trapped by substitutional Te atoms. In the second stage of the post-laser-anneal heat treatment (850°C) the formation of extended defects (dislocation loops) and a slight displacement of the Te atoms from their lattice site is observed. The later observation is obtained from the angular distribution of Fig. 6. The assumption is then made that the second drop in carrier concentration is due to the formation of Te clusters.

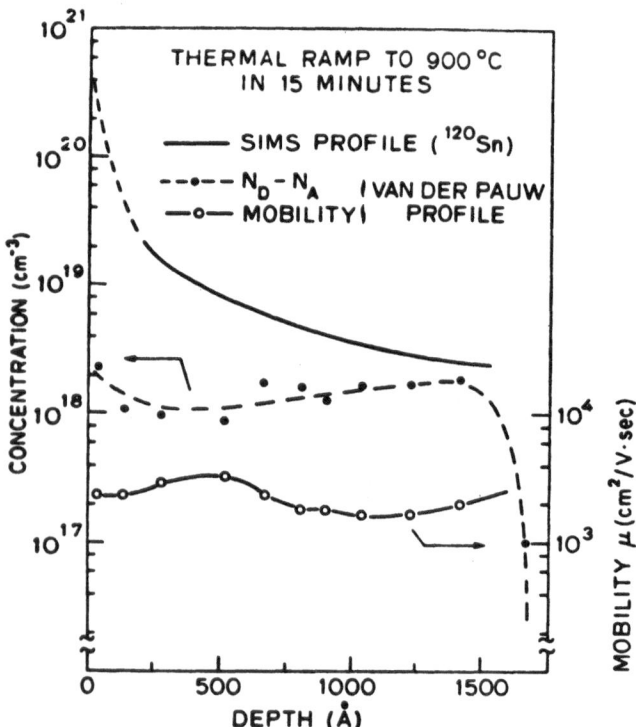

Fig. 7 SIMS and Van der Pauw stripping profiles for a sample ramped to 900°C in 15 minutes. The Hall mobility is also shown.

The behaviour of the carrier concentration and thus the behaviour of the value of contact resistance of ohmic contacts made on these layers is a serious limitation for the lifetime of a device. However this study is a model in the comprehension of ohmic contact formation. The ohmic conduction is fully explained by the high doping level. One of the studies reported here [18] has measured values of ρ_s that are exactly consistent with calculated values [3] obtained at the measured doping level for a metal−semiconductor barrier height of 0.7 eV.

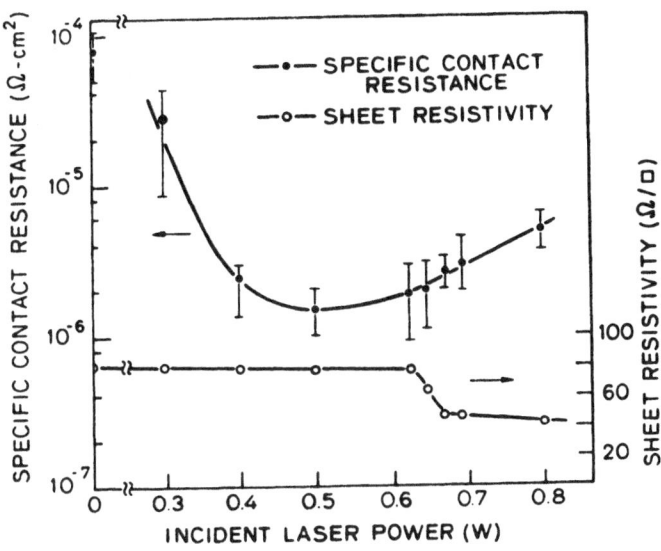

<u>Fig. 8</u> Effet of laser power on the properties of the diffused n+
layers and ohmic contacts formed on them.

5. LASER-ASSISTED DIFFUSION

Diffusion processes in GaAs are difficult due to the high vapor pressure of arsenic at diffusion temperatures. They usually
require sophisticated equipment such as sealed ampoules or controlled environment. An open tube diffusion technique is presented now
using a thin film solid source deposited on a GaAs substrate. Thermal or laser treatments or a combination of the two can be used to
control the diffusion process. This process represents a low cost,
reproducible method to obtain thin heavily doped n+ layers on semi-
insulating GaAs that is very attractive for the ohmic contact formation. A number of sources and treatments have been tried such as
an As_2Se_3 thin film source to diffuse Se using a pulsed electron
beam [20] or a Si thin film source using a pulsed Nd :YAG laser [21].
In the following the use of a SnO_2/SiO_2 source to diffuse Sn with
a combination of thermal and laser processing will be presented in
more details since the mechanism of ohmic formation has been fully

elucitated.

Relatively high carrier concentration can be achieved in GaAs when Sn impurities are diffused from a mixture of SnO_2/SiO_2 deposited as a thin film. The thermal cycle initiating the diffusion of dopants has been optimized for the best value of contact resistance with two main criteria : a high value of active carrier concentration and a short diffusion length. These conditions are met experimentally after a thermal ramp from room temperature to 900°C in 15 minutes. The resulting profile [22] is shown in Fig. 7. It corresponds to a relatively active carrier concentration profile at some $2-3\times10^{18}cm^{-3}$ to a depth of about 1600 Å. Subsequent metallization of these layers have shown an ohmic behavior without an alloying cycle. It has been shown [23] that following the thermal ramp, CW laser scans can assist the diffusion and activation of Sn. Specific contact resistance of the diffused layers obtained by this combination of processes has been measured for varying incident laser power (all the other parameters remaining constant) [24]. The results are shown in Fig. 8. Immediatly after the ramp, the contacts are already ohmic with $\rho_s \simeq 10^{-4}$ Ωcm^2. A dramatic decrease of two orders of magnitude is observed when the wafer is scanned at incident laser powers between 0.4 and 0.6 W. Specific contact resistance (ρ_s) between 1 and $2\times10^{-6}\Omega cm^2$ are obtained for this processing window while the GaAs surface remains smooth and free of visible damage. Temperature calculation [25] were carried out to define the temperature limits of the different regions. The results are shown in Fig. 9. The processing window that leaves a mirror finish on the surface corresponds to temperature between 600 and 800°C.

In order to understand the mechanism of formation of these contacts, two observations have been made; (i) the value of the carrier concentration is not high enough to be responsible for such a low contact resistance ($\rho_s = 1\times10^{-6}$ Ωcm^2) and (ii) there is an anomalously high concentration of inactive Sn atoms in the first 300 Å of the diffused layers. This suggest that a chemical reaction has taken place. Transmission Electron Microscopy (TEM) indicated precipitation in the form of isolated plates after the thermal ramp. Selected area electron diffraction patterns revealed that the precipitates were composed of a tin-arsenide compound (As_2Sn_3). It is striking to notice that the temperature induced by the laser (600°C) that is responsible for a drop of two orders of magnitude in specific contact resistance (ρ_s) correspond to the melting temperature of the compounds As_2Sn_3 (596°C). The action of the laser scan can then be interpreted as causing the melting of isolated Sn_3As_2 plates nucleated during the thermal ramp, after which they regrow into a more uniform film. This compound would then form an interface layer that reduces the barrier between the metal and the semiconductor. The low value of specific contact resistance can now be explained as coming from two contributions : the high carrier concentration (3×10^{18} cm^{-3}) in the diffused layers and the formation of the As_2Sn_3

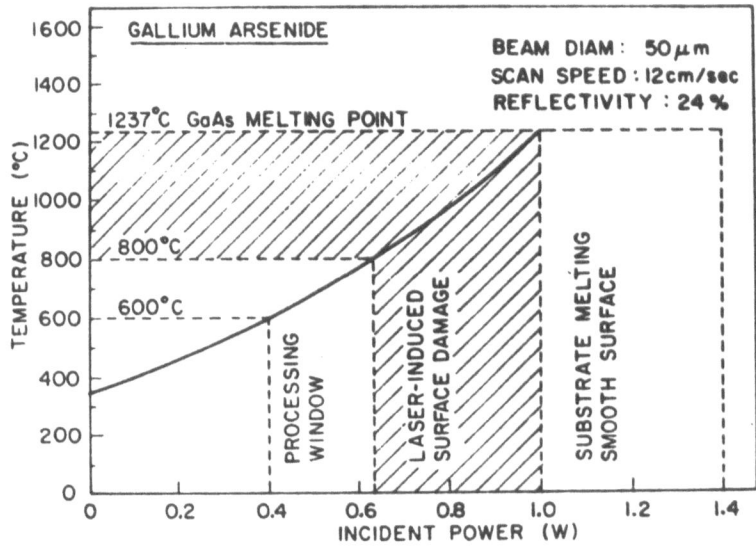

<u>Fig. 9</u> Temperature induced in the GaAs during irradiation as a
function of incident power. The different regime for ohmic contact
formation are shown.

compound at the surface that will form an interface after metallization.

To check the stability of these contacts, ageing experiments at
elevated temperatures were carried out. The degradation of contact
resistance normalized to 1 mm of contact width is plotted for two
different temperatures (320 and 400° C) in Fig. 10. It can be seen
that these contacts are a lot more stable than their Au-Ge alloyed
counterparts ([5 and 6]). These results are compatible with the
high stability of the Sn$_3$As$_2$ compound compared to the eutectic Au:Ge
one.

Finally since these results are very attractive to improve the
reliability of GaAs MESFET's, this diffusion technique has been
applied in the fabrication of such devices. It has been shown that
the process was fully compatible with current microelectronics fabrication techniques and that the resulting devices presented

544

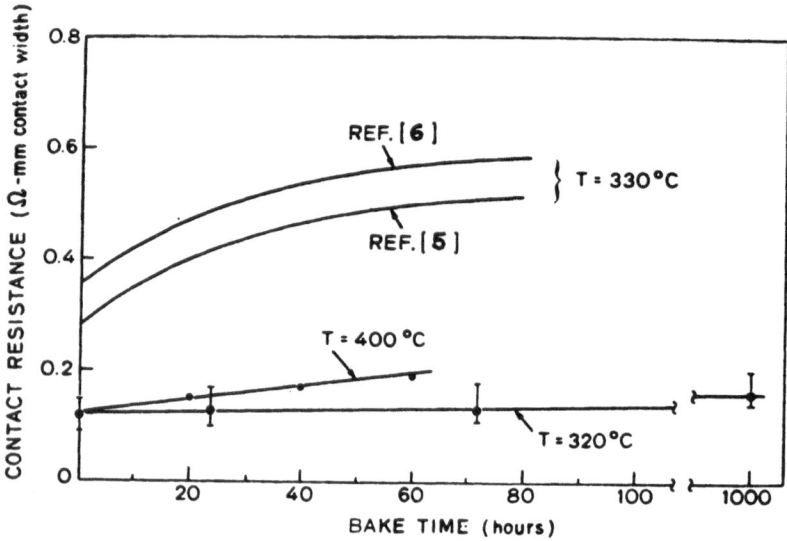

Fig. 10 High temperature stability of the planar diffused contacts.

excellent characteristics [26].

6. CONCLUSION

The beam processing technique has brought a significant improvement in the formation of ohmic contacts in GaAs. The efforts
made on the subject have shined some lights in the mechanism of
ohmic conduction. The best results are obtained when carriers
tunnel thru the top of the metal-semiconductor barrier. This indicates that both a high doping level and an interface that lowers
the barrier are required. Along this presentation it has been shown
that the stability of contacts is a very important property for
applications. The directions to follow in order to obtain a good
stability is to keep the doping level at the limit of solid solubility and to use interface compound of high melting point such as the
compound presented in the previous section. Finally the lamp irradiation technique that is currently being developed could replace
the laser to achieve high temperature in short periods of time and
make this process more accessible for device applications.

REFERENCES

1. Libov, L.D., Meskin, S.S., Nasledov, D.N., Sedov, V.E. and Tsarenkov, B.V., Instrum. Exp. Tech. USSR 4. (1975) 746.
2. Robinson, G.Y., Solid St. Electron. 18 (1975) 331.
3. Chang, C.Y., Fang, Y.K. and Sze, S.M., Solid St. Electron. 14 (1971) 541.
4. Moutou, P.C., Godard, J.J., Montel, J.M. and Dixneuf, B., Cornell Conf. Proceedings (1975).
5. Christou, A. and Sleger, K., Proc. 6th Biennal Cornell Elec. Eng. Conf. 169 (1977).
6. Macksey, H.M., Gallium Arsenide and Related compounds, edited by L.F. Eastman (Inst. of Phys. Conf. Ser. 33b, London, 1977).
7. Sze, S.M., Physics of semiconductor devices, (New York, J.Wiley, 2nd edition), p. 304.
8. Berger, H.H., Solid St. Electron 15 (1972) 145.
9. Yokoyama, N., Ohkawa, S. and Ishikawa, H., Jap. J. Appl. Phys. 14 (1975) 1071.
10. Pound, R.S., Saifi, M.A. and Hahn, W.C.,Jr., Solid St. Electron 17 (1974) 245.
11. Gold, R.B., Powell, R.A. and Gibbons, J.F., in "Laser Solid Interactions and Laser Processing" (S.D. Ferris, H.J. Leamy and J.M. Poate, eds). p. 635, New York, AIP, 1979.
12. Gold, R.B., PH. D. Thesis, Stanford University, Department of electrical Engineering, 1981.
13. Eckhardt, G., in "Laser and Electron Beam Processing of Materials" (C.W. White and P.S. Peercy, eds) (New York, Academic Press, 1980), p. 467.
14. Piotrowska, A., Guivarc'h, A. and Pelous, G. to be published in Solid St. Electron.
15. Eisen, F.H. in "Laser and Electron Beam Processing of Materials, (C.W. White and P.S. Peercy eds) (New York, Academic Press, 1980), p. 309.
16. Barnes, P.A., Leamy, H.J., Poate, J.M., Ferris, S.D., Williams, J.S. and Celler, G.K., Appl. Phys. Lett. 33 (1978) 965.
17. Mozzi, R.L., Fabian, W. and Piekarski, F.J., Appl. Phys. Lett. 35 (1979) 337.
18. Pianetta, P.A., Stolte, C.A. and Hansen, J.L., Appl. Phys. Lett. 36 (1980) 597.
19. Amano, J., Pianetta, P.A. and Stolte, C.A., Appl. Phys. Lett. 37 (1980) 948.
20. Davies, D.E., Ryan, T.G., Lorenzo, J.P., Appl. Phys. Lett. 37 (1980) 443.
21. Nissim, Y.I., Greiner, M., Falster, R.J., Gibbons, J.F., Chye, P. and Huang, C., in "Laser and Electron Beam Interactions with solids" (B.R. Appleton and G.K. Celler eds), (New York, North Holland, 1982), p. 677.
22. Nissim, Y.I., Gibbons, J.F., Evans, C.A.,Jr., Deline, V.R. and Norberg, J.C., Appl. Phys. Lett. 37 (1980) 89.

23. Nissim, Y.I., Gibbons, J.F., Magee, T.J. and Ormond, R., J. of Appl. Phys. 52 (1981) 227.
24. Nissim, Y.I., Gibbons, J.F. and Gold, R.B., IEEE Trans. Electron Devices ED 28 (1981) 607.
25. Nissim, Y.I., Lietoïla, A., Gold, R.B. and Gibbons, J.F., J. of Appl. Phys. 51 (1980) 274.
26. Dobkin, D.M., Gold, R.B., Nissim, Y.I. and Gibbons, J.F., IEDM Proc.(Washington, december 1981).

SYNTHESIS OF HIGH PURITY SEMICONDUCTING COMPOUNDS BY LASER IRRADIATION

Luc Baufay

Faculté des Sciences, Université de l'Etat,
Av. Maistriau, 23 7000 MONS, Belgium.

1. INTRODUCTION

In recent years, lasers have been proved to be useful tools in crystallizing amorphous elemental semiconductors or in recrystallizing ion-implanted region of single crystals [1]. In such cases, the same element is present before but also after the laser irradiation. This paper deals with a laser processing which is essentially different. Laser annealing experiments described here differ from most others in that the irradiation is performed on multilayered metallic films of, alternately, components A and B, in order to produce the III-V or II-IV semiconducting compound AB. For instance, by irradiating a composite sandwich film of Al and Sb, we have prepared the compound AlSb. In the same manner, we have also formed other materials : CdTe, CdSe and AlAs [2-11]. This list is not limitative and has been extended recently to different oxides (CdO, MgO, Cu_2O) [12,13].

Nevertheless, the process by which such transformation occurs has been (to us, at least) unclear. Initially, the reaction takes place over ~1000 Å thick films for laser pulse in the μs time domain. Any model describing that transformation has thus to take in account atomic displacement over several hundred Å in times of the order of 10^{-6}s. That suggests liquid diffusion but the measured laser pulse energy is insufficient to melt the semiconducting compound.

Aside from the intrinsic interest in the fundamentals of such laser processing, the potential applications of such semiconductors (for instance, AlSb and CdTe are very good candidate as photovoltaic materials [14] suggest) that the use of the laser may be of a great technological interest. For this purpose, experiments will be described in the following in order to propose a model, the quality of such laser processed materials will be tested and finally the possibility to produce large area of semiconducting thin film will be analyzed.

2. GENERAL EXPERIMENTAL CONDITIONS

2.1 Film preparation

Components A and B are successively and alternately condensed onto freshly cleaved salt or sputter-cleaned glass substrates, using a multiple e-gun source and a quartz thickness monitor at a base pressure of 10^{-6} - 10^{-7} torr, in such a way that the numbers of A and B atoms are as nearly as possible equal within the total film thickness. The number and, thus, the thickness of each elemental layer are parameters which may be tuned. In order to protect the films from air degradation, a ~250 Å thick SiO layer is sometimes evaporated on top of and, in the case of NaCl substrates, also beneath the multilayered metallic films in the same chamber. Portions of the films formed on NaCl are floatted off onto Cu grids for TEM work. These films will be in the following referred as free-standing samples. Films formed on glass substrates will be referred as supported samples.

2.2 Laser annealing

These samples are laser irradiated in air, at room temperature (except in special cases which will be mentionned) in order to achieve the phase transformation (see Fig. 1). Several lasers have been used :
- a dye laser pumped by a flash lamp; the laser pulse duration is t_p = 1.8 µs and the photon energy is hν = 2 eV;
- a ruby laser (t_p ~ 35 ns, hν ~ 1.8 eV);
- a CW krypton ion laser (all red lines, hν ~ 1.9 eV);
- a CW argon ion laser (hν is chosen to equal either 2.4 or 2.6 eV). The continuous beam can be chopped, the pulse duration is then at least 10^{-4}s.

2.3 Identification

Identification of the semiconducting compounds is obtained by a combination of techniques related to the atomic configuration (transmission electron microscopy, high resolution electron diffraction) and the physical properties of the materials (optical absorption in the visible range and electrical conductivity measurements). Only the electron diffraction patterns are reported in this section (see Fig. 2), the other characterization measurements will appear in Section 4. Note that CdTe and CdSe are known to exist under two different stable cristallographic configurations : the blende and the wurtzite structures. For single crystal, the blende type

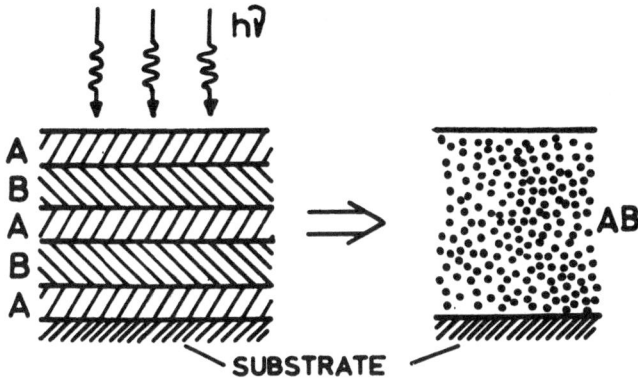

Fig. 1 Schematic representation of the laser induced transformation (from multilayered metallic to semiconductor).

(resp. wurtzite) is reported to be associated with a low temperature processing (resp. high temperature)[15] . In laser annealing experiments, the wurtzite structure appears when pulse duration is very short. Meanwhile, the transformation to the semiconductors studied here is easily evidenced by optical transparency since thin films are highly transparent in the low energy region of the visible spectrum.

3. THE NATURE OF THE TRANSFORMATION

3.1 Free-standing samples [5,6,9]

To demonstrate the intrinsic nature of such a transformation, standard 1000 Å thick, free-standing Al/Sb sandwiches films are considered here in which interaction between substrate and film can effectively be neglected. Using ruby, dye or chopped Kr$^+$ lasers, and not taking in account small differences in wavelengh

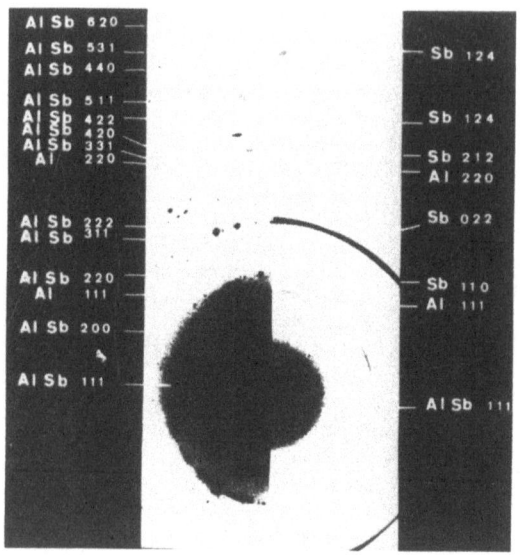

Fig. 2a Electron diffraction patterns of 1000 Å thick multilayered polycrystalline Al–Sb films and the same after laser irradiation.

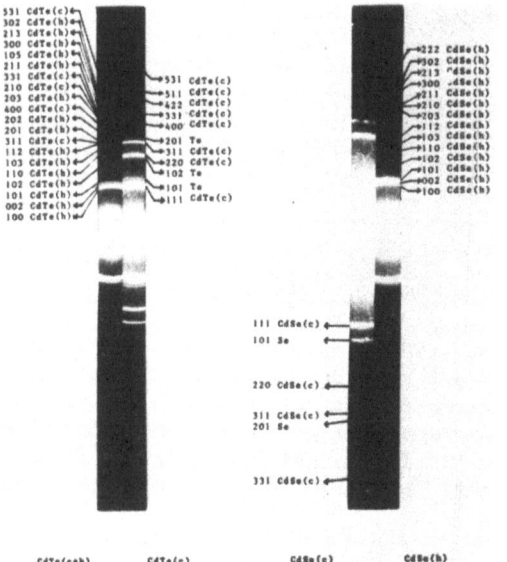

Fig. 2b Electron diffraction patterns of the two observed structures respectively for CdTe and CdSe. The hexagonal structure (or the mixture of hexagonal and cubic structure in the CdTe case) results from the dye-laser irradiation; the cubic structure results from the argon or krypton ion laser irradiation (chopped or not).

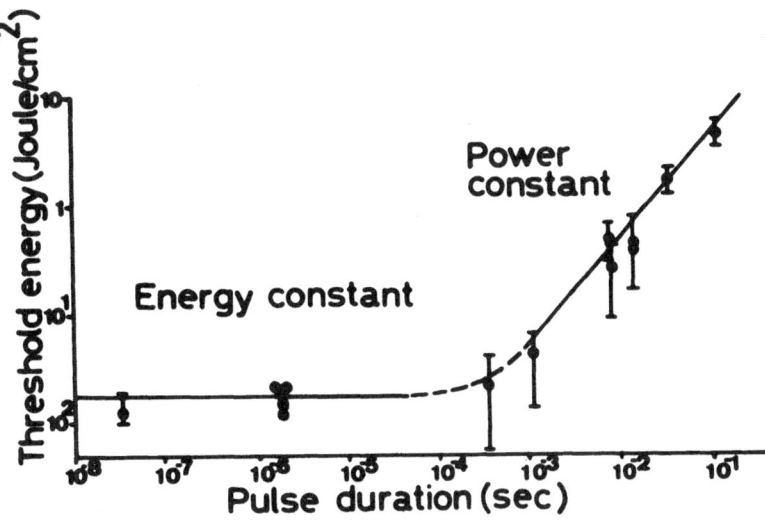

<u>Fig. 3</u> Laser pulse energy necessary as a function of pulse duration for transformation under threshold conditions (1000 Å thick Al/Sb sandwiches, free-standing samples).

dependence), the pulse energy required to just transform the film has been measured, after correction for front surface reflections, as a function of the pulse duration (see Fig. 3). For pulses shorter than 10^{-4}s, the energy threshold is constant which tells us that the pulse duration is short compared to the time over which significant thermal losses occur. That threshold energy can be thus simply related to the maximum film temperature induced via the thermal capacity of the film and no other parameters. This energy (17.6 ± 3.4 mJ/cm^2 or 12±2 kcal/mole) would allow at best to melt the film and thus its temperature could not exceed 933 K (see Fig. 4).
A similar value could be deduced from the constant power regime for pulses longer than 10^{-4} sec but we do not attempt a proper analysis here since these calculations have to take in account several parameters (thermal conductivity, ...) which are less well defined than the thermal capacity.
Fig. 5 shows the pulse threshold energy as a function of film temperature for pulse duration $\gg 10^{-4}$s, achieved by mounting the support grid in a hot air flow. In the constant power regime for pulses longer than 10^{-4} sec, i.e. an equilibrium state is reached where laser input exactly balances thermal losses, the temperature of the film is governed principally by its support and the energy vs temperature relationship should be a linear one. Thus the plot of Fig. 5 should also be linear and the extrapolation to zero energy

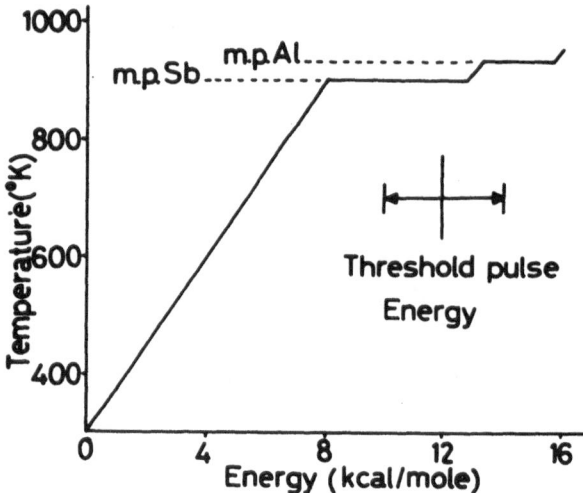

Fig. 4 Calculated temperature vs energy plot for 1000 Å thick Al/Sb film.

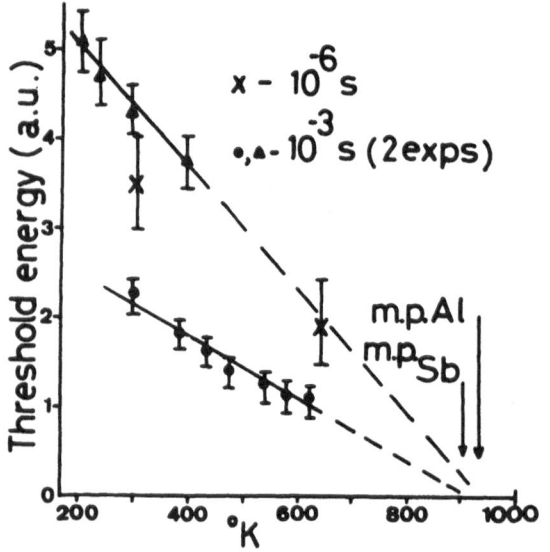

Fig. 5 Laser pulse energy necessary(as a function of film temperature)for transformation under threshold conditions.

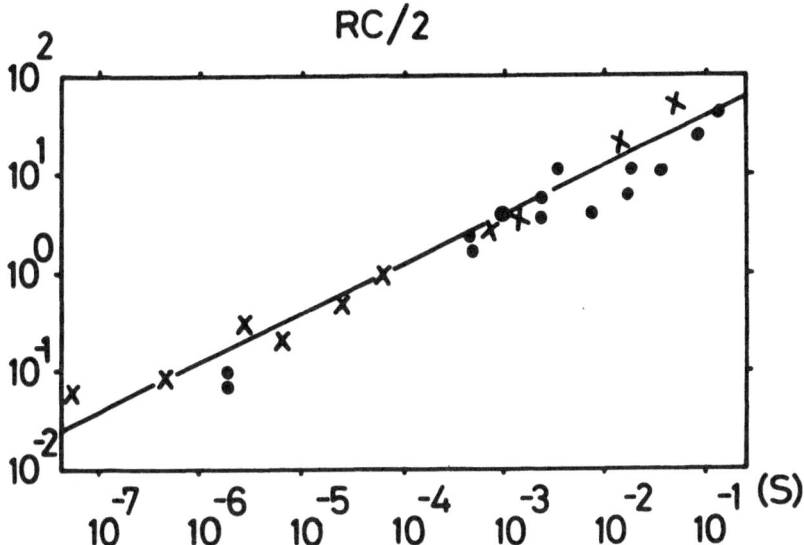

<u>Fig. 6</u> Threshold energy as a function of the pulse duration for laser (o) and capacity discharge (+) annealing in the case of supported samples.

simply indicates to which temperature the film is heated by the laser pulse. A value of 900-950 K seems reasonnable here in excellent agreement with our above calculations.

3.2 <u>Supported samples.</u>

Threshold energy has been also measured as a function of the pulse duration (see Fig. 6). Moreover we have taken advantage of the low electrical resistivity of metallic samples before transformation to simulate the temporal energy profile of the laser pulse by the electrical discharge of a capacitor through the film. In this case, the transformation also occurs and the threshold energy is measured as a function of the characteristic time of the discharge (see Fig. 6). Fig. 6 allows a comparison between these annealing

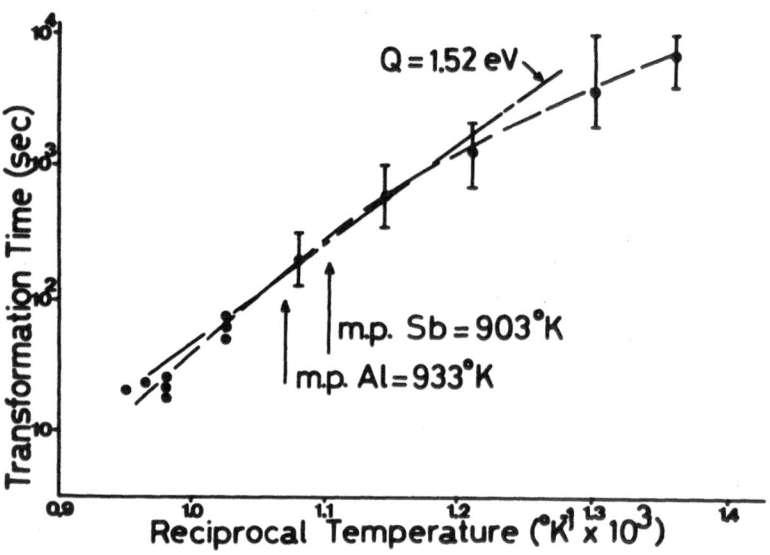

<u>Fig. 7</u> Transformation time as a function of temperature inverse
for the furnace annealing of glass supported Al/Sb films.

processes. The threshold energy is proportional to the square root
of the characteristic time of the energy pulse; that behavior is
essentially different from that corresponding to free-standing samples and shows the role of the substrate. Indeed it can be shown
that the thermal behavior of the film (its thermal capacity is
negligible compared to the substrate one) is governed by the substrate (assumed to be a semi-infinite solid) which is submitted to
the heat flux stemming from the laser pulse and the film temperature
is equal to the surface temperature of the substrate. In laser or
capacitor discharge annealing, the maximum temperature induced by
the energy pulse is calculated to be approximatively 900 K [11].

3.3 Discussion

As a conclusion to the previous experiments, the film temperature induced by the laser pulse under threshold conditions is at best
933 K, whatever the pulse duration. On the other hand, the Arrhenius plot of (inert gas) furnace annealing data for identical
supported films (Fig. 7) indicates that, at some 930 K, the complete
transformation takes place in more than 100 s. Further, it would
not be sufficient to simply melt the components in order to have
some more rapid compound formation, presumably because such formation can only occur by diffusion of (probably) Sb across the

already formed and growing solid AlSb barrier (T_{melt}=1353 K). Why that difference between laser and furnace annealing? Up to now, we have neglected any contribution arising from the heat of formation, ΔH, which is high for all materials studied. In the AlSb case, $\Delta H \simeq -22$ and -30 kcal/mole at 300 and 930 K, respectively, for the formation stemming from the liquid components. That heat of formation is very significant compared to the laser pulse energies employed. Indeed, a calculation shows that, if totally liberated without losses, ΔH will easily achieve the bizarre result of melting all the AlSb, so that the apparently low laser threshold energies measured should not at first sight present too big a problem. On the other hand, the reaction must first proceed in order to make this energy available and as shown in Fig.7, it will not have proceeded very quickly at 930 K. So assume that the laser pulse melts the Sb part plus x% of the Al in a time short compared to that in which a solid AlSb barrier of sufficient thickness can be formed. Ensuing liquid mixing and compound formation then liberates x% of 30 kcal/mole which is sufficient to melt the remaining Al and elevates the temperature of the now-composite film to or near the melting point of AlSb at which temperature we might imagine the reaction to be successfully completed. Calculation shows x ~ 25% for this to occur which would require an energy input of ~ 20 mJ/cm^2 (14 kcal/mole) for our film, in good agreement with the measured laser pulse energy (see Table 1).

TABLE 1

Systems AB	Thickness (Å)		ΔH(kcal/mol)		Model prediction		Experimental Threshold energy (mJ/cm^2)
	A	B	at 300K	at T_m	% molten	Energy (mJ/cm^2)	
AlSb	350	650	−22	−30	25	20	17.6±3.4
CdTe	390	610	−24	−31	38	13.7	11.8±0.6
CdSe	440	560	−32.6	−38	30	8.1	7.8±0.4
AlAs	430	570	−35	−	−	−	16±5

Table 1 : Showing for the systems studied the initial layer thicknesses, published values of ΔH, appropriate values of ΔH for formation from molten components at the higher T_m, % of higher T_m component melted by laser pulse according to model and the derived and measured threshold energies.

556

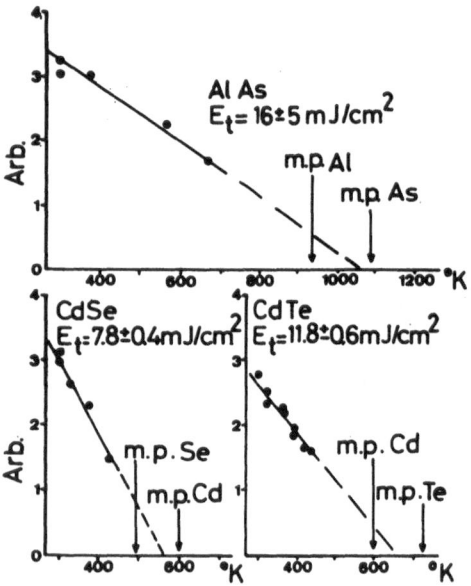

<u>Fig. 8</u> Laser pulse energy necessary as a function of film tempera-
ture for AlAs, CdTe and CdSe.

We thus conclude that our model is workable for AlSb and therefore
proceed to test its applicability to other similar systems.
Similar experiments are performed on (Cd,Te), (Cd,Se) and (Al,As)
systems in order to produce CdTe, CdSe and AlAs. Fig. 8 shows in
each case extrapolation of the threshold energy versus temperature
to temperatures commensurate with the melting points of the constituents (as in Fig.5); data similar to Fig. 3 give in each case,
a threshold energy insufficient to completely melt the film. These
values are listed in Table 1, together with the expected threshold
values calculated according to our hypothesis and in the manner
described for AlAs. Note that thermodynamic data for AlAs are incomplete and, in addition, As normally sublimes so that a model
calculation was not practicable here. For CdTe and CdSe, however,
it is fair to call the agreement between experiment and theory
excellent.

3.4 <u>Conclusion</u>

Quite clearly, the laser induced transformation observed in
the systems AlSb, AlAs, CdTe and CdSe is not simply a fast furnace
anneal, i.e., an essentially isothermal process, but one in which
the heat of formation itself plays a critical role. The function
of the laser is merely a trigger to a reaction that is to some extent self-sustaining.

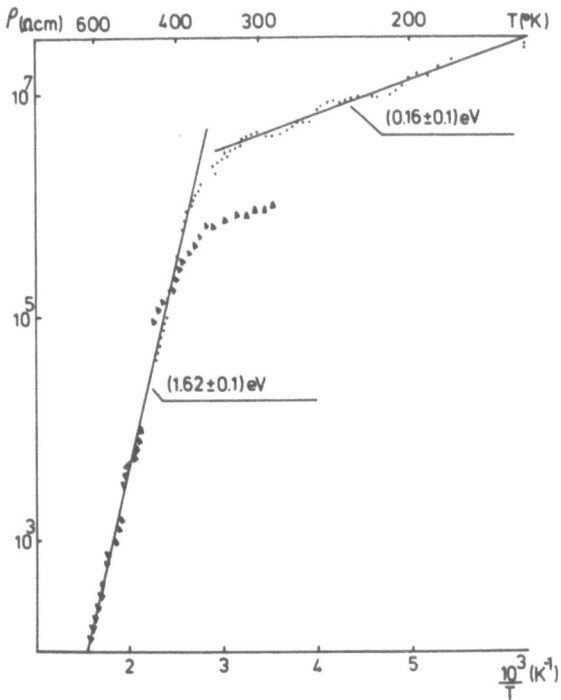

Fig. 9 Logarithm of the resistivity against the temperature inverse for CdTe. Activation energies, films composition before irradiation and resolution are indicated.

4. CHARACTERIZATION [7,8]

Large area of semiconducting compounds are obtained by scanning a continuous beam on the sample at a speed of about 1 cm/s. The beam diameter is equal to 1.6 mm and the beam power ranges from 50 to 500 W/cm^2. That section is devoted to glass-supported films of Cd/Te which are laser irradiated in order to form CdTe.

4.1 Transmission electron microscopy

1000 Å thick sample are polycrystalline with a crystallite size ranging from 100 Å to 5000 Å (for a photocell, grain size of 5000 Å to 1μ will be good enough for a 1 to 2μ thick sample).

4.2 Electrical transport measurements (Fig. 9)

The room temperature resistivity is about 10^6 Ωcm, higher than several reported values for single crystal. Plotting the logarithm of the resistivity vs temperature inverse indicates that the width of the mobility gap is 1.62 ± 0.10 eV. Worth to note is that intrinsic conduction manifests itself down to about 400 K. Below that temperature, conduction is controled by impurity activation. One single activation energy at 0.16 ± 0.10 eV is detected down to 200 K.

4.3 Optical absorption measurements (Fig. 10)

The optical transmission of 1000 Å thick sample is measured in the visible and near-infrared range. Two transitions are evidenced at 1.47 ± 0.05 eV and 2.38 ± 0.10 eV. They correspond with the fundamental absorption edge and the transition occuring between the spin-orbit split-off valence band and the conduction band at Γ [16,17], respectively.

4.4 Photocurrent measurements

The photoresponse has been measured as a function of the wavelength. The fundamental absorption edge is also evidenced in good agreement with published data.

4.5 Conclusion

The optical absorption measurements, allowing in particular to measure the spin-orbit splitting, stress the high quality of the laser-processed material. Moreover, electrical measurements seem to indicate that the forbidden gap is devoid of deep-lying impurity states and the shallow impurity concentration is estimated to be less than 10^{16} – 10^{17} cm^{-3}. This is achieved despite sample preparation conditions which would have been a priori considered to be approximate (i.e. the accuracy of the initial stoichiometric composition of the films is about 1% before irradiation and the phase transformation occurs in air or in contact with SiO layers). This means that, most probably, impurities are driven out to the grain boundaries and not frozen in the network.

5. FINAL COMMENTS

The feasibility of such synthesis is well demonstrated and, in addition, the laser processed material seems to be a high purity one. Such kind of rapid reactions can shed light upon the fundamentals of nucleation and crystal growth (i.e. possibility of choice between different crystallographic structures, purity of the

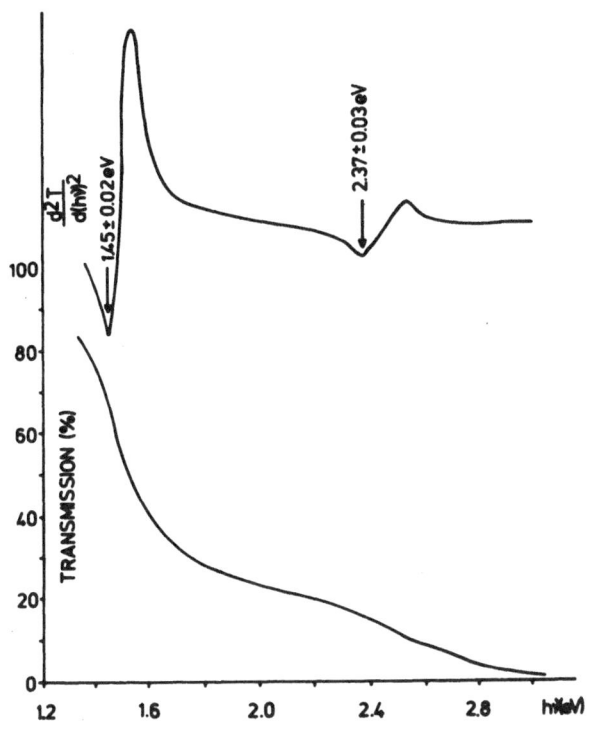

Fig. 10 Optical transmission spectrum of laser annealed CdTe films.
Structure location is indicated on the second-derivative of that
spectrum.

compounds) but that technique can be also of a great interest for
applications. For that purpose, the simple thermal model describing
the laser-initiated formation of semiconducting compounds which has
been proposed in good agreement with experiments is very helpfull,
i.e. we can *a priori* well choose the substrate with regard to its
thermal properties and to the film thickness (thicker the film, more
important the heat of formation which has to flow across the film-
substrate interface without damage). As a final conclusion, the
results presented in this paper show that the laser can be a powerful tool not only for silicium processing but also, and perhaps
essentially, for the preparation of thin film of semiconducting
compounds.

ACKNOWLEDGEMENTS

I would like to acknowledge L.D. Laude, R. Andrew and M. Wautelet for helpful discussions and for their support. I am also indebted to A. Pigeolet for his assistance and to M.C. Joliet for the TEM work. This work is supported by the National Energy Programme of the Belgian Ministry for Science Policy.

REFERENCES

1. See for instance, the Proceedings of the MRS Symposium, "Laser and Electron-Beam Solid Interactions and Materials Processing".

2. Andrew, R., Ledezma, M., Lovato, M., Wautelet, M. and Laude, L.D., Appl. Phys. Lett. 35 (5) (1979) 418.

3. Andrew, R., Baufay, L., Laude, L.D., Lovato, M. and Wautelet, M., J. de Phys. 41 (1980) C4-71.

4. Baufay, L., Failly-Lovato, M., Andrew, R., Joliet, M.C., Laude, L.D., Pigeolet, A., Wautelet, M. in Insulating Films on Semi-conductors (eds M. Schulz and G. Pensl), Springer Series in Electrophysics 7 (1981) 242.

5. Baufay, L., Andrew, R., Pigeolet, A. and Laude, L.D., Thin Solid Films 90 (1) (1982) 69.

6. Andrew, R., Baufay, L., Pigeolet, A. and Laude, L.D. in Laser and Electron-Beams Interactions with Solids, vol. 4 (eds. B.R. Appleton and G.K.Celler) (North Holland, 1982), p. 719.

7. Baufay, L., Dispa, D., Pigeolet, A. and Laude, L.D., to be published in J. of Crystal Growth.

8. Baufay, L., Dispa, D., Pigeolet, A., Joliet, M.C. and Laude, L.D., in 4th EC Photovoltaic Solar Energy Conference, eds. W.H. Bloss and G.Grassi (D. Reidel Publishing Company,1982) p. 839.

9. Andrew, R., Baufay, L., Pigeolet, A. and Laude, L.D., J. Appl. Phys. 53 (7) (1982) 4862.

10. Baufay, L., Pigeolet, A. and Laude, L.D., to be published in J. Appl. Phys, 54 (2) (1983) 660.

11. Baufay, L., Andrew, R., Pigeolet, A. and Laude, L.D. to appear in Physica.

12. Wautelet, M. and Baufay, L., Thin Solid Film, 100 (1983) L9.

13. Andrew, R., Baufay, L., Pigeolet, A. and Wautelet, M., submitted for publication in the Proceedings of the MRS Symposium of Laser Diagnostics and Photochemical Processing for Semiconductor Devices , Boston 1982.

14. See for instance J.J. Wysocki and P. Rappaport, J. Appl. Phys. 31(3) (1960) 571.

15. Abrikosov, N.Kh., Bauknia, V.F., Poretskaya, L.V., Shelimova, L.E. and Skudnova, E.V. in "Semiconducting II-VI, IV-VI and

V-VI compounds (Plenum Press 1969) p. 3.
16. Myers, T.H., Edwards, S.W. and Schitzina, J.F., J. Appl. Phys. 52 (1981) 4231.
17. Dimmock, J.O., in Proc. Intern. Conf. on II-VI Semiconducting Compounds, Ed. D.G. Thomas(New York, Benjamin, 1967) p.227.

EFFECTS OF PULSED LASER IRRADIATION ON THE ELECTRICAL PROPERTIES OF GaAs (*)

Didier PRIBAT

Thomson-CSF, Laboratoire Central de Recherches,
Domaine de Corbeville 91401 Orsay - France.

It has long been shown that pulsed laser irradiation can be used to recrystallize implanted, heavily damaged or amorphized GaAs layers. According to R.B.S. and T.E.M. measurements performed after laser irradiation (i) the implanted atoms are substitutionnaly incorporated in the matrix, and (ii) the recrystallyzed layers are generally free of extended defects. However, electrical activation of n-type implanted dopants can only be achieved at high implanted dose ($> 10^{14}$cm2) and with low yields (typically $< 60\%$). For low doses (in the range $10^{12} - 10^{13}$ at./cm2) electrical activation cannot be obtained using pulsed laser irradiation. Furthermore, in the latter situation methods generally used for crystal defect characterization (such as R.B.S. in channeling conditions) fail to give informations because of their insensivity to doses less than 10^{13} at./cm2. Consequently, electrical or optical methods are currently being used to characterize defect introduction in GaAs following low dose implantation and laser processing. The purpose of this paper is to focus on electrical measurements made after pulsed laser irradiation of both virgin and low-dose implanted GaAs material. Emphasis is put on the use of both I(V), C(V) and D.L.T.S. measurements on Schottky structures in order to characterize defect introduction. The basic physical principles of these methods are first outlined in some details. I(V), C(V) and D.L.T.S. data are then interpreted and defect introduction is characterized as a function of laser incident energy. Either surface or bulk damage effects are observed, depending on incident laser energy and wavelength. Thicknesses of damaged layers are compared to calculated melted thicknesses (according to the thermal model). A general discussion follows, concerning laser applications to low-dose implantation damage annealing in GaAs.

(*) Work partially supported by DRET.

1. INTRODUCTION

Within the last few years, a significant number of studies have been carried out on damage recovery of ion-implanted semi-conductor materials. In particular, laser annealing of implanted semiconductor layers has been an intensive area of investigation due to its potential advantages over the usual thermal annealing process [1], [2], [3], [3a]. However, laser processing techniques that have been so extensively applied to silicon have not yet been so equally successful in Gallium Arsenide.

In this material, when high dose n-type implants ($>10^{14}$At/cm^2) are pulsed-laser irradiated, very high carrier concentrations can be attained, often exceeding the maximum anticipated from thermal solubility considerations [4], [5], [6], [7]. Despite this, the electron concentration for n-type impurities never exceeds 60%, and the reported Hall mobilities are low (3 to 5 times lower than expected from layers with peak concentrations of the order of 10^{19} cm^{-3} [8]). Another discouraging factor is that no electrical activity can be obtained after pulsed irradiation of low dose implanted material (in the range $10^{12}-10^{13}$ At/cm^2) as would be of interest in forming channels in Field Effect Transistors [9], [10], [11].

Pulsed-laser irradiation in the liquid phase regime while providing layers free of extended defects [12], [12a] does leave or introduce point defects which affect the electrical properties of the processed material [13]. Compensating defect centers are created in the form of energy levels lying deep in the forbidden gap which are thought to account for the lack of electrical activity observed after irradiation of low dose implants [14]. The problem however is to find out some convenient means of characterizing those defects.

A first thing to be pointed out is that crystallographic methods, such as R.B.S. in channeling conditions which have proved to be useful in the characterization of impurity location and surface damage after laser annealing of high dose implanted GaAs [15] fail to give informations when low implanted doses (light ions, in the 100 KeV range) are of concern. A GaAs crystal having 10^{18} displaced atoms/cm^3 within the first few thousand angstroms from surface would appear as perfect from the channeling point of view. Consequently, no direct defect observation can be performed on slightly damaged layers as is the case we are dealing with.

If we except Electron Spin Resonance techniques (E.S.R.), the most sensitive methods for deep energy levels characterization are unvariably based either on electrical effects (current or charge measurements associated with a carrier accumulation in a region transitorily out of neutrality) or on optical effects (absorption or emission on localized centers). Let us note that such approaches are quite different from the R.B.S.' one, since they provide the properties (electrical or optical) of a trap energy level rather than the lattice location of the associated point defect.

We will not review in this paper all the recently developed methods for deep level characterization since they are legion [16]. We will rather focus (i) on capacitance measurementsmade on Schottky structures which provide a global means of defect characterization either at the surface or in the bulk and (ii) on Deep Level Transient Spectroscopy (D.L.T.S.) measurements which can identify and track individual defect energy levels in the forbidden gap.

These two methods will then be applied to the characterization of laser induced defects in GaAs.

2. PROPERTIES OF METAL SEMICONDUCTOR JUNCTIONS

2.1 With shallow levels only

Since all the electrical measurements presented in this paper are based on Schottky structures, it is appropriate to review first the basic properties of metal-semiconductor junctions. We will only give those formulae which are necessary for the interpretation of experimental data presented thereafter. For more details on the topic, the reader is referred to Rhoderick [17] or Sze [18].

In our case, the metal (Au or Al) due to the position of its Fermi level acts as a p^+ layer with respect to the n-doped semiconductor. On the other hand, in the semiconductor, the shallow levels introduced in the forbidden gap near the conduction band by n-type impurity incorporation are supposed to be completely ionized at normal temperature of device operation.

In the ideal case, the initial difference between the two Fermi levels is compensated by an electron transfer from the semiconductor into the metal. This electron transfer leaves behind uncompensated ionized donnors. Consequently, a positive charge distribution is built within the region of the semiconductor depleted with mobile carriers. This charge distribution induces band bending near the surface and a built-in potential within the bulk. The electric field in the depletion region compensates the flow of carriers from the semiconductor into the metal in such a way that new equilibrium conditions are created. The barrier height ϕ_{bn} is the difference between the metal work function ϕ_m and the electron affinity of the semiconductor χ :

$$\phi_{bn} = \phi_m - \chi \ .$$

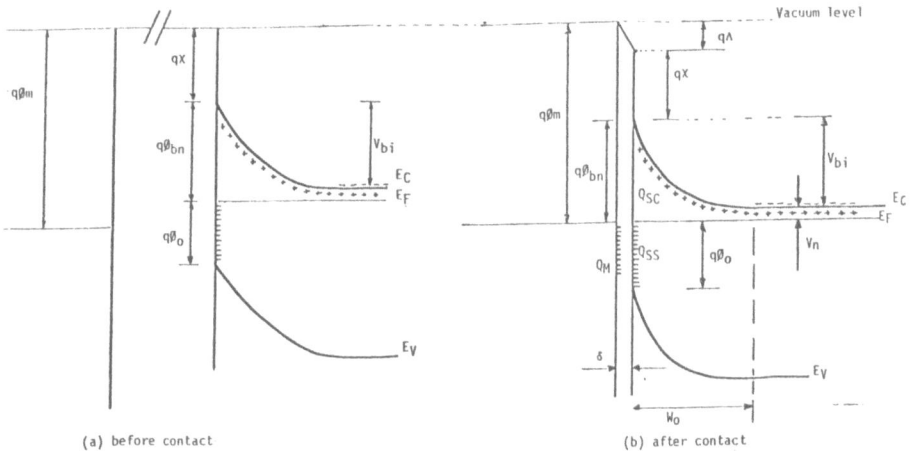

Fig. 1 Formation of a Schottky barrier between a metal and an n-type
semiconductor when the Fermi level is pinned by a high surface state
density.

Fig. 1 shows the energy band diagram of a p[+]-n junction with
zero bias applied. This band diagram has been drawn according to
the particular case where the Fermi level of the semiconductor is
pinned at the surface by a high quantity of surface states regardless of whether those states are intrinsic [19] or extrinsic [20],
[21], [22]. We introduce here the concept of equilibrium level
ϕ_o which is the level to which the surface states are filled when
the surface is in thermodynamic equilibrium. If their density is
very high, the surface states will be filled by electrons from the
bulk in such a way as to make the Fermi level of the bulk match ϕ_o
at the surface (ϕ_o behaves as a sort of "surface Fermi level").

In this situation, band bending occurs prior to metal deposition. When the metal is deposited, the Fermi level of the semiconductor relative to that of the metal has to fall an amount equal to
the contact potential qΔ(see fig. 1). If then the surface charge
density of the semiconductor is high enough to screen qΔ, the space
charge of the semiconductor remains unaffected by the deposition of
the metal. As a result the barrier height is the difference between
the conduction band minimum at the surface and the equilibrium level
ϕ_o and is therefore independant of the metal work function ϕ_m :

$$\phi_{bn} = E_g/q - \phi_o .$$

This latter situation is expected to happen when gold/(110) GaAs Schottky structures are formed, the deposition of the metal [21] , the prior oxygen adsorption [20] or the presence of surface defects [22] leading to the formation of high densities of extrinsic surface states. For the (100) surface which is of more general use throughout industry, the situation is not so clear [23] and intrinsic surface states in the gap can also account for the pinning of the Fermi level.

When a reverse bias V is applied to the structure, the surface states remain in equilibrium with the metal and they are not affected by changes in the pseudo Fermi level of the semiconductor.

Keeping in mind the above remark, and under the complete depletion approximation in the space charged layer (i.e. neglecting the Debye tail of carriers), double integration of the Poissons' equation leads to express the depleted width W as :

$$W = [2 \, \varepsilon_s (V_{bi} - V - qN_d]^{1/2} \tag{1}$$

where V_{bi} is the built-in bias voltage of the junction, V is the externally applied bias voltage, q is the electronic charge, N_d is the density of ionized impurities and ε_s is the dielectric constant of the depleted semiconductor material.

The reverse biased structure behaves as a parallel plate capacitor whose thickness is W and dielectric constant is ε_s. The capacitance per unit area is therefore :

$$C = \varepsilon_s/W = [q \, \varepsilon_s \, N_d / 2 \, (V_{bi} - V)]^{1/2} . \tag{2}$$

This capacitance decreases monotonically with increasing reverse bias voltage until electrical breakdown occurs (due to either impact ionization or tunneling through the forbidden gap).
Relation (2) can be written as :

$$C^{-2} = 2(V_{bi} - V) / q\varepsilon_s N_d \tag{3}$$

and, if N_d is constant throughout the depleted region, a straight line results from plotting C^{-2} versus V. From the intercept with the voltage axis, V_{bi} can be determined and the barrier height is derived from :

$$\phi_{bn} = V_{bi} + V_n \tag{4}$$

where V_n is (see fig. 1) the depth of the Fermi level below the conduction band. In all those derivations, the effect of image-force barrier lowering has been neglected.

Another property of importance for a Schottky junction is its current-voltage characteristic. The forward current per unit area of a Schottky diode obeying thermionic emission is given by :

$$I = I_s [\exp(qV/nkT) - 1] \qquad (5)$$

where V is the applied forward voltage and n the ideality factor. I_s is the extrapolated reverse saturation current and is related to ϕ_{bn} by :

$$I_s = A^* T^2 \exp[-(q\phi_{bn}/kT)] \qquad (6)$$

where A^* is the modified Richardson constant. As it appears in relation (6), from the extrapolated value of current density at zero voltage (I_s), the barrier height can be obtained as :

$$\phi_{bn} = kT/q [\ln(A^* T^2/I_s)] . \qquad (7)$$

2.2 With shallow and deep levels

As outlined earlier, a deep level is defined as an electronic energy level introduced deep into the forbidden gap. It results from a perturbation of the bonding structure of the host material by the presence of a lattice defect or impurity.

Generally deep levels are not fully ionized. Their charge state can be modified by electrical fields, optical illumination and temperature variations (this property is the basis of their determination).

A deep level in neutral material can behave either as a trap or a recombination center depending on the relative magnitudes of electron (hole) capture and emission rates. For example, in n-type material, as is the case we are dealing with, when an electron captured on a defect site stays there until it is re-emitted back to the conduction band, we speak of a majority carrier trap. On the other hand, if a hole is captured on the same site before the electron re-emission, then the center is classified as a recombination center.

2.2.1 Trap occupation statistics

According to the Schockley-Read model [24] a deep level can experience the following four processes :

568

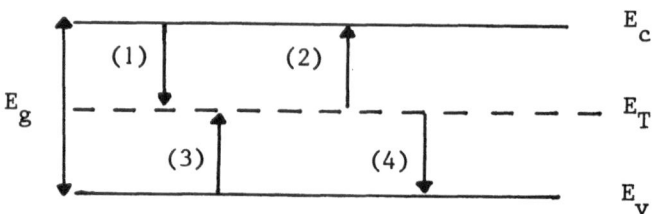

_(1) Electron capture from the conduction band.
(2) Electron emission to the conduction band.
(3) Hole capture.
(4) Hole emission.

For the sake of simplicity, we shall only consider the case of an electron trap T_D.
The energy level under the edge of the conduction band and the concentration (assumed to be constant) of T_D are respectively E_T and N_T.
A first relation we can write is :

$$N_T = N_{T^o} + N_{T^+} \tag{8}$$

where N_{T^o} is the concentration of occupied traps and N_{T^+} the concentration of ionized traps. (We suppose that the trap can only display two charge states, namely positive when the trap is empty and neutral when an electron is captured). If n_T is the number of trapped electrons, we have : $n_T = N_{T^o}$ (9) on the one hand, and :

$$n_T/N_T = f(E_T) = 1 + [\exp - \{(E_F - E_T)/kT\}]^{-1} \tag{10}$$

on the other hand, the occupation of the trap level E_T being described in terms of Fermi–Dirac statistics.
We now consider the reaction between the free electron gas and the traps :

$$e^- + T_D + \underset{e_n}{\overset{C_n}{\rightleftarrows}} T_{D^o}. \tag{11}$$

The kinetics of this reaction is described by :

$$dn/dt = - dn_T/dt = e_n N_{T^o} - C_n n N_{T^+} , \tag{12}$$

n being the free electron concentration.

Under thermodynamic equilibrium conditions (in neutral material) :

$$dn/dt = dn_T/dt = 0 \; .$$

Hence, keeping equation 9 in mind, e_n can be written as :

$$e_n = C_n \times n \; N_{T^+}/n_T \; . \tag{13}$$

From equations 8 and 13 we can derive :

$$N_{T^+} = N_T \; [\; 1 - f(E_T)] = N_T \; f(E_T) \; \exp[- (E_F - E_T)/kT \;]$$

We now describe the conduction band electrons in terms of Boltzmann statistics :

$$n = N_c \; \exp[-(E_c - E_F)/kT \;]$$

where N_c is the effective density-of-states of the conduction band.

Under the above conditions, (13) reduces to:

$$e_n = C_n \; N_c \; \exp[- \Delta E_T/kT \;] \tag{14}$$

where $\Delta E_T = E_c - E_T$.

If now we express C_n, the capture rate for free electrons as :

$$C_n = \sigma_n \; V_{th} \; ,$$

where σ_n and V_{th} are respectively the capture cross-section and the thermal velocity of the free electrons, the final form of e_n will read as :

$$e_n = \sigma_n \; V_{th} \; N_c \; \exp[- \Delta E_T/kT \;] \; . \tag{15}$$

(In the above derivation, the degenerescence factor g of the trap level has been taken equal to unity).
In fact, ΔE_T represents the free energy of ionization of the defect level which formally can be expressed as :

$$\Delta E_T = \Delta H_T - T \; \Delta S_T \; .$$

2.2.2 Capacitance Transients

Fig. 2a shows a Schottky structure built on an n-type material containing a trap level E_T. Under quiescent reverse bias V_R, all the traps above the Fermi level are empty ($x < W_1$) and generation recombination processes determine the occupation of the traps below the Fermi level ($W_1 < x < W_0$). If a positive bias pulse V_0 is now superimposed to V_R, the pseudo Fermi level for electrons in the

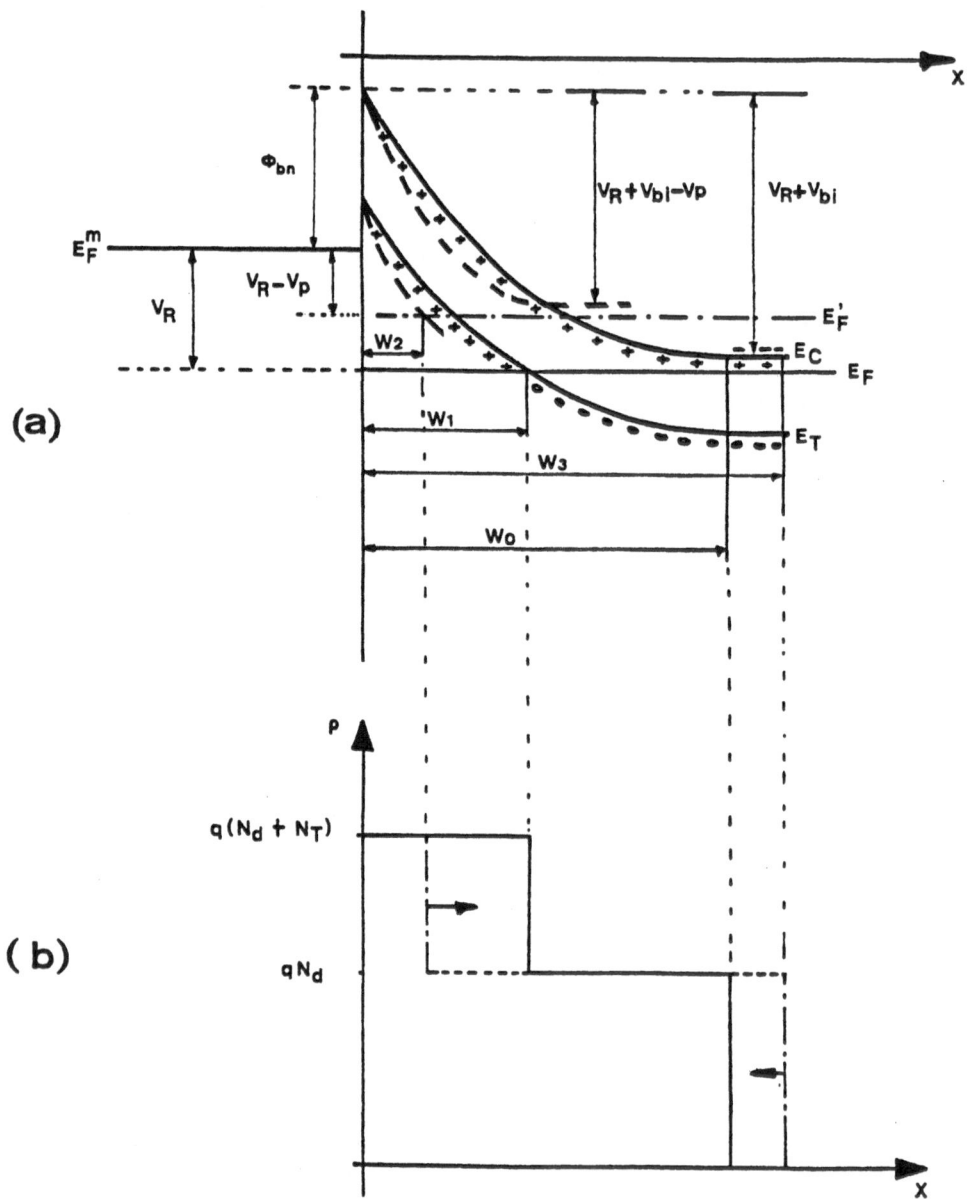

(a)

(b)

Fig. 2a   Energy band diagram of a Schottky structure with one deep donor-like level E_T. The dashed lines indicate band position at the end of the application of a positive bias pulse of magnitude V_p.
Fig. 2b   Charge density distribution in the depletion region. The arrows indicate the evolution of charge density from the end of the application of the positive bias pulse.

semiconductor will be raised, and the quiescent depletion width W_o will decrease. An amount of carriers brought into the depletion region by the positive pulse will be trapped on those empty levels between W_1 and W_2 which are now under the Fermi level. As the positive voltage is turned off the bias voltage of the diode returns to V_R, but (i) the depletion width is now increased, because, as electrons from the bulk have been trapped between W_1 and W_2, the positive charge due to uncompensated ionized donnors extends deeper in the semiconductor material (up to W_3) and (ii) the filled traps within W_2 and W_1 now emerge above the Fermi level. Consequently, reaction (11) (see Section 2.2.1) will be shifted in the direction of electron emission within the region between W_1 and W_2. Equation 12 may now be written as :

$$dn_T/dt = - e_n \, n_T \quad , \tag{16}$$

because in the depleted region the free carrier concentration has been rendered so small that capture processes are negligible. Eq. 16 canreadily be integrated to give :

$$n_T = n_{T^o} \, \exp(- e_n \, t) \tag{17}$$

where n_{T^o} is the number of traps that have been filled during the application of the positive pulse.

We will suppose in the following derivations that the pulse is long enough to fill all the traps within W_1 and W_2, so that $N_T = n_{T^o}$.

According to the above description, the depletion width of the diode will relax from its initial value just after the positive pulse (namely W_3) to the equilibrium quiescent value W_o where all the traps between W_1 and W_2 are empty. The physical reason for this is that electrons emitted from the filled traps between W_1 and W_2 (according to eq. 17) will be swept out of the depletion region and will compensate ionized impurities at the edge of this depletion region. If we suppose N_d and N_T constant and if $N_T \ll N_d$, then it can be shown [25,27a] that the depleted width relaxes exponentially with a time constant e_n^{-1}. The induced relative capacitance change (see fig. 3) is then expressed as :

$$|C(t) - C(o)| \, / \, C_o = \Delta C(t)/C_o = \Delta C(o)/C_o \, \exp(-e_n t) \tag{18}$$

where $C_o = \varepsilon_s/W_o$ is the high frequency capacitance of the junction and $\Delta C(o)$ the maximum capacitance change at $t = o$ (fig.3). $\Delta C(o)/C_o$ is related to the trap concentration by the relation [26, 27a] :

$$\Delta C(o) \, / \, C_o = N_T \, (W_1^2 - W_2^2)/2 \, N_d \, W_o^2 \quad . \tag{19}$$

572

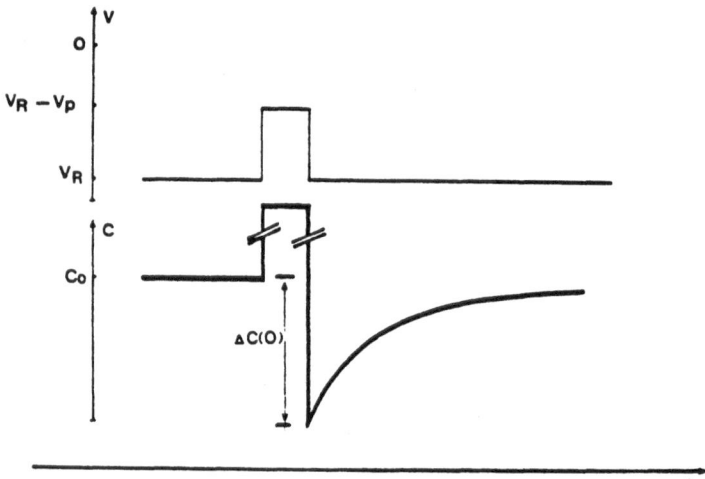

<u>Fig. 3</u> Capacitance transient resulting from the application of a
positive bias pulse V_p.

If, as supposed earlier, $N_T \ll N_d$, then relation (1) remains unaffected
and :

$$W_o = [\, 2\varepsilon_s(V_{bi} + V_R)/q\, N_d \,]^{1/2} \, ,$$

$$W_1 = W_o - [\, 2\varepsilon_s(E_F - E_T)/q^2\, N_d \,]^{1/2}$$

and :

$$W_2 = [\, 2\varepsilon_s(V_{bi} + V_R - V_p)/q\, N_d \,]^{1/2} - [\, 2\varepsilon_s(E_F - E_T)q^2\, N_d]^{1/2} \, .$$

For a positive voltage pulse which zero biases the junction (magnitude $V_R + V_{bi}$), then relation (19) is reduced to $\Delta C(o)/C_o = N_T/2N_d$,
which is the most generally used formula [16, 27] .

According to the above relations, recording of $C(t)$ at a fixed

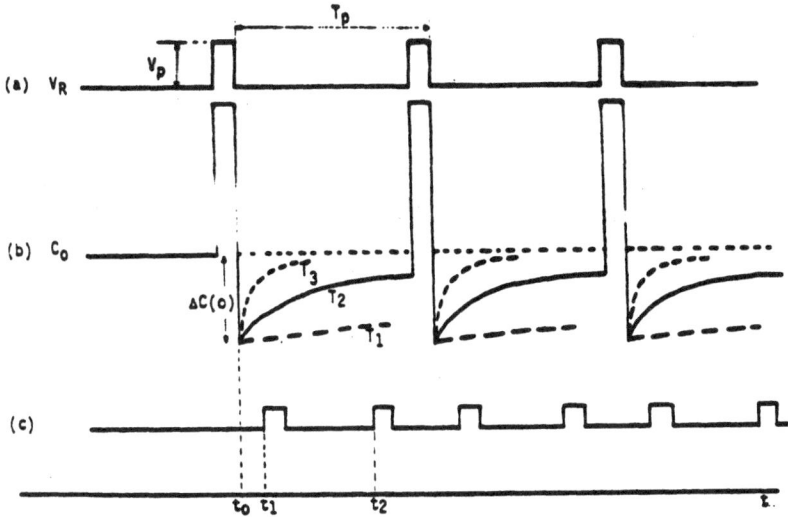

Fig. 4 Principle of D.L.T.S. signal generation :
(a) Trap filling pulse,
(b) Capacitance transient,
(c) Double box-car timing sequence.

temperature allows the determination of the trap concentration N_T.

Furthermore, by plotting $\text{Ln}|C(t)|$ versus t, a straight line results whose slope is e_n. The complete expression of e_n has already been derived (relation (15)) and, keeping in mind that N_c is proportional to $T^{3/2}$ and V_{th} to $T^{1/2}$, relation (15) can be written as :

$$e_n = BT^2 \sigma_n \exp[-\Delta E_T/kT] . \qquad (20)$$

Hence, as the temperature is varied, different values of e_n can be calculated and, from the plot of $\text{Ln}(e_n/T^2)$ versus $1/T$ the activation energy ΔE_T along with the capture cross-section σ_n can be obtained.

However, this method is time consuming, poorly accurate and does not provide ease of operation.

Emission rate and capture cross-section determinations can be made by a much more convenient filtering operation which, when first

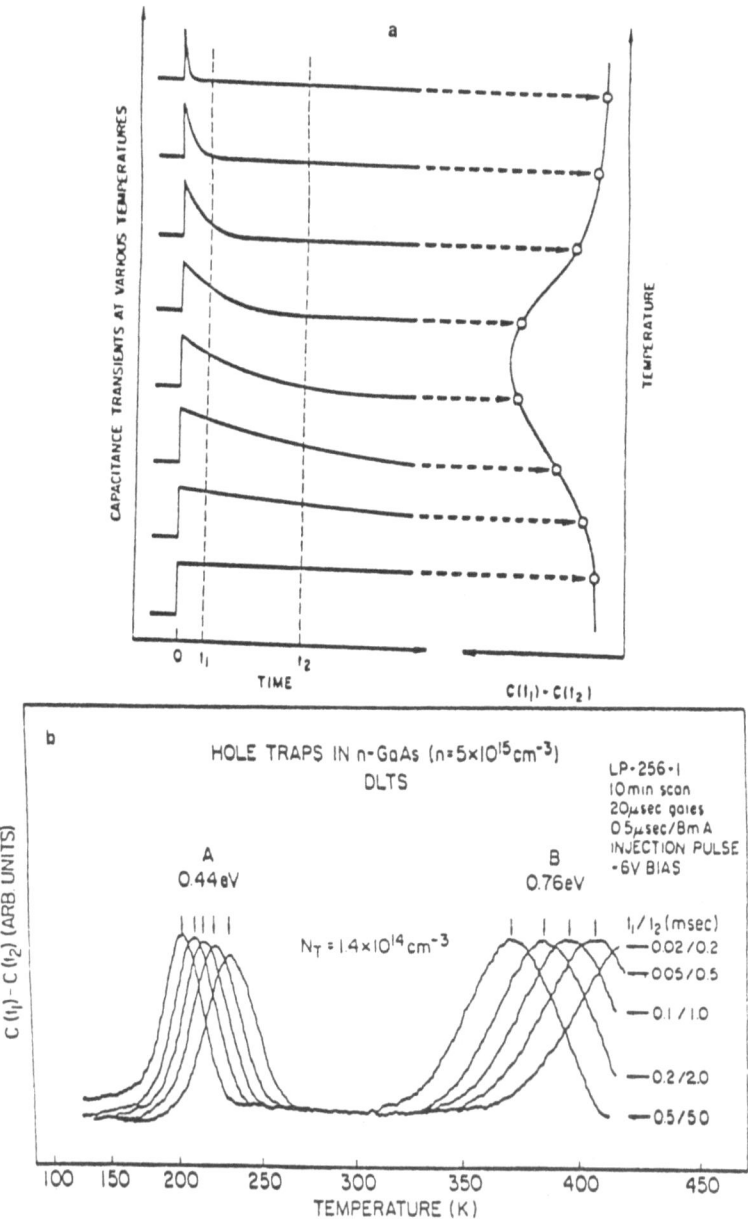

Fig. 5 (After LANG [27]).
a) Principle of generation of a D.L.T.S. Spectrum,
b) Principle of activation energy determination from window rate variation and thermal scans.

proposed by Lang in 1974, was called Deep Level Transient Spectroscopy [27].

2.2.3 Deep Level Transient Spectroscopy (D.L.T.S.)

This technique consists in bringing the emission rate e_n of a trap within a rate window, fixed by the measurement apparatus, by simply varying the temperature of the sample.

In D.L.T.S. technique the positive bias pulses (V_p) are repeatidly applied with a period T. and the capacitance values are measured at times t_1 and t_2 within the pulse period T_p, as the temperature is varied say from LN_2 temperature to 450-500 K.

Fig. 4 illustrates how the D.L.T.S. signal (namely $C(t_1) - C(t_2)$) is obtained from the differential output of a double box-car with input gates set at t_1 and t_2.

Fig. 5(a) illustrates how the D.L.T.S. signal $C(t_1) - C(t_2)$ varies with temperature. At low temperature (T_1 on Fig. 4) e_n^{-1} is large; emptying of traps will be very long and consequently the difference between $C(t_1)$ and $C(t_2)$ will be small. At high temperature (T_3 on Fig. 4), e_n^{-1} is small, emptying of traps will be accomplished within a very short time and once again $C(t_1) - C(t_2)$ is small.

Qualitatively, it is clear that $C(t_1) - C(t_2)$ goes through a maximum, when e_n^{-1} is of the order of $t_2 - t_1$. This is what we will now demonstrate.

From equ. 18 we can write :

$$C(t_1) - C(t_2) = \Delta C(o) [\exp(-e_n t_2) - \exp(-e_n t_1)] .$$

The maximum of the D.L.T.S. signal is expected to happen when :

$$d[C(t_1) - C(t_2)]/de_n = 0$$

and therefore at the maximum :

(i) $e_{no} = (t_2-t_1)^{-1} \times Ln(t_1/t_2)$, $\qquad\qquad$ (21)

(ii) $C(t_1) - C(t_2) = \Delta C_{max} = \Delta C(o)[\exp(-e_{no} t_2) - \exp(-e_{no} t_1)]$,

$\qquad\qquad\qquad\qquad\qquad\qquad\qquad\qquad\qquad\qquad\qquad\qquad$ (22)

(iii) from relation (20) e_{no} can also be expressed as :

$$e_{no} = B T_m^2 \sigma_n \exp(- \Delta E_T/kT_m) \qquad\qquad (23)$$

where T_m is the temperature at which the maximum occurs.

C_{max} and T_m are directly measured on an XY recorder, the D.L.S.T. signal being displayed on the Y channel and the temperature signal on the X one.

$C(o)$ can easily be derived from relation (22) and, as already discussed in section 2.2.2, N_T is obtained from relation (19). It is worth pointing out that C_o is the value of the capacitance at the temperature T_m of the maximum.

Inspection of relations (21) and (23) shows that at each couple of values of t_1 and t_2 corresponds a couple of values of T_m and e_{no}.

Therefore, by changing the window rate, (namely t_1, t_2 and t_2-t_1) and according to relations (21) and (23), a plot of $Ln(e_n/T^2)$ versus T^{-1} results in a straight line whose slope is $\Delta E_T/k$, assuming that σ_n is not temperature dependant. Furthermore, from the intercept with the Y axis, σ_n can also be determined.

In all the above derivations, we have considered only one majority carrier trap level, but of course the results can be extended to the more general case where several majority carrier traps are present. As $\Delta C(t) = \Sigma \Delta C_i(o) \exp -(e_{ni}.t)$, the resulting D.L.T.S. spectra will show several maxima for $\delta [C(t_1) - C(t_2)]/\delta e_{ni} = 0$ and, as previously discussed, by varying the window rate and thermally scanning the sample, the activation energy, the trap concentration and the capture cross-section of each level can be obtained. It is in that sense that the method is spectroscopic.

We did not speak up to now of minority carrier traps since only Schottky structures have been considered. However, when n-p diodes are allowed to be manufactured, minority carriers can be injected in the n-type material and minority carrier traps can be detected (Fig. 5b).

Trap profiling can also be performed by step reduction of V_R (in this case, the smaller both the positive pulse V_p and the step reduction in V_R, the more precise resolution depth is obtained). A last thing to be pointed out is that, when the depletion width is W_o (see Fig. 2a), the trap concentration is determined around W_1 if V_p is small. For graphic display of the trap concentration versus depth, this fact has to be taken into account and, as already discussed, if $N_T \ll N_d$ then :

$$W_o - W_1 = [2 \, \varepsilon_s (E_F-E_T)/q^2 \, N_d]^{1/2} .$$

The analysis of the periodic transients have been presented here as performed by a double box-car. Alternative means of filtering have been reported (lock-in amplifier, exponential correlator...) and for a review of the topic the reader is referred to a recent article of Crowell and Alipanahi [28] .

3. EXPERIMENTAL RESULTS

The results presented here were obtained using either a Q-switched Nd-YAG laser (pulse duration 20ns) or a Q-switched Ruby laser (pulse duration 25ns and $\lambda=0.69$ µm). The beam diameters were of the order of 6 mm.

The Nd-YAG laser operated either in the green ($\lambda = 0.53$ µm) or in the infrared ($\lambda= 1.06$ µm) or in a mixture of the two. After the output of the frequency doubler, the green radiation (when used alone) was separated from the infra-red by the use of two dichroic filters. The energy in the laser spots was made uniform through the use of either a light pipe (Nd-YAG laser) or a diaphragm (Ruby laser).

The substrates were n-type doped (either Te or Si, around $5\times10^{16}/cm^3$) GaAs with (100) orientation.

After irradiation, the Schottky diodes ($\phi= 340$ µm) were fabricated by evaporating a front Au contact through a mask. The back-side AuGe contact was sputter-deposited and subsequently alloyed at 400°C during 10 mn in flowing H_2.

When implanted material was under examination, the implantations were performed with Ga and As ions at the same dose which does not modify the doping impurity concentration nor the stoichiometry In this case, the initial electrical properties of the substrate must be restored by the annealing (in particular the carrier concentration).

3.1 I(V) and C(V) measurements

In an effort to clarify the situation regarding defect introduction after laser irradiation, an attempt has been made to separate surface effects from bulk effects. This separation may seem somewhat arbitrary but as we will see, at low deposited energy densities, laser effects are primarily typical surface effects. This will be illustrated by a set of two irradiations performed in green (20%) + infra-red (80%) at a total deposited energy density of respectively $0.5J/cm^2$ and $1J/cm^2$.
Fig. 6 shows the I(V) characteristics for three groups of diodes :
(i) A first group irradiated at 480 mJ/cm^2,
(ii) A reference group built on virgin material with ideality factors between 1 and 1.1,
(iii) A third group irradiated at $1J/cm^2$.

578

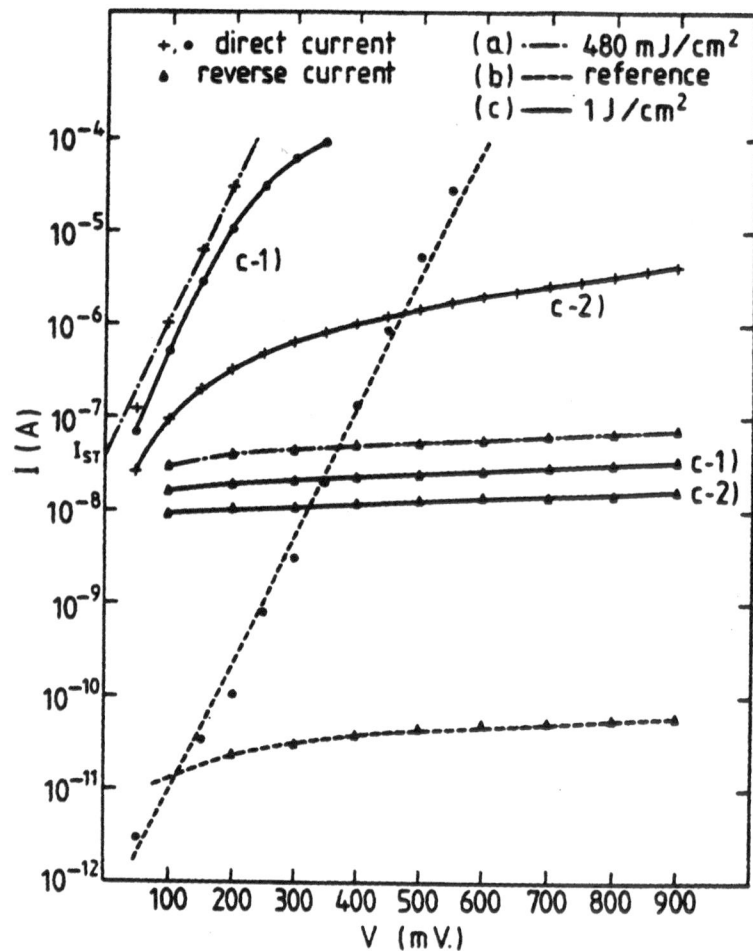

Fig. 6 I(V) characteristics of virgin and laser irradiated Schottky devices :
a) Laser irradiated (green + infra red (80%) : 480 mJ/cm^2),
b) Reference unirradiated sample,
c) Laser irradiated (green + infra red (80%) : 1J/cm^2).
From I_{ST}, barrier heights of 0.65 and 0.95 V are deduced respectively for the 0.48 J/cm^2 irradiated and reference samples.

A first thing to be pointed out is that, despite the use of a light pipe to homogenize the beam, there is an important scattering in experimental observations for the group of diodes irradiated at 1J/cm² (curves labelled C-1 and C-2). This may be attributed to some speckle effects in laser energy distribution.

Apart from that fact, it can be noticed that for a given voltage, the reference devices present some orders of magnitude lower currents than the ones irradiated at 480 mJ/cm² while exhibiting the same slope.

As shown in Fig. 6 and as will be confirmed later, this is a consequence of barrier lowering effects following laser irradiation. From Fig. 6, barrier heights of 0.95 V and 0.65 V are calculated (according to relation (7)) respectively for the reference and low energy irradiated devices.

For the high energy irradiated samples, no quantitative data regarding barrier height can be extracted from the I(V) plots. The I(V) behavior for this group of samples is found to be identical to some series resistance effect and a high resistivity layer results from laser irradiation. If not quantitative, these plots qualitatively show that the barrier height for the high energy irradiated samples is more of the order of 0.6 V than of the order of 0.9 V.

This is consistent with the idea according to which the barrier height is closely related to the surface structure through surface states. As far as melting occurs, the re-solidified surface structure (different from the original one [28a]) and consequently the barrier height should not depend on laser energy density. This simple situation is however further complicated by As and Ga losses from surface at high energy density [28b] which can induce additional surface defects.

The values of barrier heights (deduced from relation (4)) for reference and low energy irradiated samples are confirmed in Figs. 7 and 8 from the C^{-2} versus V plots. Another thing to be noticed from the examination of the latter figures is the little dependence of the C(V) and C^{-2}(V) plots on the measurement frequency (10KHz and 1MHz), indicating that few deep levels are present in the layers irradiated at 480 mJ/cm². (Indeed surface change had been evidenced following laser irradiation by condensation of water vapor on the samples which reveals the laser impact).

A completely different behavior is found to occur when the high energy irradiated devices are examined. A good straight line results from plotting C^{-2} versus V, but large apparent barrier height are found, (see Fig. 9) which can be interpreted by considering the existence of an insulating layer near the surface (at a measurement frequency of 1MHz) [29,30]. The measured capacitance C_M is

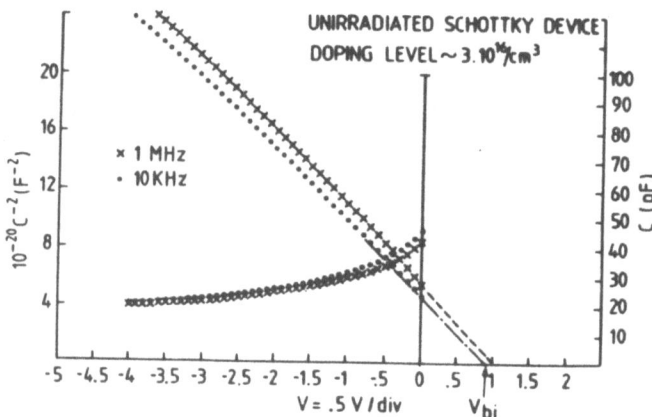

Fig. 7 C(V) and C^{-2}(V) characteristics of un-irradiated Schottky device. The barrier height determined from V_{bi} is of the order of 1V.

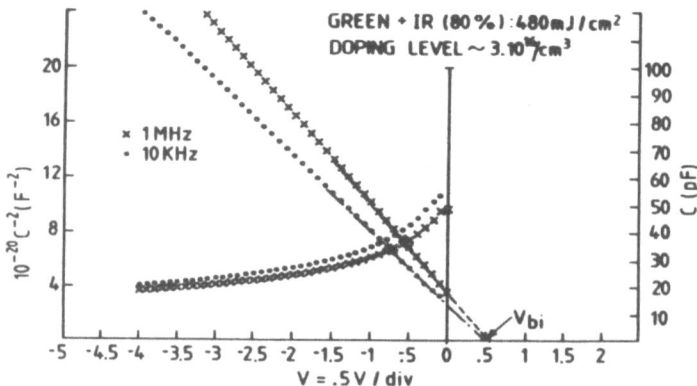

Fig. 8 C(V) and C^{-2}(V) characteristics of low energy irradiated Schottky device. The barrier height determined from V_{bi} is of the order of 0.65 V.

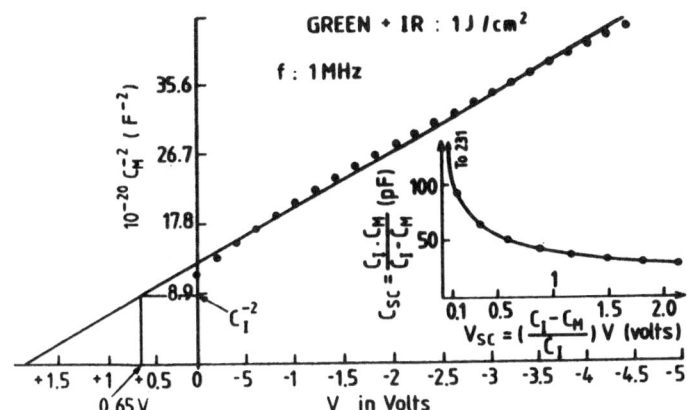

Fig. 9 $C^{-2}(V)$ characteristic of high energy irradiated Schottky device. The calculated insulating capacitance is C_I= 28pF, and the corresponding insulating thickness is ~3800 A. The plot $C_{sc}(V_{sc})$ is the characteristic of the diode corrected for the contribution of the insulating layer (see text).

such that $C_M^{-1} = C_i^{-1} + C_{sc}^{-1}$ since the capacitance C_i of the insulating layer is in series with the capacitance C_{sc} of the depleted semiconductor. The capacitance C_i can be determined from the C^{-2} (V) plot since, for a voltage equal to the built-in potential (around 0.6 V, according to the I(V) characteristics), C_{sc} is infinite and $C_M^{-1} \sim C_i^{-1}$. From the knowledge of C_i, the thickness of the insulating layer can be derived and the exact characteristics C_{sc} (V_{sc}) are obtained by taking out the contributions of C_i in the measured values C_M and V (see Fig. 9). Hence the carrier concentration profile behind the insulating layer can also be evaluated [30] .

A first conclusion can be derived from these results : (i) at low energy density, laser irradiation induces surface re-organization and barrier lowering and (ii) at high energy density a semi-insulating layer (at a measurement frequency of 1MHz) is created in the sub-surface melted region. The initially present carriers (5×10^{16}/cm^3) are compensated in this layer, indicating a trap density greater than 5×10^{16}/cm^3. Davies et al.[30a] using differential Hall measurements on highly doped epitaxial GaAs layers (~10^{18}/cm^3) also found substantial carrier loss induced by pulsed e-beam irradiation (pulse duration~50ns).

The transition from typical surface effects to bulk effects is of course gradual. However, for practical considerations, we have defined laser energy thresholds for the manifestation of bulk effects. The threshold for a particular incident laser wavelength is defined as the deposited energy density at which the $C^{-2}(V)$ plot (measured at 1MHz) give an apparent barrier height equal to the one measured on a virgin sample (~0.95V).

This barrier height of course is only apparent, because the $C^{-2}(V)$ plots convolutes the effect of barrier lowering along with the effect of the series semi-insulating capacitor. However, below the threshold, the laser effects will be considered as surface effects, whereas above the threshold they will be considered as bulk effects.

Table I gives the energy thresholds experimentally determined for green, ruby and the mixture of green + infra-red wavelengths.

Table 1

Wavelength	Threshold energy density
Green (0.53 μm)	250 mJ/cm^2
Ruby (0.69 μm)	350 mJ/cm^2
Green+I.R. (80%) (0.53 μm + 1.06 μm)	650 mJ/cm^2

Above the threshold the thickness of the insulating layer is found to increase with deposited energy density. As the defects are thought to be introduced in the re-solidified melted layers, this observation is consistent with the fact that the melted thicknesses increase with deposited energy densities [31], [32].

For implanted material, interpretation of C(V) data can be made using the same model of insulating series capacitor [29]. The thickness of the implantation induced insulating layer is found to be about 4 times larger than the (Rp + ΔRp) of the implanted species, indicating that the defect implantation tail extends deep in the material, (see Fig. 10).

A partial recovery is observed after ruby laser irradiation, but an insulating layer is still present as indicated in the $C^{-2}(V)$ plots by the large intercept on the voltage axis and the initial carrier concentration is not restored in the implanted surface

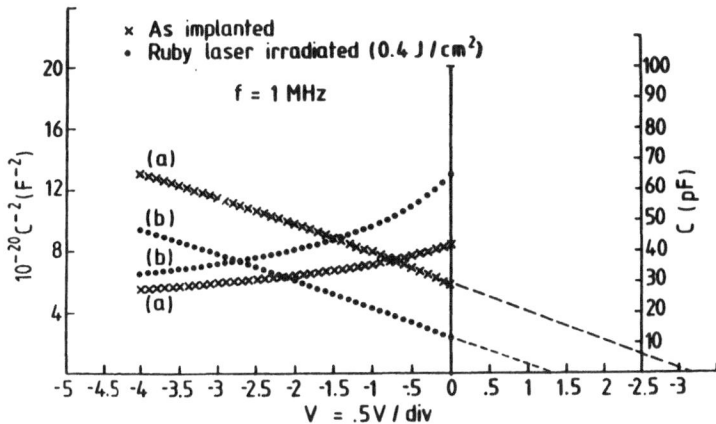

Fig. 10 C(V) and C^{-2}(V) plots on self implanted GaAs (150KeV implantation of Ga + As ions to a dose of 5×10^{11}/cm^2 each, at 200°C. Approximate Rp and ΔRp are respectively 570 Å and 210 Å [32a]).
a) As implanted : an insulating layer of the order of 2300 Å is calculated from the C^{-2}(V) plot;
b) After ruby laser irradiation (0.45J/cm^2) : Partial recovery is observed but an insulating layer of thickness~1500 Å still remains at the surface. The approximate melt depth is 2000 Å [32] .

region. Within the frame of the melting model, it seems that the liquid phase does not extend deep enough in the material to whipe out the memory of the implantation damage.

3.2 D.L.T.S. measurements

Deep level spectroscopy is performed at 1MHz. As a result, if at that frequency an insulating layer exists in the sub-surface region, defect spectroscopy will only be operative behind this insulating layer. Moreover, for slightly damaged layers, the investigated depths generally start from the zero bias width of the depleted layer, which varies with doping. To avoid capacitor breakdown at small reverse bias associated with high doping levels an upper limit of $\sim 2 \times 10^{17}$/cm^3 is set for the carrier concentration of the starting material. The corresponding limit for the zero bias width is ~1000 Å. Consequently, the very sub-surface layer in which defect concentration after laser irradiation is thought to be maximum is not investigated. The use of forward bias voltage

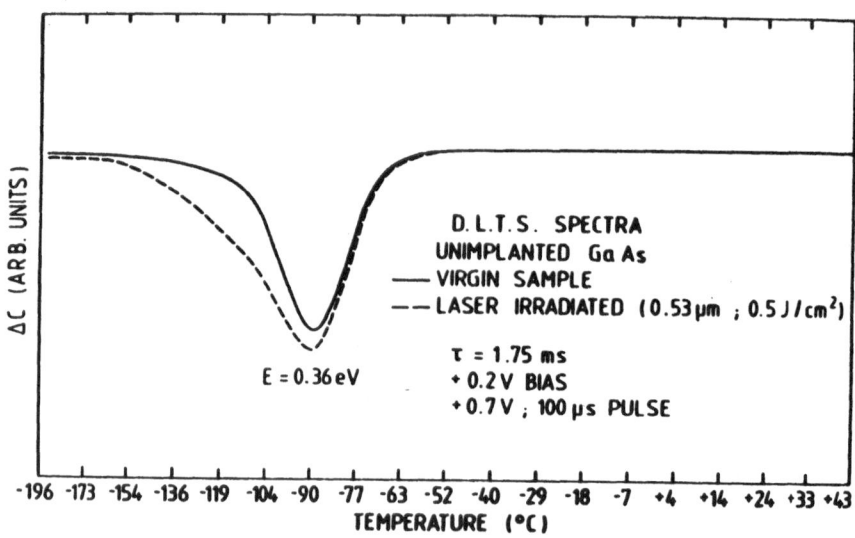

Fig. 11 D.L.T.S. Spectra in forward bias.
a) Virgin GaAs sample showing the E(0.36eV) level, $(N_T \sim 8 \times 10^{15}/cm^3)$;
b) Laser irradiated sample (green : 0.5J/cm²);
Note that the laser-induced peak (activation energy <0.36 eV) cannot be separated from the 0.36 eV center present in the starting
material.

pulses to reduce the depleted width is rendered difficult because of the presence of important shunt leakage currents in irradiated devices [14,30]. Fig. 11 shows the differences between two DLTS spectra recorded on virgin and high energy (green : 0.5 Jcm^{-2}) irradiated devices. Surprisingly, small laser-induced defect concentrations ($\sim 5 \times 10^{15} cm^{-3}$) [14, 29, 30] are found in the investigated layers, even after high energy irradiation. This is due to the fact that trap profiling can only be made behind the laser induced insulating layer whose thickness is of the order of magnitude of the melted layer. This point is not in contradiction with the reported annealing - deep in the material, behind the melted layer - of the well known El_2 trap level (0.83 eV) [29,33] when present in the starting material. Effectively, this annealing certainly takes place via a low temperature radiation-enhanced solid phase mechanism which is consistent with the observation of El_2 elimination after ion-implantation only [29].

When ion-implanted layers are irradiated, the D.L.T.S. spectra before and after irradiation show very few changes [29,34]. However for proton-implantation at low energy (30KeV) and low dose ($10^{12}/cm^2$) [35] a significant decrease in the concentration of implantation-induced defects is observed. It is generally pointed out [29], [34], [35], [36] that the implantation defects left after laser annealing are responsible for the lack of electrical activity reported for low implanted doses. It is also pointed out that carrier loss associated with laser irradiation is not important enough to totally compensate layers with peak concentrations in the range $10^{17}-10^{18}$ carriers/cm^3. Nojima [34] stated that the implantation damage (even in laser annealed layers), estimated through the density of induced localized states, saturates around $10^{18} cm^{-3}$. Thus, this point could explain why electrical activation is observed for high implantation doses with peak carrier concentrations higher than $10^{18} cm^{-3}$, whereas for low doses, peak concentrations remain weaker than $10^{18}/cm^3$ leading to a total carrier compensation.

We think here that within the frame of the melting model, which is now the most widely accepted [3a] [37], two cases have to be considered, depending on the extent of the liquid phase as compared to the implantation induced insulating layer, namely :
(i) If the melted thickness is wider than the insulating layer, then the crystal completely loses the memory of the implantation damage and obviously, the non-activation phenomenon for low doses has to be related to laser-induced defects only;
(ii) If the melted layer is thinner than the insulating layer, then both residual implantation damage and laser-induced damage will be active in compensating carriers.

It is clear then that if the extent of the liquid phase is greater than the implantation-induced insulating layer, a necessary condition for low dose activation is fullfilled. However, this condition may not be sufficient since laser-induced defects in the recrystallized layer can totally compensate out the activated dopant impurities.

If the material escapes from melting by a "soft" plasma mechanism [38] then implantation damage is likely to remain within the irradiated layer [34] , even if the "extent" of the plasma phase is larger than the implantation induced insulating layer. But in this case the repetition of the irradiation of the same layer should bring the crystal into a perfect state within a finite number of pulses. This is not observed, even after more than 25 shots of the same energy and the C(V) plots remain unchanged [39] .

4. CONCLUSION

A model was presented allowing the interpretation of I(V), C(V), $C^{-2}(V)$ and D.L.T.S. data recorded on laser irradiated, either virgin or implanted GaAs material.

In virgin material, laser irradiation induces Schottky barrier lowering. At the same time, a semi insulating sub-surface layer is built in the material (measurement frequency : 1MHz and initial carrier concentration $\sim 5 \times 10^{16} cm^{-3}$), whose thickness increases with laser energy density.

The D.L.T.S. data are representative only of the "tail" of the insulating layer and consequently indicate quite low mean defect concentrations ($\sim 5 \times 10^{1} cm^{-3}$). D.L.T.S. might not be the most interesting technique to study highly damaged material.

In implanted material, using $C^{-2}(V)$ measurements, the defects are found to extend much deeper (3 to 4 times) than the $(R_p + \Delta R_p)$ of the implanted species. Within the frame of the melting model and keeping in mind the extension depth of the implantation damage, the extent of the liquid phase in implanted material determines a necessary condition for the annealing of implantation damage even at low doses. This condition may not be sufficient since we do not know what is the ultimate carrier concentration which would be compensated out by laser irradiation itself. However, it has been demonstrated that laser-induced defects are responsible for carrier compensation at least up to $5 \times 10^{16} cm^{-3}$. This would tend to indicate that compensation of low dose first occurs due to laser-induced defects.

Further studies should include (i) irradiations on highly doped material (despite of the problems associated with capacitance measurements) to assess the upper limit for carrier compensation and (ii) low energy implantation (10-50KeV) to produce implantation-induced insulating layers thinner than the extent of the liquid phase at an adequately chosen laser energy density.

ACKNOWLEDGMENTS

The author would like to thank J.C. BOURGOIN for introducing him to D.L.T.S. techniques, J. ICOLE, S. DELAGE and D. DIEUMEGARD are gratefully acknowledged for helpful discussions and suggestions. Thanks are also due to M. CROSET for careful reading of the manuscript and to J. BOURQUIN for her attention and patience in typing the manuscript.

REFERENCES

1. A.I.P. Conf. Proc., Ser. n°50 (1979).
2. "Laser and Electron Beam Processing of Materials" Edited by C.W. White and P.S. Peercy (New York, Academic Press, 1980).
3. "Laser and Electron-Beam Solid Interactions and Materials Processing" Edited by J.F. Gibbons, L.D. Hess and T.W. Sigmon, (New York, North Holland, 1981).
3a. "Laser and electron-beam interactions with solids" Edited by B.R. Appleton and G.K. Celler, (New York, North Holland, 1982).
4. Liu, S.G., Wu, C.P., Magee, C.W. in Ref. 2, page 341
5. Sealy, B.J., Kular, S.S., Stephens, K.G., Croft, R., Palmer, A., Electr. Lett. 14, (1978) 512.
6. Pianetta, P.A., Stolte, C.A. and Hansen, J.L., in ref. 2 , page 328.
7. Lowndes, D.H., Cleland, J.W., Christie, W.H. and Eby, R.E., in ref. 3, page 223.
8. Sze, S.M. and Irvin, J.C., Solid State Electron, 11 (1965) 599.
9. Eisen, F.H. in ref. 2 page 309.
10. Gamo, K., Yuba, Y., Oraby, A.H., Murakami, K., Namba, S. and Kawasaki, Y., in ref. 2 page 322.
11. Anderson, C.L., Dunlap, H.L., Hess, L.D., Olson, G.L. and Vaidyanathan, K.V., in ref. 2 page 334.
12. See for example : Leamy, H.J., J. Vac. Sci. Technol. 18 (2) (1981) 208.
12a. Tandon, J.L., Nicolet, M.A., Tseng, W.F., Eisen, F.H., Campisano, S.U., Foti, G. and Rimini, E., Appl. Phys. Lett. 34 (1979) 597.
13. Nojima, S., J. Appl. Phys. 52 (1981) 7445.
14. Emerson, N.C. and Sealy, B.J., Electron Lett. 16 (1980) 512.
15. See for example : Campisano, S.U., Catalano, I., Foti, G., Rimini, E., Eisen, F. and Nicolet, M.A., Solid-State Electron, 21 (1978) 485.
Also : Golovchenko, J.A. and Venkatesan, T.N.C., Appl. Phys. Lett. 32 (1978) 147.
16. For a review on capacitance transient measurements, see :

Miller, G.L., Lang, D.V. and Kimmerling, L.C., Ann. Rev. Mat. Sci. 377 (1977).

17. Rhoderick, E.H., Metal semiconductor contacts (Oxford, Clarendon Press, 1978).

18. Sze, S.M., Physics of semiconductor devices, (2nd Edition, Wiley-Interscience) 1981.

19. Bardeen, J., Phys. Rev. 71 (1947) 717.

20. Spicer, W.E., Chye, P.W., Garner, C.M., Lindau, I. and Pianetta, P., Surface Sci. 86 (1979) 763.

21. Skeath, P., Su, C.Y., Hino, I., Lindau, I. and Spicer, W.E., Appl. Phys. Lett. 39 (1981) 349.

22. Allen, R.E. and Dow, J.D., Phys. Rev. B. 25 (1982) 1423.

23. Friedel, P., Thesis, Université de Paris (1982).

24. Shockley, W., Read, W.R. Jr., Phys. Rev. 87 (1952) 835.

25. Pons, D., Thesis, Université de Paris (1979).

26. Zohta, Y., Watanabe, M.O., J. Appl. Phys. 53 (1982) 1809.

27. Lang, D.V., J. Appl. Phys. 45 (1974) 3023.

27a. Lannoo, M. and Bourgoin, J.C., Point Defects in Semiconductors, Vol. 2 (Springer Verlag, in press).

28. Crowell, C.R. and Alipanahi, S., Solid State Electron, 24 (1981) 25.

28a. Zehner, D.M., White, C.W., Appleton, B.R. and Ownby, G.W., in ref. 3a page 683.

28b. de Jong, T., Wang, Z.L. and Saris, F.W., Phys. Lett. 90A (3) (1982) 147.

29. Mooney, P.M., Bourgoin, J.C. and Icole, J., in ref. 3 page 255.

30. Pribat, D., Delage, S., Dieumegard, D., Croset, M., Srivastava, P.C. and Bourgoin, J.C., to be presented in M.R.S. Meeting, Boston 1982.

30a. Davies, D.E., Lorenzo, J.P., Kennedy, E.F. and Ryan, T.G., Proceedings of the 1981 International Symposium on GaAs and related Compounds, (OISO, Japan) page 255.

31. Tsu, R., Baglin, J.E., Lasher, G.J. and Tsang, J.C., Appl. Phys. Lett. 34 (1979) 153.

32. Wood, R.F., Lowndes, D.H. and Christie, W.H., in ref. 3, page 231.

32a. Johnson, W.S., Gibbons, J.F., Projected range statistics in semiconductors, Distributed by Stanford University Bookstore (1970).

33. Yuba, Y., Gamo, K., Murakami, K. and Namba, S., Appl. Phys. Lett. 35 (1979) 156.

34. Nojima, S., Journal Appl. Phys. 53 (1982) 5028.

35. Yuba, Y., Gamo, K., Oraby, A.H., Murakami, K. and Namba, S., Nucl. Instr. and Meth. 182-183 (1981) 699.

36. Yuba, Y., Gamo, K. and Namba, S., Proceedings of the 1981 International Symposium on GaAs and Related Compounds, (OISO, Japan) page 221.

37. Rimini, E., these Proceedings.

38. Van Vechten, J.A., Tsu, R. and Saris, F.W., Phys. Lett. A 74 (1979) 442.

Also see : Van Vechten, J.A., in ref. 3a and references there in.

39. Pribat, D., unpublished results.

LASER ANNEALING OF SEMICONDUCTORS STUDIED BY MÖSSBAUER SPECTROSCOPY.

G. Langouche

Institut voor Kern- en Stralingsfysika,
Leuven University, B-3030 Leuven, Belgium.

1. INTRODUCTION

Mössbauer spectroscopy offers a wealth of information on a microscopic scale. The measured hyperfine interaction parameters at the Mössbauer nucleus reflect the bonding symmetry of the atom, the electron density at the nucleus and the coupling strength of the atom to the lattice. These Mössbauer atoms, or their parent atoms decaying to a Mössbauer atom, can be introduced as impurities in semiconductors and serve there as microscopic probes. They then reflect the electronic and vibronic properties of the surrounding lattice and are therefore very sensitive to the characteristics of the lattice site occupied by them.

In this paper we will first deal with Mössbauer spectroscopy as a method in materials research and demonstrate its capacities and limitations. In the second part we will review shortly the few extensive studies that were made on Mössbauer impurities in semiconductors, using the Mössbauer resonances of ^{57}Fe, ^{119}Sn, ^{125}Te and ^{129}I, and we will discuss the effects of laser annealing as observed by Mössbauer spectroscopy.

2. THE METHOD

Many textbooks [1] have been published in recent years explaining in great detail how Mössbauer spectroscopy works, and reviewing the major results obtained by this method. We refer the reader to these books for all details.
We only recall that the discovery by Rudolf Mössbauer in 1958 of

the fact that,when atoms are imbedded in a lattice, a part of these atomic nuclei will emit or absorb gamma radiation without generating lattice vibrations due to their recoil, gave birth to a spectroscopic method with an enormous resolving power. This recoil-free fraction, generally called the f-factor, of atoms can indeed be used in a resonant absorption experiment. This is usually done in the following way : the energy of a gamma transition between an excited nuclear state and the ground level is Doppler-shifted by giving a small relative velocity to this radioactive source with respect to an absorber containing the same isotope. By observing the amount of radiation passing through this absorber as a function of velocity, the resonant absorption curve can be scanned. As the linewidth of such a Mössbauer resonance is only limited by the Heisenberg uncertainty in the energy value of the involved nuclear levels, one can obtain a typical intrinsic resolution of 1 part in 10^{12}.

In order to observe a Mössbauer resonance with an appreciable line intensity, the f-factor must be as large as possible. As f depends on the gamma ray energy, the atomic mass and the temperature, Mössbauer spectroscopy with a measurable effect is limited to nuclear transitions to the ground state with gamma ray energies lower than 150 keV, to mass numbers larger than 40 and,for the large majority of Mössbauer transitions,to temperatures far below room temperature.

3. PARAMETERS OBSERVABLE IN MÖSSBAUER EXPERIMENTS

Three major parameters can be deduced from a Mössbauer spectrum, and all three have their own power in materials research studies. An excellent review article was recently published [2] on the use of Mössbauer spectroscopy for the characterization of solids. We will only summarize the major points.

As mentionned above, the intensity of the Mössbauer resonance is proportional to the *recoil free fraction f*. This factor probes the vibrational properties of the lattice around the Mössbauer atom. The mean square vibrational amplitude of the atoms can be derived and this ,of course, is directly related to the bond strength between this atom and its neighbours. For the case of impurity atoms in semiconductors, it is clear that different lattice sites will have different f-factors. Especially the presence of vacancies associated with the probe atom can drastically lower the f-factor and strong directional effects with respect to the crystal symmetry have been observed.

Due to the extremely high energy resolution of Mössbauer spectroscopy,one can detect nuclear hyperfine interactions induced on the nuclear energy levels by the electromagnetic interaction between this nucleus and the surrounding electrons. The two important interactions for this study are the isomer shift and the electric quadrupole interaction.

The *isomer shift* of a Mössbauer resonance, which is observed as a shift of the Mössbauer line away from zero velocity, is due to the Coulomb attraction between the nuclear charges and the electronic charges that penetrate the nuclear volume. Whenever this Coulomb interaction is different in source and absorber, a slightly different energy distance between excited and ground nuclear states will result and the Mössbauer resonance will be shifted. In practice, whenever the chemical surrounding of the Mössbauer atom is different, a different isomer shift will result. An interpretation of observed isomer shifts is hampered by the fact that the formula [3] for this isomer shift not only contains the electronic term $|\Psi_s(0)|^2 - |\Psi_a(0)|^2$, being the difference in s-electron density at the nucleus between source and absorber, but is also proportional to the nuclear factor $\Delta R/R$, ΔR being the change in nuclear radius R between the excited and the ground state. This nuclear factor is unknown and is most often estimated from semi-empirical approaches using systematics of series of chemical compounds with assumedly known electron configurations. Another complicating factor is the fact that the s-electron density is indirectly influenced by the screening action of p, d, f, ... electrons and the combined effect is often hard to predict especially when volume and pressure effects play a role, as when over- or undersized atoms are introduced in a lattice. The least we can say on impurity atoms in semiconductors is that different lattice surroundings will definitly be characterized by different isomer shifts.

The third important parameter to be derived from Mössbauer spectra is the *quadrupole splitting*, which is due to the perturbation of the nuclear levels by the electric field gradient (EFG). An EFG is a higher moment of the electron distribution which is generated at the nucleus, whenever the surrounding electron distribution is not spherical. The interaction between the nuclear quadrupole moment Q and this EFG will split the nuclear levels into a set of sublevels, the number of which depends on the nuclear spin I. A typical exemple is shown in Figure 1 for an $I = 3/2$ to $I = 1/2$ transition, where the excited state is split into two levels, and the ground state remains unsplit (all $I = 1/2$ levels have $Q = 0$). In combination with an unsplit absorber, such a source will give rise to a doublet in the Mössbauer spectrum. An intrinsic difficulty in the interpretation of Mössbauer spectra becomes immediatly apparent : the Mössbauer spectrum of a source where all the atoms occupy a unique site characterized by a quadrupole splitting is completely identical to the Mössbauer spectrum of a source where two sites are occupied, characterized by different isomer shifts. As will be mentionned later in this review, this ambiguity has often given rise to mis-interpretations of spectra.

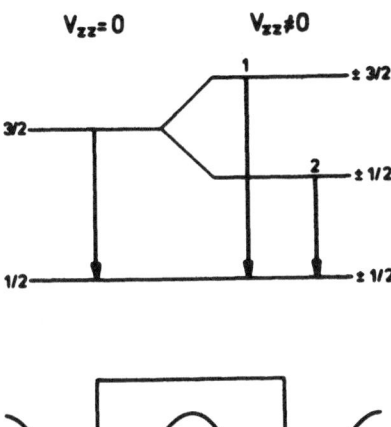

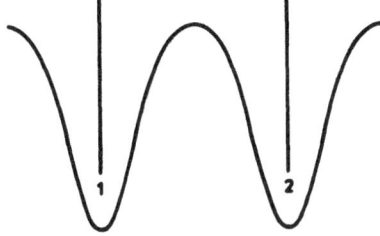

Fig. 1   Quadrupole splitting of a Mössbauer resonance (from [2])

The EFG can directly be extracted from the Mössbauer spectrum, as
the nuclear quadrupole moment is most often available from other
nuclear studies.   This EFG contains the structural information on
the electron distribution around the Mössbauer probe.   Some simple
rules can for example be applied for impurity atoms in semiconductors.   When the symmetry around the Mössbauer atom is cubic the EFG
has to be 0.   In diamond-type lattices both substitutional or regular interstitial impurities have to be characterized by unsplit
Mössbauer lines.   A non-zero EFG appears whenever the surrounding
is lower than cubic, so that sites associated with point defects,
non cubic interstitials sites or even substitutional ones which
are only slightly shifted from their equilibrium position, will be
characterized by their EFG the magnitude of which is mostly hard
to calculate as long as the exact electron configuration is unknown.

4. MÖSSBAUER PROBES

For over one hundred nuclear transitions the Mössbauer effect has been demonstrated so far. Nevertheless, not more than about ten of these have ever been used in the study of impurity atoms in semiconductors, and for only four of these (Table 1) really extensive .studies were made. Several reasons are responsible for this. Especially if one wants to introduce a radioactive parent to the Mössbauer atom as an impurity in the semiconductor host, one needs a parent with a suitable lifetime and a high specific activity. The restrictions are even more severe if one likes to use an isotope separator to implant the impurity atom, as this machine has to be dedicated to the implantation of radioactive isotopes.

In principle, a nuclear Mössbauer transition can be approached from two directions. A first way is to start from the Mössbauer nuclei in their ground state and to use them in an absorber experiment. At first sight this way looks very attractive as the studied sample is not radioactive. But there is a drawback as, due to the limited thickness allowed for such an absorber, the concentration of Mössbauer atoms has often to be higher than one desires. Much lower concentrations can be used if one approaches the Mössbauer transition from the other direction. One needs then a radioactive parent which decays via this Mössbauer transition. This parent can be a long-lived metastable state of the same Mössbauer isotope, as for example $^{125}Te^m$ or $^{119}Sn^m$, but is most often one of the neighbouring isotopes decaying via electron or positron emission. This leads to a very particular situation, which is often confusing for

Table 1 Thoroughly studied Mössbauer isotopes as impurities in semiconductors and the parent isotopes used in these experiments.

Mössbauer Isotope	Parent Isotopes	Ref.
^{57}Fe	^{57}Co	5 to 18
^{119}Sn	$^{119}Sn^m$, ^{119}Sb, ^{119}Te, ^{119}Cd, ^{119}Xe, ^{119}In	18 to 21
^{125}Te	$^{125}Te^m$, ^{125}Sb, ^{125}I	
^{129}I	$^{129}Te^m$	22 to 30

an outsider to the Mössbauer technique. The lattice location of
the impurity atom in the semiconductor is in general inherited from
the long-lived radioactive parent isotope, as the decay process via
the Mössbauer transition proceeds too fast (typical tens or hundreds
of nanoseconds) for any geometrical rearrangement to take place.
The electronic configuration, on the other hand, is typical for the
Mössbauer atom itself as even this short time is largely sufficient
for the electrons to rearrange themselves. So, looking at Table 1,
one really studies Fe on a Co site, I on a Te site, ... This situation can even be more complex when there are intermediate states
involved with long lifetimes, as in some of the ^{119}Sn studies.

5. WHAT CAN BE LEARNED FROM MÖSSBAUER SPECTROSCOPY

The major power of Mössbauer spectroscopy on impurity atoms in
semiconductors is the fact that one really introduces a microscopic
probe which senses the structure of the semiconductor host on an
atomic scale. A very high density of information is revealed by
this probe as the electronic and vibrational properties of the populated site are explored. A very high sensitivity can be obtained
as implanted doses of 10^{10} atoms/cm^2 have been shown to be still
accessible for Mössbauer spectroscopy and the method is also very
versatile as all solid materials, crystalline or amorphous, can be
approached.

Two major drawbacks, however, have seriously hindered Mössbauer
spectroscopy to become a leading spectroscopic method in this field.
First, the limited number of available Mössbauer atoms and the fact
that the major semiconductor dopants are not amongst them. Second,
once a Mössbauer spectrum is obtained, the interpretation of the measured hyperfine parameters in view of a particular microscopic
lattice model is not straightforward. Already the identification
of spectrum components can be ambiguous and has led to erroneous
interpretations. This indirect and complex way of extracting information from Mössbauer spectra has certainly not favored a generalized use of this spectroscopy method.

6. ^{57}Fe MÖSSBAUER STUDIES

More than 90% of all the Mössbauer work is making use of this
resonance because of its excellent resolution and possibility to
be used in a wide range of temperatures, even way above room temperature. It is therefore quite logical that also ^{57}Fe in semiconductors has been studied. Moreover, it has been demonstrated in recent
years [4] that Fe is one of the most common impurities in Si, introduced unintentionally but very easily up to its solubility limit

from external sources. Already in 1961, shortly after the discovery of the Mössbauer effect, studies of ^{57}Fe in Si and Ge were published [5] . Now, more than 20 years later, one has to admit that the Mössbauer spectra of ^{57}Fe in Si and Ge are still not fully understood, in spite of the wealth of available data.

In the search of the Mössbauer resonances corresponding to the most simple Fe configurations in Si, the substitutional and tetrahedral interstitial sites, one soon realized that the major part of the observed Mössbauer resonances are due to Co or Fe precipitates in Si, which are very easily formed in any diffusion experiment [6] in view of the very low solubility and very high diffusion coefficients in Si. Only at high annealing temperatures [7],the typical Co or Fe silicides are recognized while, at lower annealing temperatures lack of the exact local stoichiometry gives rise to a range of different Mössbauer spectra [8] .

Both ^{57}Co and ^{57}Fe were also implanted into Si. A doublet is observed in the Mössbauer spectrum and a controversy grew whether this spectrum is due to the population of two sites [9] or to one single quadrupole split site [10] . An unambiguous answer was obtained both for ^{57}Fe [11] and ^{57}Co [12] implantations, both in favor of the latter interpretation. The exact microscopic origin of this EFG is still not available, but several data [13] show that it belongs to Fe atoms in a surrounding which is typical for amorphous Si. In implantation experiments at lower fluences (10^{10}-10^{14} atoms/ cm^2) a drastic dose dependence is observed [14] leading to two conclusions : first, below the full amorphization limit,part of the implanted atoms occupy a regular lattice site, secondly, the fact that the amorphous doublet persists and even saturates at a constant value well below the amorphization threshold(Figure 2), supports a model in which individual implantation tracks are considered amorphous.

Laser annealing of both ^{57}Co [15] and ^{57}Fe [16] implanted silicon gives rise to complete precipitation at the surface. This is consistent with Rutherford Backscattering work [17] where precipitates at the surface, and for larger amounts of Fe, also in cell walls was observed under the form of FeSi.

Laser implantation [18] by laser irradiation of a deposited ^{57}Fe layer results in a uniform distribution of iron in the surface layer, presumably under the form of FeSi.

7. ^{119}Sn MÖSSBAUER STUDIES

Also the ^{119}Sn resonance is one of the few Mössbauer resonances that can be used at room temperature. It is different from the

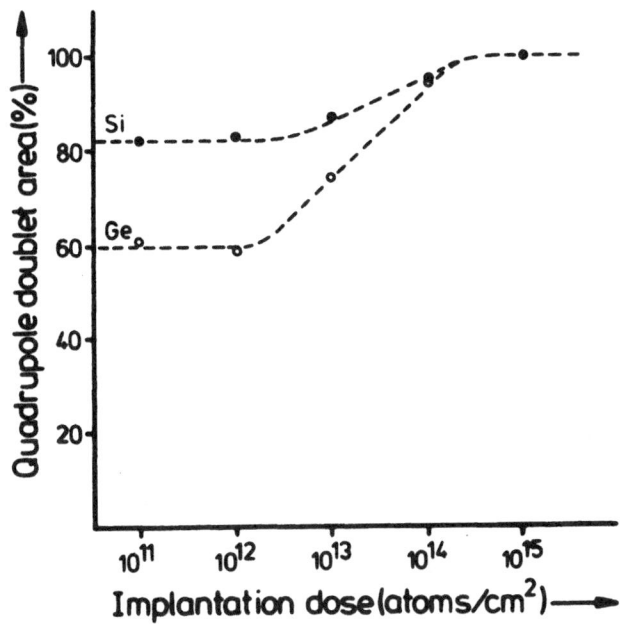

<u>Fig. 2</u> The relative area of the quadrupole doublet which is pro-
portional to the fraction of Co atoms in an amorphous surrounding,
as a function of implantation dose. The saturation behaviour at
low doses supports a model in which individual implantation tracks
are amorphous. (from [14]).

previously mentioned ^{57}Fe resonance, in two major aspects. As the
quadrupole moment of the excited Mössbauer state is rather small,
a quadrupole splitting will only give rise to a small broadening of
the Mössbauer spectrum and is therefore hard to detect. Moreover,
as the nuclear factor $\Delta R/R$ is rather small, the isomer shift scale
is very restricted and Mössbauer lines will be barely resolved.
The fact that this resonance can be approached from many parent isotopes (Table 1) made it the subject of several Mössbauer studies,
mainly by the Aarhus group [19]. While the implantation of $^{119}Sn^m$
gives rise to a 100% population of the substitutional site, several
other Mössbauer lines were discovered when other parent isotopes were
implanted. Based on the registration of hundreds of Mössbauer spectra with various parameters such as the parent isotope, the

implantation temperature, the implantation dose, the annealing temperature, post-implantations and, by comparing with the results obtained with other techniques, most of these Mössbauer resonances could be interpreted. Amongst them are interstitial Sn, Sn with an adjacent vacancy, Sn in the middle of a di-vacancy, Sn associated with oxygen, Sn trapped in dislocation loops and other configurations. The [117]In parent isotope, which as a half life of only 2.4 minutes, could be approached in an on-line production, followed by isotope separation into a Mössbauer spectrometer set up at the ISOLDE facility at CERN [20] . Using this parent and combining with the results of the [119]Sb parent, site-selective doping of III-V semiconductors could be demonstrated [21] .

Laser annealing on [119]Te implanted in Si reveals only substitutional [119]Sn [22] and, also, in laser implantation experiments on a deposited Sn layer [18] only substitutional [119]Sn is observed.

8. [125]Te and [129]I MÖSSBAUER STUDIES

Laser annealing was the determining experiment in the understanding of the Mössbauer spectrum of [125]Te and [129]I in semiconductors. While the [125]Te resonance has only a poor resolution due to the large linewidth, it has the interesting feature that it can be approached from the Te side itself, as well as from Sb and I parent isotopes. The [129]I resonances, on the other hand, offers an excellent resolution, due to the large quadrupole moment and the large $\Delta R/R$ value involved, but can only be approached from the Te parent isotope. Both resonances need low temperatures. An early study [23] after implantation of [129]Te^m in semiconductors was interpreted with about equal populations of a substitutional and interstitial site. Thermal treatment of the sample did not affect these populations. Mössbauer spectroscopy after implantation of [125]Te^m on the other hand [24] was hard to reconcile with this model, as the two [125]Te lines observed in the Mössbauer spectrum were both different from the [125]Te single lines observed after [125]Sb implantation, which was generally believed to be substitutional.

The introduction of laser annealing solved this controversy as after laser irradiation, the as-implanted spectrum of both [129]Te^m and [125]Te^m is replaced by a single line [25] corresponding to a substitutional site, consistent with the just-mentioned [125]Sb Mössbauer spectrum and supported by channeling data [26] .

The Mössbauer spectrum of the as-implanted sample, which is almost not affected by oven annealing, is now interpreted as a quadrupole splitting [27] . This EFG probably originates from threefold coordinated Te atoms, slightly shifted from the substitutional site, presumably towards a vacancy [28] . This site is very stable, in contrast with the site populated after laser annealing as the

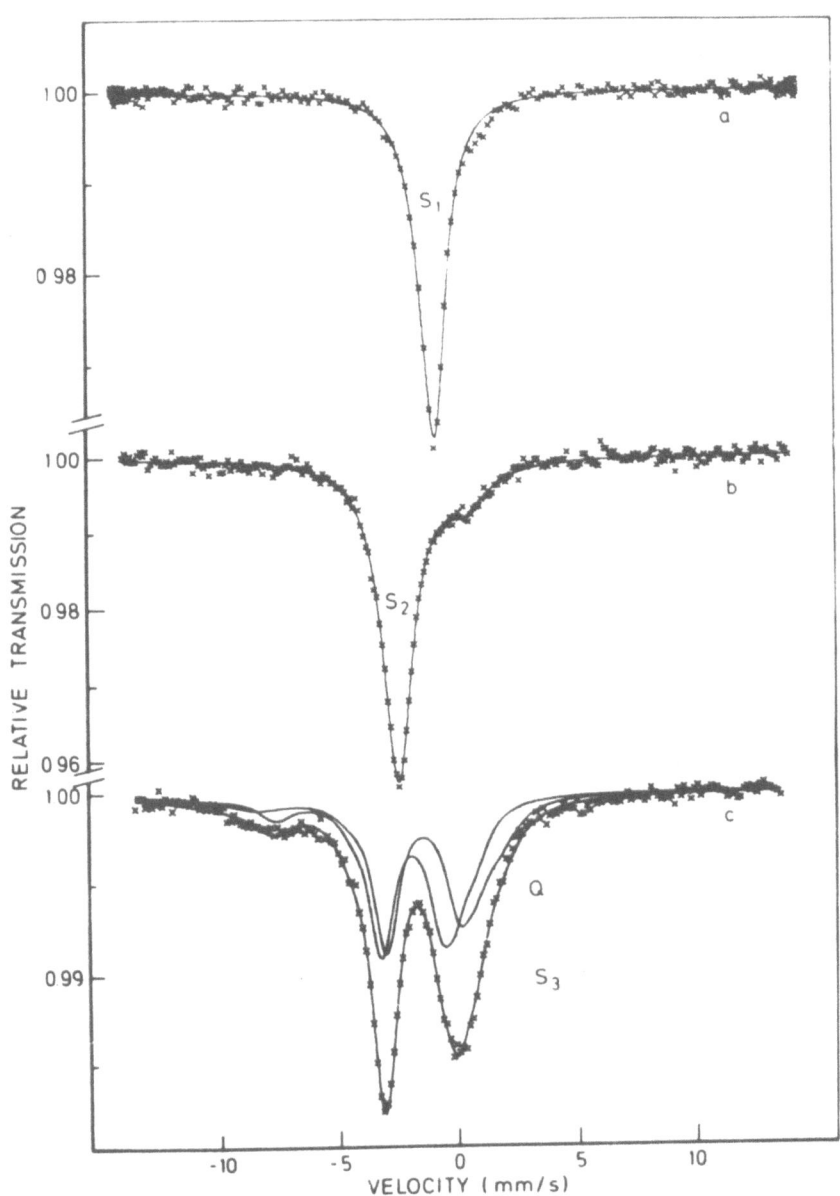

Fig. 3 Mössbauer spectra of substitutional ^{129}I in laser annealed a) heavily p-doped Si, b) compensated Si, c) heavily n-doped Si. (from [22]).

latter one relaxes immediately into the former one after a thermal treatment of the laser annealed sample, a process which can be recycled several times [29] .

A thorough study was recently undertaken by the Groningen group on the laser annealed $^{129}Te^m$ implants in Si [30] in which substitutional Te is formed. It is shown that the isomer shift of the ^{129}I Mössbauer line can take three distinct values (Figure 3) depending on the position of the Fermi-level, probably corresponding to I^0, I^+ and I^{++}.
Moreover, it is found from the Mössbauer spectrum that the I^0 state is dynamically Jahn-Teller distorted below 40 K where the I atom jumps between 4 equivalent positions, displaced by about 0.3 Å along the $<111>$ axes. This is demonstrated in a measurement of the Mössbauer f-factors along different crystallographic directions.

REFERENCES

1. Greenwood, N.W. and Gibb, T.C., "Mössbauer Spectroscopy", (London, Chapman and Hill, 1971);
   May, L., "An Introduction to Mössbauer Spectroscopy", (New York, Plenum Press, 1971);
   Gibb, T.C., "Principles of Mössbauer Spectroscopy", (London, Chapman and Hill, 1976).
2. Van Rossum, M., Prog. Crystal Growth Charact. 5 (1982) 1.
3. Shenoy, G.K. and Wagner, F.E., "Mössbauer Isomer Shifts" (Amsterdam, North-Holland, 1978).
4. Weber, E. and Riotte, H.G., Appl. Phys. Lett. 33 (2978) 433; Lee, Y.H., Kleinhenz, R.L. and Corbett, J.W., Appl. Phys. Lett. 31 (1977) 142.
5. de Coster, M., Pollak, J., Amelincks, S., "Proceedings of the Second International Conference on the Mössbauer Effect (Saclay 1961)", Ed. D.H.J. Compton and A.H. Schoen (New York, J. Wiley & Sons, 1962) p. 289.
6. Bergholz, W., J. Phys. D 14 (1981) 1099.
7. Dézsi, I., Coussement, R., Langouche, G., Molnar, B., Nagy, D.L. and de Potter, M., J. de Physique 41 (1980) C1-425.
8. Langouche, G., de Potter, M., Van Rossum, M., De bruyn, J., Dézsi, I. and Coussement, R., "Nuclear and Electron Resonance Spectroscopy Applied to Materials Science", Ed. Kaufmann & Shenoy (Amsterdam, North Holland, 1981) 353.
9. Latshaw, G.L., Russell, P.B. and Hanna, S.S., Hyp. Int. 8 (1980) 105 ;
   Weyer, G., Grebe, G., Kettschau, A., Deutsch, B.I., Nylandstedt Larsen, A. and Holck, O., J. de Physique 37 (1976) C6-893.
10. Sawicka, B., Sawicki, J. and Stanek, J. de Physique 37 (1976) C6-879.
11. Sawicka, B.B. and Sawicki, J.A., Phys. Lett. A64 (1977) 311.
12. Langouche, G., Dézsi, I., Van Rossum, M., De bruyn, J. and

Coussement, R., Phys. Stat. Sol. b89 (1978) K17.
13. Sawicki, J.A. and Sawicka, B., Phys. Stat. Sol. b80 (1977) K41.
14. Langouche, G., de Potter, M., Dézsi, I. and Van Rossum, M., Rad. Eff. Lett. 67 (1982) 101.
15. Langouche, G., de Potter, M., De bruyn, J., Van Rossum, M., Coussement, R. and Dézsi, I., J. de Physique 41 (1980) C1-421; de Potter, M., Langouche, G., De bruyn, J., Van Rossum, M., Coussement, R. and Dézsi, I., Hyp. Int. 10 (1981) 769.
16. Damgaard, S., Oron, M., Petersen, J.W., Petrikin, Y.V. and Weyer, G., Phys. Stat. Sol. a59 (1980) 63.
17. Narajan, J., J. Appl. Phys. 52 (1981) 1289; White, C.W., Narajan, J., Appleton, B.R. and Wilson, S.R., J. Appl. Phys. 50 (1979) 2967.
18. Damgaard, S., Andreasen, H., Nevolin, V.I., Petersen, J.W. and Weyer, G., "Proceedings of the International Conference on Nuclear Physics Methods in Materials Research - Darmstadt 1980" (Braunschweig, Vieweg, 1980) 432; Petrikin, Y.V., Damgaard, S., Oron, M., Peterson, J.W. and Weyer, G., J. de Physique 41 (1980) C1-423.
19. Nylandsted Larsen, A., Weyer, G. and Nanver, L., Phys. Rev. B21 (1980) 4951; Weyer, G., Nylandsted Larsen, A., Holm, N.E. and Nielsen, H.L., Phys. Rev. B21 (1980) 4939; Weyer, G., Petersen, J.W. and Damgaard, S., Hyp. Int. 10 (1981) 775.
20. Weyer, G., Damgaard, S., Petersen, J.W. and Heinemeier, J., Hyp. Int. 7 (1980) 449.
21. Weyer, G., Petersen, J.W., Damgaard, S., Nielsen, H.L. and Heinemeier, J., Phys. Rev. Lett. 44 (1980) 155.
22. Kemerink, G., Ph.D. Thesis (University of Groningen, The Netherlands).
23. Hafemeister, D.W., de Waard, H., Phys. Rev. B7 (1973) 3014.
24. De bruyn, J., Academiae Analecta 44 (1982) 47.
25. De bruyn, J., Langouche, G., Van Rossum, M., de Potter, M. and Coussement, R., Phys. Lett. 73A (1979) 356.
26. Foti, G., Campisano, S.V., Rimini, E. and Vitali, G., J. Appl. Phys. 49 (1978) 2569.
27. De bruyn, J., Coussement, R., Dézsi, I., Langouche, G. and Van Rossum, M., Hyp. Int. 10 (1981) 973.
28. Van Rossum, M., Dézsi, I., Misra, K.C., Das, T.P. and Coker, A., to be published in Phys. Rev. B.
29. Dézsi, I., Van Rossum, M., De bruyn, J., Coussement, R. and Langouche, G., Phys. Lett. 87A (1982) 193.
30. Kemerink, G.J., De Wit, J.C., de Waard, H., Boerma, D.O. and Niesen, L., Phys. Lett. 82A (1982) 255.

PARTICIPANTS

ANDREW, R.   Faculté des Sciences, Université de l'Etat,
                       Av. Maistriau, 23 7000 Mons, Belgium.

ARNONE, C.   Piazza Amendola, 31
                       90141 Palermo, Italy.

AUVERT, J.   C.N.E.T. - C.N.S.
                       38240 Meylan, France.

BAERI, P.   Istituto di Fisica dell'Università
                       57, Corso Italia, 95129 Catania, Italy.

BAHIR, G.   Solid State Institute Technion
                       Israel Institute of Technology,
                       Haifa, Israel.

BAUFAY, L.   Faculté des Sciences, Université de l'Etat,
                       Av. Maistriau, 23 7000 Mons, Belgium.

BERTOLOTTI, M.   Istituto di Fisica - Facoltà di Ingegneria
                       Università di Roma, Roma, Italy.

BILGRAM, J.   Solid State Physics Laboratory
                       Eidegenössiche Technische Hochschule Zürich
                       8093 Zürich, Switzerland.

BINDER, K.   Institut für Festkörperforschung,
                       Kernforschungsanlage, 5170 Jülich,
                       Postfach 1913, W-Germany.

BISWAS, R.   Cornell University, Lab. of Atomic and
                       Solid State Physics, Clark Hall,
                       Ithaca, New York 14853, U.S.A.

BOK, J.   Ecole Normale Supérieure, Laboratoire de
                       Physique, 24, rue Lhomond
                       75005 Paris, France.

BOYD, I.

Physics Department,
Heriot-Watt University, Riccarton, Currie,
Edinburgh EH 14 4AS, Scotland.

COMPAAN, A.

Kansas State University, Dept. of Physics
Cardwell Hall,
Manhattan, Kansas 66506, U.S.A.

DAGONNIER, R.

Faculté des Sciences, Université de l'Etat,
Av. Maistriau, 23, 7000 Mons, Belgium.

DEWEL, G.

Chimie-Physique II, Université Libre de
Bruxelles, Campus Plaine C.P. 231
1050 Bruxelles, Belgium.

DUMONT, M.

Faculté des Sciences, Université de l'Etat,
Av. Maistriau, 23, 7000 Mons, Belgium.

FOGARASSY, E.

C.R.N. Strasbourg-Cronenbourg,
Groupe P.H.A.S.E. 67037 Strasbourg, B.P.20,
France.

GASPARD, J.-P.

Université de Liège, Institut de Physique,
4000 Liège, Sart-Tilman, Belgium.

GERMAIN, P.

Groupe de Physique des Solides de l'ENS,
Tour 23, Université Paris VII,
2, Place Jussieu 75231 Paris, France.

GRIMALDI, M.G.

Istituto di Struttura della Materia
Corso Italia, 57 95129 Catania, Italy.

HARO, E.

Laboratoire de Physique des Solides,
Université de P. et M. Curie
4, Place Jussieu, Tour B, 75230 Paris,France

HANUS, J.

Faculté des Sciences de Luminy,
Depart. de Physique Case 901,
Route Léon-Lachamp,70 13288 Marseille
Cédex 9, France.

JAOUEN, H.

Institut National Polytechnique de Grenoble
Laboratoire de Physique des composants à
semiconducteurs, E.N.S.E.R.G.
Rue des Martyrs, 23 38031 Grenoble, France.

JOLIET, M.-C.

Faculté des Sciences, Université de l'Etat,
Av. Maistriau, 23 7000 Mons, Belgium.

KEILMANN, F.	Max Plank Institut für Festkorperforschung 7000 Stuttgart 80, Busnauer Strasse, 171, W-Germany.
KURZ, H.	Philips Research Lab. Vogt-Köllnstrasse, 30 2000 Hamburg 54, W-Germany.
LANGOUCHE, G.	Katholieke Universiteit Leuven, Department Natuurkunde, Celestijnenlaan,200D 3030 Leuven, Belgium.
LAUDE, L.D.	Faculté des Sciences, Université de l'Etat, Av. Maistriau, 23 7000 Mons, Belgium.
LEGROS, R.	C.N.R.S., Laboratoire de Physique des Solides, 1, Pl. A. Briand, 92190 Meudon, France.
LUCHES, A.	Physics Department, CP 193 73100 Lecce, Italy.
MARFAING, J.	Faculté des Sciences de Luminy Case 901, 13288 Marseille Cédex 2, France.
MARINE, V.	Faculté des Sciences de Luminy Case 901, 13288 Marseille Cédex 2, France.
MARINELLI, M.	Via Salvatore Talamo, 61 00177 Roma, Italy.
MAY, M.	Laboratoire d'Optique, Université Pierre et Marie Curie, Tour 13, 3ème étage, 4, Place Jussieu 75230 Paris Cédex 05, France.
METEV, S.	Sofia University Faculty of Physics 5 Anton Ivanov Blvd. BG 1126 Sofia, Bulgaria.
MULLER, J.-C.	CRN Groupe PHASE 23, rue du Loess 67037 Strasbourg, France.
MÜLLER-KRUMBHAAR, H.	Institut für Festkörperforschung 5170 Jülich, W-Germany.
NISSIM, Y.	Centre National d'Etudes des Télécommunications, 196, rue de Paris, 92220 Bagneux, France.

PIERRARD, P.

CRMC2 Case 913
13288 Marseille Cédex, France.

PIGEOLET, A.

Faculté des Sciences, Université de l'Etat,
Av. Maistriau, 23 7000 Mons, Belgium.

PREVOT, B.

Université Louis Pasteur,
Laboratoire de Spectroscopie et d'Optique
du Corps Solide, 5, rue de l'Université,
67084 Strasbourg Cédex, France.

PRIBAT, D.

Thomson-CSF, Laboratoire Central de
Recherches, Domaine de Corbeville B.P. 10
91401 Orsay, France.

RICCIARDIELLO, F.G.

Universita degli studi di Trieste
Istituto di Chimica Applicata e Industriale
Via A. Valerio, 34127 Trieste, Italy.

RIMINI, E.

Istituto di Fisica
Corso Italia, 57 95129 Catania, Italy.

ROSENBERGER, F.

Department of Physics, University of Utah,
Salt Lake City, Utah 84112, U.S.A.

RUTERANA, P.

Laboratoire de Physique du Solide
Université de Caen, 14032 Caen Cédex, France

SBAIZERO, O.

Istituto di Chimica Applicata e Industriale
via A. Valerio, 34127 Trieste, Italy.

SCHRÖTER, W.

IV Physikalisches Insitut
Bunsenstrasse 11-15, 3400 Göttingen,
W-Germany.

SIBILIA, C.

Istituto di Fisica, Facoltà di Ingegneria
Via Tiburtina, 205 00100 Roma, Italy.

SIEJKA, J.

Groupe de Physique des Solides de l'E.N.S.
2, Place Jussieu, Tour 23,
75231 Paris, France.

SCUDIERI, F.

Istituto di Fisica-Ingegneria
Università di Roma, Roma, Italy.

SMIRL, A.L.

Center for Applied Quantum Electronics,
North Texas State University, Physics Dept.
Denton, Texas 76203, U.S.A.

TROTT, G.   Kansas State University, Dept of Physics,
                              Cardwell Hall,
                              Manhattan, Kansas 66506, U.S.A.

ULBRICH, R.G.   Universität Dortmund, Institut für Physik
                              46 Dortmund-Eichlinghofen 46 DO-50
                              Postfach 500 500, W-Germany.

VAN VECHTEN, J.A.   IBM Thomas J. Watson Research Center
                              P.O. Box 218, Yorktown Heights,
                              New York 10598, U.S.A.

von ALLMEN, M.   Universität Bern, Institut für Angewandte
                              Physik, 3012 Bern Sidlerstrasse, 5,
                              Switzerland.

VOOS, M.   Groupe de Physique des Solides de l'E.N.S.
                              24, rue Lhomond, 75005 Paris Cédex 05,
                              France.

WALSER, R.   Dept. of Elec. Eng., the University of
                              Texas at Austin, Austin, Texas 78712,U.S.A.

WARTMANN, G.   Universität Essen, Fachbereich 7 Physik
                              Universitätstrasse, 5, Postfach 103764
                              4300 Essen 1, W-Germany.

WAUTELET, M.   Faculté des Sciences, Université de l'Etat,
                              Av. Maistriau, 23, 7000 Mons, Belgium.

WESTON, J.   British Aerospace, P.O. Box 5,
                              F.P.C. 67 Filton Bristol BS12 7QW, England.

YAMADA, M.   Dept. of Electronics, Faculty of Engineering
                              Kobe University, Kokkodai, Nada,
                              Kobe 657, Japan.

ZELLAMA, K.   Groupe de Physique des Solides de l'ENS
                              Tour 23, Place Jussieu
                              75221 Paris Cédex 05, France

ORGANIZING COMMITTEE :

ANDREW, R.   ⎫
DAGONNIER, R.   ⎪ Université de l'Etat
LAUDE, L.D.   ⎬ Avenue Maistriau, 23
WAUTELET, M.   ⎭ 7000 Mons, Belgium

INDEX